Mechanosensitive Ion Channels

Mechanosensitivity in Cells and Tissues

Volume 1

Series Editors

A. Kamkin
Department of Fundamental and Applied Physiology, Russian State Medical University, Ostrovitjanova Str. 1, 117997 Moscow, Russia and Institute of Physiology, Humboldt-University, Charité, Tucholskystrasse 2, 10117 Berlin, Germany

I. Kiseleva
Department of Fundamental and Applied Physiology, Russian State Medical University, Ostrovitjanova Str.1, 117997 Moscow, Russia and Institute of Physiology, Humboldt-University, Charité, Tucholskystrasse 2, 10117 Berlin, Germany

Andre Kamkin · Irina Kiseleva
Editors

Mechanosensitive Ion Channels

Forewords by Max J. Lab and Jürgen Hescheler

Springer

Editors
Andre Kamkin
Russian State Medical
University, Moscow, Russia
and Humboldt-University,
Berlin, Germany

Irina Kiseleva
Russian State Medical
University, Moscow, Russia
and Humboldt-University,
Berlin, Germany

Editorial Assistant
Ilya Lozinsky
Department of Physiology,
University of Maastricht,
The Netherlands

ISBN: 978-90-481-7638-0 e-ISBN: 978-1-4020-6426-5

Softcover reprint of the hardcover 1st edition 2008

Printed on acid-free paper

9 8 7 6 5 4 3 2 1

springer.com

Contents

Foreword

Mechanosensitivity in Cells and Tissues as an Overall Regulatory System

Kamkin and Kiseleva's assembly of leaders in the mechanosensitivity field provides an excellent broad based and up-to-date book, with chapters in the first part discussing mechanically gated channels (mechanogated), mechanosensitive channels (MGC, MSC) systems (as the editors air the nomenclature in their editorial), the second describing their signaling, and the third presenting aspects of cell mechanobiology. Conceptually starting with tension at the surface membrane, the book's theme moves on to some molecular mechanisms, and then on to MSC as initiating complex cell signal cascades, sometimes invoking signalsomes. Mechanosensitivity is also described in organs.

Mechanosensitivity does not only relate to very soft deformable tissue, but chapters also focuses on cells and tissues concerned essentially with skeletal growth and development. The state of the art covered in the book heralds, to me, a slant to mechanosensitivity in general.

Galileo remarked in 1638 that longer bones were thicker for a given structural strength (Galilei, 1638) (translation in (Crewew & da Silavio, 1939)), and this was followed in 1859 by Darwin's noting that flying ducks have underdeveloped legs as compared with terrestrial bound ones (Darwin, 1859).

These early observations were already suggesting a compensatory homeostatic feedback system applying to the whole organism. Based on the wide-ranging and authoritative chapters, I propose that mechanogated, mechanosensitive systems in biology as a whole preserves physiological, integrative homeostasis as closed feedback control systems, and Fig. 1 is indication of how this could fit into system biology. Feedback interaction is between the mechanosensor to spatiotemporal integration of systems and organs. Mechanoelectric Feedback in heart has already been similarly viewed and invokes mechanosensitivity as the detector/transducer in feedback control of electrophysiology. (Lab, 1999) Although I have applied the homeostatic feedback notion to heart, this book suggests that I can broaden this mechanical homeostatic process to biology in general, for mechanosensitivity exists throughout the phyla - bacteria to mammalian tissue.

In pursuing this notion, I used a type of meta-analyses — an accepted method to address a defined clinical topic, sometimes as a hypothesis (treatment X works). The current approach is roughly analogous, examining this book as well as the literature with the proposal in mind. My analysis roughly encompasses literature over 20 years or so, using "Entrez Pubmed," with search terms such as "Mechanotransduction,"

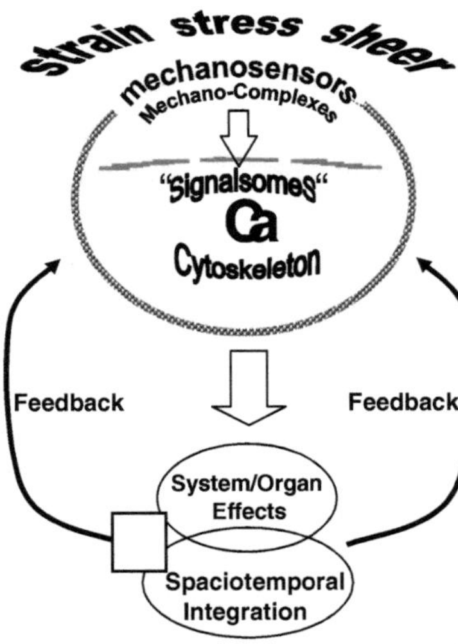

Fig. 1 Stress,strain, and sheer forces (very top) affect the membrane Mechanosensors / Mechano-complexes and then into the cell, feeding down (top block arrow - "feed-forward") down signalsomes and signal cascades often involving calcium. Importantly, influences via the cytoskeleton modulate (bottom block arrow) organ and spatiotemporal integration. These feed back ("return" - outside curved black arrows) via the same or alternate paths to modulate the consequences of the initiating mechanical changes

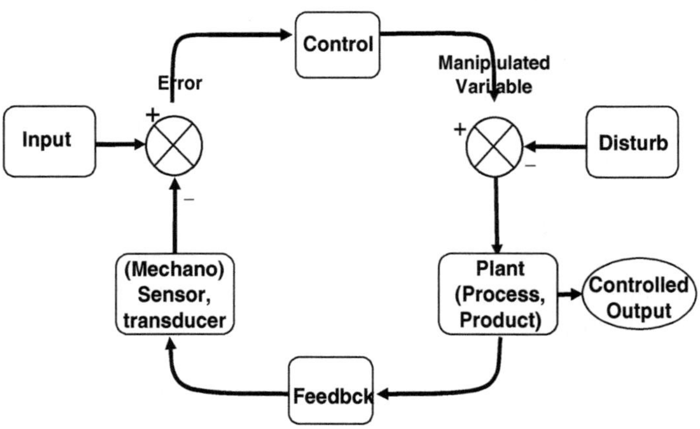

Fig. 2 Simplified block diagram of mechanosensitivity as a homeostatic feedback control system. The mechanosensors or transducers (bottom left element in loop) have mechanical inputs (top left) to a summing point (left crossed circle in loop). These via the control mechanisms and transfer functions going clockwise reach another summing point, interacting with the factors disturbing the system (top right – "disturb"). The product of these interactions (bottom left element – "plant", in loop) would produce a controlled output (elipse – bottom right). The plant/process element feeds back to influence the mechanosensor or transducer.element

"Mechanosensitivity," "Mechanogated channels," and "Stretch activated channels", and "Mechano Feedback." The findings were then organised into appropriate components of feedback control loops, see Fig. 2.

The "feedback" label in heart (Lab, 1999) was negative and when the Feedback Loop is altered (gain changes) it heralds pathology, with arrhythmia, hypertrophy.

It can also apply in the vasculature with its regulatory flow dynamics (Lehoux *et al.*, 2006). In this case an altered loop can produce atherosclerosis. Growth & proliferation, repair and regeneration can also be treated in analogous fashion, and altered loop gains here produce cancer.

Although several elegant reviews covering aspects of mechanobiology (for example (Ingber, 2003; Ingber, 2006)) offer an open system view with feed forward descriptions, my current proposal in this article is for mechanosensitivity to involve closed negative feedback systems. As a homeostatic regulatory system, it has several putative characteristics found in other regulatory systems with biophysical, biochemical and physiological regulatory functions, operating from the molecule to the organ/system.

References

Crewew H & da Silavio A (1939). *Dialogue concerning two new sciences*. Northwestern University Press.

Darwin C (1859). *On the origin of species by means of natural selection. London: John Murray* John Murray, London.

Galilei G (1638). *Discorsi e dimonstrazioni mathematichia intorno a due nuove scienze*. Lieda: Elsevier.

Ingber DE (2006). Cellular mechanotransduction: putting all the pieces together again. *FASEB J* **20**, 811–827.

Ingber DE (2003). Tensegrity II. How structural networks influence cellular information processing networks. *J Cell Sci* **116**, 1397–1408.

Lab MJ (1999). Mechanosensitivity as an integrative system in the heart: an audit. *Prog Biophysics & Molec Biol* **71**, 7–27.

Lehoux S, Castier Y, & Tedgui A (2006). Molecular mechanisms of the vascular responses to haemodynamic forces. *J Intern Med* **259**, 381–392.

Max J Lab

Imperial College London, National Heart & Lung Institute and Nanomedicine Research Laboratory — (Hammersmith Campus), Charing Cross, London W6 8RF, UK

Foreword

The book "Mechanosensitivity in Cells and Tissues: Mechanosensitive Ion Channels" edited by Andre Kamkin and Irina Kiseleva impressively demonstrates the diversity of cellular effects depending on mechanical stimuli. Coming from basic biophysical mechanisms of mechanosensitivity of the lipid bilayer, ionic channels including K^+, Cl^-, and nonselective channels in bacteria as well as in eukaryotic multicellular organisms are reviewed. Examples demonstrating the physiological role of mechanosensitive channels in cellular and organ physiology include odontoblasts and chondrocytes. Moreover information is given on mechanosensitive signaling cascades in articular chondrocytes, osteoblastic and mesenchymal stem cells as well as on in neuronal and cardiac cells where mechanosensitive cation channels are described.

When I was asked to write the foreword to this compelling book I doubted about what I as a stem cell researcher could say about mechanosensitivity of cells? But looking at the more recent studies in this field, and in particular on detailed insights in the earliest events of cellular development in the embryo – eureka - mechanical effects including migration and mechanosensitivity turn out to be ones of the most primal and archaic mechanisms of self organisation. Cellular mechanosensitivity is probably the most basic biological principle that is presumably expressed in every cellular phenotype and plays a pivotal role in manifold physiological mechanisms and especially in early embryonic development. There is compelling evidence that physical forces, including gravity, tension, compression, pressure and shear forces influence growth and remodelling in nearly all living tissues on a cellular level. Proliferation, growth, differentiation, secretion, movement, signalling as well as gene expression have been shown to be altered by applying mechanical stress directly to the respective cells. In parallel with the identification of more and more cellular targets of mechanosensitivity there is an increasing interest of basic and applied scientists in this field. Just looking at the number of publications in Pubmed[1] as a measure of the scientific impact which well reflects the interest of the scientific community in this specified research area, one can identify a dramatic rise of interest within the last year. While the field of mechanosensitivity attracted around 100 papers per annum from 2000 to 2005 the number immensely jumped to more than 450 in the year 2006.

[1] Pubmed-listed publications with the term "mechanosensitivity" in the title/abstract

This might also be due to the fact that great technological advances in areas such as nanotechnology, micromanipulation, biological imaging and computer modelling have enabled us to analyze mechanotransduction: how forces affect the biochemical activities of individual molecules, both in isolation and within living cells. Also, the detailed analysis of mechanosensitive ionic channels using 2D crystallography, electronparamagnetic resonance spectroscopy, fluorescence resonance energy transfer spectroscopy as well as computational modelling based on molecular dynamics and Brownian dynamics gave a new impact in this field. Mechano-regulation is once again becoming a central focus in fields ranging from molecular biophysics and cell biology to human physiology and clinical medicine.

The present book demonstrates the various aspects of mechanosensitivity. It thus considerably contributes to a better understanding of the primary processes in cellular mechanosensitivity and highlights the electrophysiological responses. Together with other well known signalling cascades directly coupled to the cytoskeleton and thus to cellular mechanical responses our picture on these important signalling processes becomes more and more detailed and hopefully will integrate into the complex cellular processes such as embryonic development and differentiation of cells as well into the complex physiological organic functions.

Jürgen Hescheler
Institut für Neurophysiologie. Universität zu Köln. Robert-Koch Str. 39 D-50931 Köln, Germany

Editorial

Mechanically Gated Channels and Mechanosensitive Channels

Andre Kamkin and Irina Kiseleva

Cell reaction to mechanical stress is one of the oldest cellular functions from the point of view of evolution. Response to mechanical stress in different forms is present in all organisms, from bacteria to mammals. Mechanical stimulus triggers different electrophysiological and biochemical responses in cells. It can influence physiological processes at the molecular, cellular, and systemic level. During last 20 years there was major progress in both phenomenological description of cell membrane responses to mechanical stress and in investigation of signaling pathways and mechanism underlying cellular responses to mechanical stress. Ion channels, reacting to membrane deformation, were shown to play the key role in one of the mechanisms, through which the cell responds to mechanical stimulus. This channels were originally called mechanosensitive channels (MSC). Creation of the patch clamp method allowed Guharay and Sachs (1984) to record first whole cell MSC currents.

Present book is devoted to discussion of the latest findings in the field of MSC research. The Volume was compiled by a group of scientists from all over the world, who are considered to lead the development of the field. The book comprises three parts, first discussing mechanosensitive channels, second–mechanosensitive signaling and the third–cell mechanobiology. At the same time such set up of reviews is rather conventional and is designed for the readers convenience since all presented topics intersect.

The first part presents topics dealing with mechanosensitive channels. MSC is a channel whose conductivity is a response to membrane deformation. At the same time back in 1998 Sachs and Morris noted that among MSCs under investigation there are channels, that recognize mechanical deformation as a proper physiological signal, and those, that react to mechanical stimulation with slight changes in kinetics. For the latter channels the authors introduced the term—channels with weak mechanosensitivity (Sachs and Morris, 1998). Recent findings call for new definitions and new specific terminology.

Ability of mechanosensitive ion channels to change their spatial organization from closed to open state during the transition period presently is considered to be their major distinctive characteristic, while permeability modulation of voltage-gated channels and ligand-gated channels during mechanical stress can no longer be used for defining a channel as one properly responding to mechanical stress (Sachs and Morris, 1998), although all channels that recognize mechanical deformation as a proper 2 physiological signal have been called—mechanically gated channels (MGC)—by a number of authors (Hamill and Martinac, 2001; White, 2006; Zhang and Hamill, 2000).

Many authors divide mechanically gated channels (MGC) into stretch activated channels (SACs) that underlie mechanically gated whole-cell current during cell stretching and are registered under pipette (although their identity as SACs has not been proven), and volume activated channels (VAC) for the second corresponding category of channels (Baumgarten et al, 2005).

Recent works by Honoré et al, (2006) and Sunchya & Sachs (2007) show that SICs, isolated earlier into a separate group, are most probably typical SAC channels in pre-stretch state. As for PACs, their existence remains questionable considering implications of Laplaces law. Those issues are addressed by Kamkin et al. in the first Chapter of present Volume (Kamkin et al, 2007).

Mechanosensitive channels (MSC) as a term recently is usually used to describe channels that only modulate their permeability in response to mechanical stress (Morris et al, 2006). Definition of this term in such way allows further division of MSCs into two groups—mechanosensitive voltage-gated channels (MSCVG) and mechanosensitive ligand-gated channels (MSCLG).

At the same time during preparation of Chapters of this Volume we did not insist on strict use of this novice definition. Therefore in many Chapters, including our own, traditional abbreviations are used for readers convenience and according to the context MSC can mean either MGC or MSC.

The first part of the Volume begins with a Chapter, devoted to experimental methods of studying MSC and possible mistakes in data interpretation (Kamkin et al, 2007). Authors discuss the most widely used experimental methods, which are used in the field of MSC research, and their limitations. The following Chapter deals with the role of lipid bilayer mechanics in functioning of mechanosensitive channels, that respond to membrane tension. The authors discuss the role of bilayer deformation in the context of the well known mechanosensitive channel MscL (Ursell et al, 2007). It is followed by discussion of mechanosensitive channels, which are gated by membrane tension. It goes without saying that transformation of mechanical stimulus into biological response is one of the most exiting problems in physiology, pathology and practical medicine. The authors discuss two different theoretical models of it which are used now. The first one is based on the assumption that channels are 'tethered' to cytoskeleton and/or to extracellular components, which thus exert forces on the channel, which in turn leads to gating. The second model postulates that the channel protein directly senses biophysical changes that occur within the membrane, when it is under tension (Blount et al, 2007). The discussion is continued in the following Chapter, which is describing studies of the bacterial mechanosensitive channels. It elucidates the nature of the imaged structures of the protein, the interactions between the protein and the lipid, the conformational changes, which are involved in channel gating, and the specific roles played by various protein 3 domains (Martinac and Corry, 2007). The reader, interested in bacterial MSC research, can acquire useful insight into mechanosensory transduction in simple animal models, such as the nematode *Caenorhabditis elegans* and the fruit fly *Drosophila melanogaster*

(Kourtis and Tavernarakis, 2007). Further discussion concentrates on mechanosensitive ion channels in odontoblasts, which are post-mitotic cells, involved in the dentine formation throughout the life of the tooth and which are suspected to play a role in tooth pain transmission. Two kinds of mechanosensitive K^+-channels are discussed: TREK-1 channels, belonging to the two-pore-domain potassium channel family, and high-conductance Ca^{2+}-activated potassium channels (K_{Ca}), activated by stretching of the membrane as well as osmotic shock. (Magloire et al, 2007). The discussion of the topic is further broadened by the Chapter discussing K^+-channels in articular chondrocytes, their putative roles in mechanotransduction, metabolic regulation and cell proliferation (Mobasheri et al, 2007). Recent studies have indicated that SAC, integrins, growth factor receptors, cytoskelton, extracellular matrix and other various molecules contribute to the osmotransduction. However, little is known about how these molecules sense the osmotic and mechanical stresses and how they function in order to transduce the stress downstream their signaling cascade. Separate Chapter deals with osmotransduction through volume-sensitive Cl^- channels (Niisato and Marunaka, 2007). The final Chapter in the first part of the Volume is devoted to mechanosensitive channel TRPV4. It is known that the TRPV4, a member of the TRP family, that has 6-transmembrane domains, was initially identified as a hypoosmolality-activated Ca^{2+} permeable channel. Recent findings suggest that TRPV4 along with other molecules together constitute a physiological response, such as cell volume regulation, epithelial permeability, and vascular dilatation (Suzuki, 2007).

The second part of the Volume deals with discussion of mechanotransduction, which is defined as the biochemical response of cells to mechanical stimulation (Pingguan-Murphy and Knight, 2007; Liedert et al, 2007, Calaghan, 2007). This part of the Volume describes signaling cascades, which participate in modulation of mechanosensitive channels function (Pellegrino, 2007), and discusses regulation of intracellular signaling transduction pathways by mechanosensitive ion channels (Boriek and Kumar, 2007).

Up to date several stretch-activated signaling pathways have been identified. In 2004 Lammerding et al. and in 2005 Lab described cardiac specific schemes of signal transduction pathways, which are involved in the cellular response to mechanical stimulation. Their generalizations united multiple experimental observations into a comprehensive theoretical model of mechanosensitivity, which may become very useful for generation and testing of novel hypothesis further promoting development of the field. At the same time Cingolani et al., 2005 reported their scheme of hypothetical intracellular hypertrophic pathway, triggered by myocardial stretch, and Liedert et al., (2005) presented their model of signal transduction pathways in 4 mechanotransduction in bone cells. First three Chapters of the second part of our Volume deal with those topics.

Current research in our field goes beyond investigation of the stretch-activated signaling cascades into advocating the possibility that mechanoelectro-chemical transduction forms a part of a network of mechanically linked crosstalk (Mechanically Mediated Crosstalk: MMC) (Lammerding et al, 2004; Lab, 2005). Mechanical components can hypothetically provide the bond between interactions at molecular, cellular, and macro levels to enable such crosstalk. Although our Volume focuses on stretch activated channels, stresses and strains can affect functioning of other membrane channels or receptors, which are located downstream signaling cascades, which include SACs. A cellular mechanical transformation can thus trigger several different short and/or long term ionic or other downstream responses. Several cell signaling cascades have been implied and, mostly via intracellular Ca^{2+}, can affect membrane electrophysiology. MMC can shape downstream signals leading to alterations of intracellular Ca^{2+} signaling. MMC can also span other regulatory systems and processes, such as the autonomic nervous system, and in addition, can operate through the whole heart as an integrative system (Lab, 2005).

The first Chapter of this part is devoted to mechanosensitive purinergic calcium signaling in articular chondrocytes (Pingguan-Murphy and Knight, 2007). In some respect it intersects with Chapter 7 by Mobasheri et al (2007), while focusing on the influence of physiological mechanical stimuli on intracellular Ca^{2+} signalling in articular chondrocytes and its potential role in cartilage mechanotransduction. This review examines the downstream cellular response to mechanically-activated purinergic Ca^{2+} signaling and its importance in maintenance of tissue homeostasis (Pingguan-Murphy and Knight, 2007). Next Chapter discuses involvement of mechanotransduction in various signal transduction pathways, including the activation of ion channels and other mechanoreceptors in the membrane of the bone cell, resulting in gene regulation in the nucleus (Liedert et al, 2007). Next Chapter of this part of the Volumedescribes the role of caveolae in stretch-activated signaling. The review sums up the latest data relating to the role of caveolae in Ca^{2+} regulation, role of caveolae in G protein-coupled receptor signal cascades and the role of caveolae in mechanotransduction. In this Chapter authors also discuss the relationship between mechanotransduction, caveolae and disease (Calaghan, 2007).

The role of signaling cascades in modulation of MSC in the central nervous system is discussed in terms of leech neurons model in the next Chapter. It covers multimodal activation by membrane potential, intracellular calcium and pH, as well as powerful modulation by adenosine nucleotides (Pellegrino et al, 2007).

Boriek and Kumar (2007) in their Chapter discuss the role of MSC in the regulation of different mechanosensitive signaling proteins and signaling pathways. They describe how mechanical stretch causes Ca^{2+} and other ions influxes through the MSC and how this leads to the 5 activation of signaling cascades, leading to activation of several kinases and phosphatases. They conclude that activation of these intermediate signaling molecules leads to the activation of transcription factors leading to the increased expression of mechanosensitive genes. Chapter reveals how an increased Ca^{2+} influx in response to mechanical stretch could also lead to the activation of these transcription factors via Ca^{2+}-responsive proteins, such as protein kinase C (Boriek and Kumar, 2007).

The third part of the Volume is devoted to cell mechanobiology. First Chapter of this part of the Volume describes the effects of mechanical stimulation in vertebrate hearts. It contains discussion of differences in vertebrate hearts and myocyte structures, and the effect of stretch on cardiac force and electrical activity (Shiels and White, 2007). Final Chapter of the Volume provides a throughout discription of mechanobiology of fibroblasts. The authors focus on mechanobiological responses of fibroblasts and on fibroblast specific cellular mechanotransduction (Thampatty and Wang, 2007).

The volume dwells on the major issues of mechanical stress influencing the ion channels and intracellular signaling pathways. In our opinion the book presents the latest achievements in the field and provides a broader vision of the field to experts, working on related topics of fundamental and clinical research. It can also provide a useful insight for practicing physicians, which can improve their understanding of cellular mechanisms, underlying different pathologies. We also hope that this Volume will attract more attention to the field both from researchers and practitioners and will assist to efficiently introduce it into the practical medicine.

References

Baumgarten CM (2005) Cell volume-sensitive ion channels and transporters in cardiac myocytes. In: Kohl P, Franz MR, Sachs F (ed) Cardiac Mechano-Electrical Feedback and Arrhythmias: From Pipette to Patient, Saunders, Philadelphia, pp. 21–32.

Blount P, Li Y, Moe PC, Iscla I (2007) Mechanosensitive channels gated by membrane tension: bacteria and beyond. *In this volume.*

Boriek AM and Kumar A (2007) Rregulation of intracellular signal transduction pathways by mechanosensitive ion channels. *In this volume.*

Calaghan S (2007) Caveolae: Co-ordinating centres for mechanotransduction in the heart? *In this volume.*

Cingolani HE, Aiello EA, Pérez NG, Ennis IL, Camilión de Hurtado MC (2005) The Na^+/H^+ exchanger as the main protagonist following myocardial stretch: The Anrep effect and myocardial hypertrophy In: Kamkin A and Kiseleva I (ed) Mechanosensitivity in Cells and Tissues. Academia Publishing House Ltd, Moscow, 2005: 271–290.

Guharay F and Sachs F (1984) Stretch-activated single ion channel currents in tissue cultured embryonic chick skeletal muscle. *J Physiol (Lond)* 352:685–701.

Hamill OP and Martinac B (2001) Molecular basis of mechanotransduction in living cells. *Physiol Revs* 81:685–740.

Honoré E, Patel AJ, Chemin J, Suchyna T, Sachs F (2006) Desensitization of mechano-gated K2P channels. *Proc Natl Acad Sci USA* 103(18):6859–6864.

Kamkin A, Kiseleva I, Lozinsky I (2007) Experimental methods of studying of mechanosensitive channels and possible mistakes in data interpretation. *In this volume.*

Kourtis N, Tavernarakis N (2007) Mechanosensory transduction in the nematode *Caenorhabditis elegans. In this volume.*

Lab MJ (2005) Mechanically mediated crosstalk in heart. In: Kamkin A and Kiseleva I (ed) Mechanosensitivity in Cells and Tissues. Academia Publishing House Ltd, Moscow, 2005: 58–78.

Lammerdiung J, Kamm RD, Lee RT (2004) Mechanotransduction in cardiac myocytes. Ann NY Acad Sci 1015: 53–70.

Liedert A, Claes L, Ignatius A (2007) Signal transduction pathways involved in mechanotransduction in osteoblastic and mesenchymal stem cells. *In this volume.*

Liedert A, Kaspar D, Augat P, Ignatius A, Claes L (2005) Mechanobiology of bone tissue and bone cells In: Kamkin A and Kiseleva I (ed) Mechanosensitivity in Cells and Tissues. Academia Publishing House Ltd, Moscow, 2005: 418–433.

Magloire H, Allard B, Couble M-L, Maurin J-C, Bleicher F (2007) Mechanosensitive ion channels in odontoblasts. *In this volume.*

Martinac B, Corry BA (2007) Computational approaches in studies of bacterial mechanosensitive channels. *In this volume.*

Mobasheri A, Dart C, Barrett-Jolley R (2007) Potassium ion channels in articular chondrocytes: putative roles in mechanotransduction, metabolic regulation and cell proliferation. *In this volume*.

Morris CE, Juranka PF, Lin W, Morris TJ, Laitko U (2006) Studying the mechanosensitivity of voltage-gated channels using oocyte patches. *Methods Mol Biol* 322:315–329.

Niisato N and Marunaka Y (2007) Osmotransduction through volume-sensitive Cl^- channels. *In this volume*.

Pellegrino M, Barsanti C, Pellegrini M (2007) Multimodal activation and regulation of neuronal mechanosensitive cation channels. *In this volume*.

Pingguan-Murphy B and Knight MM (2007) Mechanosensitive purinergic calcium signalling in articular chondrocytes. *In this volume*.

Sachs F and Morris CE (1998) Mechanosensitive ion channels in nonspecialized cells. *Rev Physiol Biochem Pharmacol* 132: 1–77.

Shiels HA and White E (2007) The effects of mechanical stimulation on vertebrate hearts: a question of class. *In this volume*.

Suchyna TM and Sachs F (2007) Mechanosensitive channel properties and membrane mechanics in mouse dystrophic myotubes. *J Physiol (Lond)* (in press).

Suzuki M (2007) Mechanosensitive channel TRPV4. A micro-machine converting physical force into an ion flow. *In this volume*.

Thampatty BP, Wang J H-C (2007) Mechanobiology of fibroblasts. *In this volume*.

Ursell T, Kondev J, Reeves D, Wiggins PA and Phillips R, (2007) The role of lipid bilayer mechanics in mechanosensation. *In this volume*.

White E (2006) Mechanosensitive channels: therapeutic targets in the myocardium? *Curr Pharm Des* 12(28):3645–3663.

Zhang Y and Hamill OP (2000) On the discrepancy between membrane patch and whole cell mechanosensitivity in *Xenopus* oocytes. *J Physiol (Lond)* 523.1:101–115.

Contributors

Bruno Allard University of Lyon 1, Physiologie integrative, Cellulaire et Moléculaire, Villeurbanne F-69100, The UMR CNRS 5123, Villeurbanne F-69100, France

Richard Barrett-Jolley Department of Veterinary Preclinical Sciences, Faculty of Veterinary Science, University of Liverpool, Liverpool L69 7ZJ, UK

Cristina Barsanti Laboratorio di Genetica Molecolare, Istituto G. Gaslini, largo G. Gaslini 5, I-16148 Genova, Italy

Françoise Bleicher University of Lyon 1, EA 1892-IFR 62, Développement et Régénération des Tissus Dentaires, Faculté d'Odontologie, Lyon F-69008, and INSERM/ERI 16, Lyon F-69008, France

Paul Blount Department of Physiology, University of Texas-Southwestern Medical Center, 5323 Harry Hines Blvd. Dallas, TX 75390-9040, USA, Paul.Blount@UTSouthwestern.edu

Aladin M. Boriek Department of Medicine, Baylor College of Medicine, Houston, Texas 77030, USA

Sarah Calaghan Institute of Membrane and Systems Biology, Faculty of Biological Sciences, University of Leeds, Leeds LS2 9JT, UK, S.C.Calaghan@leeds.ac.uk

Lutz Claes
Institute of Orthopedic Research and Biomechanics, University of Ulm, Helmholtzstrasse 14, 89081 Ulm, Germany

Ben Corry School of Biomedical, Biomolecular and Chemical Sciences, University of Western Australia, Crawley, WA 6008, Australia

Marie-Lise Couble University of Lyon 1, EA 1892-IFR 62, Développement et Régénération des Tissus Dentaires, Faculté d'Odontologie, Lyon F-69008, and INSERM/ERI 16, Lyon F-69008, France

Caroline Dart
School of Biological Sciences, University of Liverpool, Liverpool L69 7ZB, UK

Anita Ignatius
Institute of Orthopedic Research and Biomechanics, University of Ulm, Helmholtzstrasse 14, 89081 Ulm, Germany

Irene Iscla
Department of Physiology, University of Texas-Southwestern Medical Center, 5323 Harry Hines Blvd. Dallas, TX 75390-9040, USA

Andre Kamkin
Department of Fundamental and Applied Physiology, Russian State Medical University, Ostrovitjanova Str.1, 117997 Moscow, Russia and Institute of Physiology, Humboldt-University, Charité, Tucholskystrasse 2, 10117, Berlin, Germany, Kamkin.A@g23.relcom.ru

Irina Kiseleva
Department of Fundamental and Applied Physiology, Russian State Medical University, Ostrovitjanova Str.1, 117997 Moscow, Russia and Institute of Physiology, Humboldt-University, Charité, Tucholskystrasse 2, 10117 Berlin, Germany

Martin M. Knight
Medical Engineering Division, Department of Engineering Queen Mary, University of London, Mile End Rd., London, E1 4NS, UK, m.m.knight@qmul.ac.uk

Jané Kondev
Department of Physics, Brandeis University, Waltham MA 02454, USA

Nikos Kourtis
Institute of Molecular Biology and Biotechnology, Foundation for Research and Technology, Heraklion 71110, Crete, Greece

Ashok Kumar
Laboratory for Skeletal Muscle Physiology and Neurobiology, Jerry L. Pettis Memorial Veterans Affair Medical Center, Loma Linda, CA 92357; Loma Linda University School of Medicine, Loma Linda, CA 92350, USA,
ashok.kumar2@va.gov

Yuezhou Li
Department of Physiology, University of Texas-Southwestern Medical Center, 5323 Harry Hines Blvd. Dallas, TX 75390-9040, USA

Astrid Liedert
Institute of Orthopedic Research and Biomechanics, University of Ulm, Helmholtzstrasse 14, 89081 Ulm, Germany

Ilya Lozinsky
Department of Biomedical Engineering, Stony Brook University HSC T18-030B Stony Brook, NY 11794-8181 USA

Henry Magloire
University of Lyon 1, EA 1892-IFR 62, Développement et Régénération des Tissus Dentaires, Faculté d'Odontologie, Lyon F-69008, and INSERM/ERI 16, Lyon F-69008, France, magloire@laennec.univ-lyon1.fr

Boris Martinac
School of Biomedical Sciences, University of Queensland St Lucia, Brisbane, QLD 4072 Australia, b.martinac@uq.edu.au

Yoshinori Marunaka
Department of Molecular Cell Physiology, Graduate School of Medical Science, Kyoto Prefectural University of Medicine Kyoto 602-8566, Japan

Jean-Christophe Maurin
University of Lyon 1, EA 1892-IFR 62, Développement et Régénération des Tissus Dentaires, Faculté d'Odontologie, Lyon F-69008, and INSERM/ERI 16, Lyon F-69008, France

Ali Mobasheri
Division of Comparative Veterinary Medicine, School of Veterinary Medicine and Science, University of Nottingham, Sutton Bonington Campus, Loughborough, Leicestershire, LE12 5RD, UK, ali.mobasheri@nottingham.ac.uk

Paul C. Moe
Department of Physiology, University of Texas-Southwestern Medical Center, 5323 Harry Hines Blvd. Dallas, TX 75390-9040, USA

Naomi Niisato
Department of Molecular Cell Physiology, Graduate School of Medical Science, Kyoto Prefectural University of Medicine, Kyoto 602-8566, Japan, naomi@koto.kpu-m.ac.jp

Monica Pellegrini
Scuola Normale Superiore di Pisa, Piazza dei Cavalieri, 56127 Pisa, Italy

Mario Pellegrino
Dipartimento di Fisiologia e Biochimica,Università di Pisa, via S.Zeno 31, 56127 Pisa, Italy, marpell@dfb.unipi.it

Rob Phillips
Division of Applied Physics, California Institute of Technology, Pasadena CA 91125, USA; Kavli Nanoscience Institute, Pasadena CA 91125, USA, phillips@pboc.caltech.edu

Belinda Pingguan-Murphy
Biomedical Engineering Department, Faculty of Engineering University Malaya, Kuala Lumpur, Malaysia

Dan Reeves
Department of Physics, Brandeis University, Waltham MA 02454, USA

Holly A. Shiels
Faculty of Life Sciences, Core Technology Facility, 46 Grafton Street, University of Manchester Manchester, M13 9NT, UK

Makoto Suzuki
Department of Pharmacology. Jichi Medical University 3311-1, Yakushiji, Tochigi, 329-0498 Japan, macsuz@jichi.ac.jp

Nektarios Tavernarakis
Institute of Molecular Biology and Biotechnology, Foundation for Research and Technology, Heraklion 71110, Crete, Greece, tavernarakis@imbb.forth.gr

Bhavani P. Thampatty
MechanoBiology Laboratory Department of Orthopaedic Surgery, School of Medicine, University of Pittsburgh, Pittsburgh, PA, 15213, USA

Tristan Ursell
Division of Applied Physics, California Institute of Technology, Pasadena CA 91125, USA

James H-C. Wang
MechanoBiology Laboratory Departments of Orthopaedic Surgery and Bioengineering, School of Medicine, University of Pittsburgh Pittsburgh; PA, USA, wanghc@pitt.edu

Ed White
Institute of Membrane and Systems Biology, University of Leeds, Leeds, LS29JT, UK, e.white@leeds.ac.uk

Paul A. Wiggins
Whitehead Institute, Cambridge MA 02142, USA

Part I
Mechanosensitive Channels

Chapter 1
Experimental Methods of Studying Mechanosensitive Channels and Possible Errors in Data Interpretation

Andre Kamkin, Irina Kiseleva, and Ilya Lozinsky

Abstract In this review we discuss most widely used experimental methods of the membrane stretch, which are used for investigation of mechanosensitive channels (MSCs) by means of patch-clamp method. We address possible mistakes in interpreting the data, obtained by means of various methods of MSCs investigation. Under conditions of single channel recording we briefly analyse positive and negative pressure in terms of mechanical stimulation and demonstrate that MSC respond only to membrane tension. Resting tension of the membrane is created after suction, which is applied for the purpose of gigaseal formation. It is shown that some channels can be active at zero pressure because the seal adhesion energy produces tension. Such situation can be considered as pre-stretch. In this respect we discuss reports, showing that stretch-inactivated channels (SICs) do not imply the existence of a new type of channels, when inactivation of channel activity in response to suction can be explained by the activity of pre-stressing of stretch-activated channels (SACs). We discuss the controversy about the presence of pressure activated channels (PACs). According to the Laplace's equation, positive or negative pressures should make equal contributions to the stress. We also discuss reported methods of direct mechanical stretching of cells during whole cell recording. Discussion covers method of homogeneous stretching of a single cell by means of two patch pipettes and three types of axial stretch - by two glass capillaries, by glass stylus and by two thin carbon fibres. We briefly discuss the merits and imperfections of cell swelling. We analyse the possibilities of paramagnetic microbead method that allows the application of controlled forces to the membrane, at the level of which those mechanical forces are transmitted by integrins. We discuss possible methods of cell compression. Obviously distribution of forces is very complicated during compression and no one knows how to analyze the data in a mechanistic manner. We discuss the study of bacterial mechanosensitive channels. We discuss the limitation of the research using protein purification and functional reconstitution in planar lipid bilayers and in vesicles. Also, rarely used methods are presented.

Key words: Cells · Mechanosensitive channels · Mechanically gated channels · Membrane tension · Mechanical stimulus · Pressure · Stretch · Compression · Swelling · Paramagnetic microbeads method · Reconstituted system · Spheroplasts · Membrane blebs

A. Kamkin and I. Kiseleva (eds.), *Mechanosensitive Ion Channels.*

1.1 Introduction

Mechanosensitive channels (MSCs) are defined by the distinctive property that their gating is responsive to membrane deformation (Sachs and Morris, 1998). The designation of the channels as mechanically gated is only that the channels' open probability responds to membrane deformation. Creation of the patch clamp method allowed the first recordings of MSC currents (Guharay and Sachs, 1984). Various methods of mechanical stimulation of cells during electrical recording have been developed since then. Besides electrophysiology mechanical stimulation of cells is widely employed in combination with molecular biology and sometime Ca^{2+} signalling measurements. Electrophysiological recordings remain the best method of choice since the response is rapid and relating cause and effect is simpler. The tools for recording single channel and whole cell currents are well established, but initial and transitional states in which cellular membranes remain before and during been stressed in pipettes are less well understood. Distribution of the stress on the membrane, following various perturbations such as osmotic swelling, is also poorly understood. The use of positive or negative pressure, applied to the pipette tip as a mechanical stimulation, and data, obtained in those experiments, need special caution in handling. In addition, plenty of various methods of cell stretching have been used in recent years and their particular limitations need separate consideration. These are the issues under discussion in the present review.

1.2 Single Channel Recording

1.2.1 Positive and Negative Pressure as a Mechanical Stimulus

Single channel events elicited by application of positive or negative pressure applied to the pipette tip are readily identifiable as MSCs when the response is discrete, although in most cases the approximation of an open state is masked by a rapid buzz mode visible at higher bandwidth. MSCs have characteristically noisy open states where the channels are clearly changing state rapidly. The origin of this excess noise is not known. The assumption for interpreting data with this approach is that MSCs respond to mechanical forces in the plane of the cell membrane (membrane tension), not to hydrostatic pressure perpendicular to it (Gustin et al., 1988; Sokabe and Sachs, 1990; Sokabe et al., 1991). In cells, the definition of the mechanical membrane is fuzzy. Stress is distributed among various components associated with the cortex of the cell (Akinlaja and Sachs, 1998). The lipid bilayer is far from homogeneous in content (Lillemeier et al, 2006), let alone stress. The cytoskeleton applies forces parallel and normal to the bilayer (Suchyna and Sachs, 2007). The forces in the extracellular matrix are unknown. Thus, when for convenience the term "membrane" is used in the following discussion, the reader must recognize that it does not apply to the lipid bilayer, unless otherwise stated, only to some average various components of the cell.

Membrane tension is changed by application of negative or positive pressure to the patch pipette. The minimal information needed to address mechanics is geometry. High-resolution video imaging (Akinlaja and Sachs, 1998; Opsahl and Webb, 1994; Sokabe and Sachs, 1990; Sokabe et al, 1991; Sokabe et al, 1993; Suchyna and Sachs, 2007; Zhang et al, 2000), high-voltage electron microscopy (Ruknudin et al, 1991), atomic force microscopy (Hörber et al, 1995), and fluorescence microscopy (Disher and Mohandas, 1996; Discher et al, 1994) have been used to characterize the shape of the membrane patches (for review see Hamill, 2006). While several authors tried to calculate the membrane tension by applying the Laplace's law, in general they did not discuss or evaluate the effect of the assumption that the system is not two dimensional and not in equilibrium. However, applying the Laplace's law will provide a reasonable upper estimate of the stress. At low pressures Laplace's law becomes ineffective as the pressure goes to zero and the radius to infinity. However, the presence of a gigaseal specifies the minimum membrane tension (Opsahl and Webb, 1994) and this is in the range of half of the lytic tension and can never be reduced under equilibrium conditions (Honoré et al, 2006). Since the tension in different components of a patch is not known, the stimulus is commonly reported to be pressure, although pressure only acts when the boundary conditions convert it to a local stress and even that is a dynamic process (Suchyna et al, 2004; Suchyna and Sachs, 2007).

The pressure in the pipette can be controlled precisely (Besch et al, 2002), but how it is converted to a local stress varies in each experiment. Patch-pipette pressure is not the same as the trans-patch pressure because of stresses normal to the membrane (Suchyna and Sachs, 2007). It is generally accepted that some suction is necessary for formation of the gigohm seal (see Fig. 1.1A,B). In this case a fragment of the cellular membrane will be sucked into the pipette tip (Fig. 1.1C) (for review see Hamill, 2006).

Although numerous experiments definitely demonstrated that all cell attached patches creep upward regardless of stimulation (Suchyna and Sachs, 2007), in some cells, for example single enzimatically isolated cardiomyocytes or skeletal muscle, sometimes the gigaseal forms spontaneously (Sakmann and Neher, 1983) as expected from the large adhesion energy.

Suction used to obtain the gigaseal is sufficient to smooth out the surface folds and highresolution video images of cell-attached patches indicate an optically smooth membrane that is pulled flat (Hamill, 2006). The suction step causes the dome to move upward (Fig. 1.1B) and tension to decrease as the area of the initial concave spherical dome is larger than the disk required to span the pipette (Suchyna and Sachs, 2007). After the gigaseal forming suction it was shown that in a resting

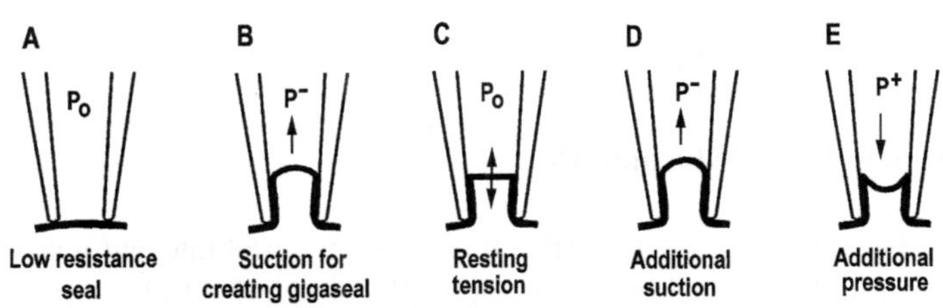

Fig. 1.1 A cartoon of patch shapes. A – Low resistance seal, B – Suction for creating gigaseal (cell-attached), C – resting tension (atmospheric pressure), D – Additional suction, E – Additional compression

patch, adhesion of the patch to the glass creates a resting tension (Akinlaja and Sachs, 1998; Opsahl and Webb1994) and indicates an optically smooth membrane (Fig. 1.1C) perpendicular to the walls of the pipette (Hamill, 2006). The patch with resting tension may be bowed in either direction by normal stresses from the cytoskeleton. The resting tension can also be increased by cyto skeletal forces bowing the patch (Honoré et al, 2006). Hamill stresses that gigaseal formation introduces significant changes in patch mechanics that can alter the mechanosensitivity of channels in the patch. The extrinsic changes in membrane geometry and structure of the cytoskeleton may have different effects on specific channels depending upon their properties (Hamill, 2006). Honore et al. demonstrated that some channels were active at zero pressure because the seal adhesion energy produces tension (Honoré et al, 2006). That data is of great importance for further discussion. The patch may actually become wrinkled as it passes through the region of zero mean curvature (Honoré et al., 2006).

In the absence of any excess membrane, pressure pulses, applied to the patch will rapidly flex the membrane and increase bilayer tension (Hamill, 2006). Please notice that the initial concave patch curvature and the increased viscoelastic relaxation time causes a delay in generating patch tension in response to a step of suction (Suchyna and Sachs, 2007). The flexing of the membrane either outward with suction (Fig. 1.1D) or inward with pressure (Fig. 1.1E) results in the activation of inward channel currents (Suchyna and Sachs, 2007; for review see Hamill, 2006).

In a number of works based on use of suction (Fig. 1.1D: P^-) and pressure (Fig. 1.1E: P^+) for changing membrane tension under pipette the authors discussed the results not only in view of stretch-activated channels (SACs) present on the membrane, but also from the point of stretch-inactivated channels (SICs) and pressure activated channels (PACs) presence. This in essence meant that the channels react to the curl of the curvature directed towards the base of the pipette (under suction) or opposite (under pressure). Until recently those views were widely discussed but then Honoré, Sachs and Suchyna on the basis of their own recent experiments without doubting the obtained results argued their interpretation, i.e. the presence of SICs and PACs (Honoré et al, 2006; Suchyna and Sachs, 2007). Honoré demonstrated that buckling of the patch increases the local curvature in both directions, but always leads to channel closing. Thus, membrane curvature, per se, cannot account for channel activation, and we favor the traditional interpretation that the primary stimulus for activation is tension, not curvature (Honoré et al, 2006).

1.2.2 SACs, SICs and PACs

1.2.2.1 Stretch-Activated Channels

MSCs have been described in several cell types from vertebrate and non-vertebrate species (Morris, 1990; Sackin, 1995). A well-characterized class of MSCs are the stretch-activated non-selective cation channels (SACs), which are activated by pipette suction (Fig. 1.1D: P^-), associated earlier with changes of the curvature of the membrane towards patch-pipette and membrane coupled cytoskeletal structures (Guharay and Sachs, 1984; Guharay and Sachs, 1985). Fred Sachs, who with his co-workers (Guharay and Sachs, 1984), was the first to characterize SACs in chick skeletal muscle, has comprehensively reviewed this work from his lab. Most studies of MSCs were focused on SACs, which are the most frequently encountered type of MSCs in recordings from membrane patches. In porcine aortic endothelial cells (Lansman et al, 1987), intact aortic endothelium of the rat (Hoyer et al, 1997), and in intact tissue preparation from porcine right atrium (Hoyer et al, 1994) cation selective SACs with a conductance for monovalent cations ranging from 20 to 40 pS have been reported.

1.2.2.2 Stretch Inactivated Channels

MSCs, which are active at resting tension and are inhibited by the application of pipette suction (Franco-Obregon, Jr. & Lansman, 1994), were isolated into a separate group and were called stretch-inactivated-channel (SICs) (Franco-Obregon & Lansman, 2002). Authors have used the *mdx* mouse model to investigate whether dystrophin contributes to MSC gating making use of the activity of single MSC. The authors showed differences (Fig. 1.2) in the suction dependence of stretch-activated (Fig. 1.2A) and stretch-inactivated (Fig. 1.2B) gating (Franco-Obregón and Lansman, 2002). According to the authors the experiments show that SACs have a very low open probability at resting tension. In response to suction, they open rapidly and then close quickly when the pressure stimulus is terminated (Fig. 1.2A). By contrast, SICs have a very high open probability at resting tension, close rapidly when suction is applied, and reopen after releasing the suction (Fig. 1.2B). This perception of SICs was accepted by scientists working in this field.

Recently Honoré et al. (2006) and later Sunchya and Sachs (2007) gave a different interpretation to the data, which was based on their own experiments. They believe that SICs do not exist and inactivation of channel activity in response to suction can be explained by the activity of pre-stressing of stretch-activated channels (Honoré et al., 2006). The origin of that phenomenon was analyzed in detail and lies in the fact that studied channels are sometimes open in absence of mechanical stimulus (Suchyna and Sachs, 2007). Moreover wild-type patches lose spontaneous

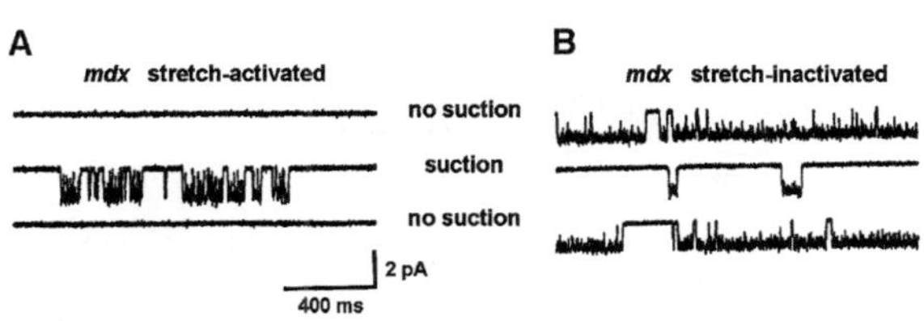

Fig. 1.2 Examples of hypothetical different types gating of MSCs. A – Stretch activated channels, B – hypothetical stretch inactivated channels in *mdx* myotubes. SICs are active at rest and close when suction is applied to the patch. Reproduced from (Franco-Obregon and Lansman, 2002) with copyright permission of the Blackwell Publishing, Journal Physiology (London), and Authors

activity over time (Morris, 2001). Thus it is possible to conclude from this data that Fig. 1.2B shows SAC in pre-stretch condition but not SIC. This was also shown for the 2P domain MSC (Honoré et al, 2006), so the presence of the phenomenology of SIC behavior does not imply the existence of a new type of channel. In this respect it is worth noting that gramicidin A (gA), which exhibits mechanosensitivity in lipid bilayers (Goulian et al., 1998; Hamill and Martinac, 2001), could behave as a stretch-activated or stretch-inactivated channel depending on the bilayer thickness (Martinac and Hamill, 2002).

1.2.2.3 Pressure Activated Channels

Köhler and colleagues reported the existence of a pressure-activated cation channel (PAC), as a novel type of a mechanosensitive channel, which has been suggested to act as a mechanosensor in aortic endothelium (Köhler et al, 2001a; 2001b). This hypothetical type of channels is activated by application of positive pressure (P^+), that changes the membrane curvature in the direction opposite to the base of the pipette (Fig. 1.3), depending on the volume of applied positive pressure (Köhler et al, 1998). The authors noted that in this tissue PACs coexist with SACs, which have been reported in porcine aortic endothelial cells (Lansman et al, 1987), in intact aortic endothelium of the rat (Hoyer et al, 1997), and in intact tissue preparation of endocardial endothelium from porcine right atrium (Hoyer et al, 1994). In that work author's show that cation currents through such SACs are sensitive to gadolinium

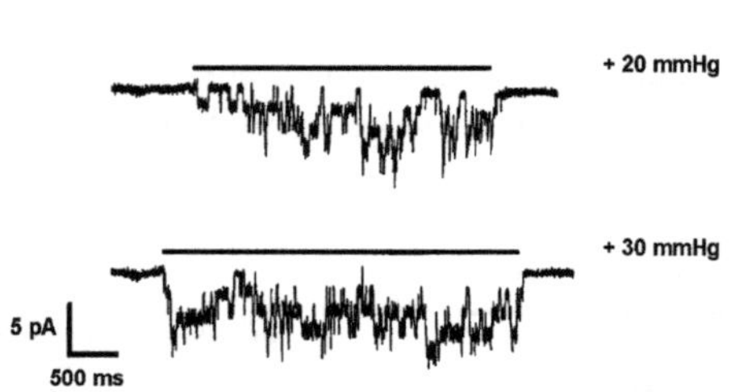

Fig. 1.3 Channel activity recording by applying positive pressure (P^+). A – Increase of spontaneous channel activity of hypothetical PAC (holding potential −80 mV in cell-attached patches) in endocardial endothelium of rat left ventricular papillary muscle by positive pipette pressure (30 mmHg) and inactivation of hypothetical PAC by negative pipette pressure (–30 mmHg). B – Gradual increase of channel activity in response to positive pipette pressure at 10, 20 and 30 mmHg, respectively. Reproduced from (Köhler et al, 1998) with copyright permission from Elsevier and Cardiovascular Research

(Hoyer et al, 1997), which is very typical for SACs and is mentioned by all the researchers working with SACs. The PAC in rat endothelium, which is also blocked by gadolinium (Köhler et al, 1998), represents however, according to the authors, another class of MSC with mechanosensitive properties different from those of SACs, as the PACs—unlike SACs—are solely activated by application of positive pressure to the plasma membrane, i.e. changing the membrane curvature in the direction opposite to the pipette base.

However, in literature there is controversy regarding existence of pressure-activated cation channels (PAC), as a novel type of mechanosensitive channel, and this view is not shared by all authors. First of all, the authors, postulating the PAC presence, believe that under pipette pressure the patch-clamped membrane becomes spherical (Köhler et al, 1998) that might cause activation of PACs. Meanwhile, a recent research in this field shows a variety of different structures, many of which are dynamic, and only some of which persist as spherical caps (Hamill, 2006; Honoré et al, 2006; Suchyna and Sachs, 2007). Secondly, based on channel activity recording in response to application of positive pressure the authors postulated that SACs and hypothetical SICs did not show any distinct response to positive pipette pressure (P^+). Therefore, the activation mechanism of pressure-activated cation channel (PAC) seems to be different from other MSC (Köhler et al, 1998). However there are no other reports showing asymmetric responses to positive and negative pressure, while several groups presented data, showing symmetric responses (Akinlaja and Sachs, 1998; Hamill, 2006; Suchina et al, 2004; Suchyna and Sachs, 2007). For example, Hamill (2006) shows for a membrane patch before (0 ms), during (50 ms), and after (250 ms) a 100-ms suction step and a similar pressure step in both cases activated a 50-pA inward current (Zhang and Hamill, 2000; Zhang et al, 2000). It was shown that both flexions of the patch outward (suction) or inward (pressure) resulted in inward current responses to suction and pressure pulses (Hamill, 2006). In the oocyte, suctions or pressures of approximately 20 mmHg produced saturating responses, so it was assumed that any channel opening caused by an increase in suction or pressure of at least 20 mmHg would involve reopening of channels that have just closed. According to the Laplace equation, positive or negative pressures should make equal contributions to the stress (Akinlaja and Sachs, 1998). Let us remind that membrane curvature, per se, cannot account for channel activation, and we favor the traditional interpretation that the primary stimulus for activation is tension, not curvature (Honore et al, 2006).

However, Suchina and Sachs have noticed that equal magnitude of negative and positive pressure steps are not equivalent in terms of a stimuli (Suchyna and Sachs, 2007). Alternating positive and negative 30 mmHg pressure steps used by them were applied to the pipette for about 6 seconds. Thus, if assumed that MSCs are gated by bilayer tension, the tension developed from positive and negative pressure is not equivalent and the system is clearly nonlinear as previously reported (Akinlaja & Sachs, 1998). It is unlikely that MSCs are responding to changes in membrane curvature since the radii of curvature of the patch are too large to generate much energy over the dimensions of a channel (Wiggins and Phillips, 2005; Markin and Sachs, 2004). The insignificant role of global curvature is also suggested by the fact that channels are not activated when the membrane wrinkles upon passing from inward to outward curvature (Honoré et al., 2006).

1.3 Whole Cell Recording

Direct mechanical stretching of a cell is the simplest form of mechanical stimulation. However from methodological point of view stretching of a cell is a very challenging and complicated procedure. Several approaches exist up to date. Choice of any of them depends on the cell type, which is used as experimental model for investigation. Stretching freshly isolated cardiomyocytes is the most challenging among them. In all cases of stretching cells it is essential to realize that the sites of stress application are different from the *in situ* cells, so that the responses may be irrelevant to normal physiology. Furthermore, the sites of stress application in cells in vivo are not known. Regardless of those considerations now we will discuss the technical aspects of methodology, which is currently used in the field.

Fabiato (1981) isolated pairs of myocytes in tandem. He mechanically fixed a force transducer to one of the cells that was damaged thereby whilst the other one remained intact. The method because of its low success rate is not widely used. Attachment of a force transducer to a single cardiomyocyte in this case by glue has low success rate because all glues, which are available on the market, damage the cell membrane.

1.3.1 Homogeneous Axial Stretch by Two Patch Pipettes

Most commonly used method of stretching of single cells is employing two patch pipettes, one of which is used for recording the whole cell current (PP), while the other patch-pipette (SP) with a wide opening lifts and stretches the cell (Fig. 1.4). This approach is difficult to implement because, in most preparations, attachment of pulling probes to the cell without producing local stress is impossible and often irreversible damage terminates the experiment.

Estimated degree of stretch (or local compression of the cell by the pipette tip) for investigated freshly isolated cardiac fibroblasts (Fig. 1.4) is reasonably high (stretch for 5 μm or local stress around the tips for 3 μm with the cell diameter at 18 μm), if guided by not only electrophysiological response—activation of MSC and inactivation of MSC (see further)—but also by calculations given by Morris and Homann (2001). According to those authors for cultivated fibroblasts, some maximum length of bilayer tether can be pulled from the cell surface using a fixed force. This identifies the magnitude of a small bilayer reservoir, amounting to 0.3–1% of the total surface area. This reservoir could buffer minor increases in membrane tension and until depleted, obviates a need for tension-sensitive surfaces area. Interactions between membrane-associated proteins and bilayer could foster micro-undulations

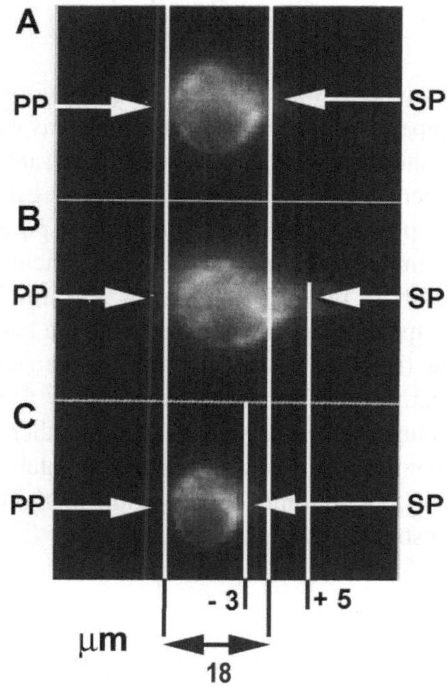

Fig. 1.4 Microphotographs of fluorescent enzymatically isolated single cardiac fibroblast in control and during mechanical deformation. PP- patch pipette, which is used for recording currents in whole cell configuration. SP- patch electrode used for mechanical deformation of the cell. A – control (fibroblast before stretching). The distance between PP and SP is 18 μm. B – fibroblast is stretched by SP by 5 μm respective to PP. SP-PP distance is increased by 28%. C – compression (implying that all responses will be dominated by local stress around the tips) of the fibroblast by means of SP by 5 μm respective to PP. SP-PP distance is decreased by 16%. Reproduced from (Kamkin et al, 2003) with copyright permission of the Naturmort Publishing House

that flattened as tension increased. Thus, artificial biological membranes stretch elastically until their area increase becomes 2–3 %.

Stretching of cells by two patch pipettes was successfully used in several cell types, for example in studies, conducted on vascular smooth muscle cells (Davis et al. 1992), single ventricular myocytes (Sasaki et al., 1992), smooth muscle cells from blood vessels and the urinary bladder (Wellner and Isenberg G, 1993, 1994, 1995) and on cardiac fibroblasts (Kamkin et al., 2003a; Kamkin et al, 2003b; Kamkin et al., 2005a).

In general, stretching of cells by two patch pipettes is analysed in detail by Sachs and Morris (1998), membrane mechanics under various types of mechanical interference with the cell was studied in detail by Morris (1990), and membrane tension is analysed in the work by Morris and Homann (2001). It can be suggested that for certain cell types like, for example, smooth muscle cells and fibroblasts, homogeneous axial stretch can be obtained by using two patch-pipettes. In most preparations it is relatively easy to fix pipettes to cells that are stretched without deformation. It was shown that in some cases auricle cardiomyocytes can be effectively stretched using this method (Zhang et al., 2000). Other cells require different stretch methods (Sachs and Morris 1998).

1.3.2 Axial Stretch by Two Glass Capillaries

Axial stretch of the sarcomeres was achieved by sucking both cell ends into the openings of two glass pipettes, distance between which was varied by piezo devices (Zeng et al, 2000). Isolated ventricular cells with clear sarcomeres were held by two concentric glass pipettes, with the inner pipette serving as a stop to prevent the cell from being sucked up by the outer pipette. The outer pipette was pulled from a glass capillary with the inner tip diameter of ~15 μm. The inner pipette was made from a glass capillary with the outer tip diameter of ~12 μm. The inner pipette was inserted into the outer pipette by a manipulator, leaving a gap of ~8 μm to the tip of the outer pipette. The tip of the outside pipette was then cut by fusion of the tip to the filament of a microforge. The cut end was lightly fire-polished so that the tips of the inside and outside pipettes were forged together and formed a cup to hold the cell. A third manipulator was used to attach a patch-pipette for whole cell electrophysiology (Fig. 1.5) recordings. Probably this method can be considered as homogeneous axial cell stretching.

1.3.3 Local Axial Stretch by Glass Stylus and Axial Stretch by Two Glass Styluses

This method is very similar to the previous one, but instead of SP uses fire-polished glass stylus (S) that adheres to the cell surface. This method of local stretch uses displacement of the glass stylus together with part of the cell surface and underlying sarcomeres (Fig. 1.6). It does not stretch the bottom of the cell that was attached to the glass bottom of the chamber where attachment was facilitated by coating it with poly-L-lysine. This method applies a component of shear stress to some components of the cell as well as axial and non axial stress (Isenberg et al., 2003; Isenberg et al, 2005; Kamkin et al., 2000; Kamkin et al., 2003c, Kamkin et al., 2005b).

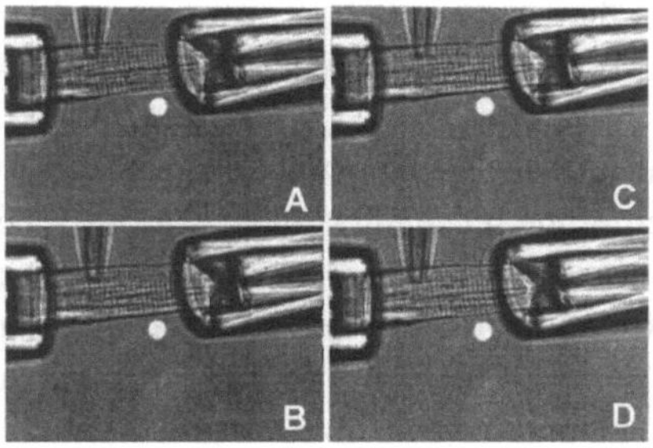

Fig. 1.5 Axial stretch. Cell was attached to two concentric pipettes (left and right) and voltage-clamped by the third (middle). Right concentric pipette moved right to stretch the cell, and patch pipette moved with local strain to reduce stress around tip. Reproduced from (Zeng et al, 2000) with copyright permission of the American Physiological Society and American Journal Physiology

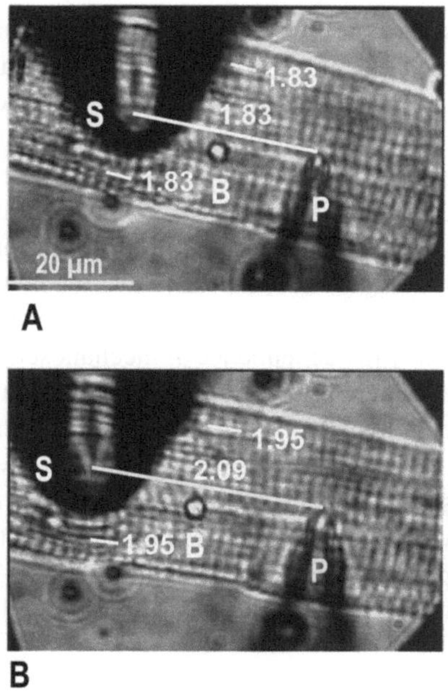

Fig. 1.6 Local axial stretch. Mechanical stimulation: Local stretch of a guinea pig ventricular myocyte. (A) before, (B) during stretch. Original microphotographs. Labels for glass stylus S, microbead B attached to the cell surface, patch pipette P, and a line connecting S over B to P. Increasing the distance S–P by 7 µm (connecting line from 31 to 38 mm or by 22%) increases distance S–B and B–P by the same extent of 22%. Reproduced from (Kamkin et al. 2000) with copyright permission from Elsevier and Cardiovascular Research

By positioning the stylus and patch-pipette ~40 µm apart before attaching them to the cell the area of stretched membrane was restricted to approximately one third of the cell length. To monitor changes of the surface membrane, microbeads of 4 µm diameter were added 20 min in advance. The cell attached glass stylus (S) is displaced from the patch pipette (P) (Fig. 1.6 : from 31 to 38 µm or by 22%). The resulting stretch increased the distance between S and the microbead (B) and the distance between B and P to the same extent (22% in Fig. 1.6 B), suggesting that the stylus did not slip on the surface of the membrane. The sarcomere length (SL), however, did not follow the glass stylus as expected. On average, SL before stretch was 1.83±0.01 µm (Fig. 1.6 A evaluations in the area on top or below the line S–B–P). During local stretch, the average SL increased to 1.92±0.08 µm or by 5%. As indicated by numbers, sarcomere length increased up to 2.09 µm between S and P in the center of the cell and only to 1.95 µm at the peripheral regions. Left to S and on the right to P, the sarcomere length did not change significantly (Fig. 1.6). The different lengthening, 5% SL versus 22% surface (microbead) and the increased S.D. of the SL are the first indication that the stylus induced surface deformation, which spread with spatial decrement.

Nevertheless the above works have certain limitation. Using the previously mentioned method (Zeng et al, 2000) the cell stretch was really axial and probably homogeneous. Sarcomere length on the background of cell stretch was the same in any area of the stretched cell. In contrast stretching the cell by one of the styluses causes local unhomogeneous stretch of cell area. Appearing stress gradients between stylus and pipette are different but homogeneous. They are the highest and most homogeneous in the center of the cell (Fig. 1.6 B) next to the drawn line (as next to the line where the sarcomere lengths are equal). But the gradients of appearing stress decreases on the background of imaginary parallel lines drawn with Δl towards upper and lower cell edges. At the same time within the boundaries of each imaginary line they are homogeneous and sarcomere lengths are equal (Fig. 1.6 B). In any case they are sufficient for producing different electro-physiological responses caused by a different degree of stretching of a local cell zone. At that there was strict dependency of inward currents amplitude on the applied stretch degree. This method allowed to investigate mechanosensitive whole-cell currents of isolated ventricular and atrial cardiomyocites in healthy hearts of mice, rats, guinea-pigs, rabbits and humans, to study the dependence of mechanosensitive whole-cell currents on the age of animals and humans and to study those currents on the background of various heart failures, e.g., hypertrophy following myocardial infarction (Kamkin et al, 2005b).

The method of stretching the cell with adhered glass stylus can be used for axial stretch of cells if the investigator uses two glass styluses (one at each end of the cell), while positioning patch electrode at the center of the cell, as it is described in the previous method (Zeng et al, 2000). This approach is much simpler to implement then stretching the cell with two capillaries.

1.3.4 Local Axial Stretch by Two Thin Carbon Fibres

Several groups succeeded in attaching a thin carbon fibre to the cell end (Fig. 1.7). When the fibre was moved together with the cell end, an almost homogenous lengthening of the sarcomeres that were far from the probes, was achieved (Le Guennec J-Y et al, 1991; White et al, 1993; White et al, 1995). However it is difficult to position the cell in such a way that both of its ends are in contact with a carbon fiber, and to achieve a good contact simultaneously between both ends of the cell

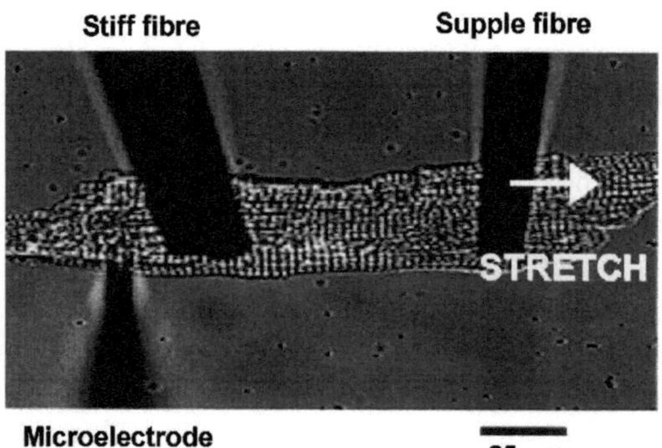

Fig. 1.7 Carbon fibre method. The microelectrode is used to record membrane potentials or currents and is placed behind the stiff fibre to protect the site of impalement during the stretch. The supple fibre is used to stretch the cell and its displacement during stimulation is a measure of active force development. Increased sarcomere spacing is used as index of stretch. Reproduced from (Belus and White, 2003) with copyright permission of the Blackwell Publishing and J Physiol (London)

and the two fibers. This method is useful for registering the contractile activity of the cell when bending one of the fibers is used as strain gauge.

1.3.5 Cell Swelling

Cell swelling by pressure inflating the cell or via exposure to hypoosmotic extracellular media can provide a three-dimensional stretch of the cell (Baumgarten and Feher, 2001; Baumgarten and Clemo, 2003; Baumgarten, 2005; Baumgarten et al, 2005; Niisato et al, 1999). The method has the problem that changes in volume are not easily related to changes in tension. The literature agrees that it is difficult to distinguish between the effects of the stretch component and other possible effects of swelling. Hu and Sachs (1996) showed that in chick heart cells direct mechanical stress activated cation currents and osmotic swelling activated anion currents. The stimuli are clearly different. The changes in electrophysiology caused by swelling cannot be compared with those due to homogenous or local stretch. Moreover, cell swelling, used by any authors, is a nonspecific stimulus (Browe and Baumgarten, 2003). Besides its mechanical effects, it reduces ionic strength, dilutes cytoplasmic ions and macromolecules, and relieves macromolecular crowding, thereby activating signaling molecules.

1.3.6 Paramagnetic Microbeads Method

Ideally, studies of stretch-activated channels would make use of the structures that normally transmit forces to cells (Browe and Baumgarten, 2003). It was supposed, that mechanical forces are transmitted by integrins (Wang et al., 1993). Glogauer and Ferrier (1998) describe a new method that uses straightforward physics to apply force to substrate-attached single cells (Glogauer et al., 1995; Glogauer and Ferrier, 1998). In this method collagen-coated magnetic ferric oxide beads ($\sim$5 μm in diameter) were attached to the dorsal surface of fibroblast via receptors of the integrin family, and a magnetic field gradient was applied to produce a force. The authors showed that applying force to beads induces membrane tension on the dorsal surface of substrate-attached fibroblasts.

This model allows the application of controlled forces to the membrane (Glogauer et al., 1995). Applied forces were parallel with the dorsal surface of the cell and had only a minimal vector component orthogonal to the dorsal surface. The total cell surface area and the percent surface area, covered by beads, were determined. Authors manipulated the magnetic force applied to cells in two different ways: 1) variation of the applied force 26, 99, or 165 dyn/cm^2 and 2) variation of the surface area of cell membrane covered with attached beads 73, 243, or 504 μm^2.

The area of cell membrane subject to the bead-induced force is significant in determining whether there will be a response. Both, (1) force application to the cells under various magnetic flux and (2) bead loading conditions demonstrated that the level of bead loading was a very important variable in determining whether the cell would react to the force. This is evident by comparing cells exposed to a 99-dyn/cm^2 force at high bead load (504 μm^2) and at low bead load (73 μm^2). Simply changing the bead load increased the proportion of responding cells (35% at the low level to almost 80% at the high level). This suggests that the area of cell membrane subject to the bead-induced force is significant in determining whether there will be a response (Glogauer et al., 1995).

Cell shape was an important criterion for determination of likelihood of response to applied force. While cells like fibroblasts in culture were spherically shaped with a round morphology and short cell processes, they responded at the low bead load (73 μm^2) at low force levels (26 dyn/cm^2). In case when the cells were long, thin, spindle-shaped with elongated filopodia, they did not produce response at such low force levels. However, when cells were loaded with high numbers of magnetic beads (504 μm^2), cell morphology was no longer an important determinant of response; both round and spindle-shaped cells exhibited responses even at low force levels (26 dyn/cm^2).

For a measured area of cell surface covered by beads of 250 μm^2 in a typical author's experiment (about 25% coverage of the dorsal cell surface) the mean number of beads per cell would be 36. This is close to 50% of the mean total bead volume per cell. For 50% bead coverage, similar calculations give a standard deviation for bead volume per cell of about 500 μm^3, or about 33% of the mean total bead volume per cell.

Authors proposed a model that accounts for possible roles of force, bead loading (force transfer sites), and actin filaments in the observed channels response to physical force application (Fig. 1.8). There is a fixed membrane tension (threshold), above which the channel will open (Glogauer et al., 1995).

Browe and Baumgarten (2003) modified the technical part of the method in the way that patch pipette could be brought up to cells (Fig. 1.9).

In conclusion it should be noted that the use of magnetic ferric oxide beads attached to the surface of the cell via receptors of the integrin family, and magnetic field gradient, applied to produce a force require certain precision of calculation. It has to do with registering of the area of cell membrane subject to the bead-induced force, cell shape, measurements, of the force, applied by the permanent magnet to a ferric oxide bead. Besides, potential effect of the magnetic field on MSCs should be taken into account.

Integrins have been suggested as a possible transducer for fluid shear stresses (Wang et al., 1993; Mazzag et al., 2003) because they extend well above the membrane and are linked to the cytoskeleton through the actin cortex. Moreover there is a common mechanism for the transduction at the cell surface of both fluid shear stresses and transverse electric fields, namely, the force exerted on an integrin. The fundamental transduction mechanism for some electric field effects may then be ultimately mechanical in nature. For electric fields it suggests that although the original stimulus is provided by an electric field acting on an integrin, subsequent steps involve mechanical coupling throughout the cell and even to the cell's neighbors (Hart, 1996).

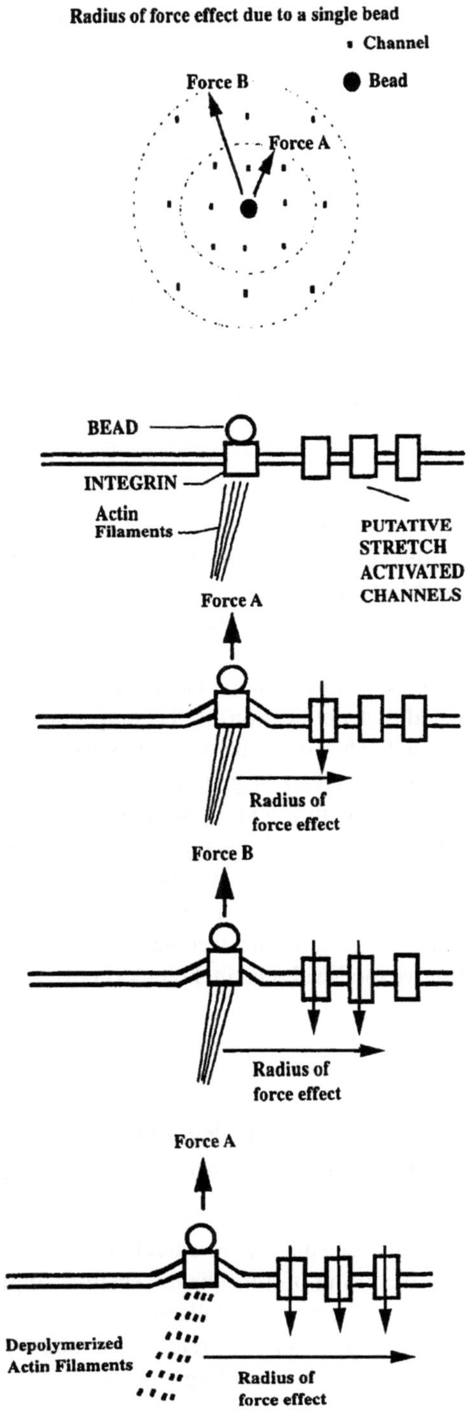

Fig. 1.8 Proposed theoretical model of effects of force, bead loading, and cytoskeleton. Applied force to membrane is translated into membrane tension, which activates channel opening. Horizontal arrows graphically indicate radius of magnetic force effect exerted on membrane by individual beads (top panel). Radius is increased with either higher force levels or by depolymerization of actin filaments with cytochalasin (bottom panel). We propose that with wider radius of force, more calcium-permeable channels are activated. Thus, after application of force level A, there is a large radius of force and many channels are activated. Reproduced from (Glogauer et al., 1995) with copyright permission of the American Physiological Society and American Journal Physiology

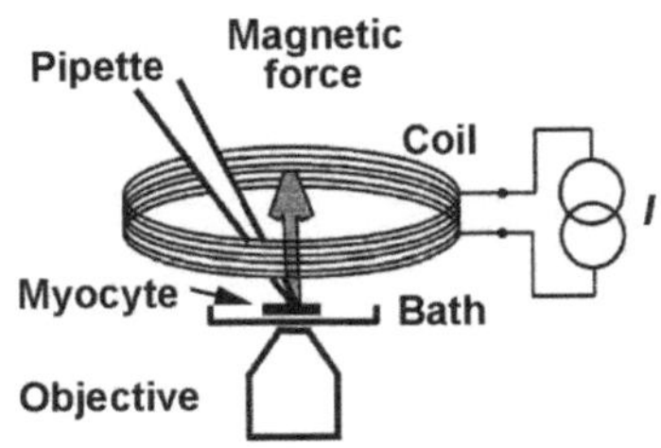

Fig. 1.9 Schematic diagram of the apparatus. A water-cooled electromagnet coil was placed directly on top of the bath. Energizing the coil with a constant current source, I, generated a magnetic flux density gradient of 2,400 G/m. The resulting force vector for each bead pointed upward toward the plane of the coil and perpendicular to the long axis of the myocyte. Reproduced from (Browe and Baumgarten, 2003) with copyright permission of the Rockfeller University Press, Journal of General Physiology and Authors

Recently, MSCs have been implied to play a role in magnetoreception. Petrov and Martinac (2007) have investigated the effect of static magnetic fields of moderate intensity on the activity and gadolinium block of MscL, the bacterial MSC of large conductance, which served as a model channel to study the basic physical principles of mechanosensory transduction in living cells. In addition it was shown that direct application of the magnetic field decreased the activity of the MscL channel. This study demonstrates for the first time that static magnetic fields can reverse the effect of gadolinium, a well-known blocker of MSCs. The results are consistent with a notion that (1) the effects of static magnetic fields on the MscL channels may result from changes in physical properties of the lipid bilayer due to diamagnetic anisotropy of phospholipid molecules and consequently (2) cooperative superdiamagnetism of phospholipid molecules under influence of static magnetic fields could cause displacement of Gd^{3+} ions from the membrane bilayer and thus remove the MscL channel block (Petrov and Martinac, 2007).

1.4 Simultaneous Recording of Single Channels and Whole-Cell Currents During Axial Stretch

The only relevant measurement of mechano-sensitive single channel activity would be a recording of a mechano-sensitive single channel in a whole cell mode, which requires a very small cell. So far there are no reports of successful single channels recordings during stretch of cardiac myocytes. However even if it was possible to perform such recordings the issue with such experiment would be that during stretch intracellular concentration of calcium is rising, which in turn would activate Ca^{2+}-sensitive K^{+} channels and/or other signaling pathways. Much worse is that the seal of a patch pipette isolates most of the stress in the patch from that in the cell.

1.5 Cell Pressing

In a different approach, whole-cell mechanosensitive currents have been evoked by pressing on a spherical cell with the tip of one pipette while voltage clamping with another (Bett and Sachs, 2000; Hu and Sachs 1994, 1995, 1996). It is known that some MSC respond when bowed toward the nucleus. Examples are found in glial cells (Bowman et al., 1992; Bowman and Lohr, 1996), smooth muscle cells (Kawahara, 1993) and endothelial cells (Marchenko and Sage, 1996).

Bett and Sachs reported currents from an approximately normal stimulus by pressing the top of a cell with the side of a patch pipette with a sinusoidal stimulus and the observed currents were phase locked to the stimulus (Bett and Sachs, 2000). This is probably the least invasive of any experiments on MSCs. Most striking in this work was the data, showing that mechanical sensitivity was not observed in resting cells, and could be detected only after exercising the cell. The sensitivity developed as an abrupt shift as though the channels served like a mechanoprotective boundary that was ruptured. This boundary would slowly recover over time with gentle stimulation causing the current to disappear, but it would be awakened with a temporary increase in the stimulus. This emphasizes that the cell is not in equilibrium and that the cytoskeleton and other components are dynamically active (Mizuno et al, 2007) so stresses are never under control. There is stress softening (Chaudhuri et al, 2007) as well as hardening and fluctuations far from equilibrium.

The studies of cardiac fibroblast responses to mechanical stress were conducted within investigating mechanosensitive whole-cell currents. Freshly isolated cardiac fibroblasts respond differently to stretching and tip pressing, when two patch-pipettes were used (Kamkin et al, 2003a,b; Kamkin et al, 2005a). Mechanical deformation of fibroblasts is shown in Fig. 1.4.

Figure 1.10 shows how net membrane currents are modulated by pipette tip pressing for 2 μm. Mechanical compression of fibroblasts (holding potential equals −45mV) causes the occurrence of inward current (compare the initial traces in Fig. 1.10 B marked by arrow and Fig. 1.10 A). Compression also increased the current amplitudes during the depolarizing clamp steps without changing their time course. At negative potentials, the currents were higher than under control conditions without compression. Hence, mechanical compression increased the membrane conductance of atrial fibroblasts. Finally, when 8 μm Gd^{3+} was added to the bath solution, not only the compression-induced currents but also a large portion of the currents in the absence of mechanical compression was blocked (Fig. 1.10 C). These findings strongly suggest that mechanical compression activates a mechanosensitive conductance in the cardiac fibroblasts.

Figure 1.11 shows membrane currents recorded under control conditions and during 2 μm lateral stretch. Stretch of the fibroblasts (holding potential equals −45mV) inhibits inward current (beginning of the traces in Fig. 1.11 B, marked by arrow). Stretch almost blocked the inward currents at negative potentials, and lowered the outward currents at positive potentials. Similar data with detailed quantitative analysis of these effects on *Shaker* are shown by Laitko and Morris (2004).

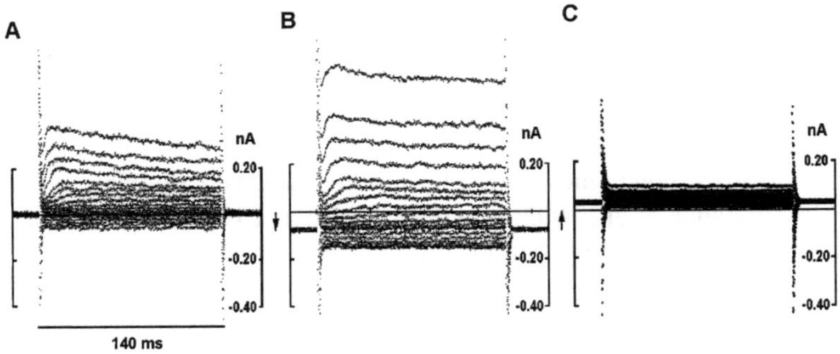

Fig. 1.10 Whole-cell currents from a cardiac fibroblast compressed by one of two hatch-pipettes for 2 μm according to Fig. 1.4 , A3 in the absence (A) and presence (B) of mechanical pressing by patch-pipette at 2 μm. (C) Effect of adding 8 μm Gd^{3+} on the whole-cell currents during sustained compression. Note: Mechanical deformation of a single enzymatically isolated cardiac fibroblast. PP- patch pipette, which is used for recording currents in whole-cell configuration. SP- patch electrode used for mechanical deformation of the cell. Sent were 20 pulses of 140 ms (0.5 Hz) that started at a holding potential of -45 mV and went to 0, −100, −90, −80, −70, −60, −50, −40, −35, −30, −25, −20, −15, −10, −5, 0, 10, 20, 30, and 50 mV. The holding potential of −45 mV was used on the assumption that it was close to the resting potential of the fibroblast. The first step went to 0 mV to have a comparison with the currents during the 16th step that went also to 0 mV. Reproduced from (Kamkin et al, 2003a) with copyright permission from Elsevier and Cardiovascular Research

These results suggest that stretch reduced the membrane conductance without significant changes of the time course of the currents (Fig. 1.11 B). The experiment was completed by addition of 8 μm Gd^{3+} to the superfusate. Gd^{3+} further reduced the currents during sustained stretch (Fig. 1.11 C). Since Gd^{3+} is known to block

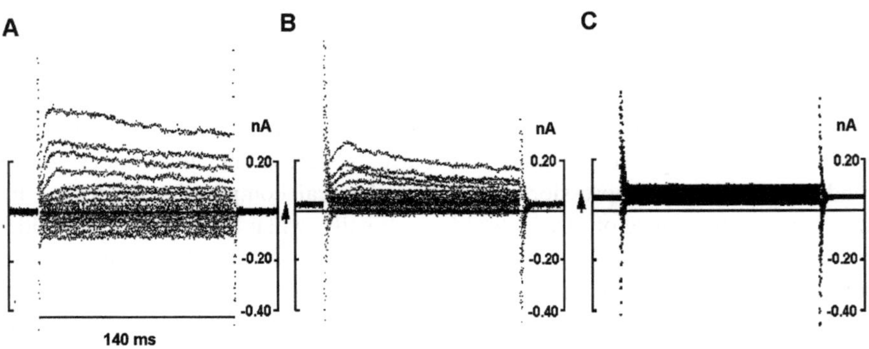

Fig. 1.11 Whole-cell currents from a cardiac fibroblast stretched by one of two hatch-pipettes according to Fig. 1.4, A2 in the absence (A) and presence (B) of mechanical stretch by patch-pipette at 2 μm. (C) Effect of adding 8 μm Gd^{3+} on the whole-cell currents during sustained compression. Series of whole-cell currents from cardiac fibroblasts in the absence (A) and presence (B) of stretch (2 μm). (C) Effect of 8 μm Gd^{3+} on the whole-cell currents during sustained stretch. Changes of the holding current are marked by arrows. Inhere condition correspond to those in Fig. 1.12. Reproduced from (Kamkin et al, 2003a) with copyright permission from Elsevier and Cardiovascular Research

mechanosensitive ion channels in a variety of cell types, these results suggest that cardiac fibroblasts express mechanosensititive ion channels that are inactivated by stretch and can be blocked by Gd^{3+}.

In case of an isolated cardiac fibroblast, which has almost spherical shape, such distinct difference in reaction to compression and stretch can be explained only by different application of the mechanical energy to the cytoskeleton, which causes different effects, possibly mediated via MSCs, but the cytoskeleton is known to be not a passive medium. In our case homogeneous axial stretch is activating MSC, when compression is suppressing their activity.

The currents at the end of the 140 ms long voltage-pulses were assembled in current-voltage relations (*I-V* curves, Fig. 1.12). Without mechanical stimulation, the *I-V* curves intersected the voltage axis at −37 mV. This zero-current potential corresponds to the normal resting membrane potential V_m (=E_m) of the non-clamped fibroblasts. Mechanical compression of isolated fibroblasts by pipette caused downward flection of the *I-V* curve and shifted V_m to more positive values (Fig. 1.12A).When freshly isolated atrial fibroblasts were stretched, the *I-V* curve was bent upward, and intersected the voltage axis at a more negative V_m (Fig. 1.12C). Both, the compression-stimulated and the stretch-reduced currents reversed their polarity close to 0 mV, as would be expected for a mechanosensitive, non-selective cation channel that can conduct Na^+, K^+ and Cs^+ ions.

Thus it was demonstrated that stretch and local stress around the tips in the form of compression have different effects on fibroblast SACs. Compression of isolated fibroblasts caused downward flection of the *I-V* curve and shifted V_m to more positive values, although stretched *I-V* curve intersected the voltage axis at a more negative V_m. Gadolinium intensifies this shift. Thus it can be assumed that compression opens SAC and stretch closes them. Probably cytoskeleton transmits the stress energy to the channel in different ways depending on which the channel either opens or closes. It should be taken into account that the cytoskeleton dynamics are much more complicated than that and they do not have to be tightly correlated in time.

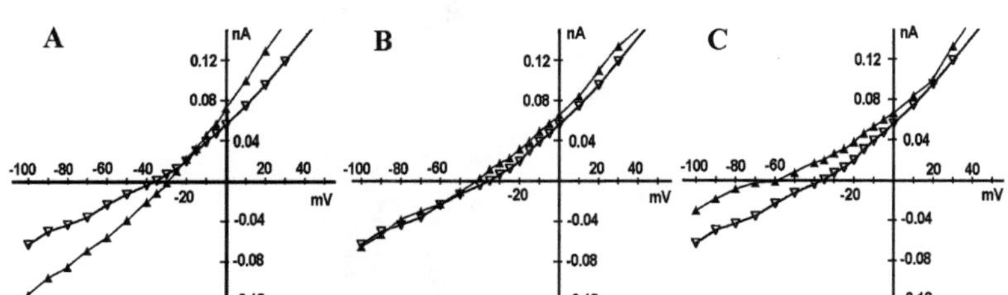

Fig. 1.12 Mechanosensitivity of membrane currents in fibroblasts that had been freshly isolated from rat atria. Current-voltage relations (*I-V* curves, current is measured at the end of the 140 ms clamp pulses). A: *I-V* curves before (empty triangles) and during 2 μm compression (filled triangles). B: Reversibility of the compression-induced changes (empty triangles before, filled triangles 2 min after stretch). Note the shift of the zero-current potential from V_m −37 to −28 mV. C: *I-V* curves before (empty triangles) and during 2 μm stretch (filled triangles). Note the shift of V_m from −37 to −60 mV. *I-V* curves A, B, C were recorded from the same cell. Reproduced from (Kamkin et al, 2003a) with copyright permission from Elsevier and Cardiovascular Research

The role of cytoskeleton can be found in different reactions of the cell towards stretch and compression of the cardiomyocytes lying edgewise, staying on the narrow side and staying on the broad side.

If we press on one part of a cell as shown in Fig. 1.13 for the cardiomyocyte the other parts of the cell will be stretched. In this case the reaction of the cell to local pressing may be the same or different from the cellular reaction to stretch, but it should be happening simultaneously with it and would be partially masked by it.

The brick-like cardiomyocytes have narrow and wide sides. Therefore, we considered two cases of cardiomyocyte compression - a compression on the narrow side and a compression on the wide side. It was not difficult to perform these experiments, because brick-like isolated cardiomyocytes, are stuck to a bottom of perfusion chamber in two positions: edgewise, staying on the narrow side or broadwise. In some experiments after seal formation and whole cell access, cardiomyocytes were rolled by the patch-pipette to attach to the glass bottom from edgewise, staying on the narrow side to broadwise. It allowed to study whole-cell current during stretch and compression on the same cardiomyocyte, placed in a different positions. Myocytes were compressed by pressing on the cell surface with the glass stylus towards the glass bottom of the chamber. Responses to stretch of cells, which were lying in two different positions, were the same. The effects of compression depended on the orientation of the myocyte, and were mediated by different mechanisms (Kamkin et al., 2005b).

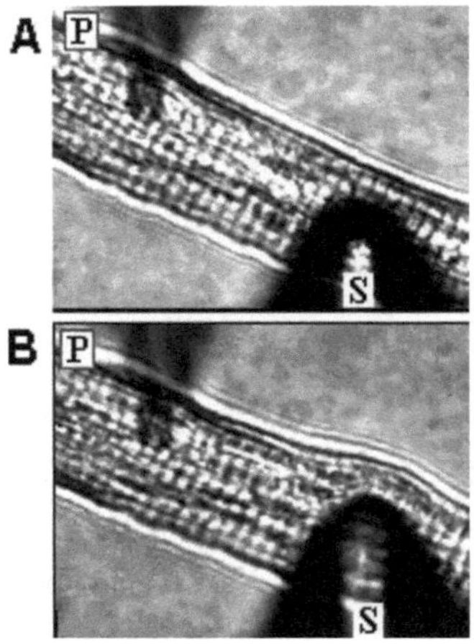

Fig. 1.13 Normal force deformation of ventricular cardiomyocyte with two glass tools. P: patch pipette fixing the myocyte to the glass coverslip. S: fire polished glass stylus. Stress caused by vertical movement of S towards the coverslip by 6 μm (C). Note: Under these conditions pressing the cell surface forms a dent in the shape of ball segment. It is due to the fact that the acting form of stylus tip is ball shaped with the diameter of about 20 μm. The stylus is brought to the cell nearly vertically which results in spherical dent in its surface 6 μm deep for this experiment. The result is that vertical pressing creates a dent on the cell surface about 900 μm^3. Reproduced from (Kamkin et al. 2005b) with copyright permission of the Academia Publishing House Ltd

We suppose that differences in cell responses to stretch and compression are determined by cytoskeleton. Cell stretching by glass stylus causes local homogeneous axial stretch of the cell, which leads to activation of certain MSC and inactivation of others, depending on their type. If we perform local compression with a glass stylus then we cause a change of membrane curvature under and around the stylus. This local change of curvature triggers an effective response of MSCs in the zone around glass stylus. Such change of membrane curvature would be quite significant and response to it would exceed the reaction of the cellular membrane to redistribution of the forces inside the cell, resulting from modest stretch of the neighboring regions of the cellular membrane. The bilayer's limited tensile strength dictates a need for surface area and, where bilayer is not highly curved, a need for mechanoprotection (Morris and Homann, 2001). Also, Honoré et al (2006) have shown that micron level curvature does not activate MSCs. But we believe that in our conditions a ball segment shaped dent is formed on the cell surface by pressing. It is related to the fact that that the acting tip of stylus ball had a diameter of approximately 20 μm. The stylus is brought to the cell nearly vertically. This causes formation of a spherical dent of 6 μm for this experiment. The result is that vertical pressing creates a dent of about 900 μm^3on the cell surface.

1.6 Reconstituted System

Reconstitution into lipid membranes simplifies the mechanics. Researchers specializing in studying reconstituted system try to find out what are the stimuli that are sensed by mechanoreceptors. Researchers investigate the molecular mechanisms, underlying activity of channels, that are gated by mechanical forces. The first way supports the idea that cytoskeletal components and extracellular matrix directly interact with fragments of the channel protein. The second way demonstrates that the channel protein directly senses changes of the properties of the membrane during tension. This approach is supported by experiments studying bacterial mechanosensitive channels, cloning and investigating them in liposomes and in blebs, which demonstrated that channels can directly sense tension within the lipid bilayer.

Recent evidence has shown that lipids are intimately involved in opening and closing the MSC of fungal, plant and animal species (Suchyna et al, 2004; Hwang et al, 2003; Andersen, 1999; Lundbaek and Andersen, 1999; Goulian, et al, 1998).

1.6.1 Bacterial Mechanosensitive Channels

The study of bacterial mechanosensitive channels demonstrated that channels can directly sense tension within the lipid bilayer. This worked well for the channels from *E. coli* (Martinac and Kloda, 2003). Three types of MSCs were identified in

E. coli, which are distinguished by their conductance, and named MscM (M for mini), MscS (S for small) and MscL (L for large) (Berrier et al, 1996). Bacterial MSCs were the only channels shown to sense bilayer tension, (Martinac et al, 1990; Martinac et al, 1987; Berrier et al, 1989; Berrier et al, 1996). The activity of MscL and MscM is not dependent on voltage. MscS displays voltage dependence (Martinac et al, 1987). In these investigations MSCs were activated by differential pressure across the patch membrane (Martinac et al, 1987).

Structural changes in MscL induced by bilayer-deformation forces were studied by combination of cysteine-scanning mutagenesis with site-directed spin labeling and electron paramagnetic resonance (EPR) spectroscopy. Those studies were complemented by analysis of channel function by means of the patch-clamp technique (Perozo et al., 2002b). The open state of MscL has a water-filled pore of >25 Å in diameter that is lined by the TM1 helices from the five subunits (Perozo et al., 2002a), and several studies show that that the channel undergoes a large conformational change when opening and closing (Sukharev et al., 2001; Gullingsrud et al., 2001; Biggin and Sansom, 2001; Betanzos et al., 2002; Colombo et al., 2003; Gullingsrud and Schulten, 2003). The study by Perozo and coworkers (Perozo et al., 2002a) demonstrates that hydrophobic mismatch is not the driving force that triggers MscL opening, although specific mismatch levels could stabilize intermediate states along the kinetic path towards the open state.

A recent study suggested two potential triggers of MscL gating by the bilayer mechanism: protein—lipid-bilayer hydrophobic mismatch and membrane curvature (Perozo et al., 2002a,b). However, it is also known that many local forces co-affect gating (Wiggins and Phillips, 2004; Wiggins and Phillips, 2005; Markin et al, 2006).

1.6.2 Purification and Functional Reconstitution in Planar Lipid Bilayers

A cloned rat epithelial Na^+ channel (ENaC) was studied in planar lipid bilayers. ENaC are highly Na^+-selective, are inhibited with high affinity by the diuretic amiloride, and are mechanosensitive. The authors studied single channel records of in vitro-translated ENaC in planar lipid bilayers (Ismailov et al, 1996a; Ismailov et al, 1996b; Ismailov et al, 1996c). Upon incorporation into planar lipid bilayers, in vitro-translated alpha-bENaC protein displayed voltage-independent Na^+ channel activity with a single-channel conductance of 40 pS, and was mechanosensitive in a sence that the single-channel open probability was maximally activated with a hydrostatic pressure gradient of 0.26 mmHg across the bilayer (Awayda et al, 1995).

Nevertheless planar bilayer has serious limitation for reconstitution and testing of the MSC activity. The measurements of membrane tension suggest that in-plane tension in the bilayer is too weak to have a significant effect on MSC gating (Dai and Sheetz, 1995). When authors inserted MSC into planar bilayer and showed activation with stress, they reported presence of an artifact. MSC does not sense tension of the planar lipid bilayer. Moreover planar lipid bilayer is under resting tension of about 1–5 dyne/cm (Elliott et al, 1983; Gruen and Wolfe, 1982; Ring, 1992; Ring and Sandblom, 1988; for review see Sachs and Morris, 1998). In this case MSC in planar lipid bilayer maintain resting activity. Therefore the results by Awayda et al. (1995), who studied ENaC in planar lipid bilayer are difficult to interpret. Finally, planar bilayers are tension clamped and cannot be used.

1.6.3 Mechanosensitive Channels in Visicles

Similar problems are caused by using visicles. Only vesicles with a fixed amount of lipid can be studied. But despite this limitation research studies targeted at the role of cytoskeleton in gating of MSC can be carried out.

1.6.4 Purification and Functional Reconstitution in Liposomes

It was demonstrated that lipid bilayer tension alone is sufficient to gate the MSCs directly, because purified MscL, MscS and other prokaryotic MSCs are still mechanically gated when reconstituted into liposomes (Sukharev et al, 1994; Häse et al, 1995; Kloda and Martinac, 2001a; Kloda and Martinac, 2001b; Kloda and Martinac, 2001c; Martinac, 2001; Perozo and Rees, 2003; Martinac et al., 1990; Martinac et al., 2004).

The membrane proteins forming specific MSCs have only recently been identified. Based on the finding that bacterial MSCs remain functional upon reconstitution into liposomes (Delcour et al., 1989) a novel strategy was developed by Sukharev et al. (Sukahrev et al, 1993; Sukahrev et al, 1994) towards molecular identification of MscL that involved detergent-solubilizing and fractionating membrane proteins, reconstituting the protein fractions in liposomes, then assaying the fractions for stretch sensitivity using the patch clamp recording. This technique was used to identify a variety of MSC proteins in bacteria and archaea (Kloda and Martinac, 2001c; Martinac and Kloda, 2003; Sukharev, 2002; Sukharev et al, 1994) and, most recently, the TRPC1 channel protein was identified as the MscCa channel in *Xenopus* oocytes (Maroto et al, 2005). A purified protein, reconstituted in pure liposomes, can retain stretch sensitivity. The technique also provided further evidence for the hypothesis that bilayer tension alone gated MSC (Martinac at al, 1990). In contrast, transient gating kinetics of MscCa are not retained in pure liposome patches expressing MscCa activity, following reconstitution of detergent-solubilized oocyte membrane protein, since it lacks an underlying cytoskeleton (for review see Hamill, 2006).

Using liposomes allows to address another old question. It is clear that stretching the bilayer will tend to decrease its lipid packing density and thickness. If the channel proteins experience a change, occurring in the membrane, it will respond with changes in the distribution between closed and open channel conformations (Hamill and Martinac, 2001; Kung, 2005; Markin and Sachs, 2004). Introducing into liposome membranes lysophospholipids and amphipathic molecules may cause local changes in tension and curvature at the lipid–protein interface and thereby shift the channel distribution (for review see Hamill, 2006). Thus we can consider the bilayer mechanism in relation to MSCs in terms of their activation by amphipaths or lysophospholipids.

Irrespectively of the limitation of using liposomes for MSC studies Morris and Homann (2001) discuss possibility of membrane surface area curves in liposomes being sufficient or not sufficient to activate MSCs. For illustration the authors consider two spherical liposomes, 2 and 20 μm in diameter, in saline.

1.7 Studies of MSCs in Situ in Spheroplasts and in Membrane Blebs

In many reviews (Hamill, 2006; Morris, 1990; Morris and Homann, 2001; Morris, 1992; Sachs, 1988; Saimi et al, 1988) it is demonstrated that MSCs of various ionic selectivities have been found in spheroplasts prepared from bacteria and fungi, in protoplasts prepared from plants, and in a multitude of animal cells of vertebrate and invertebrate origin. Macroscopic currents can be activated in spheroplasts formed from *E. coli*, yeast, and other microbial cells (Cui et al, 1995; Gustin et al, 1991; Gustin, 1988; Hamill and Martinac, 2001; Zhou et al, 1991). To form spheroplasts, the parent cells are enzymatically treated to remove the cell wall and, as a consequence, the bacterial cells assume a simple spherical geometry so that even a slight inflation causes the MSC activation (Cui C et al, 1995; Gustin et al, 1991; Gustin et al, 1988; Hamill and Martinac, 2001; Saimi et al, 1993; Zhou et al, 1991).

Two current models describe the MSC gating: the bilayer model and the more speculative, tethered model (Hamill and McBride, 1997). In the bilayer model, initially proposed for gating of MSC in Escherichia coli giant spheroplasts (Martinac et al., 1990), lipid bilayer tension alone is sufficient to gate the MS channels directly, because purified MscL, MscS and other prokaryotic MSC are still mechanosensitive when reconstituted into liposomes (Sukharev et al., 1994; Häse et al., 1995; Kloda and Martinac, 2001a; Kloda and Martinac, 2001b; Kloda and Martinac, 2001c; Martinac, 2001; Perozo and Rees, 2003). Recent demonstration that TREK and TRAAK retain stretch sensitivity in cytoskeleton-free membrane blebs indicates that they are also bilayer-gated channels (Honoré et al, 2006).

In contrast, transient gating kinetics of MscCa are not retained in blebs that lack an underlying cytoskeleton (Zhang et al, 2000). However, membrane tension measurements of cytoskeleton-free membrane blebs indicate that the in-plane tension contributes only 25% of the membrane tension value (Dai and Sheetz, 1999).

1.8 Other Methods of Investigation of Effects of Mechanical Stimulus

1.8.1 Incorporation of Cells in Agarose Gel

Isolated myocytes were embedded with an argarose gel in a thin wall polyethylene tubings. Stretching the tubing stretched the gel together with the cells (Petroff et al, 2001; Sollott and Lakatta, 1994). This method was used in combination with fluorescence indicators. Although there is an opinion that such approach can not be combined with patch clamp method and is not suitable for electrophysiology (Isenberg et al, 2005), in all probability it does not correspond to reality as the principles of using patch clamp method on isolated myocytes in argarose gel do not differ much from application of this method on cells within the brain slices.

1.8.2 Cell Stretching on Bioflex Membranes

Ventricular myocytes are seeded and attached to Bioflex® membranes covered with collagen or fibronectin. The commercial device applies a vacuum for stretching the membrane rhythmically, in both x and y to the same extent, and the attached cells are stretched together with the membrane (Husse et al, 2003; Komuro et al, 1990).

1.8.3 Fluorescence Indicators

Increments in intracellular Ca^{2+} concentration were among the first effects reported for local mechanical stretch of embryonic chick heart cells (Sigurdson et al, 1992). Without simultaneous analysis of membrane currents for possible involvement of I_{SAC}, those experiments are relatively easy to perform, and stretch induced increments in intracellular Ca^{2+} concentration were analyzed by both luminescence and fluorescence indicators (Allen and Kurihara, 1982; Blinks, 1990; Saeki et al, 1993 Calaghan and White, 1999). Increments in intracellular Ca^{2+} concentration due to axial stretch were measured in guinea-pig ventricular cardiomyocytes (Hongo et al, 1996; White et al, 1993) and discussed in context with the stretch activation of SAC (Gannier et al, 1996). Although voltage clamping is preferred to obtain quantitative data from excitable cells, current clamp experiments and the use of Ca^{2+} indicators showed stretch-induced depolarization and increases in Ca^{2+} in guinea pig heart cells (White et al. 1993). Similar results were obtained in chick heart cells (Sigurdson et al. 1992), endothelia (Diamond et al. 1994; Sigurdson et al. 1993), glia (Charles et al. 1991), osteoclasts (Xia and Ferrier 1995), nodose ganglia (Sharma et al. 1995) and epithelial cells (Boitano et al. 1994). For example, in the case of the heart cells (Sigurdson et al. 1992), mechanical deformation induced Ca^{2+} waves initiating from the site of stimulation.

1.9 Conclusion

We have considered several basic methods of membrane stretching and MSCs investigation by means of patch clamp method in various configurations. Having analyzed up to date literature we tried to discuss possible errors in interpreting data obtained by application of different methods of MSC studies. The above information allows to accept certain views and to treat some data with certain caution.

First of all, in experiments using application of positive or negative pressure by pipettes tip, the assumption for interpreting data with this approach is that MSCs respond to mechanical forces due to membrane tension, and not due to hydrostatic pressure perpendicular to membrane. Patch-pipette pressure is not the same as the trans-patch pressure, because of stresses normal to the membrane. It is generally accepted that some suction is necessary for formation of the gigohm seal. Fragment of the cellular membrane will be sucked into the pipette tip. After the suction initiating the formation of the gigaseal it was shown that in a resting patch, adhesion of the patch to the glass creates a resting tension and indicates an optically smooth membrane, perpendicular to the walls of the pipette.

The lipid bilayer is far from homogeneous in content, let alone stress. The cytoskeleton applies forces parallel and normal to the bilayer. The forces in the extracellular matrix are unknown. Thus when talking about membrane we should bear in mind that it does not apply to the lipid bilayer, unless otherwise stated, only to some average over the various components of the cell.

The flexing of the membrane either outward with suction or inward with pressure results in the activation of inward channel currents. It was shown that some channels can be active at zero pressure because the seal adhesion energy produces tension. The sharing of stress between the bilayer and the cortical cytoskeleton affects channel activity.

Honoré and colleagues (2006) and Suchyna and Sachs (2007) believe that SICs do not exist and inactivation of channel activity in response to suction can be explained by the activity of pre-stressing of stretch-activated channels (SACs). The question of PACs existence also remains a controversy. According to the Laplace's equation, positive and negative pressures should make equal contributions to the stress.

Discussing numerous methods of direct mechanical stretching of a cell and registration of stretch activated whole-cell currents, the first one to attract attention is the stretching of single cells by two patch pipettes, one of which is used for recording of whole-cell current, while the other patch-pipette with a wide opening lifts and stretches the cell. This method is efficient for certain types of cells, e.g., smooth muscle cells and fibroblasts. Three types of axial stretch are similar; stretch by two glass capillaries is complicated, but allows to perform axial stretch; stretch by glass stylus is very simple, but allows to perform only local axial stretch, and using two glass styluses to perform axial stretch makes the method more complicated. Local axial stretch by two thin carbon fibres is rather a method for registering cell contractive activity on the background of registering whole-cell currents, and using this method for cell stretching is quite problematic.

Cell swelling is a separate method that reveals certain channels. But this method applies a nonspecific stimulus. Besides its mechanical effects, it reduces ionic strength, dilutes cytoplasmic ions and macromolecules, and relieves macromolecular crowding, thereby activating signaling molecules.

Paramagnetic microbead method allows the application of controlled forces to the membrane, at which those mechanical forces are transmitted by integrins. The method requires recording of the area of cell membrane subjected to the bead-induced force and of cell shape, as well as of the value of the force, applied by the permanent magnet to a ferric oxide bead.

Cell compression poses a problem. Whole-cell mechanosensitive currents can be evoked by pressing on a spherical cell with the tip of one pipette. Obviously the stresses distribution is very complicated in compression and no one knows how to analyze the data in a mechanistic manner.

Reconstitution into lipid bilayers simplifies the mechanics. The studies of bacterial mechanosensitive channels have been very successful. Purification and functional reconstitution in planar lipid bilayers or in vesicles has certain limitations. Most preferable are vesicles, but only vesicles with a fixed amount of lipid can be studied.

Other methods with the exception of fluorescence indicators are not widely used.

At the same time taking into account all the merits and despite certain imperfections the use of the above methods contributed dramatically towards our understanding of MSC mechanisms and the new findings will give us a lot of pleasure in the future.

Acknowledgment We are grateful to Fred Sachs and Boris Martinac for discussions of these problems, as well as for the important comments on the manuscript. AK acknowledges the support of Alexander von Humboldt-Stiftung (Germany), IK is supported by travels grant of the Humboldt University of Berlin (Germany).

References

Akinlaja J and Sachs F (1998) The breakdown of cell membranes by electrical and mechanical stress *Biophys J* 75: 247–254.

Allen DG, Kurihara S (1982) The effects of muscle length on intracellular calcium transients in mammalian cardiac muscle. *J Physiol (Lond)* 327:79–94.

Andersen OS, Nielsen C, Maer AM, Lundbaek JA, Goulian M, Koeppe RE 2nd. (1999) Ion channels as tools to monitor lipid bilayer-membrane protein interactions: gramicidin channels as molecular force transducers. *Methods Enzymol* 294:208–224.

Awayda MS, Ismailov II, Berdiev BK, Benos DJ (1995) A cloned renal epithelial Na+ channel protein displays stretch activation in planar lipid bilayers. *Am J Physiol* 268(6 Pt 1): C1450–C1459.

Baumgarten CM (2005) Cell volume-sensitive ion channels and transporters in cardiac myocytes. In *Cardiac Mechano-Electrical Feedback and Arrhythmias: From Pipette to Patient*, eds. Kohl P, Franz MR, Sachs F, Saunders, Philadelphia, pp. 21–32.

Baumgarten CM, Browe DM, Ren Z (2005) Swelling- and Stretch-Activated Chloride Channels in the Heart: Regulation and Function. In: Kamkin A and Kiseleva I (ed) Mechanosensitivity in Cells and Tissues. Academia Publishing House Ltd, Moscow, 2005: pp. 79–102.

Baumgarten CM, Clemo HF (2003) Swelling-activated chloride channels in cardiac physiology and pathophysiology. *Prog Biophys Mol Biol* 82:25–42

Baumgarten CM, Feher JJ (2001) Osmosis and the regulation of cell volume. In *Cell Physiology Source Book: A Molecular Approach*, ed. Sperelakis N, Academic Press, New York, 319–355.

Belus A, White E (2003) Streptomycin and intracellular calcium modulate the response of single guinea-pig ventricular myocytes to axial stretch. *J Physiol (Lond)* 546:501–509.

Berrier C, Besnard M, Ajouz B, Coulombe A, Ghazi A (1996) Multiple mechanosensitive ion channels from *E. coli*, activated at different thresholds of applied pressure. *J Membr Biol* 151:175–187.

Berrier C, Coulombe A, Houssin C, Ghazi A (1989) A patch-clamp study of inner and outer membranes and of contact zones of *E. coli*, fused into giant liposomes. Pressure activated channels are localized in the inner membrane. *FEBS Letters* 259:27–32 .

Besch SR, Suchyna T, Sachs F (2002) High-speed pressure clamp. *Pflügers Arch - Eur J Physiol* 445(1):161–166.

Bett GCL and Sachs F (2000) Whole-cell mechanosensitive currents in rat ventricular myocytes activated by direct stimulation. *J Membrane Biol* 173: 255–263.

Biggin PC and Sansom MSP (2001) Channel gating, Twist to open. *Curr.Biol* 11: R364–R366.

Betanzos M, Chiang C-S, Guy HR, Sukharev S (2002) A large iris-like expansion of a mechanosensitive channel protein induced by membrane tension. *Nat Struct Biol* 9: 704–710.

Blinks JR (1990) Use of photoproteins as intracellular calcium indicators. *Environ Health Perspect* 84:75–81.

Boitano S, Sanderson MJ, Dirksen ER (1994) A role for Ca^{2+}-conducting ion channels in mechanically induced signal transduction of airway epithelial cells. *J Cell Sci* 107: 3037–3044.

Bowman CL and Lohr JW (1996) Mechanotransducing ion channels in C6 glioma cells. *Glia* 18(3):161–176.

Bowman CL, Ding JP, Sachs F, Sokabe M. (1992) Mechanotransducing ion channels in astrocytes. *Brain Res* 584(1–2):272–286.

Browe DM and Baumgarten CM (2003) Stretch of β1 integrin activates an outwardly rectifying chloride current via FAK and Src in rabbit ventricular myocytes. J Gen Physiol 122: 689–702.

Calaghan SC, White E (1999) The role of calcium in the response of cardiac muscle to stretch. *Prog Biophys Mol Biol* 71:59–90.

Charles AC, Merrill JE, Dirksen ER, Sanderson MJ (1991) Intercellular signaling in glial cells: calcium waves and oscillations in response to mechanical stimulation and glutamate. *Neuron* 6(6):983–992.

Chaudhuri O, Parekh SH, Fletcher DA (2007) Reversible stress softening of actin networks. *Nature* 445(7125):295–298.

Colombo G, Marrink SJ, Mark AE (2003) Simulation of MscL gating in a bilayer under stress. *Biophys J* 84: 2331–2337.

Cui C, Smith DO, Adler J (1995) Characterization of mechanosensitive channels in *Escherichia coli* cytoplasmic membrane by wholecell patch-clamp recording. *J Membr Biol* 144: 31–42.

Dai J and Sheetz MP (1995) Regulation of endocytosis, exocytosis, and shape by membrane tension. *Cold Spring Harbor Symp Quant Biol* 60: 567–571.

Dai J and Sheetz MP (1999) Membrane tether formation from blebbing cells. Biophys J 77:3363–3370

Davis MJ, Donovitz JA, Hood JD (1992) Stretch-activated single-channel and whole cell currents in vascular smooth muscle cells. *Am J Physiol* 262:C1083–C1088.

Delcour, A.H., Martinac, B., Adler, J. and Kung, C. (1989) Modified reconstitution method used in patch-clamp studies of Escherichia coli ion channels. *Biophys. J.* 56: 631–635.

Diamond SL, Sachs F, Sigurdson WJ (1994) The mechanically induced calcium mobilization in cultured endothelial cells is dependent on actin and phopholipase. *Arterioscler Thromb* 14:2000–2009.

Discher DE and Mohandas N (1996) Kinematics of red cell aspiration by fluorescence-imaged microdeformation. *Biophys J* 71:1680–1694.

Discher DE, Mohandas N, Evans EA (1994) Molecular maps of red cell deformation: hidden elasticity and "*it situ*" connectivity. *Science* 266:1032–1035.

Elliott JR, Needham D, Dilger JP, Haydon DA (1983) The effects of bilayer thickness and tension on gramicidin single-channel lifetime. *Biochim Biophys Acta* 735:95–103.

Fabiato A (1981) Myoplasmic free calcium concentration reached during the twich of an intact isolated cardiac cell and during calcium-induced release of calcium from the sarcoplasmic reticulum of a skinned cardiac cell from the adult rat or rabbit ventricle. *J Gen Physiol* 78:457–497.

Franco-Obregón A and Lansman JB (1994). Mechanosensitive ion channels in skeletal muscle from normal and dystrophic mice. *J Physiol (Lond)* 481, 299–309.

Franco-Obregón A and Lansman JB (2002) Changes in mechanosensitive channel gating following mechanical stimulation in skeletal muscle myotubes from the *mdx* mouse. *J Physiol (Lond)* 539.2, 391–407.

Gannier F, White E, Garnier D, Le Guennec JY (1996) A possible mechanism for large stretch-induced increase in $[Ca^{2+}]_i$ in isolated guinea-pig ventricular myocytes. *Cardiovasc Res* 32:158–167.

Glogauer G and Ferrier J (1998) A new method for application of force to cells via ferric oxide beads. *Pflügers Arch - Eur J Physiol* 435:320–327.

Glogauer M, Ferrier J, McCulloch CAG (1995) Magnetic fields applied to collagen-coated ferric oxide beads induce stretch-activated Ca^{2+} flux in fibroblasts. *Am J Physiol* 269: C1093–C1104.

Goulian M, Mesquita ON, Fygenson DK, Nielsen C, Andersen OS, Libchaber A (1998) Gramicidin channel kinetics under tension. *Biophys J* 74(1):328–37.

Gruen DW and Wolfe J (1982) Lateral tensions and pressures in membranes and lipid monolayers. *Biochim Biophys Acta* 688: 572–580.

Guharay F and Sachs F (1984) Stretch-activated single ion channel currents in tissue cultured embryonic chick skeletal muscle. *J Physiol (Lond)* 352:685–701.

Guharay F and Sachs F (1985) Mechanotransducer ion channels in chick skeletal muscle: The effects of extracellular pH. *J Physiol (Lond)* 363: 119–134.

Gullingsrud J and Schulten K (2003) Gating of MscL studied by steered molecular dynamics. *Biophys. J.* 85, 2087–2099.

Gullingsrud J, Kosztin D, Schulten K (2001) Structural determinants of MscL gating studied by molecular dynamics simulations. *Biophys J* 80: 2074–2081.

Gustin MC, Sachs F, Sigurdson W, Ruknudin A, Bowman C, Morris CE, Horn R (1991) Single-channel mechanosensitive currents. *Science* 253: 800.

Gustin MC, Zhou XL, Martinac B, Kung C (1988) A mechanosensitive ion channel in the yeast plasma membrane. *Science* 242: 762–765.

Hamill OP (2006) Twenty odd years of stretch-sensitive channels *Pflügers Arch - Eur J Physiol* 453:333–351.

Hamill OP and Martinac B (2001) Molecular basis of mechanotransduction in living cells. *Physiol Revs* 81:685–740.

Hamill OP and McBride DW Jr (1997) Induced membrane hypo/hyper-mechanosensitivity: a limitation of patch-clamp recording. *Annu Rev Physiol* 59: 621–631.

Hart FX (2006) Integrins may serve as mechanical transducers for low-frequency electric fields. *Bioelectromagnetics* 27(6):505–508.

Häse CC, Le Dain AC and Martinac B (1995). Purification and functional reconstitution of the recombinant large mechanosensitive ion channel (MscL) of *Escherichia coli. J Biol Chem* 270:18329–18334.

Hongo K, White E, Le Guennec JY, Orchard CH (1996) Changes in $[Ca^{2+}]_i$, $[Na^+]_i$ and Ca^{2+} current in isolated rat ventricular myocytes following an increase in cell length. *J Physiol (Lond)* 491:609–619.

Honoré E, Patel AJ, Chemin J, Suchyna T, Sachs F (2006) Desensitization of mechano-gated K2P channels. *Proc Natl Acad Sci USA* 103(18):6859–6864.

Hörber JKH, Mosbacher J, Häbele W, Ruppersberg JP, Sakmann B (1995) A look at membrane patches with scanning force microscope. *Biophys J* 68:1687–1693.

Hoyer J, Distler A, Haase W, Gogelein H (1994) Ca^{2+} influx through stretch-activated cation channel activates maxi K^+ channels in porcine endocardial endothelium. *Proc Natl Acad Sci USA* 91:2367–2371.

Hoyer J, Köhler R, Distler A (1997) Mechanosensitive cation channels in aortic endothelium of normotensive and hypertensive rats. *Hypertension* 30:112–119.

Hu H and Sachs F (1994) Effects of mechanical stimulation on embryonic chick heart cells. *Biophys J* 66:A170.

Hu H and Sachs F (1995) Whole cell mechanosensitive currents in acutely isolated chick heart cells: correlation with mechanosensitive channels. *Biophys J* 68: A393.

Hu H and Sachs F (1996) Mechanically activated currents in chick heart cells. *J Membr Biol* 154: 205–216.

Husse B, Sopart A, Isenberg G (2003) Cyclical mechanical stretch induced apoptosis in myocytes from young rats but necrosis in myocytes form old rats. *Am J Physiol Heart Circ Physiol* 285:H1521–H1527.

Hwang TC, Koeppe RE 2nd, Andersen OS (2003) Genistein can modulate channel function by a phosphorylation-independent mechanism: importance of hydrophobic mismatch and bilayer mechanics. *Biochemistry* 42(46):13646–13658.

Isenberg G, Kazanski V, Kondratev D, Gallitelli MF, Kiseleva I, Kamkin A (2003) Differential effects of stretch and compression on membrane currents and $[Na^+]_C$ in ventricular myocytes. *Progr Biophys Mol Biol* 82:43–56

Isenberg G, Kondratev D, Dyachenko V, Kazanski V, Gallitelli MF (2005) Isolated cardiomyocytes: Mechanosensitivity of action potential, membrane current and ion concentration. In: Kamkin A and Kiseleva I (ed) Mechanosensitivity in Cells and Tissues. Academia Publishing House Ltd, Moscow, 2005: pp. 126–164.

Ismailov II, Awayda MS, Berdiev BK, Bubien JK, Lucas JE, Fuller CM, Benos DJ (1996a) Triple-barrel organization of ENaC, a cloned epithelial Na^+ channel. *J Biol Chem* 271(2): 807–816.

Ismailov II, Awayda MS, Jovov B, Berdiev BK, Fuller CM, Dedman JR, Kaetzel M, Benos DJ (1996b) Regulation of epithelial sodium channels by the cystic fibrosis transmembrane conductance regulator. *J Biol Chem* 271(9):4725–32.

Ismailov II, Berdiev BK, Bradford AL, Awayda MS, Fuller CM, Benos DJ (1996c) Associated proteins and renal epithelial Na^+ channel function. *J Membr Biol* 149(2):123–132.

Kamkin A, Kiseleva I, Isenberg G (2000) Stretch-activated currents in ventricular myocytes: amplitude and arrhythmogenic effects increase with hypertrophy. *Cardiovasc Res* 48: 409–420.

Kamkin A, Kiseleva I, Yarigin V (2003) Mechanoelectrical feedback in the heart. *Monograph.* Naturmort Publishing House. Moscow. 352 pp. (Russian).

Kamkin A, Kiseleva I, Isenberg G (2003a) Activation and inactivation of a non-selective cation conductance by local mechanical deformation of acutely isolated cardiac fibroblasts. *Cardiovasc Res* 57: 793–803.

Kamkin A, Kiseleva I, Isenberg G, Wagner KD, Günther J, Theres H, Scholz H (2003b) Cardiac fbroblasts and the mechano-electric feedback mechanism in healthy and diseased hearts. *Prog Biophys Mol Biol* 82: 111–120.

Kamkin A, Kiseleva I, Isenberg G (2003c) Ion selectivity of stretch-activated cation currents in mouse ventricular myocytes. *Pflügers Arch - Europ J Physiol* 446(2): 220–231.

Kamkin A, Kiseleva I, Lozinsky I, Wagner KD, Isenberg G, Scholz H (2005a) The role of mechanosensitive fibroblasts in the heart. In: Kamkin A and Kiseleva I (ed) Mechanosensitivity in Cells and Tissues. Academia Publishing House Ltd, Moscow, 2005: pp. 203–229.

Kamkin A, Kiseleva I, Wagner KD, Scholz H (2005b) Mechano-electric feedback in the heart: Evidence from intracellular microelectrode recordings on multicellular preparations and single cells from healthy and diseased tissue. In: Kamkin A and Kiseleva I (ed) Mechanosensitivity in Cells and Tissues. Academia Publishing House Ltd, Moscow, 2005: pp. 165–202.

Kawahara K (1993) Stretch-activated channels in renal tubule. *Nippon Rinsho* 51: 2201–2208.

Marchenko SM and Sage SO (1996) Mechanosensitive ion channels from endothelium of excised rat aorta. *Biophys J* 70: A365.

Kloda A and Martinac B (2001a) Molecular Identification of a Mechanosensitive Channel in Archaea. *Biophys J* 80:229–240.

Kloda A and Martinac B (2001b). Structural and functional similarities and differences between MscMJLR and MscMJ, two homologous MS channels of *M. jannashii. EMBO J* 20: 1888–1896.

Kloda A and Martinac B (2001c) Mechanosensitive channel in *Thermoplasma* a cell wall-less Archaea: cloning and molecular characterization. *Cell Biochem. Biophys.* 34: 321–347.

Köhler R, Distler A, Hoyer J (1998) Pressure-activated cation channel in intact rat endocardial endothelium. *Cardiovasc Res* 38: 433–440

Köhler R, Grundig A, Brakemeier S, Rothermund L, Distler A, Kreutz R, Hoyer J (2001a) Regulation of pressure-activated channel in intact vascular endothelium of stroke-prone spontaneously hypertensive rats. *Am J Hypertension* 14:716–721.

Köhler R, Kreutz R, Grundig A, Rothermund L, Yagli C, Yagli Y, Pries AR, Hoyer J (2001b) Impaired function of endothelial pressure-activated cation channel in salt-sensitive genetic hypertension. *J Am Soc Nephrol* 12: 1624–1629.

Komuro I, Kaida T, Shibazaki Y et al. (1990) Stretching cardiac myocytes stimulates protooncogene expression. *J Biol Chem* 265, No.7:3595–3598.

Kung C (2005) A possible unifying principle for mechanosensation. *Nature* 436:647–654.

Laitko U and Morris CE (2004) Membrane tension accelerates rate-limiting voltage-dependent ativation and slow inactivation steps in a Shaker channel. *J Gen Physiol* 123: 135–154.

Lansman JB, Hallam TJ, Rink TJ (1987) Single stretch-activated ion channels in vascular endothelial cells as mechanotransducers. *Nature* 325:811–813

Le Guennec J-Y, White E, Gannier F, Argibay JA, Garnier D (1991) Stretch-induced increase of resting intracellular calcium concentration in single guinea-pig ventricular myocytes. *Exp Physiol* 76:975–978.

Lillemeier BF, Pfeiffer JR, Surviladze Z, Wilson BS, Davis MM (2006) Plasma membrane-associated proteins are clustered into islands attached to the cytoskeleton. *Proc Natl Acad Sci USA* 103(50):18992–18997.

Lundbaek JA and Andersen OS (1999) Spring constants for channel-induced lipid bilayer deformations. Estimates using gramicidin channels. *Biophys J* 76(2):889–895.

Markin VS and Sachs F (2004a) Thermodynamics of mechanosensitivity: lipid shape, membrane deformation and anesthesia. *Biophysical J* 86, 370A.

Markin VS and Sachs F (2004b) Thermodynamics of mechanosensitivity. *Phys Biol* 1:110–124.

Markin VS, Shlenskii VG, Saimon SA, Benos DD, Ismailov II (2006) Mechanosensitivity of gramicidin A channels in semispherical bilayer membranes at constant tension. *Biofizika* 51(6):1014–1018 (Russian).

Maroto R, Raso A, Wood TG, Kurosky A, Martinac B, Hamill OP (2005) TRPC1 forms the stretch-activated cation channel in vertebrate cells. *Nature Cell Biol* 7:1443–1446.

Martinac B (2001) Mechanosensitive channels in prokaryotes. *Cell Physiol Biochem* 11:61–76.

Martinac B (2004) Mechanosensitive ion channels: molecules of mechanotransduction. *J Cell Sci* 117:2449–2460.

Martinac B and Hamill OP (2002) Gramicidin A channels switch between stretch activation and stretch inactivation depending on bilayer thickness. *Proc. Natl. Acad. Sci. USA* 99:4308–4312.

Martinac B and Kloda A (2003) Evolutionary origins of mechanosensitive ion channels. *Prog Biophys Mol Biol* 82:11–24.

Martinac B, Adler J and Kung C (1990) Mechanosensitive ion channels of *E. coli* activated by amphipaths. *Nature* 348: 261–263.

Martinac B, Buechner M, Delcour AH, Adler J, Kung C (1987) Pressure-sensitive ion channel in *Escherichia coli. Proc Natl Acad Sci USA* 84:2297–2301

Mazzag BM, Tamaresis JS, Barakat AI (2003) A model for shear stress sensing and transmission in vascular endothelial cells. *Biophys J* 84:4087–4101.

McBride DW Jr, Hamill OP (1992) Pressure-clamp: a method for rapid step perturbation of mechanosensitive channels. *Pflügers Arch - Eur J Physiol* 421:606–612.

McBride DW Jr, Hamill OP (1993) Pressure-clamp technique for measurement of the relaxation kinetics of mechanosensitive channels. *Trends Neurosci* 16:341–345.

Mizuno D, Tardin C, Schmidt CF, Mackintosh FC (2007) Nonequilibrium mechanics of active cytoskeletal networks. *Science* 315(5810):370–373.
Morris CE (1990) Mechanosensitive ion channels. *J Membr Biol* 113:93–107.
Morris CE (1992) Are stretch-sensitive channels in molluscan cells and elsewhere physiological mechanotransdueers? *Experientia* 48: 852–858.
Morris CE and Homann U (2001) Cell surface area regulation and membrane tension. *J Membr Biol* 179(2):79–102.
Niisato N, Ito Y, Marunaka Y (1999) Activation of Cl^- channel and $Na^+/K^+/2Cl^-$ cotransporter in renal epithelial A6 cells by flavonoids: genistein, daidzein, and apigenin. *Biochem Biophys Res Commun* 254:368–371.
Opsahl LR and Webb WW (1994) Lipid-glass adhesion in giga-sealed patch clamped membranes. *Biophys J* 66:75–79.
Perozo E and Rees DC (2003) Structure and mechanism in prokaryotic mecahnosensitive channels. *Curr Opin Struct Biol* 13: 432–442.
Perozo E, Cortes DM, Sompornpisut P, Kloda A and Martinac B (2002a) Structure of MscL in the open state and the molecular mechanism of gating in mechanosensitive channels. *Nature* 418: 942–948.
Perozo E, Kloda A, Cortes DM and Martinac B (2002b) Physical principles underlying the transduction of bilayer deformation forces during mechanosensitive channel gating. *Nat Struct Biol* 9: 696–703.
Petroff MGV, Kim SH, Pepe S et al. (2001) Endogenous nitric oxide mechanisms mediate the stretch dependence of Ca^{2+} release in cardiomyocytes. *Nat Cell Biol* 3:867–873.
Petrov E and Martinac B. (2007) Modulation of channel activity and gadolinium block of MscL by static magnetic fields. *Eur Biophys J* 36(2):95–105.
Ring A (1992) Monitoring the surface tension of lipid membranes by a bubble method. *Pflü gers Arch - Eur J Physiol* 420: 264–268.
Ring A and Sandblom J (1988) Evaluation of surface tension and ion occupancy effects on gramicidin A channel lifetime. *Biophys J* 53: 541–548.
Ruknudin A, Song MJ, Sachs F (1991) The ultrastructure of patch-clamped membranes: a study using high voltage electron microscopy. *J Cell Biol* 112:125–134.
Sachs F and Morris CE (1998) Mechanosensitive ion channels in nonspecialized cells. *Rev Physiol Biochem Pharmacol* 132: 1–77.
Sachs F (1988) Mechanical transduction in biological systems. *Crit Rev Biomed Eng* 16(2):141–169.
Sackin H (1995) Mechanosensitive channels *Annu Rev Physiol* 57:333–353.
Saeki Y, Kurihara S, Hongo K, Tanaka E (1993) Tension and intracellular calcium transients of activated ferret ventricular muscle in response to step length changes. *Adv Exp Med Biol* 332:639–648.
Saimi Y, Martinac B, Delcour AH, Minorsky PV, Gustin MC, Culbertson MR, Adier J, Kung C (1993) Patch clamp studies of microbial ion channels. *Methods Enzymol* 207: 681–691
Saimi Y, Martinac B, Gustin M, Culbertson MR, Adler J, Kung C (1988) Ion channels in Paramecium, yeast and *Escherichia coli*. Trends Biochem Sci 13(8):304–309.
Sakmann B and Neher E (1983) Geometric Parameters of Pipettes and Membrane Patches. *In: Single-Channels Recording*. Plenum Press, New York. A Division of Plenum Publishing Corporation 233 Spring Street, New York, N.Y.10013. 700 pp. 637–650.
Sasaki N, Mitsuiye T, Noma A (1992) Effects of mechanical stretch on membrane currents of single ventricular myocytes of guinea-pig heart. *Jpn.J.Physiol* 42: 957–970.
Sharma RV, Chapleau MW, Hajduczok G, Wachtel RE, Waite LJ, Bhalla RC, Abboud FM (1995) Mechanical stimulation increases intracellular calcium concentration in nodose sensory neurons. *Neurosci* 66: 433–441.
Sigurdson W, Ruknudin A, Sachs F (1992) Calcium imaging of mechanically induced fluxes in tissue-cultured chik heart: role of stretch-activated ion channels. *Am J Physiol* 262: H1110–H1115.

Sigurdson WJ, Sachs F, Diamond SL (1993) Mechanical perturbation of cultured human endothelial cells causes rapid increases of intracellular calcium. *Am J Physiol* 264: H1745–H1752.

Sokabe M and Sachs F (1990). The structure and dynamics of patch clamped membrane, a study using differential interference contrast microscopy. *J Cell Biol* 111, 599–606.

Sokabe M, Nunogaki K, Naruse K, Soga H (1993) Mechanics of patch clamped and intact cell-membranes in relation to SA channel activation. *Jpn J Physiol* 43:S197–S204.

Sokabe M, Sachs F, Jing Z (1991) Quantitative video microscopy of patch clamped membranes, stress, strain, capacitance and stretch channel activation. *Biophys J* 59, 722–728.

Sollott SJ, Lakatta EG (1994) Novel method to alter length and load in isolated mammalian cardiac myocytes. *Am J Physiol Heart Circ Physiol* 267:H1619–H1629.

Suchyna TM and Sachs F (2007) Mechanosensitive channel properties and membrane mechanics in mouse dystrophic myotubes. *J Physiol (Lond)* (in press).

Suchyna TM, Tape SE, Koeppe RE, Andersen OS, Sachs F, Gottlieb PA (2004) Bilayer-dependent inhibition of mechanosensitive channels by neuroactive peptide enantiomers. *Nature* 430: 235–240.

Sukahrev SI, Martinac B, Arshavsky VY, Kung C (1993) Two types of mechanosensitive channels in the E. coli cell envelope: Solubilization and functional reconstitution. *Biophys J* 65:177–183.

Sukharev S (2002) Purification of the small mechanosensitive channel in Escherichia coli (MscS): the subunit structure, conduction and gating characteristics. *Biophys J* 83: 290–298.

Sukharev S, Betanzos M, Chiang CS, Guy HR (2001) The gating mechanism of the large mechanosensitive channel MscL. *Nature* 409: 720–724.

Sukharev SI, Blount P, Martinac B, Blattner FR and Kung C (1994) A large mechanosensitive channel in *E. coli* encoded by mscL alone. *Nature* 368: 265–268.

Wang N, Butler JP, Ingber DE (1993) Mechanotransduction across the cell surface and through the cytoskeleton. *Science.* 260: 1124–1127.

Wellner MC and Isenberg G (1993) Properties of stretch activated channels in myocytes from the guinea-pig urinary bladder. *J Physiol (London)* 466:412–425.

Wellner MC and Isenberg G (1994) Stretch effects on whole-cell currents of guinea-pig urinary bladder myocytes *J Physiol (London)* 480.3: 439–448.

Wellner MC and Isenberg G (1995) cAMP accelerates the decay of stretch-activated inward currents in guinea-pig urinary bladder myocytes. *J Physiol (London)* 482:141–156.

White E, Boyett MR, Orchard CH (1995) The effects of mechanical loading and changes of length on single guinea-pig ventricular myocytes. *J Physiol (Lond)* 482:93–107.

White E, Le Guennec IY, Nigretto JM, Gannier F, Argibay JA, Gamier D (1993) The effects of increasing cell length on auxotonic contractions: membrane potential and intracellular calcium transients in single guinea-pig ventricular myocytes. *Exp Physiol* 78: 65–78.

Wiggins P and Phillips R (2004) Analytic models for mechanotransduction: gating a mechanosensitive channel. *Proc Natl Acad Sci USA* 101(12):4071–4076.

Wiggins P and Phillips R (2005) Membrane-protein interactions in mechanosensitive channels. *Biophys J* 88: 880–902.

Xia SL, Ferrier J (1995) A calcium signal induced by mechanical pertubation of osteoclasts. *J Cellular Physiol* 167: 148–155

Zeng T, Bett GCL, Sachs F (2000) Stretch-activated whole cell currents in adult rat cardiac myocytes. *Am J Physiol* 278: H548–H557.

Zhang Y and Hamill OP (2000) On the discrepancy between membrane patch and whole cell mechanosensitivity in *Xenopus* oocytes. *J Physiol (Lond)* 523.1:101–115.

Zhang Y, Gao F, Popov V,Wan J, Hamill OP (2000) Mechanically-gated channel activity in cytoskeleton deficient blebs and vesicles from *Xenopus* oocytes. *J Physiol (Lond)* 523.1: 117–129.

Zhou XL, Stumpf MA, Hoch HC, Kung C (1991) A mechanosensitive channel in whole cells and in membrane patches of the fungus *Uromyces*. *Science* 253: 1415–1417.

Chapter 2
The Role of Lipid Bilayer Mechanics in Mechanosensation

Tristan Ursell, Jané Kondev, Dan Reeves, Paul A. Wiggins, and Rob Phillips

Abstract Mechanosensation is a key part of the sensory repertoire of a vast array of different cells and organisms. The molecular dissection of the origins of mechanosensation is rapidly advancing as a result of both structural and functional studies. One intriguing mode of mechanosensation results from tension in the membrane of the cell (or vesicle) of interest. The aim of this review is to catalogue recent work that uses a mix of continuum and statistical mechanics to explore the role of the lipid bilayer in the function of mechanosensitive channels that respond to membrane tension. The role of bilayer deformation will be explored in the context of the well known mechanosensitive channel MscL. Additionally, we make suggestions for bridging gaps between our current theoretical understanding and common experimental techniques.

Key words: Lipid bilayer mechanics · Statistical mechanics · Mechanosensitive ion channels · Membrane-protein interactions

2.1 Introduction

Cells interact with each other and with their external environment. These interactions are enabled by transmembrane proteins—machines that have evolved to allow cells to detect and respond to changes in their environment. These proteins detect external cues, such as an increase in ligand concentration or the presence of forces or voltage, and transiently alter the permeability of the cell membrane allowing ions, water, or even larger molecules to cross as well as triggering receptors for signaling (Barry and Lynch, 2005; Clapham et al, 2001). The passage of these ions (or molecules) and the triggering of receptors then leads to a series of downstream events within the cell, enabling a response to these environmental cues.

Mechanical forces and their corresponding deformations constitute one of the most important classes of external cues. Mechanosensation is a widespread phenomenon in a host of different single-celled and multicellular organisms (Fain, 2003; Gillespie and Walker, 2001; Katsumi et al, 2004; Kloda and Martinac, 2001; Nauli and Zhou, 2004; Sachs, 1991). In bacteria, experimental evidence suggests that mechanosensation arises to detect and regulate the response to changes in the osmotic environment (Pivetti et al, 2003; Chang et al, 1998; Sukharev et al, 1997).

More generally, the issue of cell shape and its attendant deformation is important not only in the context of osmotic stress and the management of physical stresses to which membranes are subjected (Morris and Homann, 2001), but also arises in context of remodeling of the cell and organelle membranes during cell division (Christensen and Strange, 2001; Kamada et al, 1995).

In multicellular organisms, mechanosensation is important in a variety of ways. One intriguing class of mechanosensors is linked to motility. For example, in nematodes like the much studied *C. elegans*, mechanosensation permits the worm to decide which way to move and may have a role in detecting body curvature, thus telling the worm when to change its wave-like shape (Gillespie and Walker, 2001). Similarly, flies have hair bristles that respond to touch (Duggan et al, 2000), while the mechanosensitive lateral-line organelles in zebrafish provide the means for detecting directional water movement in a way very similar to the workings of our inner ear (Gillespie and Walker, 2001). In each of these cases, genetics has led to the identification of a variety of genes implicated in the ability of the organism to respond to some form of mechanical stimulus. Parallel insights have been obtained in plants (*Arabidopsis* in particular), with the identification of a collection of novel proteins that also appear to be mechanosensitive (Haswell and Meyerowitz, 2006).

Mechanosensitive ion channels are a class of membrane proteins that have recently garnered significant interest. Genetic, biochemical and structural studies all conspire to make this a particularly opportune time to demand a more quantitative picture of the function of these channels. In particular, there is a growing list of success stories in which the structures of channels associated with mechanosensation have been found in both closed and open states (Bass et al, 2002; Chang et al, 1998; Perozo et al, 2002a; Perozo and Rees, 2003). In addition, functional studies that probe how gating depends upon membrane tension or external forces are beginning to make it possible to dissect the various contributions to the energetics of channel gating (Akitake et al, 2005; Chiang et al, 2004; Perozo et al, 2002b; Sukharev et al, 1999).

As a result of these studies, a number of ideas have been proposed to explain the different ways in which external force can couple to membrane-protein conformation. Two modes of action that have been hypothesized for channels are: i) cases in which physical, polypeptide linkers pull on some part of the protein resulting in gating, ii) cases in which tension in the surrounding bilayer forces the channel to open. The aim of this article is to show how statistical mechanics and simple models of bilayer elasticity can be used to glean insights into this second class of mechanosensors.

The remainder of the article is built in four main sections. In the next section, we describe how statistical mechanics can be used to analyze the probability that a two-state mechanosensitive channel is open. This discussion will include an analysis of how the external load (*i.e.* the tension) can be included in the statistical mechanical treatment of these problems. The next section considers the elastic deformations imposed on a bilayer by the presence of a transmembrane protein, and shows how these deformations result in a mechanosensitive channel acting as a bistable switch (*i.e.* a protein with two stable conformations). In the subsequent section, we discuss experimental considerations that will help form a tighter connection between theory and experimental techniques. Finally, we examine the way multiple channels in a membrane might interact through the intervening lipid bilayer and how these interactions can alter the conformational statistics of individual channels.

2.2 Statistical Mechanics of Mechanosensitive Channels

To begin, we review the application of statistical mechanics to a simple two-state mechanosensitive channel. This analysis will serve as the starting point for our subsequent, more detailed analysis which explores how bilayer elasticity can contribute to the energetics of the closed and open states of a channel.

2.2.1 Lipid Bilayer vs. Protein Internal Degrees of Freedom

One convenient scheme for characterizing the state of ion channels is to invoke the state variable σ, which is defined by $\sigma = 0$ if the channel is closed and $\sigma = 1$ if the channel is open. Our aim is to compute the open probability P_{open} which, in terms of our state variable σ, can be written as $\langle\sigma\rangle$, where $\langle\cdots\rangle$ denotes an average. When $\langle\sigma\rangle \approx 0$, this means that the probability of finding the channel open is low. Similarly, when $\langle\sigma\rangle \approx 1$, this means that it is almost certain that we will find the channel open. To evaluate these probabilities we need to invoke the Boltzmann distribution, which tells us that the probability of finding the system in a state with energy $E(\sigma)$ is $p(\sigma) = e^{-\beta E(\sigma)}/Z$, where Z is the partition function defined by $Z = \sum_\sigma e^{-\beta E(\sigma)}$, $\beta = 1/k_B T$, k_B is Boltzmann's constant, and T is the temperature in degrees Kelvin.

On the level of a single channel, we introduce ϵ_{closed} and ϵ_{open} for the energies of the closed and open states, respectively, as shown in Fig. 2.1. These energies contain contributions from deformations of the surrounding lipid bilayer as well as internal protein energetics; however, they do not contain the tension-dependent driving force which we will address separately. The state variables can be used to write the channel energy (in the absence of tension) as

$$E(\sigma) = (1-\sigma)\epsilon_{closed} + \sigma\epsilon_{open}. \tag{2.1}$$

With these energies in hand, we can assign weights to the different states as shown schematically in Fig. 2.1. Within this scheme, the probability that the channel is open is given by $\langle\sigma\rangle$ and can be computed as $\langle\sigma\rangle = \sum_{\sigma=0}^{1} \sigma p(\sigma)$, where $p(\sigma)$ is the probability of finding the channel in state σ. To compute these probabilities, we invoke the Boltzmann distribution, and evaluate the partition function given by

$$Z = \sum_{\sigma=0}^{1} e^{-\beta E(\sigma)} = e^{-\beta\epsilon_{closed}} + e^{-\beta\epsilon_{open}}. \tag{2.2}$$

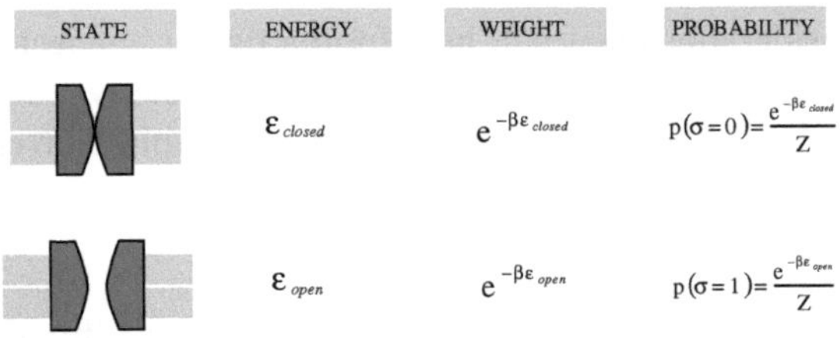

Fig. 2.1 States, Boltzmann weights and corresponding probabilities for a two-state ion channel. The two different states have different energies and the probability of these different states is determined by the Boltzmann distribution

As a result, we see that the open probability can be written as

$$\langle \sigma \rangle = \frac{e^{-\beta\epsilon_{open}}}{e^{-\beta\epsilon_{closed}} + e^{-\beta\epsilon_{open}}} = \frac{1}{1 + e^{\beta(\epsilon_{open} - \epsilon_{closed})}}. \tag{2.3}$$

This expression is relatively sterile in the absence of some term that tunes the energies of the open and closed states to reflect the impact of external driving forces. In fact, one of the most remarkable features of ion channels is that the probability of being in different states can be tuned by external factors such as ligand concentration, the application of a voltage, or application of tension in the surrounding membrane. In general, this formalism can account for any of these driving forces, but we will restrict our attention to the important case of mechanosensitive channels, where the key driving forces are mechanical. In this case, gating occurs when the energy balance between the open and closed states is altered by membrane tension.

To give the origin of membrane tension a physical meaning, we introduce the notion of a "loading device", which we define as the external agent acting on a lipid bilayer to alter its tension. As depicted by hanging weights on the bilayer in Fig. 2.2, we can make a toy model of how changes in bilayer geometry are coupled to the energy of this loading device. The point of introducing this hypothetical situation is to enforce the idea that, in our statistical mechanical treatment of this problem, the loading device is an important part of the overall free energy budget of the system. As a result, when we write down the partition function for a problem involving a channel and a deformable membrane, we have to account for the internal protein energetics, the deformation energy of the lipid bilayer, and the energy associated with the loading device itself. In particular, we note that an increase in the membrane area will lead to a lowering of the weights depicted in Fig. 2.2 and a corresponding decrease in the energy of the loading device. Of course, the application of tension in real membranes is not performed by hanging weights, but through techniques such as micropipette aspiration (Goulian et al, 1998; Rawicz et al, 2000). Nevertheless, the concept of hanging weights brings the importance of the energy of the loading device into sharp focus.

For the case of tension-activated ion channels, the open probability, $\langle \sigma \rangle$, is dictated by a competition between the energetic advantage associated with reduction in the energy of the loading device and the energetic cost of the open state due to

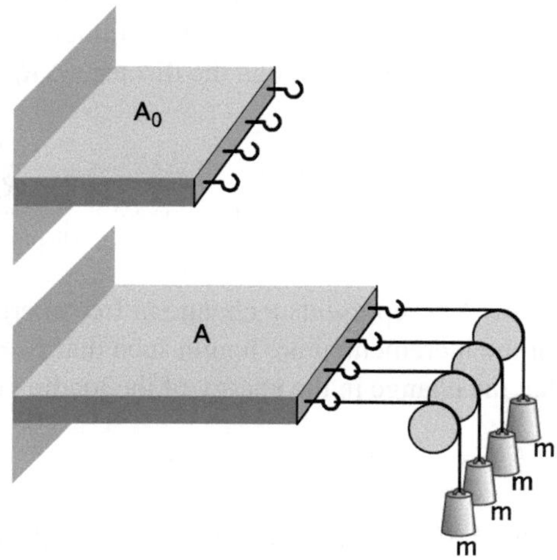

Fig. 2.2 Energy of the loading device for membrane deformation. This figure compares the unloaded and loaded membrane and shows how membrane deformation results in a *lowering* of the potential energy of the loading device. In this hypothetical experiment, the tension (force per unit edge length) in the membrane is given by $\tau = mg/\Delta l$ where Δl is the distance between two consecutive hooks, and g is the acceleration due to gravity

both the internal protein energetics and the energetics of membrane deformation. Following up on the idea of Fig. 2.2, but now with special reference to the case of a mechanosensitive ion channel, Fig. 2.3 shows how the opening of the channel results in a reduction of the energy of the loading device.

The total area of the bilayer is constant (to within a few percent), and as a result, when the channel opens and the radius gets larger the weights in our hypothetical loading device are lowered by some amount, which lowers the potential energy. The greater the weights, the larger the change in potential energy. The notion of weights is a simple representation of externally applied forces on the membrane. If we imagine a finite membrane with fixed area as shown in Fig. 2.3, when the channel opens, the outer radius will change as $\Delta R_{out} = (R/R_{out})\Delta R$, where R is the closed channel radius, ΔR is the change in channel radius upon opening, R_{out} is the outer radius of the membrane when the channel is closed, and ΔR_{out} is the increase in the outer radius of the membrane when the channel opens. We are interested in evaluating the change in potential energy of the loading device (*i.e.* the dropping of the weights) as

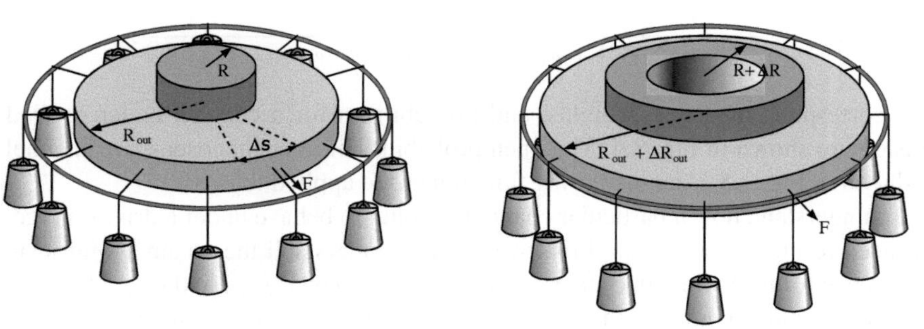

Fig. 2.3 Schematic of how channel opening results in a relaxation in the loading device. For simplicity, we represent the loading device as a set of weights attached to the membrane far from the channel. When the channel opens, these weights are lowered, and the potential energy of the loading device is decreased

a result of channel opening. To do so, we compute the work associated with the force F, which is most conveniently parameterized through a force per unit length (the tension, τ) acting through the distance ΔR_{out} as shown in Fig. 2.3. This results in

$$\Delta G_{tension} = \underbrace{\tau \Delta s}_{\text{force on arc}} \times \underbrace{\frac{R}{R_{out}} \Delta R}_{\text{displacement of arc}} \times \underbrace{\frac{2\pi R_{out}}{\Delta s}}_{\text{number of arcs}} . \tag{2.4}$$

where ΔG represents a change in free energy. We have introduced the variable Δs for the increment of arc length such that $\tau = F/\Delta s$. Given these definitions, we see that the change in the energy of the loading device is given by

$$\Delta G_{tension} = -\tau 2\pi R \Delta R. \tag{2.5}$$

In light of our insights into the energy of the loading device, we introduce the energy as a function of the applied tension τ, which is given by

$$E(\sigma, \tau) = (1 - \sigma)\epsilon_{closed} + \sigma \epsilon_{open} - \sigma \tau \Delta A. \tag{2.6}$$

The term $-\sigma \tau \Delta A$ favors the open state and reflects the fact that the energy of the loading device is lowered in the open state. In fact, this term reveals that any increase in protein area is energetically favored when membrane tension is present, which could imply hidden mechanosensitivity in other classes of ion channels and receptors—a subject we will touch upon later in this review.

To compute the open probability of the channel in the presence of applied tension, we need to once again evaluate the partition function $Z = \sum_\sigma e^{-\beta E(\sigma)}$. Using the energy given in eqn. 2.6, we find

$$Z = e^{-\beta \epsilon_{closed}} + e^{-\beta(\epsilon_{open} - \tau \Delta A)}. \tag{2.7}$$

This permits us to write down the open probability directly as

$$P_{open} = \frac{e^{-\beta(\epsilon_{open} - \tau \Delta A)}}{e^{-\beta(\epsilon_{open} - \tau \Delta A)} + e^{-\beta \epsilon_{closed}}} = \frac{1}{1 + e^{\beta(\epsilon_{open} - \epsilon_{closed} - \tau \Delta A)}}. \tag{2.8}$$

The corresponding states, weights, and probabilities for a channel under applied tension are shown in Fig. 2.4. The open probability of a mechanosensitive channel is shown in Fig. 2.5 as an increasing function of the applied tension.

To understand how a particular channel is going to behave under a driving force, we need to know two things. First, we need to understand the channel's intrinsic preference for each of its two states, which is encoded by ϵ_{closed} and ϵ_{open}. Second, we need to understand how the external driving force alters the relative energies of these different states. With these two quantitative measurements in hand, statistical mechanics allows us to compute the behavior of the channel under a range of driving forces. To make further progress, we need to examine the microscopic origins of ϵ_{closed} and ϵ_{open}. Intriguing recent experiments suggest that these energies are driven in large measure by membrane deformations.

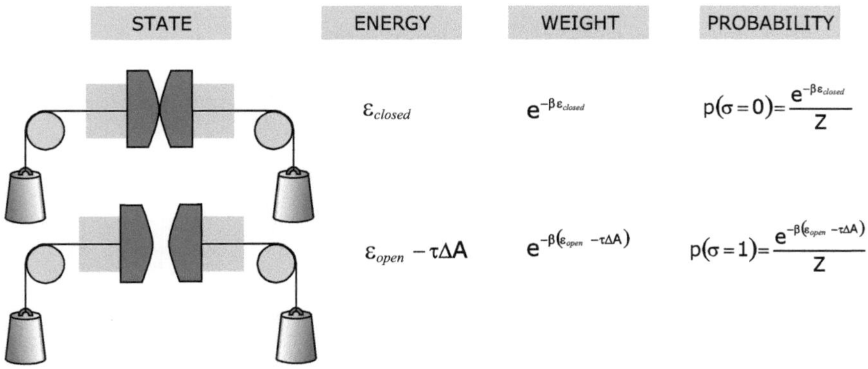

Fig. 2.4 States, weights and corresponding probabilities for a two-state mechanosensitive channel under load

2.3 Bilayer Free Energy and Gating of a Mechanosensitive Channel

The abstract formalism of the previous section leaves us poised to examine mechanosensation to the extent that we can understand the physical origins of ϵ_{closed} and ϵ_{open}. The main idea of this part of the review is to show how simple models of the elastic properties of lipid bilayers can be used to determine the bilayer's contribution to ϵ_{closed} and ϵ_{open}. One of the key clues that hints at the importance of membrane deformation in dictating channel gating is the data shown in Fig. 2.6. In particular, this plot shows how the open probability depends upon the lipid carbon tail length. This data strongly suggests that the energetics of the surrounding membrane is an important part of the overall free energy budget of channel gating (also see (Martinac and Hamill, 2002 and the informative review by Jensen and Mouritsen, 2004).

The parameters ϵ_{closed} and ϵ_{open} can each depend on some combination of the energetics of protein conformation, membrane deformation, and hydration energy. Our

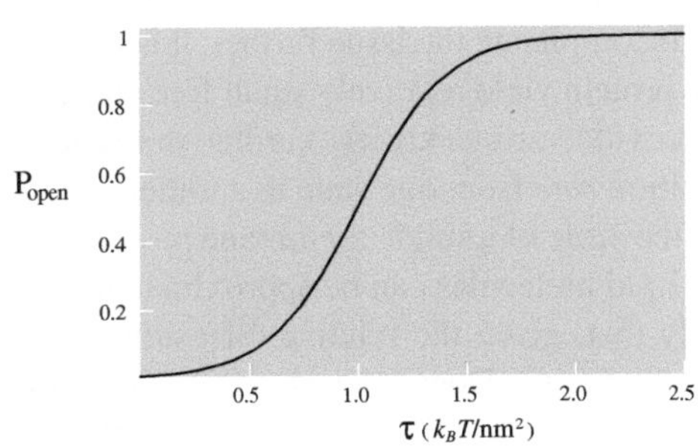

Fig. 2.5 Ion channel open probability as a function of applied tension. The plot shows $P_{open} = \langle\sigma\rangle$ as a function of the applied tension τ. The parameters used in the plot for a model mechanosensitive channel are $\epsilon_{open}-\epsilon_{closed} = 10\,k_BT$ and $\Delta A = 10\,\text{nm}^2$. The critical tension is $1.0\,k_BT/\text{nm}^2$, corresponding to $P_{open} = 1/2$. For reference, the tension can be rewritten as 1 pN/nm $\simeq 0.25\,k_BT/\text{nm}^2$

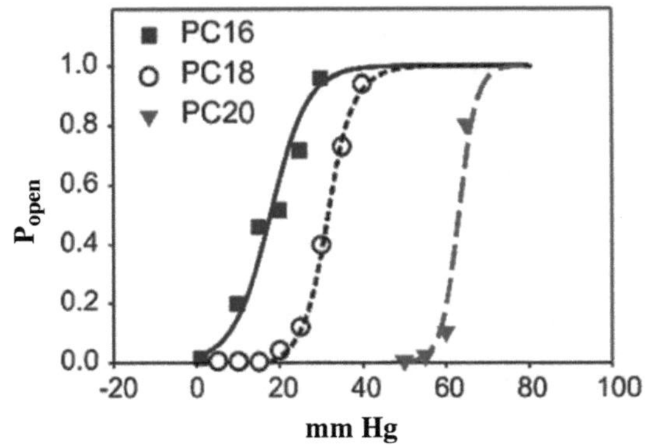

Fig. 2.6 Ion channel open probability for different lipids. The graph shows the open probability of the mechanosensitive channel MscL as a function of the applied pressure across the bilayer for three different lipid tail lengths. Pressure difference is related to bilayer tension via a constant, and hence this suggests that bilayer thickness (for carbon tail lengths of 16, 18 and 20) affects channel function. Figure adapted from Perozo et al [2002b]

strategy is to use the tools of continuum mechanics to calculate how the deformation of lipids surrounding a protein and the applied tension work in concert to affect the channel's preference for a particular state (Dan and Safran, 1998; Huang, 1986; Nielsen et al, 1998; Wiggins and Phillips, 2005). Unfortunately, relatively little is known about how the internal rearrangements of the protein and the hydration energy of the channel pore contribute to the overall free energy balance (Anishkin et al, 2005; Yoshimura et al, 1999). This ignorance is in part due to a lack of general rules that tell us how internal rearrangements translate into changes in protein energy. Further, the lack of crystal structures in the open and closed states of many channel proteins means we cannot be sure where each residue moves, which are exposed to the surrounding lipids and which are facing the hydrated internal pore. It is also difficult for molecular dynamics to comment on the energies associated with the internal movements of the protein (Elmore and Dougherty, 2001, 2003; Gullingsrud et al, 2001; Gullingsrud and Schulten, 2003) because the all-atom energies of these simulations are very large in comparison to the changes in free energy, and hence it is difficult to distill relatively small free energy changes in the background of large energy fluctuations. To complicate the issue further, it is also possible that the internal movements of the protein yield relatively small free energy changes between the two states, but may provide various kinetic hurdles in the form of energy barriers, which affect the transition *rate* from one state to another.

It is reasonable on the scale of a single membrane protein to ask whether a bilayer composed of discrete lipid molecules can be approximated as a continuum material. We argue heuristically that, given the relative diffusion coefficients of membrane proteins ($D \sim 0.1 - 1\mu\text{m}^2/s$) (Doeven et al, 2005; Gambin et al, 2006; Guigas and Weiss, 2006) and lipids ($D \sim 10\mu\text{m}^2/s$) (Kahya et al, 2003), in the time it takes a transmembrane protein to diffuse one lipid diameter, many lipids will have exchanged places near the protein, in a sense averaging out the discreteness of the lipid molecules. Additionally, the transition time for protein conformational change ($\sim 5\ \mu s$) (Shapovalov and Lester, 2004) is slow compared to lipid diffusion. Hence, we argue the bilayer can be approximated as a continuous material in equilibrium with well-defined elastic properties (Harroun et al, 1999). Further, we choose to formulate our analysis in the language of continuum mechanics, rather than lateral pressure profiles (Cantor, 1999).

Approximating the membrane as a continuum material (Dan and Safran, 1998; Harroun et al, 1999; Helfrich, 1973; Huang, 1986; Nielsen et al, 1998; Wiggins and Phillips, 2004, 2005), we will concentrate our analysis on how the mechanical properties and deformations of lipids affect the energy balance of the protein, and how tension can play the role of a driving force for gating the channel. In particular, the mechanosensitive channel of large conductance (MscL) is one of the best characterized mechanosensitive channels. Additionally, a combination of X-ray crystallography and electron paramagnetic resonance studies have yielded insights into the structures of both the closed and open states of MscL (Chang et al, 1998; Perozo et al, 2002a, 2001). One of the outcomes of this structural analysis is the idea that the structure can be roughly approximated as a cylinder, making it amenable to mechanical modeling. MscL exemplifies many of the characteristics one might call "design principles" for a mechanosensitive channel (Wiggins and Phillips, 2004, 2005), such as change in hydrophobic thickness, a change in radius, and sensitivity to membrane curvature. In the remainder of the review, we will lay the foundation for a continuum mechanical understanding of how lipid deformations and tension work together to give a switchable channel.

2.3.1 The Case Study of MscL

In the prokaryotic setting, the physiological purpose of MscL is thought to be an emergency relief valve under conditions of hypoosmotic shock (Chang et al, 1998; Pivetti et al, 2003; Sukharev et al, 1997), whereby the osmotic pressure difference between the inside of a cell and the environment translates into increased membrane tension. The channel responds by gating and non-selectively releasing osmolytes to the environment until the internal and external pressures are equilibrated (Sukharev et al, 2001, 1999). This presents us with (at least) two key questions. First, what gives MscL its ability to "sense" tension in the membrane? Second, what role is the lipid bilayer playing in the gating transition?

We will argue that the answers to these questions are found in the properties of a lipid bilayer and the geometrical features of the channel as revealed in Table 2.1. In particular, the bilayer has four key elastic properties that give it the ability to transduce tension and resist deformation by a transmembrane protein. The most striking elastic feature is the in-plane fluidity of the bilayer, which, in the absence of cytoskeletal interactions, results in equalization of tension throughout the membrane. This means that any in-plane stress (*i.e.* tension) on the membrane is felt everywhere equally. Hence, in the case of MscL, an increase in tension is applied uniformly to the outer edge of the protein, essentially trying to "pull" the channel open. We argue it is this "pulling" which constitutes the driving force for channel gating. However, this driving force is competing with the energetic cost to gate the

Table 2.1 MscL geometrical and bilayer elastic parameters. $^{(*)}$ These parameters depend on the elastic properties of the bilayer, in particular the bilayer bending modulus (κ_b), the bilayer area stretch modulus (K_A), and the leaflet hydrophobic thickness (l)

Parameter	Value	Source
Closed height	3.8 nm	Chang et al [1998]
Closed radius	2.5 nm	Chang et al [1998]
Open height	2.5 nm	Perozo et al [2002a]
Open radius	3.5 nm	Perozo et al [2002a]
Measured ΔA^*	$20\,\text{nm}^2$	Chiang et al [2004]
Measured ΔG^*	$51\,k_B T$	Chiang et al [2004]
Calculated ΔG^* (at critical tension)	$\sim 55\,k_B T$	this article
Critical Tension*	$\sim 2.5\,k_B T/\text{nm}^2$	Chiang et al [2004]
Lytic Tension*	$\sim 3.5\,k_B T/\text{nm}^2$	Rawicz et al [2000]
Bending Modulus (κ_b)	$\sim 20\,k_B T$	Niggemann et al [1995] Rawicz et al [2000]
Area Stretch Modulus (K_A)	$\sim 60\,k_B T/\text{nm}^2$	Rawicz et al [2000]
Leaflet Thickness (l)	1.75 nm	Rawicz et al [2000]

channel due to internal conformational changes within the protein and deformations of the surrounding lipid.

Three other properties give the membrane the ability to store energy elastically upon deformation. First, each leaflet of the membrane resists changes in the angle between adjacent lipid molecules, leading to bending stiffness of the membrane (Nielsen et al, 1998; Harroun et al, 1999; Helfrich, 1973; Huang, 1986; Dan and Safran, 1998; Wiggins and Phillips, 2005). Second, the membrane has a preferred spacing of the lipid molecules in-plane and will resist any changes in this spacing due to external tension (Dan and Safran, 1998; Rawicz et al, 2000). Third, the membrane has a well-defined equilibrium hydrophobic thickness which, when given an embedded protein of a different hydrophobic thickness, leads to energetically costly 'hydrophobic mismatch' (Dan and Safran, 1998; Harroun et al, 1999; Nielsen et al, 1998; Wiggins and Phillips, 2004, 2005).

The competition between the driving force and the energetic cost to gate the channel hints at a set of design principles that dictate how the channel behaves as a bistable switch. If we neglect the molecular details of MscL, its conformational change can be characterized by a set of simple changes in geometrical parameters. In particular, in our coarse-grained description we will think of the gating transition as being accompanied by changes in height, radius and protein angle, all of which couple to various modes of membrane deformation as shown in Fig. 2.7. The central question becomes, is deformation of the lipids surrounding the protein a major player in gating energetics? Indeed, experiments have already suggested that the gating characteristics are intimately linked to the hydrophobic mismatch between the protein and bilayer as was shown in Fig. 2.6 (Jensen and Mouritsen, 2004; Martinac and Hamill, 2002; Perozo et al, 2002b). It is the goal of the following sections to build up a theoretical framework for understanding the various kinds of bilayer deformation around a transmembrane protein and to describe how these

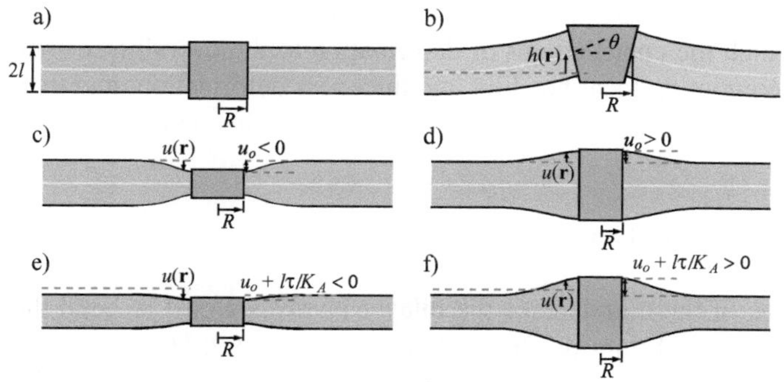

Fig. 2.7 Modes of bilayer deformation. Differences in the equilibrium shape of the membrane and an embedded protein give rise to local deformations. a) The undeformed state is a transmembrane protein with zero hydrophobic mismatch and a flat midplane. Even an initially undeformed membrane exhibits tension dependence since tension induces bilayer thinning. b) An angled protein induces midplane bending, characterized by the function $h(\mathrm{r})$ and the boundary slope θ. As tension increases, the most preferred energetic state is $\theta = 0$. c) A membrane protein that is thinner than the equilibrium thickness of the bilayer compresses the bilayer causing local area expansion and bending of each leaflet, characterized by the function $u(\mathrm{r})$. d) A membrane protein that is thicker than the equilibrium thickness of the bilayer stretches the bilayer causing local area reduction and bending of each leaflet. e) An increase in tension will decrease the energetic cost of a membrane protein that is thinner than the equilibrium thickness of the bilayer, as the membrane thins and approaches zero hydrophobic mismatch. f) Likewise, an increase in tension will increase the energetic cost of a membrane protein that is thicker than the equilibrium thickness of the bilayer

deformations contribute to the overall free energy budget associated with the gating of MscL (and probably other channels as well).

2.3.2 *Bilayer Deformation, Free Energy and the Role of Tension*

To investigate the contribution of membrane deformation to channel gating in mechanosensitive channels, we put our ignorance of the internal protein energetics aside and focus on the response of the membrane. The point of this analysis is to see how large the membrane contributions are to the free energy of channel gating, and to examine how they compare to the measured values. A mechanosensitive channel must resist the driving force due to tension to exhibit the properties of a bistable switch. As we will demonstrate in this section, deformation of the surrounding lipids can provide this resistance, and almost certainly does in the case of MscL, given our knowledge of the open and closed structures and the body of experimental data describing the interactions between lipids and MscL (Perozo et al, 2002b; Powl et al, 2003; Sukharev et al, 1999).

The deformations that a transmembrane protein induces can be most broadly split into two main classes: those that deform the midplane of the bilayer, and those that deform the bilayer leaflet thickness. If the deformation is not too severe, these two types of deformation are independent of one another (Wiggins and Phillips, 2005). Fig. 2.7 shows these two classes of deformation and the simple model idealizations implied by elastic descriptions. The basic structure of the models we consider are those in which the contributions of deformation to the overall free energy are obtained by computing local bending and thickness deformation, and then summing over the contributions from all the area elements making up the bilayer.

2.3.2.1 Midplane Deformation

Deformation of the midplane of the bilayer involves a cost to bend the midplane from its flat, equilibrium position (Harroun et al, 1999; Helfrich, 1973; Wiggins and Phillips, 2005). We use the function $h(\mathrm{r})$ to denote this change in height of the bilayer midplane as a function of the position r as shown in Fig. 2.7. The energy cost associated with bending the membrane away from its flat configuration can be written as

$$G_{\text{bend}}^{(\text{mid})} = \frac{\kappa_b}{2} \int \left(\nabla^2 h(\mathrm{r}) - C_o\right)^2 \mathrm{d}^2\mathrm{r}, \tag{2.9}$$

where the bilayer bending modulus $\kappa_b \simeq 20\, k_B T$ (Niggemann et al, 1995; Rawicz et al, 2000) and C_o is the midplane spontaneous curvature. Throughout the review the gradient operator is defined by $\nabla = (\partial/\partial x, \partial/\partial y)$, and the Laplacian operator by $\nabla^2 = \partial^2/\partial x^2 + \partial^2/\partial y^2$, in Cartesian coordinates. In general, bilayers with symmetric leaflet compositions have zero midplane spontaneous curvature. Tension also plays a role in the energetics of midplane deformation because any bend in the midplane results in a reduction in the projected area of the membrane, which couples directly to an increase in the energy of the loading device. This effect is quite intuitive when one considers deformations of a macroscopic membrane under tension and results in a contribution to the free energy of the form

$$G_{\text{ten}}^{(\text{mid})} = \frac{\tau}{2} \int (\nabla h(\mathrm{r}))^2 \, \mathrm{d}^2\mathrm{r}, \tag{2.10}$$

where the tension, τ, ranges from zero up to the nominal membrane lytic tension of $\sim 3.5\, k_B T/\mathrm{nm}^2$ [1] (Rawicz et al, 2000). In general, the elastic parameters we use are representative of a typical phosphatidylcholine (PC) lipid. Thus the total energy expended to deform the midplane over an area A is

$$G^{(\text{mid})} = \int_A \left(\frac{\tau}{2}(\nabla h(\mathrm{r}))^2 + \frac{\kappa_b}{2}(\nabla^2 h(\mathrm{r}))^2\right) \mathrm{d}^2\mathrm{r}. \tag{2.11}$$

The logic behind this kind of analysis is to find the free energy minimizing function $h(\mathrm{r})$. One way to carry out this minimization is by solving a partial differential equation that is generated by formally minimizing the free energy. An alternative

[1] The lytic tension of a bilayer is technically a dynamic quantity (Evans et al, 2003), however, we quote the lytic tension as the tension at which bilayer lysis is a rapid, spontaneous process.

Table 2.2 Typical free energies for midplane deformation. The first row indicates how tension leads to an increase in deformation energy. The second and third rows show the sensitivity to the boundary slope. The fourth row indicates how protein radius changes deformation energy. The last row is a comparison with the known radius change and critical tension of MscL

Fixed Parameters	Dynamic Parameter	Free Energy Difference
$R = 3$ nm, $\theta = 0.5$	$\tau = 0 \rightarrow 2\,k_B T/\text{nm}^2$	$10\,k_B T$
$R = 3$ nm, $\tau = 2\,k_B T/\text{nm}^2$	$\theta = 0 \rightarrow 0.5$	$10\,k_B T$
$R = 3$ nm, $\tau = 2\,k_B T/\text{nm}^2$	$\theta = 0 \rightarrow 0.8$	$26\,k_B T$
$\theta = 0.5$, $\tau = 2\,k_B T/\text{nm}^2$	$R = 3 \rightarrow 6$ nm	$14\,k_B T$
$\theta = 0.8$, $\tau = 2.5\,k_B T/\text{nm}^2$	$R = 2.5 \rightarrow 3.5$ nm	$13\,k_B T$

(and approximate) scheme to be explored later in this section is to make a guess for the functional form of $h(\mathrm{r})$ and to minimize with respect to some small set of parameters. This approach is called a variational method and can be quite useful for developing intuition.

In the midplane-deforming model, the protein can dictate the slope of the membrane at the protein-lipid interface which, in addition to the protein radius, will determine the deformation energy. The length scale over which the membrane returns to its unperturbed state is given by $\sqrt{\kappa_b/\tau}$ and the energy for this type of deformation is

$$G^{(\text{mid})}(R,\tau) = \theta^2 \pi R \sqrt{\kappa_b \tau} \frac{K_0(R\sqrt{\tau/\kappa_b})}{K_1(R\sqrt{\tau/\kappa_b})}, \tag{2.12}$$

where R is the radius of the protein, θ is the slope of the membrane at the protein-lipid interface as shown in Fig. 2.7, and K_i are modified Bessel functions of the second kind of order i (Turner and Sens, 2004; Wiggins and Phillips, 2005). Given a protein with a particular radius and fixed boundary slope, an increase in tension will make any deformation *more* costly. Hence, for midplane deformation, increased tension prefers a flatter membrane and/or smaller protein radius. To get a feel for the energy scale of this deformation several examples for different parameter values are summarized in Table 2.2.

With the contribution to the free energy difference arising from midplane deformation in hand, we can now explore the competition between applied tension and the energetics of membrane deformation in dictating channel gating. The key to understanding the interplay between tension and deformation energetics lies in the scaling of these two effects with protein radius. The midplane deformation energy scales roughly linearly with the radius of the protein and is unfavorable. On the other hand, the term proportional to the applied tension scales as the square of the protein radius and favors the open state. If we fix the membrane slope, then the energy of a midplane deforming protein as a function of protein radius and tension is

$$G(R,\tau) \simeq \underbrace{G^{(\text{mid})}(R,\tau)}_{\text{membrane}} - \underbrace{\tau \pi R^2}_{\text{loading device}}. \tag{2.13}$$

As tension increases, the potential energy of the loading device will eventually overcome the deformation energy and a larger protein radius will be the preferred state. Indeed, midplane deformations have been hypothesized to be an important functional mechanism of MscL (Turner and Sens, 2004). One of the uncertainties that accompanies a model of this type is the fact that there is some function that connects the slope of the membrane at the protein-lipid interface (θ) with the current radius of the channel, that is, there is some unknown function $\theta(R)$ (Spencer and Rees, 2002). Future experiments will be necessary to further clarify this point. If we make the simplest approximation that $\theta(R) =$ constant and look at two reasonable values of $\theta = 0.6$ and $\theta = 0.8$ (Turner and Sens, 2004), using eqn. 2.13 and the parameters in Table 2.1, we find the rather small critical tensions $\sim 0.004\,k_BT/\mathrm{nm}^2$ and $\sim 0.06\,k_BT/\mathrm{nm}^2$, respectively, compared to the known critical tension of MscL at $\sim 2.5\,k_BT/\mathrm{nm}^2$ (Anishkin et al, 2005; Chiang et al, 2004). Though we have shown that midplane deformations are capable of endowing a channel protein with bistability, the scale of the critical tension and the free energy difference between conformations indicates that, at least for MscL, an additional kind of deformation might be important as well.

2.3.2.2 Thickness Deformations

We have examined how protein conformation can alter midplane bending of the surrounding lipid bilayer and how this deformation energy penalizes the open state by virtue of its larger radius. A second major class of deformations are those that bend and compress a single leaflet of the membrane (Aranda-Espinoza et al, 1996; Huang, 1986; Nielsen et al, 1998; Wiggins and Phillips, 2005) and can be thought of as imposing a local thickness on the lipid bilayer that is different from its equilibrium value, as illustrated in Fig. 2.7. This kind of deformation relies on the fact that most proteins are rigid in comparison to the flexibility of a lipid molecule. Hence, when trying to match the hydrophobic region of the protein to the hydrophobic core of the bilayer, it is the lipid that will undergo the vast majority of the deformation. For the calculations considered here, we assume that leaflet deformations are symmetric: whatever happens to the top leaflet is mirrored in the bottom leaflet. The deformation is measured as the deviation of the equilibrium position of the lipid head-groups by the function $u(\mathrm{r})$ at each position r on the membrane as was introduced schematically in Fig. 2.7. The bending energy takes the form

$$G_{\text{bend}}^{(\text{leaf})} = \frac{\kappa_b}{4} \int (\nabla^2 u(\mathrm{r}) - c_o)^2 \, \mathrm{d}^2\mathrm{r}, \tag{2.14}$$

where $\kappa_b \simeq 20\,k_BT$ is the bending modulus of a bilayer (Niggemann et al, 1995; Rawicz et al, 2000), equal to approximately twice the bending modulus of a leaflet, and the spontaneous curvature of the leaflet, c_o, characterizes the leaflet's natural tendency for a curved state at a hydrophobic-hydrophilic interface (Dan and Safran, 1998). For many bilayer forming lipids, such as phosphatidylcholines, the spontaneous curvature is small (Boal, 2002). In addition to bending, matching the hydrophobic regions of the protein and bilayer necessarily means the bilayer will change in thickness, giving rise to a bilayer energy penalty of the form

$$G_{\text{comp}}^{(\text{leaf})} = \frac{K_A}{2} \int \left(\frac{u(\mathrm{r})}{l} \right)^2 \mathrm{d}^2\mathrm{r}, \tag{2.15}$$

where $l \simeq 1.75$ nm is the leaflet hydrophobic thickness, and due to membrane volume conservation, the bilayer area stretch modulus, $K_A \simeq 60\, k_B T/\mathrm{nm}^2$, is associated with this deformation (Rawicz et al, 2000). Yet another contribution to the free energy of deformation in those cases where the membrane thickness is perturbed is a local change in the area per lipid as the bilayer thickness varies around the protein. Membrane volume conservation arises because the membrane is roughly forty times more resistant to volume change than area change (Seemann and Winter, 2003; Tosh and Collings, 1986). As a result, if a transmembrane protein locally thins the bilayer, lipids will suffer an area expansion in a way that conserves volume. Similarly, if the protein locally thickens the bilayer, the area per lipid will locally decrease. This implies that the area change near the protein is proportional to the compression $u(\mathrm{r})$, and the work done on the bilayer is the integrated area change multiplied by tension

$$G_{\text{ten}}^{(\text{leaf})} = \tau \int \frac{u(\mathrm{r})}{l} \mathrm{d}^2\mathrm{r}, \tag{2.16}$$

where τ is the externally applied bilayer tension. Hence, u less than zero corresponds to a reduction in the energy of the loading device. All of these contributions can be added up to yield the free energy cost associated with thickness variations of the two leaflets that can be written as

$$G^{(\text{leaf})} = \int_A \left(\frac{K_A}{2} \left(\frac{u}{l} \right)^2 + \frac{\tau u}{l} + \frac{\kappa_b}{2} \left(\nabla^2 u \right)^2 \right) \mathrm{d}^2\mathrm{r}. \tag{2.17}$$

In elastic models of this type, the protein dictates the degree of hydrophobic height mismatch, $u(R) = u_o$, and the angle at which the leaflet contacts the protein at the interface between the protein and the surrounding lipids. Far from the protein, we expect the bilayer to be flat and slightly thinner in accordance with the applied tension, *i.e.* $|\nabla u(\infty)| = 0$ and $u(\infty) = -\tau l/K_A$, respectively. In the case of a cylindrical protein we make the further simplifying assumption that the angle is zero (*i.e.* $|\nabla u(R)| = 0$) (Huang, 1986). The hydrophobic mismatch itself depends on membrane properties; changes in membrane thickness are linearly related to the hydrophobic mismatch by $u_o = d/2 - l$, where d is the hydrophobic thickness of the protein. Unlike midplane deformation, the length scale at which the leaflet returns to its unperturbed state, λ, depends only on fixed elastic parameters of the membrane given by

$$\lambda = \left(\frac{l^2 \kappa_b}{K_A} \right)^{\frac{1}{4}} \simeq 1\,\mathrm{nm}. \tag{2.18}$$

The deformation energy due to thickness variation in the surrounding lipids induced by the protein can be written in a simple form when the radius of the protein is larger than λ (which is the case for MscL) as

$$G^{(\mathrm{leaf})}(R,\tau) = \pi\kappa_b\left(\frac{u_o}{\lambda}+\frac{\tau}{K_A}\frac{l}{\lambda}\right)^2\left(1+\sqrt{2}\frac{R}{\lambda}\right). \tag{2.19}$$

The deformation energy scales linearly with protein radius and depends quadratically on the hydrophobic mismatch, u_o (Wiggins and Phillips, 2005), making the overall deformation energy particularly sensitive to the hydrophobic mismatch, and hence leaflet thickness l [3]. The deformation energy is fairly insensitive to changes in stretch stiffness, K_A (*i.e.* most terms in the energy are sublinear), and generally insensitive to changes in the bending modulus since $G \propto \kappa_b^{1/4}$. Additionally, given the actual values of the elastic parameters, one finds that the leaflet free energy scales roughly *linearly* with tension, due to the very small value of τ/K_A. Like midplane deformation, we see that thickness deformation prefers a smaller protein radius. On the other hand, in the midplane case, tension always increases the deformation energy around a channel while in the case of lipid bilayer thickness variations, the tension can either increase or decrease the deformation energy depending on the sign of the hydrophobic mismatch. In fact, since the hydrophobic mismatch can be either positive or negative (*i.e.* the protein can be thicker or thinner than the bilayer), tension will increase the deformation energy around a protein that is thicker than the membrane (*e.g.* the closed state of MscL) and decrease the deformation energy around a protein that is thinner than the membrane (*e.g.* the open state), as was shown in Fig. 2.7.

One of the beautiful outcomes of this simple thickness variation elastic theory is that the total free energy as a function of protein radius can be written in the simple form

$$G(R,\tau) = \underbrace{G^{(\mathrm{leaf})}(R,\tau)}_{\text{membrane}} - \underbrace{\tau\pi R^2}_{\text{loading device}}, \tag{2.20}$$

which is reminiscent of classical nucleation theory and results in free energy profiles as shown in Fig. 2.8. At zero tension, the deformation clearly prefers a smaller protein radius, limited only by the steric constraints of the protein structure, which means that there is a certain minimum radius that the protein can adopt. As the tension increases, the quadratic dependence of the driving force on radius will eventually overcome the linear dependence of the deformation energy, leading to a preference for the open state (corresponding to larger R). We introduce a "hard wall" potential at the open radius which provides a severe energy penalty for radii

[3] The concept of hydrophobic mismatch is valid when the hydrophobic regions of the protein and the bilayer strongly interact, however, this concept has its limits based on the chemistry between the lipids and the transmembrane region of the protein (Markin and Sachs, 2004, Lee, 2003, 2005), and eventually this condition will be broken if the mismatch is too large (Wiggins and Phillips, 2005).

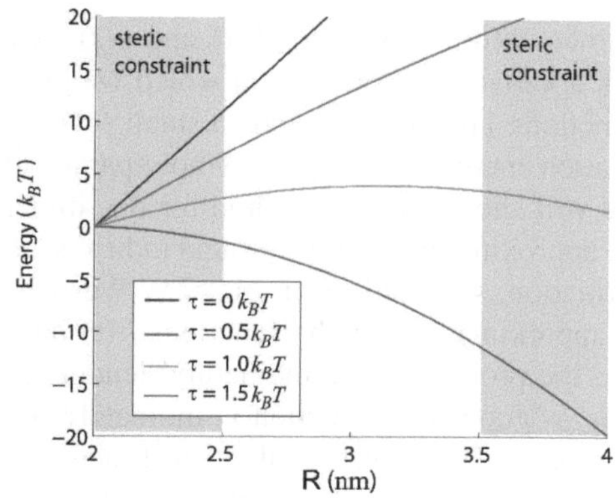

Fig. 2.8 Thickness deformation and tension induced energy of a MscL-like channel. Competition between the cost of deforming the lipid surrounding a protein and the benefit of opening a pore under tension leads to a bistable switch. At zero tension, the cost of deformation favors a small protein radius, limited only by the steric constraints of the protein. As tension increases, the benefit to opening a pore is comparable to the energetic cost to deform the lipids surrounding the protein, and a larger protein radius is now possible. At high tension, the potential energy of the loading device far outweighs the deformation cost and a larger protein radius is favored, again limited by the steric constraints of the protein

larger than the open state radius and argue that this approximation captures the idea that opening the channel any further would lead to some degree of energetically costly denaturation. This model also captures the correct scale for the critical gating tension which is on the order of $1\,k_BT/\text{nm}^2$.

It is of interest to compare the energy scale implied by this elastic model to measured values. The free energy change of MscL gating was measured to be $\simeq 51\,k_BT$ using native bacterial membranes (Anishkin et al, 2005; Chiang et al, 2004). If one uses the independently measured geometrical properties of the channel, contained in Table 2.1, and elastic properties of pure bilayers (in the text) to calculate the free energy of the closed and open states, their difference is approximately $55\,k_BT$ at the critical tension of $2.5\,k_BT/\text{nm}^2$. Though very encouraging, this close correspondence depends sensitively upon the choice of hydrophobic mismatch, as dictated by the channel structure and bilayer thickness.

2.3.3 Approximating Bilayer Deformation: The Variational Approach

In previous sections, we performed cursory derivations of the energy functionals which govern membrane shape for both midplane and membrane thickness deformations. In order to extract meaning from these energies, we had to minimize the free energy functionals of eqns. 2.11 and 2.17 with respect to membrane shape. To solve the full problem, the conventional scheme (used to obtain the earlier quoted results) is to use the calculus of variations to derive a corresponding partial differential equation in the unknown deformation fields $h(\mathrm{r})$ and $u(\mathrm{r})$. A useful and intuitive alternative is to adopt a variational approach in which we guess a family of solutions (called 'trial functions') that depend upon a small set of parameters and then minimize the deformation energy with respect to those parameters.

For simplicity, we will showcase this method for one-dimensional membranes which amounts to the approximation that the protein radius is larger than the natural length-scale of deformation, schematized in Fig. 2.9. We will use the variational approach to find an approximation for the functions $h(\mathrm{r})$ and $u(\mathrm{r})$ with their related energies, and in the process derive the natural length-scale of deformations in both cases. Picking a 'good' trial function is intimately related to the success of the variational approach. The choice of the trial function is often dictated by what we know about the character of the solution. In this case, we know that in the near-field the protein is locally disturbing the bilayer by inducing bending or hydrophobic mismatch. In the far-field, these disturbances should decay back to a flat bilayer. Keeping in mind that most of the energy cost is stored in the local disturbance around the protein, we want a trial function that has locally varying character around the protein and then a simple decay far from the protein. Such a trial function (call it $f(x)$) could be constructed using a local disturbance, $g(x)$, within a decaying envelope

$$f(x) = g(x)e^{-x/\lambda}. \tag{2.21}$$

The constant λ is an as-yet undetermined natural length scale of deformation and will emerge from the minimization process itself. Further, this choice of an exponential envelope essentially guarantees that the membrane returns to its unperturbed state far from the protein.

As a practical tool for calculation, our choice of $g(x)$ should have enough parameters to reproduce the given boundary conditions. In addition, we want to choose $g(x)$ such that the free energy is a simple function of these parameters. The power of the variational approach is that once we have written the energy in terms of these variational parameters, the best version of $f(x)$ is, by definition, the one that minimizes the energy. Thus, for instance, if the trial function has two free parameters a and b, $f(x; a, b)$, finding the best trial function amounts to solving a system of algebraic equations defined by

$$\frac{\partial}{\partial a} G[f(x; a, b)] = 0 \quad \text{and} \quad \frac{\partial}{\partial b} G[f(x; a, b)] = 0, \tag{2.22}$$

where the brackets indicate the energy, G, is calculated using the trial function f. This variational strategy can also be used as the basis of numerical approaches in which the membrane deformation is represented using finite elements, for example. In this case, the trial functions permit us to determine the energy to an arbitrary degree of accuracy. Our strategy in the remainder of this section is to use the simplicity of the variational approach to find approximate energies for the midplane and thickness deformations imposed by membrane proteins.

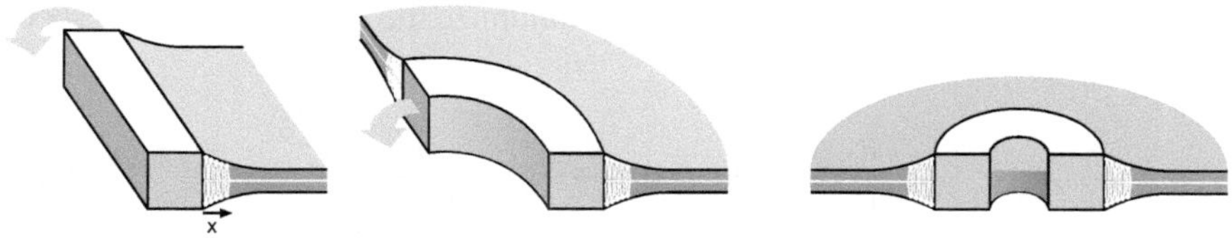

Fig. 2.9 Protein-induced line tension. Deformation of the membrane around an ion channel can be described using a line tension. This line tension is obtained by solving for a one-dimensional deformation energy per unit length and then imposing that energy around the circumference of the channel. The diagrams above show the sequential wrapping of this one-dimensional line tension around a cylindrical channel

2.3.3.1 Variational Approach for Midplane Deformations

Our goal is to obtain an approximate expression for the one-dimensional energy due to midplane bending given by

$$G^{(\mathrm{mid})} = 2\pi R \int_0^\infty \left(\frac{\tau}{2} \left(\frac{\mathrm{d}}{\mathrm{d}x} h(x) \right)^2 + \frac{\kappa_b}{2} \left(\frac{\mathrm{d}^2}{\mathrm{d}x^2} h(x) \right)^2 \right) \mathrm{d}x. \tag{2.23}$$

The presence of the $2\pi R$ in this expression is due to the fact that we are computing the energy per unit length for a deformed bilayer, as shown in Fig. 2.9, and must then multiply by the length (the circumference) of deformed material.

The strategy employed in the variational approach is to plug the trial function into the free energy functional and compute the resulting energy, which depends upon the parameters in the trial function. Our trial function has the form

$$h(x) = g(x) e^{-\left(\frac{x}{\lambda}\right)}. \tag{2.24}$$

The choice of $g(x)$ can be made based upon the boundary conditions. In particular, at the boundary of the protein, we require that

$$\frac{\mathrm{d}}{\mathrm{d}x} h(x)|_{x=0} = \theta, \tag{2.25}$$

which tells us that we can make the choice $g(x) = \text{constant}$. Applying this boundary condition yields the functional form

$$h(x) = -\theta \lambda e^{-\left(\frac{x}{\lambda}\right)}, \tag{2.26}$$

where the only remaining undetermined parameter is the length scale, λ. This trial function can be plugged into eqn. 2.23 and the integral is easily evaluated to yield the free energy

$$G^{(\mathrm{mid})}(\lambda) = \frac{\pi}{2} R \theta^2 \left(\tau \lambda + \frac{\kappa_b}{\lambda} \right). \tag{2.27}$$

The next step in the variational strategy is to minimize the free energy with respect to λ,

$$\frac{\partial}{\partial \lambda} G^{(\text{mid})}(\lambda) = 0 \quad \rightarrow \quad \lambda = \sqrt{\frac{\kappa_b}{\tau}}, \tag{2.28}$$

which upon substitution yields

$$G^{(\text{mid})} = \theta^2 \pi \kappa_b \frac{R}{\lambda} = \theta^2 \pi R \sqrt{\kappa_b \tau}. \tag{2.29}$$

This is precisely the asymptotic ($R\sqrt{\tau/\kappa_b} > 1$) form of eqn. 2.12 for midplane bending energy, and our minimization correctly defines the natural length-scale of midplane deformation.

2.3.3.2 Variational Approach for Membrane Thickness Deformations

A similar analysis can be made for the one-dimensional deformations induced by hydrophobic mismatch. In this case, the free energy functional in the absence of tension can be written as

$$G^{(\text{leaf})} = 2\pi R \int_0^{\infty} \left(\frac{K_A}{2} \left(\frac{u(x)}{l} \right)^2 + \frac{\kappa_b}{2} \left(\frac{\mathrm{d}^2}{\mathrm{d}x^2} u(x) \right)^2 \right) \mathrm{d}x. \tag{2.30}$$

We adopt the same functional form for the trial function, namely,

$$u(x) = g(x) e^{-\left(\frac{x}{\lambda}\right)}. \tag{2.31}$$

In this case, we specify two boundary conditions in the near-field; there is a hydrophobic mismatch which demands

$$u(R) = u_o, \tag{2.32}$$

and the leaflet has a particular slope at the membrane interface, which we will set to zero,

$$\frac{\mathrm{d}}{\mathrm{d}x} u(x)|_{x=0} = 0. \tag{2.33}$$

In order to accommodate these two boundary conditions, $g(x)$ must have two free parameters. As a result, we pick the simplest function which has two degrees of freedom, namely a line, and hence set $g(x) = ax/\lambda + b$, where a and b are constants. Applying the two near-field boundary conditions constrains the trial function to the form

$$u(x) = u_o \left(1 + \frac{x}{\lambda} \right) e^{-\left(\frac{x}{\lambda}\right)}, \tag{2.34}$$

where λ is a free parameter with respect to which the energy must be minimized. Using this trial function, the free energy can be written as a simple expression of the form

$$G^{(\text{leaf})}(\lambda) = \pi \kappa_b u_o^2 R \left(\frac{5}{4} \frac{K_A}{\kappa_b l^2} \lambda + \frac{1}{4\lambda^3} \right). \tag{2.35}$$

Minimizing the energy with respect to λ gives

$$\frac{\partial}{\partial \lambda} G^{(\text{leaf})}(\lambda) = 0 \quad \rightarrow \quad \lambda = \left(\frac{3}{5}\right)^{\frac{1}{4}} \left(\frac{\kappa_b l^2}{K_A}\right)^{\frac{1}{4}} \tag{2.36}$$

which upon substitution gives the membrane thickness energy

$$G^{(\text{leaf})} = \left(\frac{5}{3}\right)^{\frac{3}{4}} \left(\frac{K_A}{\kappa_b l^2}\right)^{\frac{3}{4}} \pi \kappa_b u_o^2 R. \tag{2.37}$$

Again, the variational approach has reproduced the correct asymptotic form of the energy with a small multiplicative error (see eqn. 2.19); the exact asymptotic result has $\sqrt{2}$ instead of $\left(\frac{5}{3}\right)^{\frac{3}{4}}$, introducing an error of $\sim 4\%$.

Finally, there are many forms of $u(x)$ which yield roughly the same *energy*, but how does the exact deformation *shape* compare with our minimized trial function? Here too, the variational approach gives a trial function that nearly matches the exact result as shown in Fig. 2.10.

2.3.4 Distilling the Design Principles

Having explored how midplane bending, thickness variation and area change are coupled to tension and the geometric features of the MscL channel, can we distill general rules for what makes a membrane protein mechanosensitive? One simple statement is that under tension an increase in protein area is always favored, regardless of bilayer elastic properties, because an increase in area lowers the potential

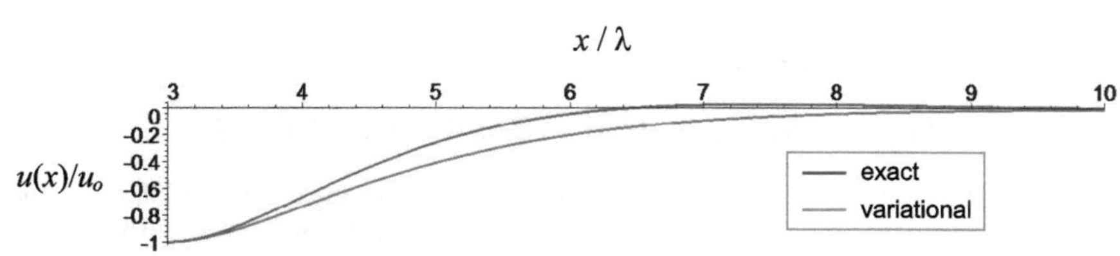

Fig. 2.10 Comparison of the exact and variational solutions for the thickness deformations around a protein. The variational approach generates an approximation to $u(x)$ which is close to the exact solution. The protein radius is $R/\lambda = 3$

energy of the loading device. Conversely, both midplane and thickness deformations prefer a smaller channel, because a larger radius results in a larger annulus of deformed lipid and hence a larger free energy penalty (except in the case where the spontaneous curvature favors a larger radius (Wiggins and Phillips, 2005)). With the area change preferring a larger radius and deformation preferring a smaller radius, we have the necessary energetic competition that ultimately leads to bistability. This also means the sign of the free energy change due to deformation (midplane or thickness) must be positive. Hence, the channel is going from a closed state with less deformed lipid surrounding it, to an open state with more deformed lipid surrounding it. The contributions to the free energy budget of a mechanosensitive protein, like MscL, due to channel-area change and membrane deformations are shown in Fig. 2.11. The basic point of this picture is to show how various contributions to the free energy scale with the radius (R) and the elastic parameters.

Midplane deformation is the deformation which depends most simply on membrane properties, since it is only linked to the bending modulus. Additionally, its tension dependence is such that the cost of the deformation always increases with tension and angle, hence we know that tension in addition to preferring a larger protein, also wants a more cylindrical protein in the case of midplane bending. This allows us, within the limitations of our theory, to put an upper bound on the cost of midplane deformations. Taking the lytic tension as an upper bound, a nominal bending modulus of $20\,k_BT$, and $\theta = 0.6$ as a reasonable value of the membrane slope (Turner and Sens, 2004), the maximum energetic cost of deformations for a protein of radius R (in nm) is $\simeq R \times 9\,k_BT/\text{nm}$.

Thickness deformation depends on all the elastic parameters; bending modulus, area stretch modulus, and membrane thickness. The tension dependence of thickness deformation energy is also more complex, though a general principle does emerge. We know that tension can increase or decrease the overall thickness deformation energy, but the general principle is that it always prefers the protein to have the

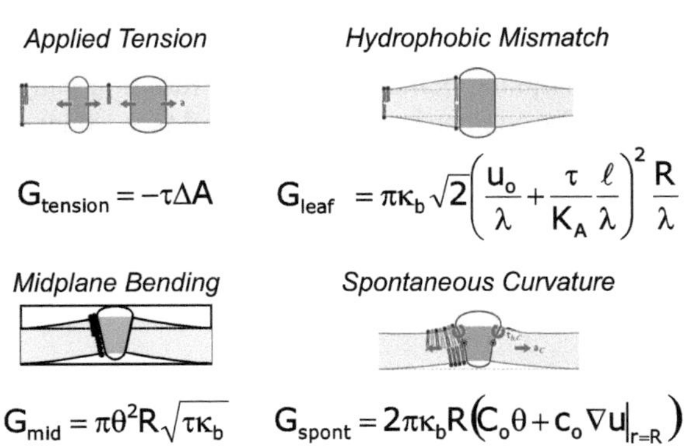

Fig. 2.11 Contributions to the free energy. This figure shows how the different modes of deformation contribute to the overall free energy budget of the membrane-protein system. The energies are written asymptotically to show their dominant scaling with the relevant parameters. For the sake of simplicity, we did not address how spontaneous curvature factors into the free energy budget. However, a thorough discussion of both midplane and leaflet spontaneous curvature energy contributions are found in (Wiggins and Phillips, 2005)

same hydrophobic thickness as the bilayer, though the bilayer thickness is itself decreased as tension increases. The other important feature to note is that a decrease in the thickness of a transmembrane protein is always accompanied by an increase in the area of the membrane surrounding the protein due to volume conservation of the membrane. This change in membrane area is *indistinguishable* from a change in protein area. Indeed for MscL, the measured area change is probably a mix of a change in the areal footprint of the protein, and a local increase in the membrane area surrounding the protein, together giving the measured value of $\sim 20\,\text{nm}^2$. An estimate of the upper bound of leaflet deformations is made by assuming the maximum $u_o = 0.5$ nm, then the maximum change in free energy for a protein of radius R (in nm) is $\simeq R \times 22\,k_BT/\text{nm}$ at zero tension with the given elastic parameters (see Table 2.1). This illustrates that while both midplane and thickness deformations are important factors in determining the preferred protein conformation, thickness deformations are generally associated with a slightly higher energy scale.

2.4 Experimental Considerations

Much of our knowledge of the function of mechanosensitive channels, including MscL, comes from detailed electrophysiology studies where gating of the channel is monitored by sharp differences in the ion flux through a membrane patch (Anishkin et al, 2005; Chiang et al, 2004; Perozo et al, 2002b; Sukharev et al, 2001, 1997, 1999). A small voltage ($\sim$ 50mV) is applied across a patch of membrane at the tip of a micropipette. As a function of pressure difference, channel opening events are recorded as stochastic changes in patch current by an ammeter with picoamp (pA) sensitivity. This truly amazing single-molecule spectroscopy technique allows the experimenter to adjust the voltage as well as the pressure difference across the membrane as shown in Fig. 2.12. The pressure difference across the membrane translates into a lateral membrane tension (via the Laplace-Young Relation), responsible for gating the mechanosensitive channel. However, there are two serious problems with this method when probing the mechanisms of mechanosensitive channels.

Arguably, the most serious problem is that often *pressure difference* (J/m^3) across the membrane is taken to be the input variable of prime importance, when in fact *tension* (J/m^2) is the membrane parameter which governs mechanosensitive gating. Pressure difference is linearly related to tension via the radius of curvature of the membrane, hence in principle the fix is straightforward - image the membrane patch (see Fig. 2.12). While certainly not impossible (Moe and Blount, 2005, Sukharev et al, 1999), the membrane patch can be difficult to image due to its small size and the fact that it is inside the micropipette. A recent study (Moe and Blount, 2005) demonstrated the importance of measuring tension in lieu of pressure difference. It was shown that using the standard methods for creating "identical" micropipettes, the measured characteristics of a channel varied significantly. However, when the membrane patch was imaged and tension used as the principle input variable, the same data collapsed to within a few percent of each other. In general, if one could perfectly control the size and shape of the micropipette tip used for contacting

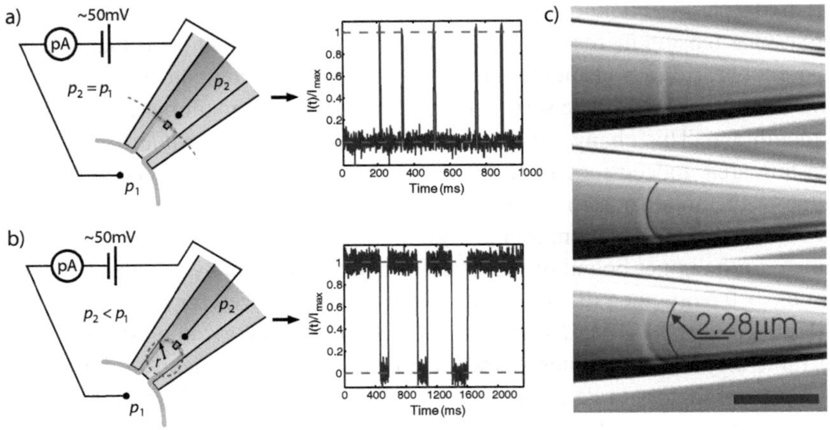

Fig. 2.12 Measurement of tension vs. pressure difference in an electrophysiological experiment. A channel protein (small blue rectangle) is embedded in a membrane patch (green). A potential of roughly 50mV is applied across the sealed membrane patch, and channel opening events are measured by an ammeter (circle) with picoamp (pA) sensitivity. a) At low pressure difference, the tension in the patch is low, the mechanosensitive channel is in the closed conformation, and the patch has a very large radius of curvature. The plot to the right shows normalized channel current as a function of time for a simulated channel; the open state has low occupation at low tension. b) At high pressure difference, the tension in the patch is high, the mechanosensitive channel will occupy the open state, and the radius of curvature (r) is on the order of microns. The plot to the right shows the open state has high occupation at high tension. c) Optical micrograph of vertically oriented membrane patch at low (top) and high (middle and bottom) pressure differences, illustrating the decrease in the radius of curvature with increase pressure difference (from (Moe and Blount, 2005)). The scale bar is 5μm

and sealing the membrane patch, all measurements would be related by a single constant (the radius of curvature). However, variations in micropipette shape and size, as well as variations in how the membrane contacts the pipette tip all lead to potentially large variations in the perceived gating characteristics of the channel. Additionally, it is difficult to compare the wealth of quantitative data coming from electrophysiology studies to theoretical models when pressure difference, instead of tension, is used as the principle input variable. Tension is routinely measured in micropipette aspiration experiments (Rawicz et al, 2000), and in fact, single-channel electrophysiology recordings are possible in such a setup (Goulian et al, 1998) using ion channels with conductances *lower* than MscL. Hence, this technique might provide a useful way to apply known membrane tension to reconstituted MscL channels in well characterized membranes.

With tension being used as the variable of prime importance, electrophysiology is poised to put the continuum mechanical view to the test, elucidating the role of lipids in ion channel function. In particular, the elastic properties of many lipids have been measured (Rawicz et al, 2000), enabling a careful examination of the dependence of gating energy on lipid carbon chain length. The simple continuum view we set forth here predicts a quadratic dependence of the lipid thickness deformation energy on hydrophobic mismatch, which is directly linked to carbon chain length. This, of course, has implications for both the function of various transmembrane proteins, and comments meaningfully on the ability of bilayer thickness to segregate proteins in biological membranes.

A second class of intriguing experiments concerns the mechanosensitivity of other ion channels and receptors, generally regarded not to be mechanosensitive (Calabrese et al, 2002; Gu et al, 2001). This is both interesting from a functional standpoint, in an effort to understand the full physiological effects of these proteins, and as a tool for understanding structural features such as the motions of transmembrane helices. Performing a similar experiment where lipid carbon chain length is varied around a voltage-gated ion channel (for example) could reveal hidden mechanosensitivity, and energetic analysis from such an experiment could comment on the degree of height and area change during the gating transition.

The second problem facing a complete understanding of the function of mechanosensitive channels is that for many such channels volumetric flow, and not ion flux, is the relevant physiological parameter[2]. Hence, ion flux is used as a surrogate measurable in place of the true physiological output of the channel. One could argue that ion flux is proportional to volumetric flow, however this assumes that the way ions flow through the channel pore is identical to the way water flows through the pore. Experiments have elucidated the roughly ohmic nature of mechanosensitive channels (Perozo et al, 2002b; Cruickshank et al, 1997) at low voltage ($\lesssim$ 80mV), however we know essentially nothing about how a pressure gradient across the membrane translates into a volumetric flow. Even the simplest continuum approximation (Hagen-Poiseuille flow) would predict a non-linear function relating the area of the channel pore to the volumetric flow, in contrast to the (roughly) linear relationship between ion flux and channel pore area as predicted by Ohm's Law (Hille, 1968). It would be of considerable physical and physiological interest to expand our understanding of fluid flow at the molecular level, by measuring the relationship between pressure gradient and volumetric flow through a large-pore channel like MscL.

2.5 Cooperativity and Interaction between Transmembrane Proteins

One intriguing consequence of the deformations induced in membranes by ion channels is that channels will interact. These interactions can lead to cooperativity in the gating of neighboring channels and can also induce spatial ordering of the proteins. These interactions can be thought of as arising from two different effects: those of *elastic* origin and those of *thermal* origin. The elastic forces are purely an enthalpic

---

[2] The mechanosensitive bacterial channel MscS (Pivetti et al, 2003; Levina et al, 1999; Bass et al, 2002) is another example. Although, there are also mechanosensitive channels that appear to be highly ion selective, such as the bacterial mechanosensitive ion channel MscK (Li et al, 2002) and the K2P family of mammalian mechanosensitive channels (Maingret et al, 1999; Maingret et al, 2000; Franks and Honore, 2004; Lauritzen et al, 2005).

effect coming from a minimization of the deformation energy around two proteins separated by a given distance. The thermal forces are entropically driven by the thermal fluctuations of the membrane and are analogous to Van der Waals forces.

2.5.1 Enthalpic Interactions

As discussed above, proteins which change the membrane thickness or bend the membrane midplane produce deformations which extend anywhere from a few nanometers (thickness) up to tens of nanometers (midplane) from the protein edge. As two proteins approach each other, their respective deformation fields overlap resulting in a deformation profile between them that is different than either of them produce separately. In this case, the total deformation energy of the system is dependent on the separation between the two proteins and results in an interaction potential which is dependent upon the conformation of the proteins. These forces arise purely from the mechanical attributes of the deformed membrane and have no entropic component. We know that midplane and thickness deformations are independent, and hence there are distinct interactions due to midplane and thickness deformations.

Pairwise interactions due to midplane deformation using eqn. 2.11 have previously been calculated for a variety of membrane curvature environments and protein shapes at zero tension (Chou et al, 2001). Using a bilayer bending modulus of $\sim\ 100\,k_BT$, attractive interactions of order $\sim\ 1-5\,k_BT$ were found when the proteins were separated by 1–2 protein radii (which we estimate to be 5–10 nm measured center-to-center for a typical transmembrane protein). If we adjust the energy scale to be consistent with a phosphatidylcholine bilayer bending modulus of $\sim\ 20\,k_BT$ this lowers the interaction energetics to $\sim\ 0.5-3\,k_BT$. These interactions tend to be long-ranged with a power-law decay of $1/r^4$ (Goulian et al, 1993). Simple pairwise interaction will be inadequate to describe the nature of interactions between more than two proteins. This arises because one protein can shield other proteins from feeling the deformation of a neighboring protein, and hence interactions are not (in general) pairwise additive. Apart from direct numerical simulation, there are few analytical (theoretical) tools which allow one to study how many interacting proteins in close proximity behave as a group (Harroun et al, 1999).

Like midplane deformations, the thickness deformation fields extending from the edges of two proteins will overlap and interact as the proteins come into close proximity (Aranda-Espinoza et al, 1996; Dan et al, 1993; Ursell et al, 2007). We provided evidence that lipids likely influence the function of MscL through thickness deformations and once again we will appeal to MscL as a case study for interacting membrane proteins. The short-range nature of thickness deformations (essentially exponential decay) means there is no power-law asymptotic formula for their interaction, though these interactions were numerically explored for all possible conformations of two MscL proteins as shown in Fig. 2.13. As we saw with single proteins, the energetic scale of thickness interactions is generally higher than with midplane

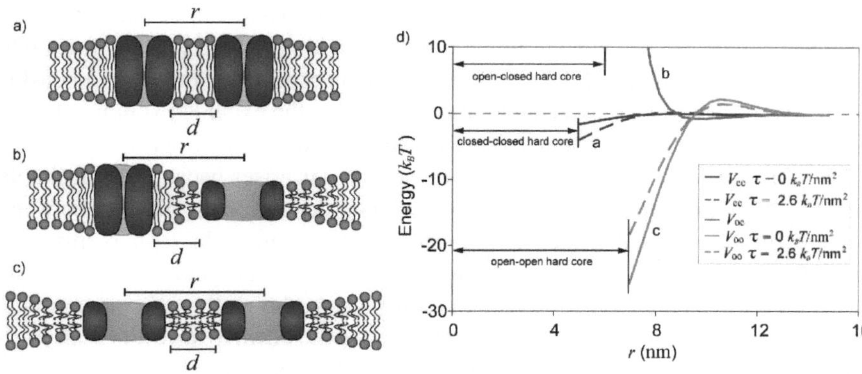

Fig. 2.13 Conformation-dependent interactions between two MscL channels. As two MscL channels (blue) come close to each other, regions of deformed lipids (green) overlap leading to deformation induced interactions. a) Deformation surrounding two closed MscL channels. b) Deformation surrounding a closed and an open MscL channel. c) Deformation surrounding two open MscL channels. The relative sizes of the open channel, the closed channel, and the lipids are roughly correct. d) Interaction potentials for the three configurations shown in a, b and c. External tension weakens the interaction between two open channels (V_{oo}) and strengthens the interaction between two closed channels (V_{cc}), but has almost no effect on the interaction between an open and closed channel (V_{oc}). The open-open and closed-closed interactions are both more strongly attractive than the open-closed interaction, indicating that elastic potentials favor interactions between channels in the same state. The 'hard core' distance is where the proteins' edges are in contact

deformations, and can vary greatly depending on the hydrophobic mismatch. The leaflet interactions between two MscL proteins are appreciable when they are within several nanometers of each other, and ranged from $\sim 2 - 25\, k_B T$ depending on the protein conformations and the tension in the membrane. This kind of short-ranged interaction might play an important role in membrane protein function (Botelho et al, 2006; Goforth et al, 2003 Molina et al,, 2006), given the nominal density of transmembrane proteins in biological membranes leads to spacings on the order of 10-100 nm (Mitra et al, 2004).

Additionally, the interactions due to thickness variations can be either attractive or repulsive depending on the shape of the proteins. The general principle that emerges is that 'like' proteins attract and 'unlike' proteins repel (in contrast to electrostatics) as shown in Fig. 2.13 (Ursell et al, 2007). Proteins whose hydrophobic mismatch has the same sign (*i.e.* both are taller or both are shorter than the membrane) lead to net attractive interactions; proteins with opposite signs of hydrophobic mismatch lead to repulsive interactions. Later in the article, we will demonstrate that these conformation-dependent interactions can communicate state information between two proteins, leading to cooperative channel gating.

2.5.2 Entropic Interactions

A second class of forces between membrane proteins arise due to membrane fluctuations. Like most entropic forces, the thermal interactions between transmembrane proteins are fairly weak, on the order of a few $k_B T$. Two fluctuation-induced forces have been studied in some detail in the literature; a long-ranged Casimir force due to the surface fluctuations of the membrane (Goulian et al, 1993;Park and Lubensky, 1996), and a very short-ranged depletion force due to the excluded volume of lipid molecules between two membrane proteins (Sintes and Baumgartner, 1997).

The Casimir force between two membrane proteins arises because the available spectrum of fluctuations of the membrane-midplane depend on the distance between two proteins. Entropically, the membrane-protein system seeks to maximize the number of available modes of fluctuation and hence an energetic potential exists between two transmembrane proteins in a fluctuating, thermally active membrane. Through a series of elegant calculations, this force was shown to have a $1/r^4$ asymptotic form, where r is the center-to-center distance between two cylindrical proteins (Goulian et al, 1993; Park and Lubensky, 1996). If we presume that it is approximately correct for small separations ($r \simeq 2R$), this implies an attractive potential with an energy scale of $\sim 1\, k_B T$.

Lateral density fluctuations of lipids in the membrane also lead to entropic forces between proteins. Using Monte Carlo simulations, these entropic depletion forces (also called 'excluded volume forces') between cylindrical proteins were shown to be appreciable only when the proteins' edges were within $\sim$ 1 lipid molecular diameter (Sintes and Baumgartner, 1997). For cylindrical proteins, with diameters on the order of $\sim 1 - 2$ nm, direct edge contact resulted in a favorable interaction with an energy scale of $\sim 2\, k_B T$.

2.5.3 Protein Conformations Affected by Interaction

As noted above, the elastic interactions between ion channels such as MscL depend upon protein conformation. In earlier sections of the paper, we established that the equilibrium conformations of the channel are entirely determined by the free energy difference between the two states. As a result, elastic interactions which change the energy of a two-channel system will affect the probability that we measure any one channel in the open state (Ursell et al, 2007). In fact, electrophysiology (see Fig. 2.2) is well suited to such measurements where the total amount of time spent in the open state divided by the total measurement time *is* the open probability.

The free energy difference between the open and closed states of a single channel is roughly $50\, k_B T$ (Chiang et al, 2004), which implies the energy scale for two channels is roughly $100\, k_B T$. We have also seen that two MscL channels in proximity have interactions on an energy scale of roughly $\sim 20\, k_B T$ as shown in Fig. 2.2. Two open channels have a strong, favorable interaction that can significantly alter the open probability of a given channel relative to the isolated channel value as shown in Fig. 2.14. Such interactions also affect channel 'sensitivity', defined by the derivative of the P_{open} curve with respect to tension (Ursell et al, 2007), which quantifies how responsive the channel is to changes in the driving force, in this case tension. The full-width at half maximum of this peaked function is a measure of the

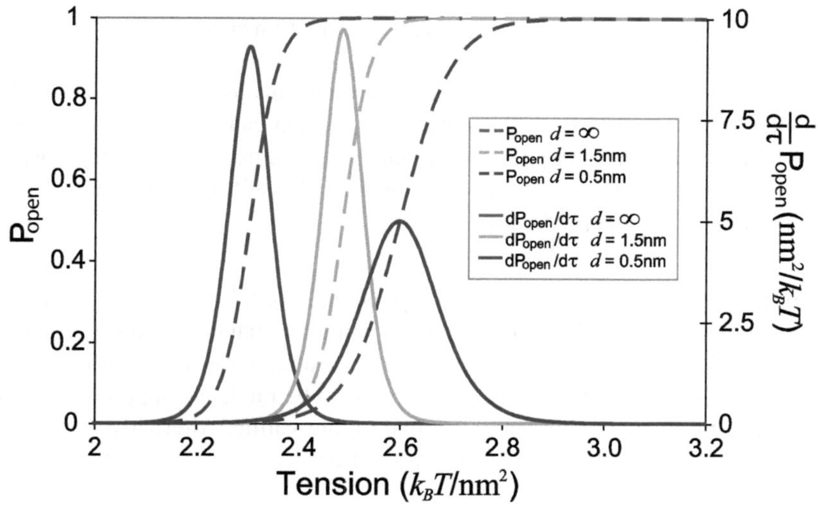

Fig. 2.14 Conformational statistics of interacting MscL proteins. Interactions between neighboring channels lead to shifts in the probability that a channel will be in the open state (dashed lines). The sensitivity and range of response to tension, $dP_{open}/d\tau$, are also affected by bilayer deformations (solid lines). P_{open} and $dP_{open}/d\tau$ are shown for separations of 0.5 nm (red) and 1.5 nm (green) with reference to non-interacting channels at $d = \infty$ (blue). Interactions shift the critical gating tension for the closest separation by $\sim 12\%$. Additionally, the peak sensitivity is increased by $\sim 90\%$ from $\sim 5\text{nm}^2/k_BT$ to $\sim 9.5\text{nm}^2/k_BT$

range of tension over which the channel has an appreciable response. In general, the area under the sensitivity curve is equal to one, hence increases in sensitivity are always accompanied by decreases in range of response, as demonstrated by the effects of the beneficial open-open interaction on channel statistics. The critical gating tension and sensitivity are essentially the key properties which define the transition to the open state, and are analogs to the properties which define the transition of *any* two-state ion channel. Hence, the elastic interactions can affect channel function on a fundamental level.

2.6 Conclusion

The goal of this article is to take stock of the role of lipid bilayer deformations in mechanosensation. More precisely, we have argued that the lipid bilayer is not a passive bystander in the energetics of channel gating. As a result, by tuning membrane properties it is possible to alter channel function. We have emphasized two broad classes of membrane deformation that are induced by the presence of a transmembrane protein: i) deformation and bending of the midplane of the lipid bilayer, ii) variations in the thickness of the lipid bilayer that are induced by hydrophobic mismatch. As a result of these deformations, there is a free energy cost to changing the radius of a channel since the open state implies a larger annulus of deformed material and hence a higher free energy. This deformation energy competes with the energetic relaxation of the loading device.

One of the key reasons for performing theoretical analyses like those described here is that they permit us to sharpen the questions that can be asked about a given biological problem. This sharpness is ultimately most meaningful if it is translated into precise experimental predictions. The theoretical results described here suggest a variety of experimental predictions.

- *Dependence of gating tension on hydrophobic mismatch.* Previous work has already shown that lipid bilayer tail lengths can alter channel gating by changing the hydrophobic mismatch. To more precisely examine this relationship, careful measurements of the membrane tension need to be made, as opposed to pipette pressures, to elucidate the energetics underlying gating. Alternatively, mutagenesis could be used to explore the same effect by changing the hydrophobic thickness of the protein.
- *Hidden mechanosensitivity in other classes of channels and receptors.* The results described here have been applied to the case study of MscL. However, we argue that any transmembrane protein that varies its radius or hydrophobic thickness upon conformational change will exhibit mechanosensitivity. Furthermore, the way tension affects the function of these proteins might help elucidate the classes of structural changes that occur during their conformational change.
- *Cooperative gating of channels.* As a result of the elastic deformations induced in the lipid bilayer by mechanosensitive channels, nearby channels can communicate their conformational state, resulting in cooperative gating. This cooperativity should be observable in electrophysiology experiments as a change in the critical tension and channel sensitivity with an increase in channel density.

Shortcomings of the Theory. Obviously, the use of simple ideas from elasticity theory to capture the complex process of mechanosensation provides a caricature of the real process. One signature of the shortcomings of this kind of approach is the fact that single amino acid substitutions can completely alter the properties of certain proteins (Yoshimura et al, 1999, 2004). This serves as a warning of the pitfalls of models that ignore atomic-level details and their impact on biological function. A second class of complaint that can be registered against the models described here is that we have ignored material heterogeneity. In particular, biological membranes are built up of a broad range of different lipids and are riddled with membrane proteins. As a result, it is not clear if an elastic description like that described here is appropriate, and if it is, how to select the relevant material parameters.

Regardless of the difficulties highlighted above, it is clear that the emergence of an increasing number of structures of ion channels coupled with functional studies of these proteins has raised the bar for what should be expected of theoretical models of channel function. The central thesis of the work described here is that the presence of the lipid bilayer provides another way in which these systems can be manipulated.

Acknowledgment We are grateful to a number of people who have given us both guidance and amusement in thinking about these problems, as well as insightful comments on the manuscript: Doug Rees, Olaf Anderson, Fred Sachs, Evan Evans, Cathy Morris, Sergei Sukharev, Mathew Turner, Eduardo Perozo, Nily Dan, Fyl Pincus, Liz Haswell and Pierre Sens. JK acknowledges the support of National Science Foundation grant No. DMR-0403997 and is a Cottrell Scholar of Research Corporation. RP acknowledges the support of the National Science Foundation grant No. CMS-0301657. TU and RP acknowledge the support of the National Science Foundation CIMMS Award ACI-0204932 and NIRT Award CMS-0404031 as well as the National Institutes of Health Director's Pioneer Award. DR acknowledges the support of National Science Foundation grant No. DGE-0549390. PAW is supported by the Whitehead Institute for Biomedical Research.

Accession Numbers The primary accession numbers (in parentheses) from the Protein Data Bank (http://www.pdb.org) are: mechanosensitive channel of large conductance (2OAR; formerly 1MSL).

References

Akitake B, Anishkin A, Sukharev S (2005) The"dashpot" mechanism of stretch-dependent gating in MscS. J Gen Physiol 125(2):143–154

Anishkin A, Chiang CS, Sukharev S (2005) Gain-of-function mutations reveal expanded intermediate states and a sequential action of two gates in MscL. J Gen Physiol 125(2):155–170

Aranda-Espinoza H, Berman A, Dan N, Pincus P, Safran S (1996) Interaction between inclusions embedded in membranes. Biophys J 71(2):648–656

Barry PH, Lynch JW (2005) Ligand-gated channels. IEEE Trans Nanobioscience 4(1):70–80

Bass RB, Strop P, Barclay M, Rees DC (2002) Crystal structure of *Escherichia coli* MscS, a voltage-modulated and mechanosensitive channel. Science 298(5598):1582–1587

Boal D (2002) Mechanics of the cell. Cambridge University Press, New York, 1st edition

Botelho AV, Huber T, Sakmar TP, Brown MF (2006) Curvature and hydrophobic forces drive oligomerization and modulate activity of Rhodopsin in membranes. Biophys J 91(12): 4464–4477

Calabrese B, Tabarean IV, Juranka P, Morris CE (2002) Mechanosensitivity of n-type calcium channel currents. Biophys J 83(5):2560–2574

Cantor RS (1999) Lipid composition and the lateral pressure profile in bilayers. Biophys J 76(5): 2625–2639

Chang G, Spencer RH, Lee AT, Barclay MT, Rees DC (1998) Structure of the MscL homolog from *Mycobacterium tuberculosis*: A gated mechanosensitive ion channel. Science 282(5397): 2220–2226

Chiang CS, Anishkin A, Sukharev S (2004) Gating of the large mechanosensitive channel in situ: Estimation of the spatial scale of the transition from channel population responses. Biophys J 86(5):2846–2861

Chou T, Kim KS, Oster G (2001) Statistical thermodynamics of membrane bending-mediated protein-protein attractions. Biophys J 80(3):1075–1087

Christensen M, Strange K (2001) Developmental regulation of a novel outwardly rectifying mechanosensitive anion channel in *Caenorhabditis elegans*. J Biol Chem 276(48):45024–45030

Clapham DE, Runnels LW, Strubing C (2001) The TRP ion channel family. Nat Rev Neurosci 2 (6):387–396

Cruickshank CC, Minchin RF, Le Dain AC, Martinac B (1997) Estimation of the pore size of the large-conductance mechanosensitive ion channel of *Escherichia coli*. Biophys J 73(4): 1925–1931

Dan N, Pincus P, Safran SA (1993) Membrane-induced interactions between inclusions. Langmuir 9:2768–2771
Dan N, Safran SA (1998) Effect of lipid characteristics on the structure of transmembrane proteins. Biophys J 75(3):1410–1414
Doeven MK, Folgering JH, Krasnikov V, Geertsma ER, van den Bogaart G, Poolman B (2005) Distribution, lateral mobility and function of membrane proteins incorporated into giant unilamellar vesicles. Biophys J 88(2):1134–1142
Duggan A, Garcia-Anoveros J, Corey DP (2000) Insect mechanoreception: What a long, strange TRP it's been. Curr Biol 10(10):R384–387
Elmore DE, Dougherty DA (2001) Molecular dynamics simulations of wild-type and mutant forms of the *Mycobacterium tuberculosis* MscL channel. Biophys J 81(3):1345–1359
Elmore DE, Dougherty DA (2003) Investigating lipid composition effects on the mechanosensitive channel of large conductance (MscL) using molecular dynamics simulations. Biophys J 85(3): 1512–1524
Evans E, Heinrich V, Ludwig F, Rawicz W (2003) Dynamic tension spectroscopy and strength of biomembranes. Biophys J 85(4):2342–2350
Fain GL (2003) Sensory Transduction. Sinauer Associates, Sunderland
Franks NP, Honore E (2004) The TREK K2P channels and their role in general anaesthesia and neuroprotection. Trends Pharmacol Sci 25(11):601–608
Gambin Y, Lopez-Esparza R, Reffay M, Sierecki E, Gov NS, Genest M, Hodges RS, Urbach W (2006) Lateral mobility of proteins in liquid membranes revisited. Proc Natl Acad Sci USA 103(7):2098–2102
Gillespie PG, Walker RG (2001) Molecular basis of mechanosensory transduction. Nature 413 (6852):194–202
Goforth RL, Chi AK, Greathouse DV, Providence LL, Koeppe 2nd RE, Andersen OS (2003) Hydrophobic coupling of lipid bilayer energetics to channel function. J Gen Physiol 121(5): 477–493
Goulian M, Mesquita ON, Fygenson DK, Nielsen C, Andersen OS, Libchaber A (1998) Gramicidin channel kinetics under tension. Biophys J 74(1):328–337
Goulian M, Pincus P, Bruinsma R (1993) Long-range forces in heterogenous fluid membranes. Europhys Letters 22(2):145–150
Gu CX, Juranka PF, Morris CE (2001) Stretch-activation and stretch-inactivation of Shaker-IR, a voltage-gated K+ channel. Biophys J 80(6):2678–2693
Guigas G, Weiss M (2006) Size-dependent diffusion of membrane inclusions. Biophys J 91(7): 2393–2398
Gullingsrud J, Kosztin D, Schulten K (2001) Structural determinants of MscL gating studied by molecular dynamics simulations. Biophys J 80(5):2074–2081
Gullingsrud J, Schulten K (2003) Gating of MscL studied by steered molecular dynamics. Biophys J 85(4):2087–2099
Harroun TA, Heller WT, Weiss TM, Yang L, Huang HW (1999) Theoretical analysis of hydrophobic matching and membrane-mediated interactions in lipid bilayers containing Gramicidin. Biophys J 76(6):3176–3185
Haswell ES, Meyerowitz EM (2006) MscS-like proteins control plastid size and shape in *Arabidopsis thaliana*. Curr Biol 16(1):1–11
Helfrich W (1973) Elastic properties of lipid bilayers: Theory and possible experiments. Z Naturforsch [C] 28(11):693–703
Hille B (1968) Pharmacological modifications of the sodium channels of frog nerve. J Gen Physiol 51(2):199–219
Huang HW (1986) Deformation free energy of bilayer membrane and its effect on Gramicidin channel lifetime. Biophys J 50(6):1061–1070
Jensen MO, Mouritsen OG (2004) Lipids do influence protein function - the hydrophobic matching hypothesis revisited. Biochim Biophys Acta 1666(1-2):205–226
Kahya N, Scherfeld D, Bacia K, Poolman B, Schwille P (2003) Probing lipid mobility of raft-exhibiting model membranes by fluorescence correlation spectroscopy. J Biol Chem 278 (30):28109–28115

Kamada Y, Jung US, Piotrowski J, Levin DE (1995) The protein kinase C-activated MAP kinase pathway of *Saccharomyces cerevisiae* mediates a novel aspect of the heat shock response. Genes Dev 9(13):1559–1571

Katsumi A, Orr AW, Tzima E, Schwartz MA (2004) Integrins in mechanotransduction. J Biol Chem 279(13):12001–12004

Kloda A, Martinac B (2001) Molecular identification of a mechanosensitive channel in archaea. Biophys J 80(1):229–240

Lauritzen I, Chemin J, Honore E, Jodar M, Guy N, Lazdunski M, Jane Patel A (2005) Cross-talk between the mechano-gated K2P channel TREK-1 and the actin cytoskeleton. EMBO Rep 6 (7):642–648

Lee AG (2003) Lipid-protein interactions in biological membranes: A structural perspective. Biochim Biophys Acta 1612(1):1–40

Lee AG (2005) How lipids and proteins interact in a membrane: A molecular approach. Mol Biosyst 1(3):203–212

Levina N, Totemeyer S, Stokes NR, Louis P, Jones MA, Booth IR (1999) Protection of *Escherichia coli* cells against extreme turgor by activation of MscS and MscL mechanosensitive channels: Identification of genes required for MscS activity. Embo J 18(7):1730–1737

Li Y, Moe PC, Chandrasekaran S, Booth IR, Blount P (2002) Ionic regulation of MscK, a mechanosensitive channel from *Escherichia coli*. Embo J 21(20):5323–5330

Maingret F, Fosse M, Lesage F, Lazdunski M, Honore E (1999) TRAAK is a mammalian neuronal mechano-gated K+ channel. J Biol Chem 274(3):1381–1387

Maingret F, Patel AJ, Lesage F, Lazdunski M, Honore E (2000) Lysophospholipids open the two-pore domain mechano-gated K+ channels TREK-1 and TRAAK. J Biol Chem 275(14): 10128–10133

Markin VS, Sachs F (2004) Thermodynamics of mechanosensitivity. Phys Biol 1(1-2):110–124

Martinac B, Hamill OP (2002) Gramicidin A channels switch between stretch activation and stretch inactivation depending on bilayer thickness. Proc Natl Acad Sci USA 99(7):4308–4312

Mitra K, Ubarretxena-Belandia I, Taguchi T, Warren G,Engelman DM (2004) Modulation of the bilayer thickness of exocytic pathway membranes by membrane proteins rather than cholesterol. Proc Natl Acad Sci USA 101(12):4083–4088

Moe P, Blount P (2005) Assessment of potential stimuli for mechano-dependent gating of MscL: Effects of pressure, tension, and lipid headgroups. Biochemistry 44(36):12239–12244

Molina ML, Barrera FN, Fernandez AM, Poveda JA, Renart ML, Encinar JA, Riquelme G, Gonzalez-Ros JM (2006) Clustering and coupled gating modulate the activity in KcsA, a potassium channel model. J Biol Chem 281(27):18837–18848

Morris CE, Homann U (2001) Cell surface area regulation and membrane tension. J Membr Biol 179(2):79–102

Nauli SM, Zhou J (2004) Polycystins and mechanosensation in renal and nodal cilia. Bioessays 26(8):844–856

Nielsen C, Goulian M, Andersen OS (1998) Energetics of inclusion-induced bilayer deformations. Biophys J 74(4):1966–1983

Niggemann G, Kummrow M, Helfrich W (1995) The bending rigidity of phosphatidylcholine bilayers: Dependences on experimental method, sample cell sealing and temperature. J Phys II France 5:413–425

Park JM, Lubensky TC (1996) Interactions between membrane inclusions on fluctuating membranes. J Phys I France 6:1217–1235

Perozo E, Cortes DM, Sompornpisut P, Kloda A, Martinac B (2002a) Open channel structure of MscL and the gating mechanism of mechanosensitive channels. Nature 418(6901):942–948

Perozo E, Kloda A, Cortes DM, Martinac B (2001) Site-directed spin-labeling analysis of reconstituted MscL in the closed state. J Gen Physiol 118(2):193–206

Perozo E, Kloda A, Cortes DM, Martinac B (2002b) Physical principles underlying the transduction of bilayer deformation forces during mechanosensitive channel gating. Nat Struct Biol 9(9):696–703

Perozo E, Rees DC (2003) Structure and mechanism in prokaryotic mechanosensitive channels. Curr Opin Struct Biol 13(4):432–442

Pivetti CD, Yen MR, Miller S, Busch W, Tseng YH, Booth IR, Saier Jr. MH (2003) Two families of mechanosensitive channel proteins. Microbiol Mol Biol Rev 67(1):66–85

Powl AM, East JM, Lee AG (2003) Lipid-protein interactions studied by introduction of a tryptophan residue: The mechanosensitive channel MscL. Biochemistry 42(48):14306–14317

Rawicz W, Olbrich KC, McIntosh T, Needham D, Evans E (2000) Effect of chain length and unsaturation on elasticity of lipid bilayers. Biophys J 79(1):328–339

Sachs F (1991) Mechanical transduction by membrane ion channels: A mini review. Mol Cell Biochem 104(1-2):57–60

Seemann H, Winter R (2003) Volumetric properties, compressibilities and volume fluctuations in phospholipid-cholesterol bilayers. Zeitschrift fur physikalische Chemie 217:831–846

Shapovalov G, Lester HA (2004) Gating transitions in bacterial ion channels measured at 3 microsecond resolution. J Gen Physiol 124(2):151–161

Sintes T, Baumgartner A (1997) Protein attraction in membranes induced by lipid fluctuations. Biophys J 73(5):2251–2259

Spencer RH, Rees DC (2002) The alpha-helix and the organization and gating of channels. Annu Rev Biophys Biomol Struct 31:207–233

Sukharev S, Betanzos M, Chiang CS, Guy HR (2001) The gating mechanism of the large mechanosensitive channel MscL. Nature 409(6821):720–724

Sukharev SI, Blount P, Martinac B, Kung C (1997) Mechanosensitive channels of *Escherichia coli*: The MscL gene, protein, and activities. Annu Rev Physiol 59:633–657

Sukharev SI, Sigurdson WJ, Kung C, Sachs F (1999) Energetic and spatial parameters for gating of the bacterial large conductance mechanosensitive channel, MscL. J Gen Physiol 113(4): 525–540

Tosh RE, Collings PJ (1986) High pressure volumetric measurements in dipalmitoylphosphatidylcholine bilayers. Biochim Biophys Acta 859(1):10–14

Turner MS, Sens P (2004) Gating-by-tilt of mechanically sensitive membrane channels. Phys Rev Lett 93(11):118103

Ursell T, Huang KC, Peterson E, Phillips R (2007) Cooperative gating and spatial organization of membrane proteins through elastic interactions. PLoS Comput Biol 3(5):e81

Wiggins P, Phillips R (2004) Analytic models for mechanotransduction: Gating a mechanosensitive channel. Proc Natl Acad Sci USA 101(12):4071–4076

Wiggins P, Phillips R (2005) Membrane-protein interactions in mechanosensitive channels. Biophys J 88(2):880–902

Yoshimura K, Batiza A, Schroeder M, Blount P, C Kung (1999) Hydrophilicity of a single residue within MscL correlates with increased channel mechanosensitivity. Biophys J 77(4):1960–1972

Yoshimura K, Nomura T, Sokabe M (2004) Loss-of-function mutations at the rim of the funnel of mechanosensitive channel MscL. Biophys J 86(4):2113–2120

Chapter 3
Mechanosensitive Channels Gated by Membrane Tension

Bacteria and Beyond

Paul Blount, Yuezhou Li, Paul C. Moe, and Irene Iscla

Abstract What are the stimuli that are sensed by mechanoreceptors? Researchers have now begun to address this question and determine the molecular mechanisms underlying channels that are gated by mechanical forces. Two quite different models now exist. The first is that the channels are 'tethered' to cytoskeleton and/or extracellular components, which thus exert forces on the channel that lead to gating. The second model predicts that the channel protein directly senses biophysical changes that occur within the membrane when it is under tension. Several lines of evidence indicate that many putative mechanosensitive channels are indeed tethered by other proteins, however in many instances the exact role this tethering plays in mechanosensing has yet to be fully clarified. On the other hand, the cloning and study of bacterial mechanosensitive channels demonstrated that channels can directly sense tension within the membrane. Evidence obtained from several of the more complex eukaryotic mechanosensory systems suggests that a number of eukaryotic channels from divergent families similarly sense tension within the membrane.

Key words: Stretch-activated channels · Mechanosensitive channels · Membrane tension · Amphipaths · Bacterial channels · MscL · MscS · K_{2P} channels · TRP channels · Channel reconstitution

3.1 Introduction

Mechanosensitive channels are essentially ubiquitous in all biological life. While the most obvious function is to serve as sensors for the human sense of touch, they also are implicated in the senses of hearing and balance, cardiovascular regulation, and sensing and relieving cellular stresses including volume and turgor effected by changing osmolarity.

There are two models or designs that can serve as a basis for the fundamental mechanism of how mechanosensitive (MS) channels sense mechanical forces.

A. Kamkin and I. Kiseleva (eds.), *Mechanosensitive Ion Channels.*

One of the models holds that intracellular (e.g. cytoskeletal components) and extracellular (e.g. extracellular matrix) tethers interact directly with portions of the channel protein, and it is these tethers that draw the channel components in different directions upon mechanical perturbation, thus supplying the energy for channel gating. Several older studies supported this model. Indeed, early publications suggested that the most likely source for the energy required to gate a mechanosensitive channel would be through proteins probably much larger than normal channels, such as cytoskeletal elements (Guharay and Sachs, 1984; Sachs, 1986). In addition, a spring/tether system was consistent with the biophysical analyses of hair cells of the inner ear (see (Ricci et al., 2006) for a discussion of current possibilities). Finally, genetic studies using the nematode *C. elegans* found that both cytoskeletal elements and extracellular matrix are associated with mechanical stimuli sensed by specific sensory neurons (Tavernarakis and Driscoll, 1997). However, another model exists: the direct detection of pressure across, or tension within the cellular membrane. How can one support the latter hypothesis? The ultimate proof would lie in the purification of the channel to apparent homogeneity and reconstitution into synthetic membranes, thus demonstrating that only the protein and force acting on a lipid membrane are needed. However, many channels cannot survive such harsh treatments as solubilization and reconstitution. Hence, more indirect means are necessary. The functional expression of the channel in heterologous systems that presumably do not contain the same tethering proteins, or the observation that the pharmacological disruption of cytoskeletal proteins does not have a negative effect on membrane-stretch activation, could also provide evidence that tethering proteins are not critical. Finally, the application of amphipaths, which intercalate asymmetrically into the membrane bilayer and disrupt its normal lateral pressure profile, may affect mechanosensitive channel gating, thus providing substantial support for the channel directly detecting biophysical changes in the membrane due to stress.

The best studied channels that directly detect membrane stretch are from the bacterium *E. coli*. In this system it has been shown that chemicals that disrupt the normal lateral pressure profile within a membrane modulate bacterial MS channel activity (Martinac et al., 1990; Perozo et al., 2002b). Further, one of the channels have been shown to be functional when heterologously expressed either in yeast membranes (Sukharev et al., 1994b) or in a cell-free system (Sukharev et al., 1994a), and the purification and reconstitution of two microbial channels have not only demonstrated that only the protein and a lipid bilayer are required for MS channel function, but have also allowed for the study of these channels in a synthetic membrane system (Moe and Blount, 2005). One interpretation of the microbial channel studies would be that these sensors, and their mechanisms of detecting mechanical force, are unique to the microbial world; however, studies of mammalian mechanosensors and channels demonstrate that at least some of these channels are not negatively affected by cytoskeletal disruption, are modulated by amphipaths, or can survive reconstitution. Hence, the data now suggest that there are common and shared mechanisms with channels as divergent as members of the two-pore K^+ (2PK) and transient receptor potential (TRP) channel families.

3.2 Bacterial Mechanosensitive Channels

Nearly 20 years ago, Dr. Ching Kung's group utilized an approach to generate giant spheroplasts from *E. coli* to obtain cells large enough to be amenable to the classic patch clamp approach. When the native membranes were subjected to this assay, the most obvious channel activities noted were those activated by differential pressure across the patched membrane (Martinac et al., 1987).

The activities observed in bacterial native membranes were robust and of relatively large conductance; these properties suggested the physiological function of what is now termed a "biological emergency release valve". At this time, it was known that bacteria, including *E. coli*, are quite proficient at adapting to large changes in osmotic environment. When at high osmolarity, bacterial pumps accumulate compatible solutes, such as proline, K^+ and quaternary amines such as betaine, to serve as osmoprotectants and maintain cell turgor, which is required for normal cellular division (see (Poolman et al., 2002) and (Wood, 2006) for reviews). But microbes must also survive rapid changes in osmolarity and unlike increases in osmolarity, a sudden decrease in the osmotic environment due to rain or other acute insult could be life threatening. Early studies demonstrated that upon such an osmotic 'downshock', proline (Britten and McClure, 1962) and other solutes (Tsapis and Kepes, 1977) are jettisoned from the cell. From studies of *E. coli* null mutants, it is now clear that mechanosensitive channels are the conduit through which these solutes leave the cell; bacteria that do not contain these channels are far less likely to survive an acute osmotic downshock (Levina et al., 1999).

Although only one channel activity type was noted in the initial reports on bacterial channels (Martinac et al., 1987; Martinac et al., 1990), it is now clear that there are at least four molecular entities native to the *E. coli* membrane that encode mechanosensitive channel activity. First, there is MscL (for mechanosensitive channel of large conductance), which normally opens for only a few tens of milliseconds and has a conductance on the order of 3.6 nanosiemens (nS). A smaller conducting channel activity (about 1nS) opens at stimuli less than that required for MscL and has been referred to as MscS (S for small). However, we now know that what was referred to as MscS activity in early studies is in reality two independent activities generated from independent genes: *mscS* and *mscK*, with the MscS activity showing a greater propensity for desensitization (Levina et al., 1999; Li et al., 2002). Finally, there is a fourth mechanosensitive channel activity of yet smaller conductance (about 300pS), called MscM (M for mini), whose encoding gene has yet to be identified, less frequently observed, not well characterized, and will not be further discussed in this chapter (but see (Berrier et al., 1996) for the best description to date of the MscM activity). The MscL and MscS proteins appear to be redundant in function and the most important in surviving an osmotic downshock; the double mutant in *E. coli* is fragile to such insult and the vast majority, $\sim$90%, will lyse with an osmotic downshock over a couple-hundred milliosmolar (Levina et al., 1999).

Because of their abundance, the MscS and MscL channels are routinely observed in essentially every patch-clamped native membrane. A typical trace is shown in

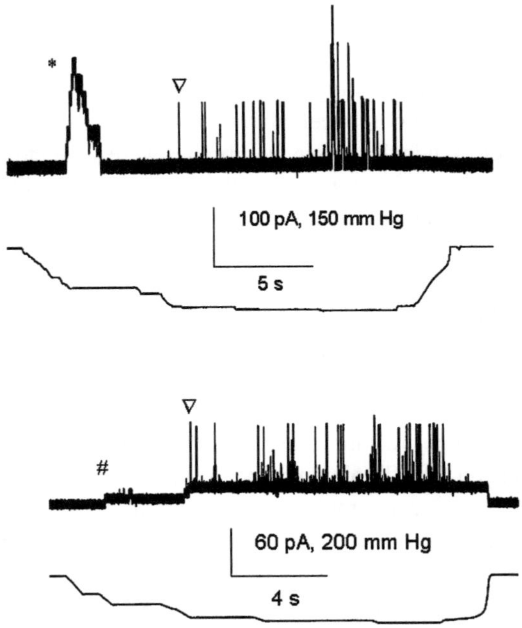

Fig. 3.1 Typical current traces of MscS (top) and MscK (bottom) in the presence of MscL. Recordings were generated from patches derived from *E. coli* giant spheroplast at −20 mV. Strain PB113 *(Δ mscS, Δ mscK)* with *mscS* expressed in trans (top panel), and strain MJF451 (Δ*mscS*) (bottom panel), were utilized for the study of MscS and MscK, respectively. MscS (*), MscK (#) and MscL (∇) activities, and the pressure traces (lower trace in each panel) are shown

Fig. 3.1(Top). In an *mscS*-null mutant, MscK activity is seen in about 20% of the patches, and is much less prone to inactivate, as shown in Fig. 3.1 (Bottom).

3.2.1 The Bacterial MscL Channel

MscL was the first of the bacterial mechanosensitive channels to be cloned and sequenced. Beyond that, MscL was the first channel from any organism that was definitively demonstrated to directly produce a mechanosensitive channel activity. To date, it remains the best characterized mechanosensitive channel. Researchers have the advantage of a crystal structure and detailed models for how the channel senses and responds to mechanical forces (see (Blount et al., 2007) for a more detailed review of many of these models).

3.2.1.1 From the *mscL* Gene to the MscL Protein Structure and Gating Models

The gene encoding the bacterial channel MscL activity was originally identified by a relatively novel method. It was discovered that when *E. coli* membrane proteins were solubilized from the membrane, fractionated by a filtration column and the resulting fractions reconstituted into patchable liposomes, both the MscL and MscS activities were still present (Sukharev et al., 1993). Hence, the laborious technique of reconstitution and patch clamp was used as an assay for the biochemical enrichment of the molecule that underlies the MscL activity (Sukharev et al., 1994a). Briefly, *E. coli* membrane proteins were solubilized in mild detergent, chromatographically fractionated by size and each fraction assayed by patch clamp. Subsequent chromatographic fractionations of pooled samples by other columns, and repeating the assay, led to additional enrichment. Using SDS-PAGE to analyze the protein composition of the fractions, one band correlated well with the MscL activity. Microsequencing identified the N-terminal of a protein that had been only partially sequenced previously, but this was enough to identify its location on the chromosome and thus facilitate cloning of the entire region. Once the full sequence was obtained, it predicted a small transmembrane protein that was only 136 amino acids in length. Using a PhoA-fusion approach (Boyd et al., 1987), it was determined that there are two transmembrane domains, and that the N- and C-termini are located in the cytoplasm (Blount et al., 1996a). Early studies indicated, we now know incorrectly, that a homohexamer was formed (Blount et al., 1996a; Saint et al., 1998).

A major advance in the study of MscL came when a crystal structure was derived for a homologue from *M. tuberculosis* (Chang et al., 1998) (Fig. 3.2). The structure resolved the uncertainty of the stoichiometry of the complex; it was undoubtedly a homopentamer, not a hexamer. The pentameric composition was later confirmed by independent approaches (Sukharev et al., 1999a). As anticipated (Blount et al., 1996a), the crystal structure suggested that each subunit within the complex has

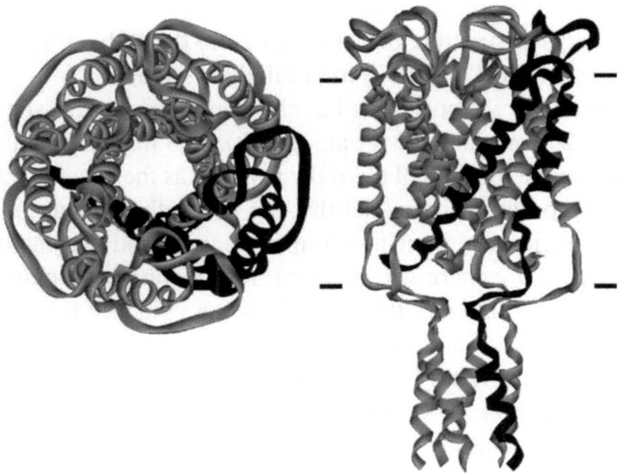

Fig. 3.2 Schematic representation of MscL based on the *M. tuberculosis* crystal structure, which some evidence now suggests is in a 'nearly closed' state. Shown are the periplasmic (*left*) and side (*right*) views. In the view from the side the approximate position of the membrane is shown. Note that it is the first transmembrane domains (TMD1) that line the pore of the channel and the second transmembrane domains (TMD2) surround and are in contact with the membrane. A single subunit is highlighted for clarity

two α-helical transmembrane domains. Because there is only a small opening at the pore (~4 Å), the channel appeared to be in a closed, or nearly closed, conformation.

The MscL channel is highly conserved, almost ubiquitous within the bacterial kingdom, and many homologues of the *E. coli* MscL channel had been studied and shown to encode mechanosensitive channel activity (Moe et al., 1998). However, at the time the crystal structure was published, no functional information existed for the *M. tuberculosis* homologue (Tb-MscL). Essentially all structure/function experiments until this time had been performed on the *E. coli* MscL (Eco-MscL). A subsequent study demonstrated that the Tb-MscL was indeed functional, but not in a normal physiological range, at least for *E. coli*; the Tb-MscL channel needed far more energy to gate than Eco-MscL, and the former did not rescue an 'osmotic-lysis' phenotype of the MscS/MscL double mutant in *E. coli* (Moe et al., 2000). On the other hand, experiments in which analogous residues were similarly mutated suggested that the Eco- and Tb-MscL functioned by similar mechanisms (Moe et al., 2000). In all, because of the marked functional differences, one must be cautious when directly correlating functional studies performed with Eco-MscL with properties of the Tb-MscL structure.

The structure of Tb-MscL has inspired researchers to predict the structure of closed state of the Eco-MscL channel (Fig. 3.3). The predictions of the relative location of specific residues in the closed structure have been supported by electron paramagnetic resonance (EPR) spectroscopy in combination with site-directed spin labeling (SDSL) (Perozo et al., 2001). However, one should note that some evidence suggests a slightly altered model around the constriction point of the pore for the "fully closed" *E. coli* MscL channel (Bartlett et al., 2004; Iscla et al., 2004; Levin and Blount, 2004; Li et al., 2004; Bartlett et al., 2006). Several experiments using relatively independent approaches performed with the *E. coli* MscL channel now suggest that the crystal structure does not represent the fully closed state found in membranes (see (Blount et al., 2007) for a full discussion).

A detailed model for the open state has also been proposed, as shown in Fig. 3.3 (Sukharev et al., 2001a; Sukharev et al., 2001b). The published models are quite detailed, predicting the location of all of the residues as the channel transits through different states. Aspects of the model that are generally agreed upon in the field are that the transmembrane domains tilt upon opening and that the channel opens not unlike the iris of a camera with the first transmembrane domain (TMD1) lining the majority of the lumen of the pore. However, evidence for alternative views for the precise locations of and interactions between the TMDs (e.g. see (Li et al., 2004)), and the exact residues lining the pore (see (Perozo et al., 2002a; Bartlett et al., 2004; Bartlett et al., 2006)) have more recently emerged. This model also proposes a structure for the N-terminal domain (that was not solved in the crystal structure) and possible function as a "second gate" (Sukharev et al., 2001a; Sukharev et al., 2001b); another study also predicts that the C-terminal bundle remains immobile upon channel gating (Anishkin et al., 2003). Again, the proposed model for these regions does not satisfy all of the data and have yet to be fully confirmed. For a more complete discussion of these structural issues, see (Blount et al., 2007).

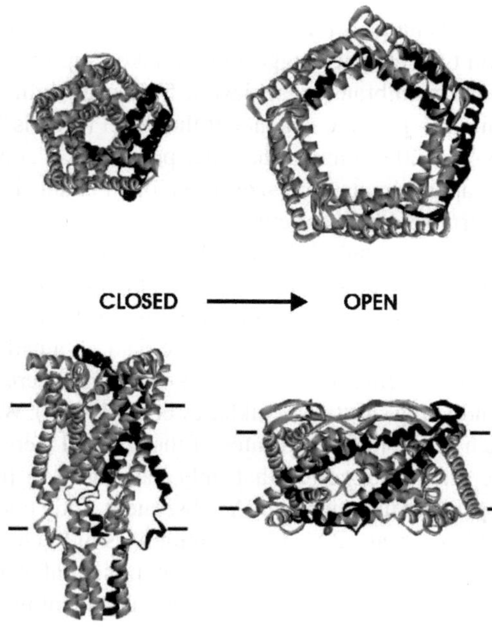

Fig. 3.3 One proposed model for the structure of the *E. coli* MscL protein (Sukharev et al., 2001a; Sukharev et al., 2001b). The periplasmic (*top*) and lateral (*bottom*) views of the model are shown. The approximate position of the membrane is shown. Note the flattening of the open structure in comparison with the closed conformation due to the tilting of the transmembrane domains. Although the C-terminal domain is shown to separate, a more recent proposal suggests that the bundle may remain intact (Anishkin et al., 2003). A single subunit is highlighted in black in each figure

3.2.1.2 MscL as a Sensor

Given that the MscL channel survives reconstitution (Sukharev et al., 1993; Sukharev et al., 1994a; Sukharev et al., 1994b) and even can be purified to apparent homogeneity and yet retain functionality (Häse et al., 1995; Blount et al., 1996a), it seemed likely that the channel sensed either tension within the membrane or pressure across it. Experiments conducted in yeast demonstrated that a yeast mechanosensitive channel can indeed sense membrane tension (discussed in Section 3 below). Hence, it was assumed the same would be found for the MscL channel. By imaging the patch within a given membrane, and measuring the pressure across the membrane with a pressure transducer, one can easily calculate the tension within the membrane by a simple equation, Laplace's law: Tension in the membrane equals the pressure across it times the radius of curvature divided by 2. The probability of channels opening within a patch (P_O) relative to the amount of tension fit a Boltzmann relationship well. All of the data obtained in early studies were consistent with MscL sensing tension within the membrane rather than pressure across it, and the authors assumed this was the case, although they did not formally demonstrate it (Sukharev et al., 1999b; Chiang et al., 2004). A more recent study, using a similar approach but analyzing patched membranes that achieved quite different radii upon stimulation, has now definitively demonstrated that the MscL channel does indeed sense tension within the membrane, not the pressure across it (Moe and Blount, 2005).

Given that the channel senses tension within the membrane, the spatial and energetic parameters of channel gating can be calculated by using the Boltzmann plots of P_O versus tension. For MscL, such an analysis has been used (Sukharev et al., 1999b), updated (Chiang et al., 2004), and utilized for some of the more sensitive mutant channels (Anishkin et al., 2005). The current derived values for opening the wild type channel suggest that it takes approximately 7 to 13 dynes/cm^2 of tension in the membrane to achieve a 50% probability of channel gating, the energy required to gate the channel at this level (ΔE) is 51 ± 13 *kT*, and the change in area (ΔA) is 20 ± 5 nm^2. The latter parameter is consistent with current predictions for the approximate pore size of the open channel. Finally, analysis was performed to determine which structural transition was sensitive to membrane tension. The channel does not only have two states, fully open and closed, but has several partially open states, or substates, that can be measured ((Sukharev et al., 1999b; Sukharev et al., 2001a), but see (Pivetti et al., 2003) for a more detailed analysis at a much higher temporal resolution). The data obtained for the major substates easily fit a simple scheme: C1$\leftrightarrow$S2$\leftrightarrow$S3$\leftrightarrow$S4$\leftrightarrow$O5 where S2 to O5 are progressively higher conducting substates (Sukharev et al., 1999b). When the rate constants for achieving each of the major substates of the channel were plotted against membrane tension, the results suggested that only the closed to first substate conversion is strongly dependent upon tension. A subsequent study confirmed and extended these findings (Chiang et al., 2004) by examining single-channel traces and demonstrating that activation was essentially monoexponential, supporting the hypothesis that gating is largely due to a single tension-dependent molecular change in confirmation; all other conformational changes are essentially independent of stimulus. Put another way, once the channel has initiated ion permeation, the progression to the different substates, and even closure of the channel, occurs independent of external stimulus.

Although it is now agreed that MscL senses tension in the membrane, the precise biophysical changes that induce gating are only now being elucidated. For example, one could imagine that thinning of the membrane leading to 'hydrophobic mismatch' between the membrane and the protein could serve as a major cue. One study tested such an hypothesis by reconstituting MscL into synthetic phosphatidylcholine (PC) containing mono-unsaturated chains of 14, 16, 18 (which is average length of a biological membrane) 20 and 22 carbons; chain lengths greater or lesser than this could not be analyzed for technical reasons (Perozo et al., 2002b). The data indicated that shortening the chain length, and thus increasing hydrophobic mismatch, did indeed decrease the activation threshold, while increasing the chain length increased the activation threshold. However, even when reconstituted in the shortest carbon lengths, the channel did not gate spontaneously. Hence, it appears that, for MscL, hydrophobic mismatch can modulate the gating, but is not the major cue in its activation.

A second biophysical change that occurs upon membrane stretch is a change in the lateral pressure profile. Computational estimates (Cantor, 1999) as well as a form of computer modeling known as molecular dynamic simulations (Gullingsrud and Schulten, 2004) suggest that there are literally hundreds of atmospheres of pressure within a bilayer. Amphipaths, non-lipid or lipid-like compounds with both a hydrophilic and hydrophobic nature, intercalate into the membrane surfaces, perturb the normal lateral pressure profiles of the membranes and cause stresses and can induce altered cellular shapes (Deuticke, 1968; Sheetz et al., 1976). An asymmetric acute application, or presence of a membrane potential in combination with a charged amphipath, can lead to asymmetric intercalation, or preferential concentration of such compounds into one leaflet of the bilayer, thus leading not only to a disruption of the lateral pressure profile, but also to mis-shaping of the membrane. These amphipaths are categorized as either cup-formers (inner leaflet) or crenators (outer leaflet) depending upon which leaflet the amphipath is concentrated within. One study tested the effects of amphipaths on the purified and reconstituted Eco-MscL; channel activity and structure were monitored by patch clamp and EPR, respectively (Perozo et al., 2002b). The findings demonstrated that lysophospholipids, which are lipid-like amphipaths that contain only one fatty acid chain, could gate the Eco-MscL channel. In a subsequent study, the channel was gated in this fashion in a reconstituted system, and SDSL and EPR was used to monitor the labeled mutant channels (Perozo et al., 2002a); this study led to one of the models for structural rearrangements occurring upon Eco-MscL gating (see (Blount et al., 2007) for a more detailed discussion of the current structural gating models for this channel). From these studies, it appeared that the asymmetry between the lateral pressure profiles of the two leaflets, which will thus lead to a curvature of the membrane, activates the channel. Using computer modeling and molecular dynamic simulations one group has studied the effect of addition of non-bilayer-forming lipids, to effect a curvature of and stress within the membrane, on the model of the closed Eco-MscL channel (Meyer et al., 2006). Although only a single simulation of 9.5 nS was performed, movements of the channel consistent with the initiation of gating were observed. While molecular dynamic simulations compute the atomic forces between each atom in a simulation, another form of computer modeling referred to as the finite element method (FEM), and is often used in engineering, bases its computations on highly idealized geometries and the mechanical properties of the elements in the simulation. Molecular dynamic simulations do not rule out the possibility that curvature plays a role in the forces sensed by MscL; however, results from FEM suggest that it is actually the membrane tension, not curvature, that is the best stimulus (Tang et al., 2006). This finding is consistent with experimental data which show that, in a visualized patch the MscL channel appears to gate after the curvature of the membrane has nearly achieved its smallest value (Moe and Blount, 2005).

MscL appears to sense its lipid environment. Hence, it seems likely that the second transmembrane domain, TMD2, which faces the lipid bilayer, is vitally important for sensing membrane tension. Perhaps one of the studies that most directly tests this hypothesis is one in which Eco-MscL was divided in half so TMD1 and TMD2 were expressed independently (Park et al., 2004) Expression of the TMD1 alone formed spontaneously gating channels, albeit of varying conductance, while TMD2 segments were completely silent as assayed by patch clamp. The co-expression of the two domains formed a mechanosensitive channel with a conductance similar to the wild type channel. Hence, the data from this study suggested that TMD1 plays the largest role in forming the channel lumen while TMD2 is vital for correct assembly and, more importantly, sensing the lipid environment. In another study, tryptophan or tyrosine residues were substituted into regions of the transmembrane domains close to the lipid-aqueous interface (Chiang et al., 2005). This latter study relied heavily on the observation that addition of such aromatic residues can decrease the ability of a transmembrane peptide to tilt within the membrane

(Ozdirekcan et al., 2005). The results are consistent with gating of the channel being linked with TMD tilt; however, one must be careful of over-interpretation: first, previous biochemical studies were performed on peptides that spanned the membrane only once, and it is unclear that such tryptophan "capping" would strongly influence a transmembrane element that is not in isolation in the membrane and is presumably strongly associating with another transmembrane domain (TMD1); second, the specificity of the phenotypes to the tryptophan substitutions was not demonstrated (e.g. a previous study (Levin and Blount, 2004) demonstrated that a C at position F93 yields a similar decreased-sensitivity phenotype as the Y or W substitutions studied).

Could there be 'specific' interactions between lipid headgroups and the protein that are important or allow the channel to be more sensitive to membrane perturbation? Some studies implied that this was the case. In one instance, a random mutagenesis study noted that loss-of-function-effecting mutations often occurred near the proposed site of interaction with membrane lipid headgroups, suggesting such specific interactions may occur (Maurer and Dougherty, 2003). Another study isolated intragenic suppressors for a gain-of-function mutant that effects a slowed-growth phenotype, I41N, and found that some of the suppressing mutations were clustered on the periplasmic side of the transmembrane domains, again near the predicted headgroups (Yoshimura et al., 2004). The authors of this latter study performed an asparagine scan of the region and found that mutation of many of the residues predicted to face the lipid led to channels that were less functional. While the interpretation was reasonable and attractive, some predictions have yet to be fulfilled by other studies. For example, a similar scanning of the entire transmembrane region did not find altered activity when the same residues were mutated to cysteine rather than asparagine (Levin and Blount, 2004). In addition, another study looking for suppressors of other gain-of-function mutants did not show the same clustering of residues (Li et al., 2004), but note that this inconsistency may simply be due to the fact that only partial suppressors were selected in the latter study, and neither study saturated their screen. Residues residing near the proposed lipid-aqueous interface at the cytoplasmic region of the protein have also been implicated as contributing to the sensitivity of the MscL sensor. For example, tryptophan mutagenesis and trp fluorescence spectroscopy revealed a relatively nonspecific association between the protein and uncharged lipids, but a highly specific binding with anionic lipids. Three positively charged residues in a cluster near the cytoplasmic end of TMD2 have been implicated (Powl et al., 2005b). The authors conclude that anionic lipids bind to basic residues at the cytoplasmic region of TMD2. Analysis of mutants was consistent with the notion that the binding was to a specific site containing basic residues within this subdomain, RKK starting at position 104 in the *E. coli* MscL. Further manipulation of this cationic "pocket" was found to perturb lipid association and yield a gain-of-function phenotype, presumably reflecting a conformational disruption of the protein. Unfortunately, some aspects of the data are difficult to interpret because of the use of *Mycobacterium tuberculosis* MscL channel, which, as discussed above, has been shown to be less functional in heterologous and reconstitution systems (Moe et al., 2000), and the fact that activity of mutated channels was inferred from whole-cell physiological approaches rather than direct electrophysiological measurements. A further complication is that the proposed anionic-lipid binding site, RKK, is not conserved; for example, it is REE in the *Pseudomonas fluorescence* MscL, which has been shown to have similar channel properties and mechanosensitivity as the *E. coli* MscL (Moe et al., 1998). Interestingly, this cationic pocket is proximal to, and included in, a charged domain that has been postulated to function as a pH sensor. Originally, as a control for a series of unrelated experiments, a modest change in activity in wild type MscL was noted when pH was varied (Iscla et al., 2004). A subsequent, much more detailed study determined that the MscL cytoplasmic residues identified in the tryptophan mutagenesis studies, and other charges just distal to TMD2, were involved in this pH-dependent channel modulation. A likely possibility seemed that pH changed the interaction of positively charged residues with negatively charged lipids, thus changing the sensitivity of the channel (Kloda et al., 2006). Although the patch clamp results are clear, what, if any, physiological role this pH modification plays has yet to be determined. A summary of relative location of the residues implicated in protein-lipid interactions within a current model for the closed structure of the Eco-MscL channel is shown in Fig. 3.4.

While many of the studies summarized in the previous paragraph imply that interactions between the MscL protein and specific lipid headgroups exist, the evidence is largely indirect and, in essentially all cases, relies heavily on mutations

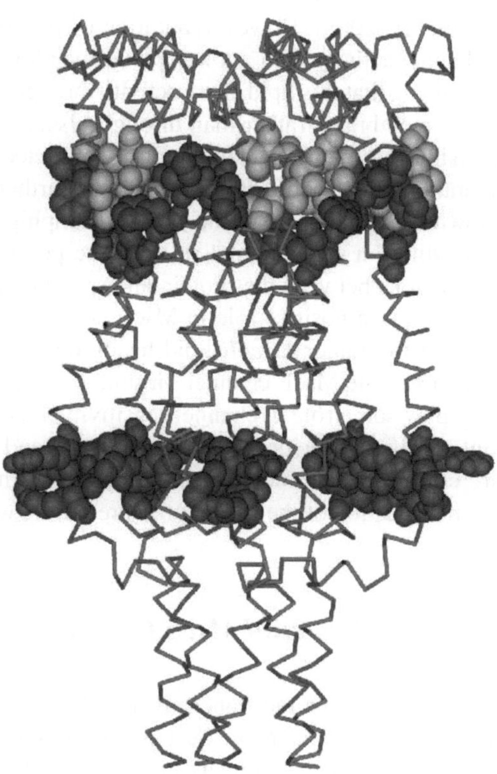

Fig. 3.4 Predicted protein-lipid interactions in the MscL channel. Residues implicated in lipid binding are shown in CPK in a side view of the channel. Mutation of residues lining the periplasmic rim of the channel yielded a loss-of-function phenotype, presumably through disruption of lipid binding (Yoshimura et al., 2004). These residues form the periplasmic end of TMD1 (gray), and TMD2 (black). The more cytoplasmic residues were identified by tryptophan fluorescence spectroscopy, which revealed heterogeneity in lipid binding to the channel (Powl et al., 2005a). This charged pocket lies at the beginning of a charge cluster, RKKEE, postulated to form a pH sensing domain that regulates the channel sensitivity (Kloda et al., 2006)

in the protein which, given the diversity of channel phenotypes observed in numerous studies, may be misleading. On the other hand, if the interactions are truly specific, then one would anticipate that changing lipid headgroups would have influences at least as profound as mutation of the protein. To test this hypothesis, one group assessed the influence of changing lipid composition on channel function (Moe and Blount, 2005). Lipids with phosphatidylcholine headgroups, which are not synthesized by bacteria, were used as a standard. To test the hypothesis that negatively charged lipids would interact with the cytoplasmic cationic 'pocket' discussed above or other regions of the protein, a generic negatively charged lipid, phosphatidylserine, was added. Surprisingly, there was no effect on the activation threshold or tension required to achieve a probability of opening of 50%. However, because phosphatidylserine is not found in *E. coli*, it was possible that an interaction could be more specific with a lipid whose headgroup was found in the *E. coli* envelope. Therefore, the major negatively charged *E. coli* lipid headgroup, phosphatidylglycerol, was tested; however, here again no measurable effect was observed. The authors then directed their efforts to study another major lipid found in the *E. coli* cytoplasmic membrane, phosphatidylethanolamine. This lipid was of particular interest because molecular dynamic simulations had indicated numerous hydrogen bond formations between phosphatidylethanolamine headgroups and the MscL protein (Elmore and Dougherty, 2001). The simulations also suggested that while the structure of MscL in phosphatidylethanolamine lipids is similar to the crystal structure, the structure in phosphatidylcholine is significantly different (Elmore and Dougherty, 2003). However, addition of phosphatidylethanolamine actually led to a channel activity with a *lower* rather than higher sensitivity to membrane tension, suggesting that this lipid alters activity through changes in the biophysical properties of the membrane, rather than through MscL-lipid specific interactions. Indeed, increasing phosphatidylethanolamine is predicted to increase thickness and change the lateral pressure profile of the membrane; furthermore, addition of this lipid has been shown to partially reverse the effects of amphipaths (Perozo et al., 2002b). The lipid-reconstitution study cannot rule out the possibilities that there are non-functional interactions between the protein and specific lipids, or that some minor lipid headgroup plays a positive role in MscL channel function, however it does appear that none of the major *E. coli* lipid headgroups specifically interact *functionally* with residues of the MscL channel. In sum, it appears that while hydrophobic mismatch may play some role in channel sensitivity, there are no obvious functionally important specific interactions with headgroups, and that the true stimuli for the channel are changes in the biophysical properties of the membrane, presumably centering on modification of the lateral pressure profile within the membrane structure.

3.2.2 The Bacterial MscS Channel and its Homologues

MscL was the first mechanosensitive channel to be defined at a genetic and molecular level and remains the best characterized of this class of molecules. However, in the very first experiments in which patch clamp was applied to native *E. coli* membranes it is likely that it is the MscS channel whose activity was first observed and characterized (Martinac et al., 1987; Martinac et al., 1990). There have been many comparisons and contrasts made between these two channels. A common and perhaps distant evolutionary origin between the family of MscS and MscL mechanosensitive channels has been proposed (Kloda and Martinac, 2001; Kloda and Martinac, 2002; Martinac, 2004). This is largely because a few channels from archaea share apparent sequence homology with MscL but are clearly related to MscS in structure; thus these chimera-like channels could be potential 'missing links' between the MscL and MscS families, assuming the limited homology is not convergent evolution. On the other hand, when one looks at the features of the current structural models, including subunit stoichiometry, there are no obvious similarities between MscS and MscL. The MscS family appears to be large and is much more diverse and complex than the highly homologous MscL family. In addition, while MscL is almost exclusively found in bacteria and archaea (with a single possible exception of *Neospora*), MscS is found in *S pombe* as well as many plants (see (Haswell et al., 2007) for a review). Although MscS is not as well characterized as MscL, and given the disparity between the two families, the parallels of how they sense and respond to mechanical force is truly impressive.

3.2.2.1 The MscS Activity, Protein Structure and Gating Models

The *mscS* gene was discovered by classical microbial genetics. The finding was somewhat serendipitous, starting when an *E. coli* mutant strain, generated by UV mutagenesis, was isolated that showed impaired growth in the presence of both, high K^+ and betaine or proline (both are compatible solutes that are transported into the cell to maintain turgor in a high-osmotic environment) (McLaggan et al., 2002). The lesion was identified as a missense mutation within a gene called *kefA* (also called *aefA*). Realizing that the phenotype had some of the characteristics of what one may anticipate from a dysfunctional bacterial mechanosensitive channel, the authors further characterized it, as well as many of its homologues. Ultimately this led to a gene, *yggB*, that correlated well with MscS activity (Levina et al., 1999). Investigation of the channel encoded by the wild type *kef*A led to the discovery that the activity was only observed in high K^+ buffer, which suggested its current name, MscK (for K^+ regulated) (Li et al., 2002), while the same publication suggested YggB be renamed MscS; these names have since been in general use.

The MscS and MscK activities are of similar conductance and both more sensitive to membrane stretch than is the MscL channel. Hence, many of the early studies characterizing the voltage-dependent nature (Martinac et al., 1987) and inactivation properties (Koprowski and Kubalski, 1998) of "mechanosensitive channels of *E. coli*" probably were characterizing a combination of MscS and MscK. More recent studies in which a *mscS/mscK/mscL* null mutant strain of *E. coli* was used for *mscS* expression allowed for the analysis of MscS in isolation (Akitake et al., 2005; Sotomayor et al., 2006). These latter works serve as the current definitive studies on the properties of the MscS activity. The findings include the observation that MscS exhibits essentially voltage-independent activation by tension, but strong voltage-dependent inactivation under depolarizing conditions. In addition, the channel appears to respond preferentially to acute stimuli but inactivates prior to opening if the stimulus is applied slowly over the course of several seconds. Finally, the study estimated the energy, area, and gating charge for the closed-to-open transition at 24 kT, 18 nm^2, and +0.8, respectively (Akitake et al., 2005).

The structure of *E. coli* MscS was solved to 3.9 Å resolution by X-ray crystallography (Bass et al., 2002); residues 27 to 280 of the total 287 were resolved (Fig. 3.5). As expected from the hydropathy plot, three helical TMDs were found at the N-terminal region of the protein. The channel, however, appeared to be a homoheptamer (Bass et al., 2002), rather than the hexamer predicted from crosslinking experiments (Sukharev, 2002) or the homopentamer of MscL (Chang et al., 1998). TMD3, which is rich in glycine and alanine, appears to form the pore. A glycine at position 113 induces a turn within the α-helical domain realigning the helix along the presumed membrane/cytoplasmic interface. Distal to this structure is a region that is relatively high in β-sheet character. The cytoplasmic region forms a cage with 7 pores, or portals, at the subunit interfaces, and all of the subunits interact to terminate in a short β-barrel-like 'crown' at the extreme C-terminal end. The unique cytoplasmic cage-like structure has been proposed to act as a molecular sieve that determines the size and perhaps selectivity of the pore. The size of the pores in the crystal structure is as follows: the pore formed by TMD3 is ∼11Å, the portals in the cytoplasmic cage are ∼14Å, and the extreme C-terminal β-barrel is ∼8Å. Given the size of the potential pores, it was presumed that the channel may be in an open state. However, molecular dynamic simulations suggested that a hydrophobic barrier may deter permeation (Anishkin and Sukharev, 2004). These data have led to the speculation that the channel is in an inactivated or desensitized state, which may indeed be a low-energy state, especially for a channel no longer under the

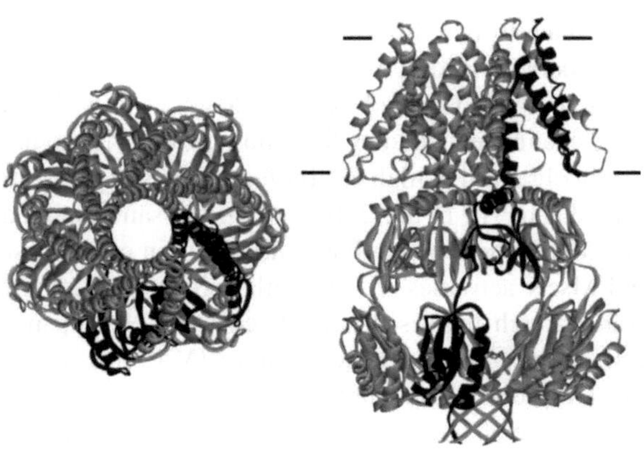

Fig. 3.5 The structure of the MscS channel. The model shown is derived from X-ray crystallographic data (Bass et al., 2002). The top (*left*) and side (*right*) views are shown. Each of the seven identical subunits passes through the membrane (indicated in the right panel) three times. The complex contains a 'cage-like' structure in the cytoplasm. A single subunit is highlighted in black in each figure. This structure is currently thought to be in an open or inactivated state

constraints of the lateral pressures of a biological membrane. Further investigation will be required to determine if the structure is open, partially open or inactivated.

The constriction defining the pore of the MscS channel is composed of a region of TM3 in which glycine and alanine residues appear to be tightly packed, with the closest association between glycine 108 and alanine 106. One model for the closure of MscS has been derived from mutagenesis experiments in which each of the glycines in this region has been substituted with serine (Edwards et al., 2005). The model proposes that the TMD3s become more vertical or normal to the membrane plane by sliding along each other and rotating. The resulting change in the packing of the small amino acids, glycine and alanine, in this region would then lead to a channel with a smaller, closed pore. The leucines at positions 105 and 109 would come into closer proximity, thus forming a tighter constriction point and more efficient hydrophobic barrier leading to a non-conducting conformation.

Structural changes are also predicted to occur in the cytoplasmic cage-like structure of MscS. One study found Ni^{2+} binding to MscS poly-histidine tagged at the C-terminal inhibited gating, suggesting movement in this region (Koprowski and Kubalski, 2003). Another study found that deletions at the C-terminal end of the protein are poorly tolerated, often yielding channels that are not expressed well in the membrane (Schumann et al., 2004). A more quantitative determination of the proximity of specific residues within the cage-like structure of the closed MscS channel has been obtained by disulfide trapping and crosslinking studies (Miller et al., 2003). The data imply that, when closed, the C-terminal cage region of the protein assumes a much more compact structure than that observed in the solved crystal structure. Hence, the transmembrane pore may not be the only permeation barrier in the closed state of the channel; the portals and β-barrel-like crown may also collapse into ion-impermeable conformations. This hypothesis has been referred to as the "Chinese-lantern" model by analogy with lanterns whose intensity is adjusted by either collapsing or expanding the lamp (Edwards et al., 2004).

Hence, one current model for MscS gating, presented in Fig. 3.6, is as follows. TMD1 and TMD2 interact with the membrane and thus constitute the primary sensors of membrane tension and modulation by voltage potential. These domains may move upon gating, flapping out like wings upon stimulation (Bass et al., 2002). Consistent with this hypothesis, placing a negative charge at position 40, which is just below the lipid-aqueous interface on the periplasmic side, should increase the probability of the extension of the these 'wings', and indeed does effect a gain-of-function phenotype (Okada et al., 2002); on the other hand, data derived from a recent asparagine scan was interpreted to suggest that this movement is trivial or non-existent (Nomura et al., 2006). Similar to MscL, in the open structure the orientation of the transmembrane domains forming the pore may adjust, becoming more tilted relative to the normal. This reorientation may be at least partially shielded from the membrane, so one of the modulators of MscL activity, thinning of the membrane, may play less of a role in MscS gating, but this has yet to be determined experimentally. Also in contrast with MscL, the structural rearrangements in MscS gating are not limited to the transmembrane domains. The potential collapse of the cytoplasmic cage domain may define, at least in part, the potential permeation barriers in the

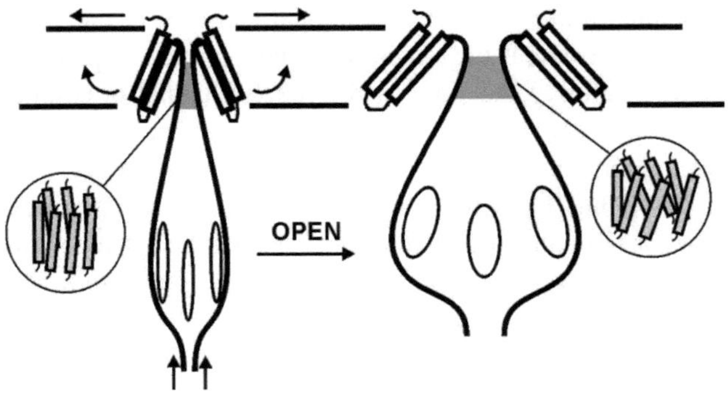

Fig. 3.6 A cartoon depicting a current model for the structural rearrangements thought to occur upon gating of the MscS channel. The 3 TMDs are within the bilayer depicted by grey horizontal cylinders. Note that there are nine pores: one in the bilayer formed by TM3, seven at the interface of the subunits within the cytoplasmic "cage", and one at the extreme cytoplasmic tip. As discussed in text, some evidence suggests that each of these may constrict upon closure of the channel. The seven TMD3's that form the pore are depicted as the circled inset for each structure to emphasize the predicted change in tilt of these domains in the different states

closed structure. While Fig. 3.6 and the discussion above describes some of the current views, one must appreciate that less is known of MscS than MscL, and the models for its gating are still in flux; for a more complete discussion of the evidence for and against aspects of gating models for MscS, see (Blount et al., 2005).

3.2.2.2 MscS as a Sensor

Given the large structural differences between MscL and MscS, it becomes a legitimate issue whether the two channels sense the same stimuli and have analogous structural responses. Similar to MscL (Häse et al., 1995; Blount et al., 1996a), MscS has been purified to apparent homogeneity, reconstituted, and shown to be functional, thus demonstrating that only the reconstituted protein and a lipid bilayer are necessary for channel activity; no other proteins are required (Okada et al., 2002; Sukharev, 2002). Hence, MscS responds directly to mechanical force. In addition, the observation that the ratio of the threshold pressure required to activate MscL versus MscS is a constant (Blount et al., 1996a; Blount et al., 1996b), and the reproducibility of the Boltzmann curves of the probability of opening versus tension (Sukharev, 2002), both suggest that the channel, like MscL, senses tension in the membrane rather than pressure across it (see Section 2.1.2, above), although this has not yet been formally demonstrated. Also, as described in the previous section, there are limited similarities in which the pore domains twist and tilt upon gating. There is evidence that MscS is probably modulated by amphipaths. A very early study, prior to the identification of multiple channels, determined the effect of addition of amphipaths on bacterial mechanosensitive channels (Martinac et al., 1990). This study found that addition of either positively charged, negatively charged or uncharged amphipaths could increase the sensitivity of the channel to stimuli. After the activation of the channel by an amphipath of one charge, its replacement with an amphipath of the opposite charge would partially reverse the effects prior to activation, presumably because the amphipaths would partition within opposing sides of the membrane, thus canceling the effect. As stated previously, because of its abundance and sensitivity, and also given the conductance of the channels reported in the study, it is likely that the channel that was modulated by the amphipaths in this study was MscS (although MscK modulation cannot be totally ruled out). A more recent study of MscS under high hydrostatic pressure suggested that it is lateral compression of the bilayer that is intimately involved in the expansion of the channel area as the channel opens (Macdonald and Martinac, 2005). Hence, it does appear that MscS senses physical changes in the membrane.

There are far fewer mutagenic and computational studies, as well as total mutants, for MscS relative to MscL. Hence, the residues at the lipid-aqueous interfaces and any specific protein-lipid interactions are less clear. In one recent study, similar to MscL, asparagine scanning in the regions of TMD1 and 2 has led to identification of functional residues and a prediction of where in the protein the lipid-aqueous interface occurs (Fig. 3.7) (Nomura et al., 2006). To date, no studies have been performed in which the lipid composition of the reconstituted channel has been manipulated and any differences measured.

3.2.2.3 MscS Homologues

Homologues of MscS are essentially ubiquitous throughout the microbial world. For example, several channels from Archaea, including MscMJ, MscMJLR and MscTA, have been functionally characterized providing evidence for a wide spread of the bacterial type MS channels in this separate phylogenetic domain (Kloda and

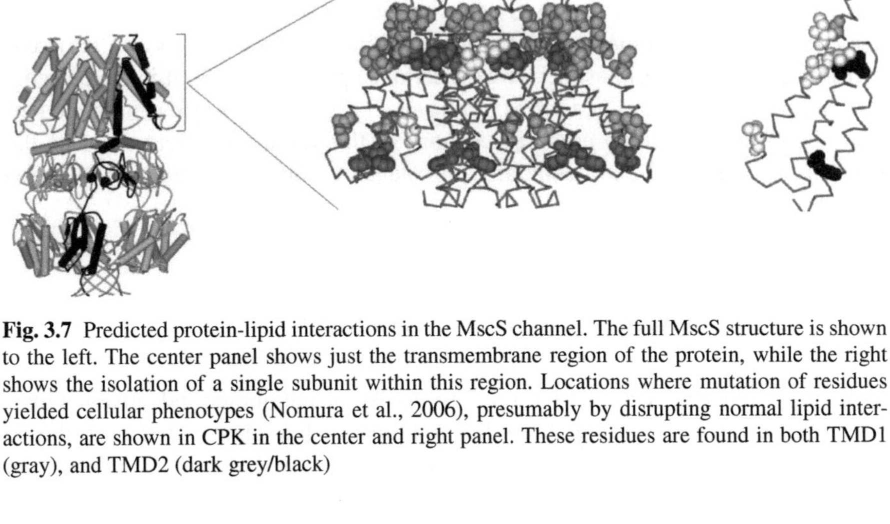

Fig. 3.7 Predicted protein-lipid interactions in the MscS channel. The full MscS structure is shown to the left. The center panel shows just the transmembrane region of the protein, while the right shows the isolation of a single subunit within this region. Locations where mutation of residues yielded cellular phenotypes (Nomura et al., 2006), presumably by disrupting normal lipid interactions, are shown in CPK in the center and right panel. These residues are found in both TMD1 (gray), and TMD2 (dark grey/black)

Martinac, 2002; Martinac and Kloda, 2003). Often a single organism will contain numerous homologues. Even *E. coli* appears to have a handful of MscS-like proteins. Often these homologues are much larger and more complex than the relatively stream-lined MscS. In these instances, the MscS structure appears to be conserved at the C-terminal end of the protein while additional transmembrane and other domains are found at the N-terminal. Of at least 4 putative homologues in the *E. coli* genome, channel activity has only been measured for one: MscK. The *mscK* gene, relative to MscS, is predicted to have eight additional transmembrane domains and a large periplasmic region (McLaggan et al., 2002). A gain-of-function mutation revealed a phenotype that was only observed in the presence of relatively high potassium concentration (McLaggan et al., 2002), and electrophysiological characterization (Li et al., 2002) strongly implied an allosteric interaction in which this ion is sensed, and is required for normal gating of the channel.

MscS homologues are also found in several eukaryotic organisms including *S pombe*, as well as essentially in all plants (see (Haswell et al., 2007)). Unfortunately little is known of these channels, and their function is inferred only by their homology to MscS. Similar to bacteria, a single plant species may contain numerous homologues within its genome, and most are much more complex structurally than *E. coli* MscS. Mutagenic studies in plants suggest that these homologues play a role in organelle size, density and perhaps the division machinery (Haswell and Meyerowitz, 2006); however, the exact role these homologues play in this process, and even whether they encode mechanosensitive channel activities that are linked to this function, have yet to be determined.

The diversity and complexity of this gene family implies multiple forms of regulation. The observation that MscK requires high concentrations of potassium or ammonium in the media or buffer to gate tempts one to believe that the additional, normally silent homologues in *E. coli* do indeed encode mechanosensitive channel activities, but we are just not clever enough to realize their requirements for gating. Thus far, none of these homologues have been reconstituted into artificial membranes, nor have they been tested in the presence of amphipaths. Because their structural design appears to overlap with MscS, it is supposed that, although they may be highly regulated, they will sense tension in the membrane. On the other hand, the additional domains may be giving a hint that the channels are 'tethered' by cytoplasmic and periplasmic (or extracellular) proteins. Thus, it is conceivable that, although the streamlined forms of this channel primarily sense membrane tension, the more structurally complex versions of this family may rely more on the tethering to intra- and extra-cellular components. Clearly, more work in the area is needed.

3.3 *S. cerivicea* Mechanosensitive Channels

As mention above in Section 2.1.2, pressure across the membrane and tension within it are related, but not the same. According to Laplace's law, tension in the membrane equals the pressure across it times the radius of curvature divided by 2. In fact, the first study to use this simple relationship to determine if a specific mechanosensitive channel was sensing pressure across the membrane or tension within it used a whole-cell patch clamp approach and was performed on a yeast channel (Gustin et al., 1988). In contrast to measuring the patch geometry, as has been performed with MscL and MscS (described above), the authors measured the size of the cell, and gated the channels with positive pressure. Three cells of very different sizes, and thus diameter and radius of curvature, were assayed. When the probability of opening, P_o, was plotted against the pressure, three distinct curves were found. However, when the P_o was plotted against the calculated tension, the data merged to form a single curve. Thus, this yeast channel appears, unambiguously, to sense membrane tension.

Calcium flux into the yeast cytosol has been shown to be elicited by hypotonic shock as well as other stresses (Batiza et al., 1996; Matsumoto et al., 2002). Other phenotypic evidence suggests that Ca^{2+} plays a role in mating; in wild type cells, exposure to the mating pheromone α factor stimulates Ca^{2+} influx and thereby increases Ca^{2+} concentration in the cytoplasm. In selective drop-out (SD) medium with low Ca^{2+}, cells with deletions of *mid1* or *cch1* grow less vigorously and exhibit lower Ca^{2+} influx (Muller et al., 2001), and mat a strains are killed by α factor (Iida et al., 1994; Fischer et al., 1997). Hence it is thought that Mid1p and Cch1p form a calcium channel or transporter. Interestingly, functional expression of Mid1 in Chinese hamster ovary cells appeared to lead to a mechanosensitive channel activity (Kanzaki et al., 1999). Although the conductance of this channel is similar to the mechanosensitive channel in the native plasmid membrane found in whole-cell patch clamp (32 and 40 pS respectively), spheroplasts from the *mid1* knockout strain appear to still retain the previously characterized channel (Palmer et al., 2004). Hence, more research is needed to determine if Mid1p functions as a mechanosensitive channel when expressed in its native environment, and if so, whether it has similar functional characteristics to MscL and MscS (e.g. senses membrane tension and sensitive to amphipaths).

3.4 Mammalian Two-Pore Domain Potassium (K_{2P}) Channels

Voltage-gated and inward rectifying potassium channels are normally tetrameric complexes in which each subunit contains a region of the protein that serves as the potassium filter known as the pore or "P" domain. Two-pore domain potassium channels (K_{2P}) are also K^+ selective channels, however each subunit of the K_{2P} channels has a structure characterized by two, rather than one, P domains. These channels contain four transmembrane domains, with both the carboxyl- and amino-termini cytoplasmic, and one P domain between TMD1 and 2, and the other between TMD3 and 4 (Lesage and Lazdunski, 2000). Assuming that four P domains make a functional complex, these channels are homo or hetero dimers (Kang et al., 2004). Electrophysiologically, K_{2P} show a small time- and voltage-dependence of their activation, and they are largely insensitive to common K^+ channel blockers. Functionally, the K_{2P} channels are responsible for the 'leak' or background currents often found in cells when investigated by electrophysiological means. These channels appear to play an essential role in the regulation of the cell membrane potential and its excitability. Presumably, in times of stress these channels will release potassium from the cell, thus preventing an undesirable depolarization that potentially could allow harmful levels of Ca^{2+} into the cell cytoplasm. Such a protective role has been recently supported by a study demonstrating that a mouse null for a member of this channel family, which can be stimulated by acidic environments (Trek-1), is more sensitive to neuronal death due to acute anoxia (Buckler and Honore, 2005).

Although K_{2P} channels were discovered less than a decade ago, 15 channel subunits have already been identified in humans (Lesage and Lazdunski, 2000; Goldstein et al., 2005). All known K_{2P} subunits have the same structural scheme, but in general they share very low sequence identity, which is reflected in their functional differences as well as the diversity of stimuli regulating their activities. They can be functionally divided in five groups: two weak inward rectifiers TWIK-1 and TWIK-2 (Lesage et al., 1996; Chavez et al., 1999); alkaline-activated K^+ channels TALK-1 and TALK-2 (Girard et al., 2001; Han et al., 2003; Duprat et al., 2005); THIK-1 and THIK-2 inhibited by halothane (Rajan et al., 2001); and the acid sensitive channels TASK-1, TASK-2 and TASK-3 (Chapman et al., 2000; Kim et al., 2000; Rajan et al., 2000; Talley et al., 2000; Bayliss et al., 2001; Niemeyer et al., 2001; Patel and Honore, 2001; Czirjak and Enyedi, 2002; Talley and Bayliss, 2002), and three lipid- sensitive mechanosensitive channels, Trek-1, Trek-2 and TRAAK (Maingret et al., 1999; Bang et al., 2000; Maingret et al., 2000b; Patel et al., 2001). These latter channels are also sensitive to a wide variety of physical and chemical factors including heat, pH, membrane stretch, hypoxia, neurotransmitters as well as volatile and local anesthetics (Duprat et al., 2000; Maingret et al., 2000a; Koh et al., 2001; Patel et al., 2001; Patel and Honoré, 2002; Kim, 2003; Heurteaux et al., 2004; Goldstein et al., 2005).

3.4.1 The Mechano-gated K^+ Channels TREK-1, TREK-2 and TRAAK Sense Changes in the Lipid Bilayer

The K_{2P} mechanosensitive channels, Trek-1 (also known as $K_{2p}2.1$ and KCNK2), Trek-2 (also known as $K_{2p}10.1$ and KCNK10) and TRAAK (also known as $K_{2p}4.1$ and KCNK4) are reversibly activated by membrane stretch. Removal of the cytoskeleton actually increases channel activity, suggesting not only that tethering of the channel plays no role in its gating, but that the cytoskeleton actually absorbs much of the energy that would normally stretch the membrane and provide the stimulus for channel gating. Although none of these channels have been purified and reconstituted into synthetic lipids, it does appear that membrane tension is a normal stimulus for these channels because they are activated by several amphipathic compounds including arachidonic acid; notably, activation can occur in excised patches. Volatile and local anesthetics, which also are thought to intercalate into the membrane, can also influence gating (Patel et al., 1999).

Unlike the bacterial mechanosensitive channels, which appear to not distinguish between cup-formers or crenators, these channels appear to be activated by crenators and inhibited by cup-formers. Strong activation by lysophospholipids and other amphipaths serves as perhaps the strongest evidence to date that these channels indeed sense membrane tension in a mechanism similar to that of MscL and MscS.

3.5 Mammalian Transient Receptor Potential (TRP) Channels

Transient Receptor Potential (TRP) Channels were first identified as playing a role in *Drosophila* vision and have since grown into a large family of related channels (see (Minke, 2006) for review). The TRP channel family now appears to contain a very large number of members that are largely cationic-specific and allow Ca^{2+} to cross the membrane. Many of these channels apparently play roles in sensory systems. The physiological roles of many of the members of this family, as well as the stimuli that they are sensing, are just now being elucidated. Members, including TRPV1 (Caterina et al., 1997) and TRPV2 (Caterina et al., 1999), TRPM8 (McKemy et al., 2002; Peier et al., 2002), and TRPV3 (Smith et al., 2002; Xu et al., 2002) have been implicated in sensing temperature, although the mechanism, whether it be by temperature-dependent protein folding or sensing temperature-dependent changes in the membrane, have yet to be determined. Several members of this family have also been implicated in responses to osmotic cell swelling and/or whole-animal osmotic regulation and thirst including TRPV1 (actually an apparent splice variant of this channel, see (Sharif Naeini et al., 2006)), TRPV4 (Liedtke et al., 2000; Strotmann et al., 2000; Vriens et al., 2004), TRPV2 (Muraki et al., 2003), TRPC1 (Chen and Barritt, 2003), TRPM3 (Grimm et al., 2003) and TRPM7 (Numata et al., 2006). Some pharmacological evidence does suggest that these channels indeed sense changes in the membrane (e.g. see (Spassova et al., 2006) for a study of TRPC6 and (Numata et al., 2006) for TRPM7). In some instances, the channels have been shown to function in excised patches, which would be consistent with 'tethers' not playing a significant role. But, the best evidence to date that at least some members of this family do not require any cytoskeleton or extracellular tethers to function comes from studies of a channel from *Xenopus* oocytes. Here, a technique was developed in which hypotonic stress was used to induce 'blebbing' of the plasma membrane and the isolation of plasma membrane vesicles. The blebs and vesicles are apparently devoid of any cytoskeletal elements, yet maintain the mechanosensitive channel activity (Zhang et al., 2000). This group further demonstrated that the channel can be solubilized, biochemically enriched, reconstituted, and yet remains functional. Using this approach, in conjunction with Western blotting, the channel was identified as an orthologue of the mammalian TRPC1 channel (Maroto et al., 2005). Although the channel was not purified to apparent homogeneity, it does not seem likely that critical interactions with other cellular components would survive solubilization and reconstitution, thus making it very probable that only the channel and membrane are required for this mechanosensitive channel activity.

3.6 Other Potential Tension-Modulated Channels

Members of the BK channel family are also strong candidates for physiologically relevant mechanosensitive channels. Activity from members of this family have been found in many organs, are regulated by internal Ca^{2+} levels and thought to play a role in a variety of cellular functions. In several cell types, including osteoblast-like cells (Davidson et al., 1990; Allard et al., 2000), renal cells (Pacha et al., 1991), smooth cells (Kirber et al., 1992; Dopico et al., 1994), neuroepithelium (Mienville et al., 1996) and skeletal muscle cells (Mallouk and Allard, 2000), it appears that membrane stretch can activate or modulate activity independent of changing Ca^{2+} levels. More recently, a member of this family, SAKCa, has been cloned from embryo chick ventricular myocytes, expressed, and shown to have mechanosensitive channel properties and pharmacology (Tang et al., 2003; Sokabe et al., 2004). Significantly, SAKCa activity has been shown to be modulated by amphipaths (Qi et al., 2005).

There are additional candidate tension-gated channels. One of the most studied is the epithelial Na^+ channel (ENaC) family. Members of this family were first identified by a touch-insensitive genetic screen in *C. elegans*. Much of the genetic evidence suggested that, within this organism, the channel functions through a tether mechanism (Tavernarakis and Driscoll, 1997); the original screen identified not only transmembrane proteins (*mec*-4 and *mec*-10), now known to be related to mammalian ENaC and to form channels, but also extracellular and intracellular cytoskeletal proteins that appear to play some role in the mechanosensory process (Tavernarakis and Driscoll, 1997). Some studies have suggested that the homologous mammalian ENaC channel may be modulated by membrane stretch (e.g. (Ji et al., 1998)); the mechanical modulation may reflect a release of the channels from a Ca^{2+} block (Ismailov et al., 1997). In addition, the ENaC channel has recently been shown to be modulated by pharmacological agents that would influence the membrane structure (Awayda et al., 2004). Some evidence also exists that the antimicrobial peptides alamethicin (Opsahl and Webb, 1994) and gramicidin A (Martinac and Hamill, 2002), which are easily reconstituted, also may detect membrane tension. Finally, the mammalian polycystins 1 and 2, encoded by Pkd1 and pkd2 respectively, which are found in the cilia of kidney epithelium, are also candidates for mechanosensitive channels (Nauli et al., 2003); however, their putative activities and molecular mechanisms have not yet been well studied.

In addition to mechano-activated channels, membrane stretch or osmotic pressures have also been shown to modulate channels that are clearly gated by other means, including the voltage-gated shaker K^+ channel (Gu et al., 2001; Laitko and Morris, 2004; Laitko et al., 2006) and the ligand-gated NMDA receptors (Paoletti and Ascher, 1994); in the latter case, lysophospholipids and arachidonic acid have both been shown to modulate activity (Casado and Ascher, 1998), suggesting that these channels are indeed sensing their membrane environment. In these cases it remains unclear what, if any, physiological role this mechano-modulation plays.

3.7 Conclusions

Many channels are now emerging as candidates for gating, or modulating their gating, in response to biophysical changes occurring within the lipid bilayer upon membrane tension. The ultimate experiment to determine that a protein or protein complex is truly sensing its membrane environment is to reconstitute it into synthetic membrane and demonstrate function. However, many proteins do not survive the harsh treatments required for such an approach. An additional or alternative approach that is often used is to modify the lipid membrane *in situ* with lysophospholipids or amphipathic compounds that modify the membrane lateral pressure profile and observe potential changes in channel activity. Fig. 3.8 diagrams and describes how such compounds may disrupt normal membrane bilayer structure.

One of the outstanding features of the list of candidates for channels that sense membrane tension is the diversity of its members. Even the two most characterized members, MscL and MscS, both from bacteria, are quite different structurally. Therefore, the search for a conserved motif or domain that encodes the ability to sense membrane tension may be futile. On the other hand, there must be some general principles underlying the coupling of membrane tension and changes in protein conformation. Such principles may even extend to some proteins where 'tethering' is thought to be critical. One recent proposal, which may be a unifying theory, predicts that many of these proteins may indeed be sensing their lipid environment, and the tethers simply displace the protein from a preferred lipid-protein interaction by pulling (Kung, 2005). Although there may not be obvious structural homology between the members of this diverse class of proteins sensitive to membrane stresses, there should be specific or gross structural features or interactions that encode this ability. Thus far, MscL remains the best characterized member of the class of mechanosensory molecules. From this model system, proposals for how a protein senses and responds to its membrane environment abound. One posit is that it is the transient exposure of hydrophobic residues to an aqueous environment that is the primary energy barrier for channel gating; in essence, the protein 'tears' before the

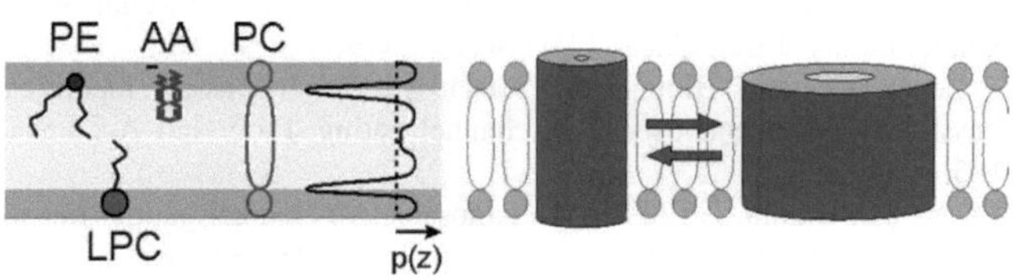

Fig. 3.8 Membrane pressure profiles affect the gating of Mechanosensitive channels. Many mechanosensitive channels gate in response of tension transmitted through the cell membrane. Each membrane is formed by bilayer forming lipids like phosphatidylcholine (PC) that have a characteristic pressure profile, shown toward the middle of the figure, with negative pressures just under the headgroups, and the largest pressures about 1/3rd and 2/3rds across the membrane. Insertion of crenators like arachidonic acid (AA), conical lipids like lysophosphatidylcholine (LPC) or lipids with different head group sizes phosphatidylethanolamine (PE) into the lipid bilayer (see left side of panel) change the membrane curvature and pressure profiles, thus affecting Mechanosensitive channel gating (right side of panel)

membrane does (Blount and Moe, 1999). Another group has utilized a detailed computational approach to determine the free energy changes induced by lipid-protein interactions and correlated this to free energy differences between conductance states (Wiggins and Phillips, 2005). The challenge for the future will be to determine if such theories and approaches will lead to models that will not only present an explanation for the mechanosensitive channel activity of MscL or other observed activities, but also be able to predict which transmembrane proteins are prime candidates for mechanosensitive proteins simply by their specific or general structural features.

Acknowledgment The authors are supported by Grant I-1420 of the Welch Foundation, Grant FA9550-05-1-0073 of the Air Force Office of Scientific Review, Grant 0655012Y of the American Heart Association—Texas Affiliate, and Grant GM61028 from the National Institutes of Health.

References

Akitake, B., Anishkin, A., and Sukharev, S. (2005) The "dashpot" mechanism of stretch-dependent gating in MscS. J Gen Physiol 125:143–154.

Allard, B., Couble, M. L., Magloire, H., and Bleicher, F. (2000) Characterization and gene expression of high conductance calcium-activated potassium channels displaying mechanosensitivity in human odontoblasts. J Biol Chem 275:25556–25561.

Anishkin, A., Chiang, C. S., and Sukharev, S. (2005) Gain-of-function mutations reveal expanded intermediate states and a sequential action of two gates in MscL. J Gen Physiol 125: 155–170.

Anishkin, A., Gendel, V., Sharifi, N. A., Chiang, C. S., Shirinian, L., Guy, H. R., and Sukharev, S. (2003) On the conformation of the COOH-terminal domain of the large mechanosensitive channel MscL. The Journal of general physiology 121(3):227–244.

Anishkin, A., and Sukharev, S. (2004) Water dynamics and dewetting transitions in the small mechanosensitive channel MscS. Biophys J 86:2883–2895.

Awayda, M. S., Shao, W., Guo, F., Zeidel, M., and Hill, W. G. (2004) ENaC-membrane interactions: regulation of channel activity by membrane order. J Gen Physiol 123:709–727.

Bang, H., Kim, Y., and Kim, D. (2000) TREK-2, a new member of the mechanosensitive tandem-pore K^+ channel family. J Biol Chem 275:17412–17419.

Bartlett, J. L., Levin, G., and Blount, P. (2004) An *in vivo* assay identifies changes in residue accessibility on mechanosensitive channel gating. Proc Natl Acad Sci U S A 101: 10161–10165.

Bartlett, J. L., Li, Y., and Blount, P. (2006) Mechanosensitive channel gating transitions resolved by functional changes upon pore modification. Biophys J 91:3684–3691.

Bass, R. B., Strop, P., Barclay, M., and Rees, D. C. (2002) Crystal structure of *Escherichia coli* MscS, a voltage-modulated and mechanosensitive channel. Science 298:1582–1587.

Batiza, A. F., Schulz, T., and Masson, P. H. (1996) Yeast respond to hypotonic shock with a calcium pulse. J Biol Chem 271:23357–23362.

Bayliss, D. A., Talley, E. M., Sirois, J. E., and Lei, Q. (2001) TASK-1 is a highly modulated pH-sensitive 'leak' K^+ channel expressed in brainstem respiratory neurons. Respiration physiology 129:159–174.

Berrier, C., Besnard, M., Ajouz, B., Coulombe, A., and Ghazi, A. (1996) Multiple mechanosensitive ion channels from *Escherichia coli*, activated at different thresholds of applied pressure. J Membr Biol 151:175–187.

Blount, P., Iscla, I., Li, Y., and Moe, P. C. (2005). The bacterial mechanosensitive channel MscS and its extended family. In Bacterial channels and their eukaryotic homologues, A. Kubalski, and B. Martinac, eds. (Washington, D.C., ASM Press).

Blount, P., Iscla, I., Moe, P. C., and Li, Y. (2007). MscL: The bacterial mechanosensitive channel of large conductance. In Mechanosensitive Ion Channels (Volume 58 Current Topics in Membranes series), O. P. Hamill, ed. (St. Louis, MO, Elsievier Press) pp 202–233.

Blount, P., and Moe, P. (1999) Bacterial mechanosensitive channels: integrating physiology, structure and function. Trends in Microbiol 7:420–424.

Blount, P., Sukharev, S. I., Moe, P. C., Schroeder, M. J., Guy, H. R., and Kung, C. (1996a) Membrane topology and multimeric structure of a mechanosensitive channel protein of *Escherichia coli*. EMBO J 15:4798–4805.

Blount, P., Sukharev, S. I., Schroeder, M. J., Nagle, S. K., and Kung, C. (1996b) Single residue substitutions that change the gating properties of a mechanosensitive channel in *Escherichia coli*. Proc Nat Acad Sci USA 93:11652–11657.

Boyd, D., Manoil, C., and Beckwith, J. (1987) Determinants of membrane protein topology. Proc Nat Acad Sci USA 84:8525–8529.

Britten, R. J., and McClure, F. T. (1962) The amino acid pool in *Escherichia coli*. Bacteriol Rev 26:292–335.

Buckler, K. J., and Honore, E. (2005) The lipid-activated two-pore domain K+ channel TREK-1 is resistant to hypoxia: implication for ischaemic neuroprotection. The Journal of physiology 562:213–222.

Cantor, R. (1999) Lipid composition and the lateral pressure profile in bilayers. Biophys J 76:2625–2639.

Casado, M., and Ascher, P. (1998) Opposite modulation of NMDA receptors by lysophospholipids and arachidonic acid: common features with mechanosensitivity. The Journal of physiology 513 (Pt 2):317–330.

Caterina, M., Rosen, T., Tominaga, M., Brake, A., and Julius, D. (1999) A capsaicin-receptor homologue with a high threshold for noxious heat. SO - Nature 1999 Apr 1;398(6726): 436–41 398.

Caterina, M. J., Schumacher, M. A., Tominaga, M., Rosen, T. A., Levine, J. D., and Julius, D. (1997) The capsaicin receptor: a heat-activated ion channel in the pain pathway. Nature 389:816–824.

Chang, G., Spencer, R. H., Lee, A. T., Barclay, M. T., and Rees, D. C. (1998) Structure of the MscL homolog from *Mycobacterium tuberculosis*: A gated mechanosensitive ion channel. Science 282:2220–2226.

Chapman, C. G., Meadows, H. J., Godden, R. J., Campbell, D. A., Duckworth, M., Kelsell, R. E., Murdock, P. R., Randall, A. D., Rennie, G. I., and Gloger, I. S. (2000) Cloning, localisation and functional expression of a novel human, cerebellum specific, two pore domain potassium channel. Brain Res Mol Brain Res 82:74–83.

Chavez, R. A., Gray, A. T., Zhao, B. B., Kindler, C. H., Mazurek, M. J., Mehta, Y., Forsayeth, J. R., and Yost, C. S. (1999) TWIK-2, a new weak inward rectifying member of the tandem pore domain potassium channel family. J Biol Chem 274: 7887–7892.

Chen, J., and Barritt, G. J. (2003) Evidence that TRPC1 (transient receptor potential canonical 1) forms a Ca(2+)-permeable channel linked to the regulation of cell volume in liver cells obtained using small interfering RNA targeted against TRPC1. The Biochemical journal 373:327–336.

Chiang, C. S., Anishkin, A., and Sukharev, S. (2004) Gating of the large mechanosensitive channel *in situ*: estimation of the spatial scale of the transition from channel population responses. Biophys J 86:2846–2861.

Chiang, C. S., Shirinian, L., and Sukharev, S. (2005) Capping Transmembrane Helices of MscL with Aromatic Residues Changes Channel Response to Membrane Stretch. Biochemistry 44:12589–12597.

Czirjak, G., and Enyedi, P. (2002) TASK-3 dominates the background potassium conductance in rat adrenal glomerulosa cells. Molecular endocrinology (Baltimore, Md 16: 621–629.

Davidson, R. M., Tatakis, D. W., and Auerbach, A. L. (1990) Multiple forms of mechanosensitive ion channels in osteoblast-like cells. Pflugers Arch 416:646–651.

Deuticke, B. (1968) Transformation and restoration of biconcave shape of human erythrocytes induced by amphiphilic agents and changes of ionic environment. Biochim Biophys Acta 163:494–500.

Dopico, A. M., Kirber, M. T., Singer, J. J., and Walsh, J. V., Jr. (1994) Membrane stretch directly activates large conductance Ca(2+)-activated K+ channels in mesenteric artery smooth muscle cells. Am J Hypertens 7:82–89.

Duprat, F., Girard, C., Jarretou, G., and Lazdunski, M. (2005) Pancreatic two P domain K+ channels TALK-1 and TALK-2 are activated by nitric oxide and reactive oxygen species. The Journal of physiology 562:235–244.

Duprat, F., Lesage, F., Patel, A., Fink, M., Romey, G., and Lazdunski, M. (2000) The neuroprotective agent riluzole activates the two P domain K(+) channels TREK-1 and TRAAK. Molecular Pharmacology 57:906–912.

Edwards, M. D., Booth, I. R., and Miller, S. (2004) Gating the bacterial mechanosensitive channels: MscS a new paradigm? Current Opinion in Microbiology 7:163–167.

Edwards, M. D., Li, Y., Kim, S., Miller, S., Bartlett, W., Black, S., Dennison, S., Iscla, I., Blount, P., Bowie, J. U., and Booth, I. R. (2005) Pivotal role of the glycine-rich TM3 helix in gating the MscS mechanosensitive channel. Nat Struct Mol Biol 12:113–119.

Elmore, D. E., and Dougherty, D. A. (2001) Molecular dynamics simulations of wild-type and mutant forms of the Mycobacterium tuberculosis MscL channel. Biophys J 81: 1345–1359.

Elmore, D. E., and Dougherty, D. A. (2003) Investigating lipid composition effects on the mechanosensitive channel of large conductance (MscL) using molecular dynamics simulations. Biophys J 85:1512–1524.

Fischer, M., Schnell, N., Chattaway, J., Davies, P., Dixon, G., and Sanders, D. (1997) The Saccharomyces cerevisiae CCH1 gene is involved in calcium influx and mating. FEBS Lett 419:259–262.

Girard, C., Duprat, F., Terrenoire, C., Tinel, N., Fosset, M., Romey, G., Lazdunski, M., and Lesage, F. (2001) Genomic and functional characteristics of novel human pancreatic 2P domain K(+) channels. Biochemical and biophysical research communications 282: 249–256.

Goldstein, S. A., Bayliss, D. A., Kim, D., Lesage, F., Plant, L. D., and Rajan, S. (2005) International Union of Pharmacology. LV. Nomenclature and molecular relationships of two-P potassium channels. Pharmacol Rev 57:527–540.

Grimm, C., Kraft, R., Sauerbruch, S., Schultz, G., and Harteneck, C. (2003) Molecular and functional characterization of the melastatin-related cation channel TRPM3. J Biol Chem 278:21493–21501.

Gu, C. X., Juranka, P. F., and Morris, C. E. (2001) Stretch-activation and stretch-inactivation of Shaker-IR, a voltage-gated K+ channel. Biophys J 80:2678–2693.

Guharay, F., and Sachs, F. (1984) Stretch-activated single ion channel currents in tissue-cultured embryonic chick skeletal muscle. J Physiol 352:685–701.

Gullingsrud, J., and Schulten, K. (2004) Lipid bilayer pressure profiles and mechanosensitive channel gating. Biophys J 86:3496–3509.

Gustin, M. C., Zhou, X. L., Martinac, B., and Kung, C. (1988) A mechanosensitive ion channel in the yeast plasma membrane. Science 242:762–765.

Han, J., Kang, D., and Kim, D. (2003) Functional properties of four splice variants of a human pancreatic tandem-pore K+ channel, TALK-1. Am J Physiol Cell Physiol 285: C529–538.

Häse, C. C., Le Dain, A. C., and Martinac, B. (1995) Purification and functional reconstitution of the recombinant large mechanosensitive ion channel (MscL) of *Escherichia coli*. J Biol Chem 270:18329–18334.

Haswell, E. S. (2007). MscS-like proteins in plants. In Mechanosensitive Ion Channels (Volume 58 Current Topics in Membranes series), O. P. Hamill, ed. (St. Louis, MO, Elsievier Press).

Haswell, E. S., and Meyerowitz, E. M. (2006) MscS-like proteins control plastid size and shape in Arabidopsis thaliana. Curr Biol 16:1–11.

Heurteaux, C., Guy, N., Laigle, C., Blondeau, N., Duprat, F., Mazzuca, M., Lang-Lazdunski, L., Widmann, C., Zanzouri, M., Romey, G., and Lazdunski, M. (2004) TREK-1, a K+ channel involved in neuroprotection and general anesthesia. Embo J 23:2684–2695.

Iida, H., Nakamura, H., Ono, T., Okumura, M. S., and Anraku, Y. (1994) MID1, a novel Saccharomyces cerevisiae gene encoding a plasma membrane protein, is required for Ca2+ influx and mating. Molecular and cellular biology 14:8259–8271.

Iscla, I., Levin, G., Wray, R., Reynolds, R., and Blount, P. (2004) Defining the physical gate of a mechanosensitive channel, MscL, by engineering metal-binding sites. Biophys J 87(5):3172–3180.

Ismailov, II, Berdiev, B. K., Shlyonsky, V. G., and Benos, D. J. (1997) Mechanosensitivity of an epithelial Na+ channel in planar lipid bilayers: release from Ca2+ block. Biophys J 72:1182–1192.

Ji, H. L., Fuller, C. M., and Benos, D. J. (1998) Osmotic pressure regulates alpha beta gamma-rENaC expressed in Xenopus oocytes. The American journal of physiology 275:C1182–1190.

Kang, D., Han, J., Talley, E. M., Bayliss, D. A., and Kim, D. (2004) Functional expression of TASK-1/TASK-3 heteromers in cerebellar granule cells. The Journal of physiology 554:64–77.

Kanzaki, M., Nagasawa, M., Kojima, I., Sato, C., Naruse, K., Sokabe, M., and Iida, H. (1999) Molecular identification of a eukaryotic, stretch-activated nonselective cation channel. Science 285:882–886.

Kim, D. (2003) Fatty acid-sensitive two-pore domain K^+ channels. TRENDS in Pharmacological Sciences 24:648–654.

Kim, Y., Bang, H., and Kim, D. (2000) TASK-3, a new member of the tandem pore K(+) channel family. J Biol Chem 275:9340–9347.

Kirber, M. T., Ordway, R. W., Clapp, L. H., Walsh, J. V., Jr., and Singer, J. J. (1992) Both membrane stretch and fatty acids directly activate large conductance Ca(2+)-activated K^+ channels in vascular smooth muscle cells. FEBS Lett 297:24–28.

Kloda, A., Ghazi, A., and Martinac, B. (2006) C-terminal charged cluster of MscL, RKKEE, functions as a pH sensor. Biophys J 90:1992–1998.

Kloda, A., and Martinac, B. (2001) Molecular identification of a mechanosensitive channel in archaea. Biophys J 80:229–240.

Kloda, A., and Martinac, B. (2002) Common evolutionary origins of mechanosensitive ion channels in Archaea, bacteria and cell-walled Eukarya. Archaea 1:35–44.

Koh, S. D., Monaghan, K., Sergeant, G. P., Ro, S., Walker, R. L., Sanders, K. M., and Horowitz, B. (2001) TREK-1 Regulation by Nitric Oxide and cGMP-dependent Protein Kinase. The Journal of Biological Chemistry 276:44338–44346.

Koprowski, P., and Kubalski, A. (1998) Voltage-independent adaptation of mechanosensitive channels in *Escherichia coli* protoplasts. J Membr Biol 164:253–262.

Koprowski, P., and Kubalski, A. (2003) C termini of the *Escherichia coli* mechanosensitive ion channel (MscS) move apart upon the channel opening. J Biol Chem 278:11237–11245.

Kung, C. (2005) A possible unifying principle for mechanosensation. Nature 436(7051):647–654.

Laitko, U., Juranka, P. F., and Morris, C. E. (2006) Membrane stretch slows the concerted step prior to opening in a Kv channel. J Gen Physiol 127:687–701.

Laitko, U., and Morris, C. E. (2004) Membrane tension accelerates rate-limiting voltage-dependent activation and slow inactivation steps in a Shaker channel. J Gen Physiol 123:135–154.

Lesage, F., Guillemare, E., Fink, M., Duprat, F., Lazdunski, M., Romey, G., and Barhanin, J. (1996) TWIK-1, a ubiquitous human weakly inward rectifying K+ channel with a novel structure. Embo J 15:1004–1011.

Lesage, F., and Lazdunski, M. (2000) Molecular and functional properties of two-pore-domain potassium channels. American Journal of Physiology - Renal Fluid & Electrolyte Physiology 279:F793-F801.

Levin, G., and Blount, P. (2004) Cysteine scanning of MscL transmembrane domains reveals residues critical for mechanosensitive channel gating. Biophys J 86:2862–2870.

Levina, N., Totemeyer, S., Stokes, N. R., Louis, P., Jones, M. A., and Booth, I. R. (1999) Protection of *Escherichia coli* cells against extreme turgor by activation of MscS and MscL mechanosensitive channels: identification of genes required for MscS activity. EMBO J **18**: 1730–1737.

Li, Y., Moe, P. C., Chandrasekaran, S., Booth, I. R., and Blount, P. (2002) Ionic regulation of MscK, a mechanosensitive channel from *Escherichia coli*. EMBO J **21**:5323–5330.

Li, Y., Wray, R., and Blount, P. (2004) Intragenic suppression of gain-of-function mutations in the Escherichia coli mechanosensitive channel, MscL. Mol Microbiol **53**:485–495.

Liedtke, W., Choe, Y., Marti-Renom, M. A., Bell, A. M., Denis, C. S., Sali, A., Hudspeth, A. J., Friedman, J. M., and Heller, S. (2000) Vanilloid receptor-related osmotically activated channel (VR-OAC), a candidate vertebrate osmoreceptor. Cell **103**:525–535.

Macdonald, A. G., and Martinac, B. (2005) Effect of high hydrostatic pressure on the bacterial mechanosensitive channel MscS. Eur Biophys J **34**:434–441.

Maingret, F., Fosset, M., Lesage, F., Lazdunski, M., and Honore, E. (1999) TRAAK is a mammalian neuronal mechano-gated K+ channel. J Biol Chem **274**:1381–1387.

Maingret, F., Lauritzen, I., Patel, A., Heurteaux, C., Reyes, R., Lesage, F., Lazdunski, M., and Honore, E. (2000a) TREK-1 is a heat-activated background K(+) channel. EMBO J **19**: 2483–2491.

Maingret, F., Patel, A., Lesage, F., Lazdunski, M., and Honore, E. (2000b) Lysophospholipids open the two-pore domain mechano-gated K(+) channels TREK-1 and TRAAK. J Biol Chem **275**:10128–10133.

Mallouk, N., and Allard, B. (2000) Stretch-induced activation of Ca(2+)-activated K(+) channels in mouse skeletal muscle fibers. Am J Physiol Cell Physiol **278**:C473–479.

Maroto, R., Raso, A., Wood, T. G., Kurosky, A., Martinac, B., and Hamill, O. P. (2005) TRPC1 forms the stretch-activated cation channel in vertebrate cells. Nat Cell Biol **7**:179–185.

Martinac, B. (2004) Mechanosensitive ion channels: molecules of mechanotransduction. J Cell Sci **117**:2449–2460.

Martinac, B., Adler, J., and Kung, C. (1990) Mechanosensitive ion channels of *E. coli* activated by amphipaths. Nature **348**:261–263.

Martinac, B., Buechner, M., Delcour, A. H., Adler, J., and Kung, C. (1987) Pressure-sensitive ion channel in *Escherichia coli*. Proc Nat Acad Sci USA **84**:2297–2301.

Martinac, B., and Hamill, O. P. (2002) Gramicidin A channels switch between stretch activation and stretch inactivation depending on bilayer thickness. Proc Natl Acad Sci USA **99**:4308–4312.

Martinac, B., and Kloda, A. (2003) Evolutionary origins of mechanosensitive ion channels. Progress in biophysics and molecular biology **82**:11–24.

Matsumoto, T. K., Ellsmore, A. J., Cessna, S. G., Low, P. S., Pardo, J. M., Bressan, R. A., and Hasegawa, P. M. (2002) An osmotically induced cytosolic Ca^{2+} transient activates calcineurin signaling to mediate ion homeostasis and salt tolerance of Saccharomyces cerevisiae. J Biol Chem **277**:33075–33080.

Maurer, J. A., and Dougherty, D. A. (2003) Generation and evaluation of a large mutational library from the *Escherichia coli* mechanosensitive channel of large conductance, MscL - Implications for channel gating and evolutionary design. J Biol Chem **278**:21076–21082.

McKemy, D. D., Neuhausser, W. M., and Julius, D. (2002) Identification of a cold receptor reveals a general role for TRP channels in thermosensation. Nature **416**:52–58.

McLaggan, D., Jones, M. A., Gouesbet, G., Levina, N., Lindey, S., Epstein, W., and Booth, I. R. (2002) Analysis of the kefA2 mutation suggests that KefA is a cation-specific channel involved in osmotic adaptation in *Escherichia coli*. Molec Microbiol **43**:521–536.

Meyer, G. R., Gullingsrud, J., Schulten, K., and Martinac, B. (2006) Molecular Dynamics study of MscL interactions with a curved lipid bilayer. Biophys J.

Mienville, J., Barker, J. L., and Lange, G. D. (1996) Mechanosensitive properties of BK channels from embryonic rat neuroepithelium. The Journal of membrane biology **153**:211–216.

Miller, S., Edwards, M. D., Ozdemir, C., and Booth, I. R. (2003) The closed structure of the MscS mechanosensitive channel - Cross-linking of single cysteine mutants. J Biol Chem **278**:32246–32250.

Minke, B. (2006) TRP channels and Ca^{+2} signaling. Cell calcium 40:261–275.
Moe, P., and Blount, P. (2005) Assessment of Potential Stimuli for Mechano-Dependent Gating of MscL: Effects of Pressure, Tension, and Lipid Headgroups. Biochemistry 44:12239–12244.
Moe, P. C., Blount, P., and Kung, C. (1998) Functional and structural conservation in the mechanosensitive channel MscL implicates elements crucial for mechanosensation. Molec Microbiol 28:583–592.
Moe, P. C., Levin, G., and Blount, P. (2000) Correlating a protein structure with function of a bacterial mechanosensitive channel. J Biol Chem 275:31121–31127.
Muller, E. M., Locke, E. G., and Cunningham, K. W. (2001) Differential regulation of two Ca(2+) influx systems by pheromone signaling in Saccharomyces cerevisiae. Genetics 159:1527–1538.
Muraki, K., Iwata, Y., Katanosaka, Y., Ito, T., Ohya, S., Shigekawa, M., and Imaizumi, Y. (2003) TRPV2 is a component of osmotically sensitive cation channels in murine aortic myocytes. Circ Res 93:829–838.
Nauli, S. M., Alenghat, F. J., Luo, Y., Williams, E., Vassilev, P., Li, X., Elia, A. E., Lu, W., Brown, E. M., Quinn, S. J., Ingber, D. E., and Zhou, J. (2003) Polycystins 1 and 2 mediate mechanosensation in the primary cilium of kidney cells. Nature genetics 33:129–137.
Niemeyer, M. I., Cid, L. P., Barros, L. F., and Sepulveda, F. V. (2001) Modulation of the two-pore domain acid-sensitive K+ channel TASK-2 (KCNK5) by changes in cell volume. J Biol Chem 276:43166–43174.
Nomura, T., Sokabe, M., and Yoshimura, K. (2006) Lipid-protein interaction of the MscS mechanosensitive channel examined by scanning mutagenesis. Biophys J 91:2874–2881.
Numata, T., Shimizu, T., and Okada, Y. (2006) TRPM7 is a stretch- and swelling-activated cation channel involved in volume regulation in human epithelial cells. Am J Physiol Cell Physiol.
Okada, K., Moe, P. C., and Blount, P. (2002) Functional design of bacterial mechanosensitive channels. Comparisons and contrasts illuminated by random mutagenesis. J Biol Chem 277:27682–27688.
Opsahl, L. R., and Webb, W. W. (1994) Transduction of membrane tension by the ion channel alamethicin. Biophys J 66:71–74.
Ozdirekcan, S., Rijkers, D. T., Liskamp, R. M., and Killian, J. A. (2005) Influence of flanking residues on tilt and rotation angles of transmembrane peptides in lipid bilayers. A solid-state 2H NMR study. Biochemistry 44:1004–1012.
Pacha, J., Frindt, G., Sackin, H., and Palmer, L. G. (1991) Apical maxi K channels in intercalated cells of CCT. The American journal of physiology 261:F696–705.
Palmer, C. P., Batiza, A., Zhou, X. L., Loukin, S. H., Saimi, Y., and Kung, C. (2004). Ion channels of microbes. In Cell signalling in prokaryotes and lower metazoa, I. Fairweather, ed. (Lkuver Academic Publishers), pp. 325–345.
Paoletti, P., and Ascher, P. (1994) Mechanosensitivity of NMDA receptors in cultured mouse central neurons. Neuron 13:645–655.
Park, K. H., Berrier, C., Martinac, B., and Ghazi, A. (2004) Purification and functional reconstitution of N- and C-halves of the MscL channel. Biophys J 86:2129–2136.
Patel, A., and Honoré, E. (2002) The TREK two P domain K^{+} channels. J Physiol 539.3:647.
Patel, A., Honore, E., Lesage, F., Fink, M., Romey, G., and Lazdunski, M. (1999) Inhalational anesthetics activate two-pore-domain background K+ channels. Nature Neuroscience 2:422–426.
Patel, A. J., and Honore, E. (2001) Molecular physiology of oxygen-sensitive potassium channels. European Respiratory Journal 18:221–227.
Patel, A. J., Lazdunski, M., and Honoré, E. (2001) Lipid and mechano-gated 2P domain K^{+} channels. Current Opinion in Cell Biology 13:422–427.
Peier, A. M., Moqrich, A., Hergarden, A. C., Reeve, A. J., Andersson, D. A., Story, G. M., Earley, T. J., Dragoni, I., McIntyre, P., Bevan, S., and Patapoutian, A. (2002) A TRP channel that senses cold stimuli and menthol. Cell 108:705–715.
Perozo, E., Cortes, D. M., Sompornpisut, P., Kloda, A., and Martinac, B. (2002a) Open channel structure of MscL and the gating mechanism of mechanosensitive channels. Nature 418:942–948.

Perozo, E., Kloda, A., Cortes, D. M., and Martinac, B. (2001) Site-directed spin-labeling analysis of reconstituted Mscl in the closed state. J Gen Physiol 118:193–206.
Perozo, E., Kloda, A., Cortes, D. M., and Martinac, B. (2002b) Physical principles underlying the transduction of bilayer deformation forces during mechanosensitive channel gating. Nature Struct Biol 9:696–703.
Pivetti, C. D., Yen, M. R., Miller, S., Busch, W., Tseng, Y. H., Booth, I. R., and Saier, M. H. (2003) Two families of mechanosensitive channel proteins. Microbiol Mol Biol R 67:66–85.
Poolman, B., Blount, P., Folgering, J. H., Friesen, R. H., Moe, P. C., and van der Heide, T. (2002) How do membrane proteins sense water stress? Molec Microbiol 44:889–902.
Powl, A. M., East, J. M., and Lee, A. G. (2005a) Heterogeneity in the Binding of Lipid Molecules to the Surface of a Membrane Protein: Hot Spots for Anionic Lipids on the Mechanosensitive Channel of Large Conductance MscL and Effects on Conformation. Biochemistry 44:5873–5883.
Powl, A. M., Wright, J. N., East, J. M., and Lee, A. G. (2005b) Identification of the Hydrophobic Thickness of a Membrane Protein Using Fluorescence Spectroscopy: Studies with the Mechanosensitive Channel MscL(,)(1). Biochemistry 44:5713–5721.
Qi, Z., Chi, S., Su, X., Naruse, K., and Sokabe, M. (2005) Activation of a mechanosensitive BK channel by membrane stress created with amphipaths. Mol Membr Biol 22: 519–527.
Rajan, S., Wischmeyer, E., Karschin, C., Preisig-Muller, R., Grzeschik, K. H., Daut, J., Karschin, A., and Derst, C. (2001) THIK-1 and THIK-2, a novel subfamily of tandem pore domain K+ channels. J Biol Chem 276:7302–7311.
Rajan, S., Wischmeyer, E., Xin Liu, G., Preisig-Muller, R., Daut, J., Karschin, A., and Derst, C. (2000) TASK-3, a novel tandem pore domain acid-sensitive K+ channel. An extracellular histiding as pH sensor. J Biol Chem 275:16650–16657.
Ricci, A. J., Kachar, B., Gale, J., and Van Netten, S. M. (2006) Mechano-electrical transduction: new insights into old ideas. The Journal of membrane biology 209:71–88.
Sachs, F. (1986) Biophysics of mechanoreception. Membrane biochemistry 6:173–195.
Saint, N., Lacapere, J. J., Gu, L. Q., Ghazi, A., Martinac, B., and Rigaud, J. L. (1998) A hexameric transmembrane pore revealed by two-dimensional crystallization of the large mechanosensitive ion channel (MscL) of Escherichia coli. J Biol Chem 273: 14667–14670.
Schumann, U., Edwards, M. D., Li, C., and Booth, I. R. (2004) The conserved carboxyl-terminus of the MscS mechanosensitive channel is not essential by increases stability and activity. FEBS Lett (in press).
Sharif Naeini, R., Witty, M. F., Seguela, P., and Bourque, C. W. (2006) An N-terminal variant of Trpv1 channel is required for osmosensory transduction. Nat Neurosci 9:93–98.
Sheetz, M. P., Painter, R. G., and Singer, S. J. (1976) Biological membranes as bilayer couples. III. Compensatory shape changes induced in membranes. J Cell Biol 70:193–203.
Smith, G. D., Gunthorpe, M. J., Kelsell, R. E., Hayes, P. D., Reilly, P., Facer, P., Wright, J. E., Jerman, J. C., Walhin, J. P., Ooi, L., Egerton, J., Charles, K. J., Smart, D., Randall, A. D., Anand, P., and Davis, J. B. (2002) TRPV3 is a temperature-sensitive vanilloid receptor-like protein. Nature 418:186–190.
Sokabe, M., Naruse, K., and Qiong-Yao, T. (2004) A new mechanosensitive channel SAKCA and a new MS channel blocker GsTMx-4. Nippon yakurigaku zasshi 124:301–310.
Sotomayor, M., Vasquez, V., Perozo, E., and Schulten, K. (2006) Ion Conduction through MscS as Determined by Electrophysiology and Simulation. Biophys J.
Spassova, M. A., Hewavitharana, T., Xu, W., Soboloff, J., and Gill, D. L. (2006) A common mechanism underlies stretch activation and receptor activation of TRPC6 channels. Proc Natl Acad Sci U S A 103:16586–16591.
Strotmann, R., Harteneck, C., Nunnenmacher, K., Schultz, G., and Plant, T. (2000) OTRPC4, a nonselective cation channel that confers sensitivity to extracellular osmolarity. Nature Cell Biology 2:695–702.

Sukharev, S. (2002) Purification of the small mechanosensitive channel of *Escherichia coli* (MscS): the subunit structure, conduction, and gating characteristics in liposomes. Biophys J 83:290–298.

Sukharev, S., Betanzos, M., Chiang, C., and Guy, H. (2001a) The gating mechanism of the large mechanosensitive channel MscL. Nature 409:720–724.

Sukharev, S., Durell, S., and Guy, H. (2001b) Structural models of the MscL gating mechanism. Biophys J 81:917–936.

Sukharev, S. I., Blount, P., Martinac, B., Blattner, F. R., and Kung, C. (1994a) A large-conductance mechanosensitive channel in *E. coli* encoded by *mscL* alone. Nature 368:265–268.

Sukharev, S. I., Martinac, B., Arshavsky, V. Y., and Kung, C. (1993) Two types of mechanosensitive channels in the *Escherichia coli* cell envelope: solubilization and functional reconstitution. Biophys J 65:177–183.

Sukharev, S. I., Martinac, B., Blount, P., and Kung, C. (1994b) Functional reconstitution as an assay for biochemical isolation of channel proteins: Application to the molecular identification of a bacterial mechanosensitive channel. Methods: A Companion to Methods in Enzymology 6:51–59.

Sukharev, S. I., Schroeder, M. J., and McCaslin, D. R. (1999a) Stoichiometry of the large conductance bacterial mechanosensitive channel of E. coli. A biochemical study. J Membr Biol 171:183–193.

Sukharev, S. I., Sigurdson, W. J., Kung, C., and Sachs, F. (1999b) Energetic and spatial parameters for gating of the bacterial large conductance mechanosensitive channel, MscL. J Gen Physiol 113:525–540.

Talley, E. M., and Bayliss, D. A. (2002) Modulation of TASK-1 (Kcnk3) and TASK-3 (Kcnk9) potassium channels: volatile anesthetics and neurotransmitters share a molecular site of action. J Biol Chem 277:17733–17742.

Talley, E. M., Lei, Q., Sirois, J. E., and Bayliss, D. A. (2000) TASK-1, a two-pore domain K+ channel, is modulated by multiple neurotransmitters in motoneurons. Neuron 25:399–410.

Tang, Q. Y., Qi, Z., Naruse, K., and Sokabe, M. (2003) Characterization of a functionally expressed stretch-activated BKca channel cloned from chick ventricular myocytes. The Journal of membrane biology 196:185–200.

Tang, Y. W., Cao, G., Chen, X., Yoo, J., Yethiraj, A., and Cui, Q. (2006) A finite element framework for studying the mechanical response of macromolecules: Application to the gating of the mechanosensitive channel MscL. Biophys J.

Tavernarakis, N., and Driscoll, M. (1997) Molecular modeling of mechanotransduction in the nematode *Caenorhabditis elegans*. Ann Rev Physiol 59:659–689.

Tsapis, A., and Kepes, A. (1977) Transient breakdown of the permeability barrier of the membrane of *Escherichia coli* upon hypoosmotic shock. Biochim Biophys Acta 469:1–12.

Vriens, J., Watanabe, H., Janssens, A., Droogmans, G., Voets, T., and Nilius, B. (2004) Cell swelling, heat, and chemical agonists use distinct pathways for the activation of the cation channel TRPV4. Proc Natl Acad Sci U S A 101:396–401.

Wiggins, P., and Phillips, R. (2005) Membrane-protein interactions in mechanosensitive channels. Biophys J 88:880–902.

Wood, J. M. (2006) Osmosensing by bacteria. Sci STKE 2006:pe43.

Xu, H., Ramsey, I. S., Kotecha, S. A., Moran, M. M., Chong, J. A., Lawson, D., Ge, P., Lilly, J., Silos-Santiago, I., Xie, Y., DiStefano, P. S., Curtis, R., and Clapham, D. E. (2002) TRPV3 is a calcium-permeable temperature-sensitive cation channel. Nature 418:181–186.

Yoshimura, K., Nomura, T., and Sokabe, M. (2004) Loss-of-function mutations at the rim of the funnel of mechanosensitive channel MscL. Biophys J 86:2113–2120.

Zhang, Y., Gao, F., Popov, V. L., Wen, J. W., and Hamill, O. P. (2000) Mechanically gated channel activity in cytoskeleton-deficient plasma membrane blebs and vesicles from Xenopus oocytes. The Journal of physiology 523 Pt 1:117–130.

Chapter 4
Computational Studies of the Bacterial Mechanosensitive Channels

Ben Corry and Boris Martinac

Abstract Bacterial mechanosensitive (MS) channels were first documented in giant spheroplasts of *Escherichia coli* during a survey of the bacterial cell membrane by the patch clamp some twenty years ago. Two major events that greatly advanced and kept the research on bacterial MS channels at the forefront of the MS channel research field include: (i) cloning of MscL and MscS, the MS channels of Large and Small conductance, and (ii) solving their 3D crystal structure. In addition to advancing further experimental studies of the bacterial MS channels by enabling the use of new techniques, such as EPR and FRET spectroscopy, these events also enabled theoretical approaches to be employed. In this chapter we will review recent computational approaches used to elucidate the molecular dynamics of MscL and MscS, which has significantly contributed to our understanding of basic physical principles of the mechanosensory transduction in living organisms.

Key words: MS channels · Patch clamp · Bilayer model · Mechanosensory transduction · EPR spectroscopy · FRET · Molecular dynamics · Brownian dynamics

4.1 Introduction

Studies of mechanosensitive (MS) channels in bacteria were facilitated by the advent of the patch-clamp technique (Hamill et al., 1981), which has removed the constraints of being able to study electrophysiologically only cells that are large enough to be impaled with glass microelectrodes. Development of a "giant spheroplast" preparation, which was essential for the examination of a bacterial cell membrane by the patch clamp technique (Ruthe and Adler, 1985; Martinac et al., 1987), led to the discovery of MscS and MscL, two of the several types of MS channels existing in bacteria (Berrier et al., 1996; Martinac, 2001; Blount et al., 2007). Both channels were cloned several years after their discovery (Sukharev et al., 1994; Levina et al., 1999). Given that bacteria can be grown in large quantities delivering milligram amounts of channel proteins, 3D structures of both MscL (Chang et al., 1998) and MscS (Bass et al., 2002) became available just a few years after their cloning (Fig. 4.1). This opened a window of opportunities to conduct studies relating

A. Kamkin and I. Kiseleva (eds.), *Mechanosensitive Ion Channels.*

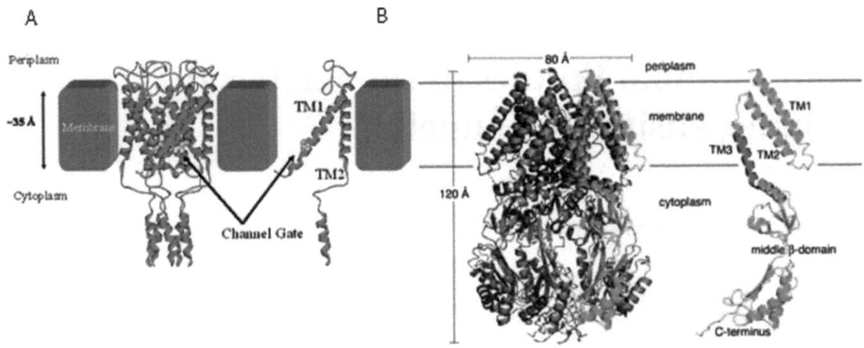

Fig. 4.1 3D crystal structure of MscL and MscS. **(A)** MscL channel homopentamer (left) and a channel monomer (right) from *M. tuberculosis* (Chang et al., 1998). The thickness of the membrane bilayer (shown as solid blocks) is approx. 3.5 nm. The channel gate is formed by a group of amino acids at the cytoplasmic end of the TM1 transmembrane domain (modified from Oakley et al., 1999). Figure based on model 1MSL in Protein Data Bank (http://www.rcsb.org/pdb). **(B)** The crystal structure at 3.9 Å resolution of MscS from E. coli showing the channel homoheptamer (left) and a monomer (right) (Bass et al., 2002) viewed by PyMol19. Residues 27 to 280 were resolved. Secondary structural domains and the position of the TM3 transmembrane helix are indicated in the diagram of the monomer. A conserved structural motif of glycine and alanine residues in the pore-lining transmembrane helix TM3 essential for MscS gating is highlighted in red (reproduced from Martinac, 2005)

the structure and function of these MS channels by employing a whole array of structural methods and techniques including 2D electron crystallography (Saint et al., 1998), electronparamagnetic resonance (EPR) spectroscopy (Perozo et al., 2001; Perozo et al., 2002a, 2002b; Tsai et al., 2005), fluorescence resonance energy transfer (FRET) spectroscopy (Corry et al., 2005) as well as computational modelling based on molecular dynamics (Gullingsrud et al., 2001; Gullingsrud and Schulten, 2003; Sotomayor and Schulten, 2004; Sotomayor et al., 2006; Meyer et al., 2006) and Brownian dynamics (Vora et al., 2006). Together, all these approaches have provided a wealth of information on the structural determinants of gating and conduction of the two MS channels. In this review we have summarized the key theoretical findings on the structure and function of MscL and MscS. See the Chapter by Paul Blount and colleagues in this volume for general information on bacterial MS channels not covered here.

4.2 Molecular Dynamics Studies of MscL

The determination of the 3D structure of the MscL homologue from *Mycobacterium tuberculosis* (Tb-MscL) by X-ray crystallography showed that the MscL protein forms a homopentameric channel (Chang et al., 1998) (Fig. 4.1A). Each subunit of the pentamer consists of two transmembrane α-helices TM1 and TM2 and a third cytoplasmic α-helix. The TM1 transmembrane domain (residues 15–43) is connected to TM2 domain (residues 69–89) by a periplasmic loop which extends into the pore region and lines the external face of the channel. The TM2 helix is continued by a second loop (residues 90–101) leading to the cytoplasmic helix (residues 102–115) at the C-terminal end of the protein facing the cytoplasm. Each TM1 helix is in contact with two TM1 helices from adjacent subunits and two TM2 helices of the channel pentamer, one from the same subunit and the other from the neighboring subunit. The helices are slanted and the five TM1 helices join together at the cytoplasmic side of the pentamer (Fig. 4.1A). The ion conduction pathway of the MscL channel is formed by the TM1 transmembrane helices with their hydrophilic residues lining the channel pore. The pore is approximately 18 Å in diameter at the periplasmic side but narrows to an occluded apex at the cytoplasmic side (approximately 2Å in diameter). The tight constriction, which is believed to act as the channel gate, is formed by hydrophobic residues (5 Val and 5 Ile) that are highly conserved among bacterial members of the MscL family (Oakley et al., 1999). The open channel conductance of the pore, on the other hand, suggests a much wider channel in the order of 30Å in diameter which implies a very large conformational change upon channel gating (Cruickshank et al., 1997).

In order to conduct molecular simulations of an ion channel, it is crucial to have atomic resolution structures of the pore. Thus, a large number of simulation studies aimed primarily at unravelling the gating pathway of the channel and how the protein detects changes in membrane tension or composition have emerged since the publication of the Tb-MscL crystal structure. Bacterial MS channels directly sense tension developed in the bilayer alone (Markin & Martinac 1991), and this mechanosensitivity is maintained when the channel has been reconstituted into artificial liposomes in which no other proteins or cytoskeleton are present (Perozo et al, 2002; Delcour et al, 1989; Berrier et al 1989; Häse et al, 1995). In particular, evidence suggests that MS channels respond only to tension in the plane of the membrane and not that perpendicular to it (Gustin et al, 1998; Sokabe & Sachs 1990). The two main questions that computational studies have aimed to elucidate, therefore, is how the lateral tension in the membrane is sensed and transformed into structural changes, and what are the conformational changes involved in moving from the closed to open states of the channel.

MS channel proteins could in principle sense bilayer tension by interacting with the membrane in a number of ways. These interactions could simply be Van der Waals contacts between the non-polar regions of the protein and lipid, or could rely on specific hydrogen bonds forming between parts of the protein and the lipid head groups. EPR studies of MscL demonstrated that hydrophobic mismatch, although not the driving force that triggers MscL opening, could stabilize intermediate conformational states of the channel along the kinetic path towards the open state (Perozo et al., 2002a, 2002b). The same studies suggested further that addition of conical lipids, such as lysophosphatidylcholine (LPC), to one monolayer of liposomes reconstituted with MscL channels created local stresses in the lipid bilayer leading to redistribution of the transbilayer pressure profile sufficient to open the channel. MscL could sense these stresses and open in the absence of externally applied membrane tension.

A number of computational studies have been undertaken to help determine if the coupling between the membrane and the MscL protein is simply a result of van der Waals interactions and hydrophobic matching, or whether hydrogen bonds with the lipid headgroups are important. In an analytic study of the energetics of channel opening, Wiggins and Phillips (2004) proposed the competition between the hydrophobic mismatch of the protein and the applied membrane tension result in a bistable system representing the open and closed states of the channel. This model suggests that hydrophobic matching may be all that is required to induce gating within the protein. Furthermore it is able to make some predictions about how altering the composition of the lipids or protein can influence the hydrophobic contacts and thus alter the energetics of channel gating. Similarly the influence of membrane curvature can be examined within this framework.

Atomistic simulations have been used to help understand why changing the nature of the lipid headgroups or the length of the lipid chains alters the pressures required to initiate channel gating. Elmore and Dougherty (2001) examined the Tb-MscL in a POPE bilayer using molecular dynamics (MD) simulations. A very large number of hydrogen bonds were formed between the lipid headgroups and the protein, primarily with either the C-terminal domain or a localised region of the periplasmic loop suggesting that these portions of the protein may be important for sensing changes in membrane tension. Exchanging the POPE bilayer for POPC resulted in many less hydrogen bonds forming, which provides a possible mechanism for explaining the different gating tensions observed in membranes of different lipid composition (Elmore and Dougherty, 2003). Although positing hydrogen bonds with the lipid headgroups as being important for linking the protein to the membrane appears contrary to the notion that hydrophobic mismatch is the cause of channel gating it is possible that these two effects could reinforce each other. Both these links between the protein and membrane would be affected in complimentary ways by membrane thinning. The more specific nature of the hydrogen bonds, however, may impose more conformational restrictions upon the protein during gating than would generic hydrophobic contacts.

An alternative explanation for the influence of lipid composition on MscL gating has also been presented, in which the differing pressure profiles of the lipids could influence the forces exerted upon the MscL protein. Gullingsrud and Schulten used MD simulations to determine the lipid bilayer pressure profiles in POPE and POPC membranes (Gullingsrud and Schulten, 2004). A non uniform pressure profile exists across the membrane as the hydrophilic headgroups are squeezed together to avoid water contacting the hydrophobic tails while maintaining a nearly constant volume in the lipid. In POPC bilayers the lateral pressure is concentrated near the headgroups, but stretching the membrane concentrates the pressure closer to the start of the aliphatic chain, a phenomenon that also arises when switching to POPE. Furthermore, if the length of the lipid tails is shortened in MD simulations this results in a thinning of the protein created by pore constriction and helix kinking to avoid a hydrophobic mismatch (Elmore and Dougherty, 2003).

In parallel to the studies examining the protein-lipid interactions, a number of computational investigations have taken place to examine the conformational changes involved in gating. The closed state of the pore can be expected to be relatively stable, as the free energy difference between the closed and open states of the protein is estimated to be $\sim$17 kT (Martinac, 2001). Therefore, it is not surprising that when the protein is placed in a bilayer with no applied tension or force, the transmembrane helices all remain relatively static and the pore remains occluded in MD simulations (Colombo et al., 2003; Gullingsrud et al., 2001; Elmore and Dougherty, 2001). The extramembrane regions of the protein (C-terminal domain and periplasmic loop) do display more mobility, but it is not clear that these represent functional motions.

Large conformational changes can be observed in MD simulations, however, when the protein is influenced by an external force. In particular, it is hoped that external forces that mimic those presented to the protein by membrane tension may be able to initiate the large structural changes involved in gating within the nanosecond timescales achievable in molecular simulations. Many simulations have been carried out to investigate the nature and sequence of these conformational changes, differing most notably in the way that the external force is applied to the protein. Gullingsrud and Schulten applied surface tension to the bare Tb-MscL protein (ie not in a lipid bilayer) (Gullingsrud et al., 2001) as well as direct, predominantly radial forces to selected regions of the Eco-MscL protein that would contact the lipid to mimic the membrane pressure profiles (Gullingsrud and Schulten, 2003). Bilston and Mylvaganam (2002) examined direct force application to the outer TM2 helix and anisotropic pressure coupling also to a bare Tb-MscL protein. Colombo et al (2003) applied pressure to the membrane around Tb-MscL, and finally Kong et al. (2002) used targeted molecular dynamics to force a conformational change of Eco-MscL from the crystal structure to a pre-determined end point. Not surprisingly, the different methods of force application resulted in differing conformational changes, although some common features are evident.

In all cases the application of force resulted in a flattening of the protein in the direction of the pore axis due to an increased tilt in the TM1 and TM2 helices, an increase in the in-plane area of the protein, as well as a shortening of the pore, all of which are in agreement with EPR studies (Perozo et al., 2002b) and results obtained by engineering disulfide bonds between subunits (Betanzos et al., 2002). This type of motion reproduces the models that suggest an iris like opening of the pore. In most cases some degree of opening of the pore constriction was observed during the simulations, provided that the strength of the applied force or pressure was above a critical point. In these cases water enters the entire length of the pore. The force or pressure required to observe this was always higher than that used experimentally to open the pore otherwise conformational changes would not occur during the short length of the simulations (typically <10ns).

Given the strength of the forces used in these molecular simulations, some caution must be taken in assuming that the conformational changes seen represent those taking place in reality. The influence of the external force on the structural changes observed can be seen from the fact that the channel opening events and their ordering differ among the various simulations. The first study of Gullingsrud et al. (2001) showed two distinct conformational changes: a widening of the extracellular end of the pore created by a retraction of the periplasmic loop, followed by a simultaneous shift in the angle of the TM1 and TM2 helices. Their second study, however, did not show much movement by the periplasmic loop (Gullingsrud and Schulten, 2003). Instead, the initial motion was a widening of the cytoplasmic end caused by a tilting of the transmembrane helices, followed by a simultaneous expansion of both the TM1 and TM2 helices that forced the hydrophobic constriction apart. Colombo et al (2003) saw similar helix tilting, but noted that the TM1 helix only moved after the TM1 had made space for it and also that the tilt of the TM2 helix was greater than that of the TM1 helix unlike in the simulations of Gullingsrud et al. (2001).

Due to the short length of the simulations employed, none of the studies discussed so far claim to witness the entire gating process. Kong et al on the other hand (Kong et al., 2002) do reach a final open state due to the targeted nature of their simulations in which they force the simulation to end at a pre-determined open state. Some caution must be used, however, in interpreting the sequence of opening events seen in this study. As the driving force in this simulation is the distance of the atoms from their final location, the resulting order of events is likely to favor large changes before small ones and may not reproduce the sequence undergone in reality. Furthermore, the structural changes will be highly dependent upon the conformation chosen to represent the final open state given that there is no detailed experimental evidence to support this choice.

The radial forces applied in all these studies are used to speed up conformational changes that may be associated with gating, but it is possible that such forces bias protein motions. For example, radial movements may take place sufficiently quickly such that movements in other directions do not have time to take place. To avoid these difficulties, Meyer et al. (2006) examined the conformational changes displayed in MD simulations of Eco-MscL embedded in a curved bilayer composed of single and double tailed lipids without the application of any external force (Fig. 4.2). This is meant to reproduce the conditions seen in patch clamp experiments in which the bilayer is bent through suction, although the degree of bending is again necessarily more in the simulation than in reality. Although the conformational changes seen in this 9.5ns simulation were not as large as in some of those using external forces and channel opening was not seen, some interesting motions possibly associated with channel gating were observed. In particular, the

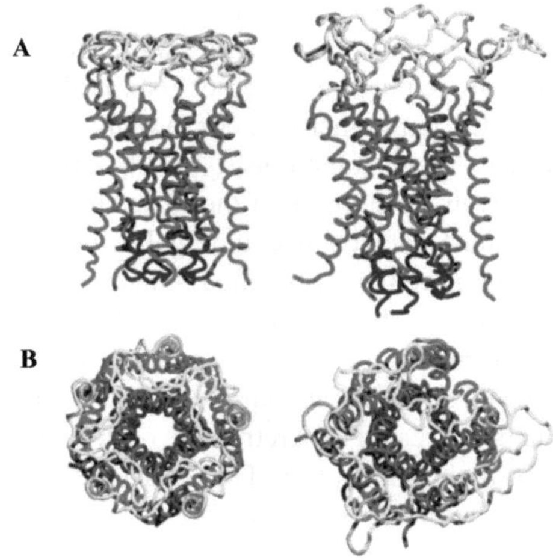

Fig. 4.2 Conformational changes of MscL in a curved bilayer studied by molecular dynamics simulation. **(A)** Side view and **(B)** top view of MscL at the beginning (left) and end (right) of the 9.5ns simulation. The C-terminus has been cut-off in the simulation and is not shown. (Reproduced from Meyer et al., 2006).

periplasmic loop was seen to be particularly mobile, something that may have been overlooked in previous simulations. There is evidence that this loop is important in channel gating as cutting it with proteases (Ajouz et al., 2000) or reconstituting protein without this connection between the C- and N-terminal halves of the protein (Park et al., 2004) resulted in functional channels with increased pressure sensitivity. This loop may possibly have to move first to allow the transmembrane helices to adjust to pressure.

Two alternatives to direct simulations of the protein have also been utilised to witness channel openings: normal mode analysis in which local fluctuations of the protein about a minimum energy conformation are extrapolated to show large scale conformational changes, and continuum models in which a non-atomistic description of the system is used. Some common motions were extracted from a normal mode analysis starting from a number of different closed and open structures (Valadie et al., 2003). Most notable was a twist and tilt of the helices, in particular TM1, as suggested previously. Notably the top half of the TM2 helix was seen to be less mobile than the bottom which resulted in a kinking of this helix, a property that had also been suggested in some of the MD studies (Meyer et al., 2006; Kong et al., 2002). Another result in common with previous studies included the lack of interaction between the C-terminus and the remainder of the protein and the lack of motion seen in the central gating residues. In order to witness large scale motions in a short time span, Tang et al (2006) developed a continuum modelling approach in which the lipid bilayer and protein are represented as elastic sheets and rods rather than as atomistic structures. Using this simplified model a widening of the pore is observed upon membrane stretching created by helix tilting and radial movement caused by a thinning of the membrane, but no significant opening was seen with membrane bending.

4.3 Computational Studies of MscS

The mechanosensitive channel of small conductance, MscS, is an archetypal mechanosensor. Unlike MscL proteins, which are almost exclusively found in bacterial cells, MscS-like proteins are widely spread among Gram-negative and Gram-positive bacteria, archaea, fungi and plants where they function in cellular processes underlying osmoregulation and growth (Martinac, 2006). They constitute a large sub-family of MS channels (Martinac and Kloda, 2003; Pivetti et al., 2003).

Shortly after MscS was cloned (Levina et al., 1999) its structure was solved by X-ray crystallography at a resolution of 3.9 Å (Bass et al., 2002). MscS forms a homoheptameric channel with a diameter of 80 Å and a length of 120 Å (Fig. 4.1B). Each subunit contains three transmembrane domains, with N-termini facing the periplasm and large C-termini extending into the cytoplasm. The transmembrane helices TM1 and TM2 are considered to constitute the sensors for membrane tension and voltage (Bass et al., 2002; Bezanilla & Perozo, 2002), whereas TM3 helices line the channel pore and facilitate the channel opening by slight iris-like rotations and tilting (Edwards et al., 2005) and possibly unkinking (Sotomayor et al, 2007).

A Gly-Ala pattern in this helix faces the pore and appears to be an important structural motif supported by the fact that it is highly conserved in the MscS sub-family of MS channel proteins (Martinac and Kloda, 2003; Pivetti et al., 2003).

Computational studies have hoped to help answer a number of unresolved questions relating to this channel. The first of these is surprisingly simple: what is the conduction state of the crystallized protein? It is obvious from the published structure that there is a large non-occluded pore passing across the membrane (Fig. 4.3) indicative of an open ion conductive pore. Noting this, Bass et al (2002) suggested that the protein has most likely been imaged in an open state. This is somewhat surprising, as the closed state of MscS is usually favoured by ~7 kcal/mol in membranes in the absence of lateral tension, but as the protein was crystallized from detergent micelles rather than in lipid this may have altered the resting state of the channel. Closer inspection of the structure, however, indicates that the narrowest region of the pore has a diameter of only ~3.5 Å, and is surrounded by two rings of non-polar leucine residues (L105 and L109). It has been suggested that such narrow hydrophobic pores can prevent the passage of ions without presenting a physical occlusion.

One way to try and ascertain the conduction state of the imaged structure is to try to determine the conductance of the protein directly from simulations and compare these to known experimental values. Although this approach is appealing, it suffers one major problem: it is computationally very demanding to conduct atomistic simulations for long enough to witness many ions passing through the pore as is required to accurately determine the channel conductance. To overcome these problems, mesoscopic simulation techniques have often been employed in which some of the atoms in the system are not treated explicitly in order to reduce the

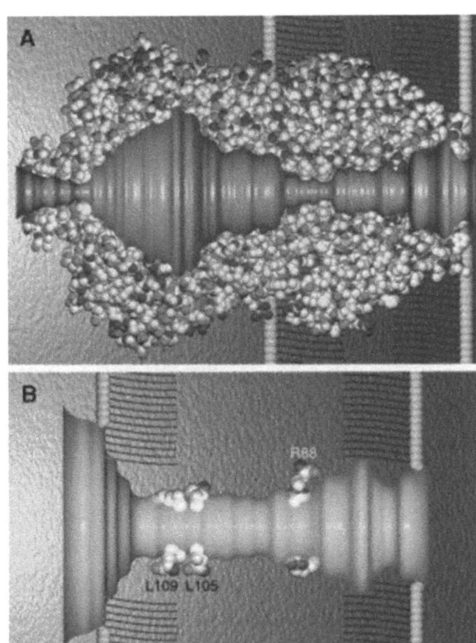

Fig. 4.3 MscS channel model used for Brownian dynamics simulations of the conduction properties of the channel. (A) Schematic of the MscS crystal structure with the top half removed to reveal the pore. **(B)** Schematic of the transmembrane region of an expanded pore structure. The highlighted residues, starting from the intracellular side, are residues L109 and L105 that line the expanded part of the channel, and R88, the only charged amino-acid residue lining the pore. (Reproduced from Vora et al., 2006)

computational demand and allow the current to be determined explicitly. In two such studies of the MscS channel (Sotomayor et al., 2006; Vora et al., 2006), the protein, lipid and water are treated as continuous dielectric media (Fig. 4.3). Ions are treated explicitly and move under the electrostatic force induced by the membrane potential, partial charges of the protein atoms and other ions, as well as experiencing scattering and frictional forces to replicate the collisions with water molecules or the protein boundary. Both studies showed that when the protein is held in its crystal structure the current (carried almost exclusively by anions) is much below recorded single open channel currents, supporting the notion that this does not represent the open state of the pore. Simulations using widened structures demonstrate that only minimal structural changes are required to create a highly conductive pore (Sotomayor et al., 2006; Vora et al., 2006).

Recently, all atom molecular dynamics simulations have also been used to estimate the conductance of the pore. Spronk et al (2006) witnessed ionic currents in their MD simulations of the MscS pore under large applied potentials and Sotomayer et. al. (2007) also measured currents in a range of situations. Some caution should be applied when directly interpreting the conductance measured in these high voltage simulations as most channel I-V curves become non-linear under large applied potentials, but the picture emerging is consistent with other results. When the protein is held constrained near the imaged structure, the conductance is much smaller (~200ps) than that measured experimentally for the open state of the channel (~1000ps) (Sotomayor et al. 2007). Furthermore, the current is carried almost exclusively by anions, rather than being only slightly anion selective as is found in patch clamp recordings (Martinac et al., 1987; Sukharev et al., 1993). Together these results support the idea that the pore has been crystallized in neither the fully open nor fully closed states, and more likely represents some intermediate or inactive configuration.

The likely conductance state of the crystallized structure can also be probed by a less direct method that examines the behaviour of water in the pore. A number of molecular dynamics investigations of model pores (Beckstein et al., 2001; Beckstein and Sansom, 2003; Beckstein and Sansom, 2004), carbon nanotubes (Hummer et al., 2001; Wan et al., 2005), and the nicotinic acetylcholine receptor (Corry, 2006; Beckstein and Sansom, 2006) have shown that water tends to evacuate the hydrophobic regions of such pores under certain conditions. In particular, if the radius of the pore is under a certain critical value that depends on its particular geometry and surface character, then the pore is often empty. But, if the radius is increased or a more polar surface is introduced then water will fully hydrate the channel. This critical radius is around 4–4.5 Å for the model pores (Beckstein and Sansom, 2004) and nicotinic receptor (Corry, 2006), roughly the radius required to allow three shells of water to enter at which stage the average number of hydrogen bonds per water molecule can approximate that found in the bulk (Anishkin and Sukharev, 2004). If water evacuates a region of a pore it is unlikely that an ion will traverse through it. In these conditions the ion will have to be stripped of most of its hydration shell (some waters may be pulled through the channel with the ion) without being able to compensate this energy with coordination to any polar groups in the protein as is the case, for example, in the selectivity filter of potassium channels.

As the radius of the hydrophobic constriction in the MscS pore lies below the critical radius for hydration seen for other pores, it is possible that a similar hydrophobic gating mechanism is at play in MscS. Initial molecular dynamics studies of this pore indicated that water does indeed evacuate the constricted region of the MscS pore when it is held constrained near the crystal structure (Sotomayor and Schulten, 2004; Anishkin and Sukharev, 2004). Adding a polar gain of function mutation L109S at this constriction leads to stable hydration of the pore and significantly reduces the barrier to ion permeation (Anishkin and Sukharev, 2004), demonstrating that a change in the surface character of the pore can significantly alter its characteristics.

If an evacuated pore can be equated with a non-conducting state of the channel, then these results support the notion that the crystal structure of the protein represents a non- or low- conducting state. A further complication has emerged from a more recent molecular dynamics study that showed the evacuation of the narrow region of the pore seen in earlier simulations is voltage dependent (Spronk et al., 2006), a result similar to that seen in model hydrophobic pores (Dzubiella et al. 2004). This result has significant implications, suggesting that application of a large enough membrane potential may result in leakage currents in hydrophobically gated pores. It should be noted, however, that a large electric field was required to observe these results. When a more modest fields of 20–50 mV/nm (transmembrane potential ~220–550 mV) were applied the pore did not remain fully hydrated suggesting that the MscS pore could still use a hydrophobic gating mechanism to prevent ion conduction under physiological conditions.

If the imaged protein structure is neither in the open or closed states an obvious question that simulations can help to answer is exactly what the open and closed states look like and how do they differ from the crystallized structure. When no harmonic restraints are applied to the protein in MD simulations to keep it near the imaged structure, the pore has been seen to constrict and become physically occluded (Sotomayor and Schulten 2004; Spronk et al. 2006; Sotomayor et al. 2007). These results support the idea that the imaged state of the channel does not represent the closed state of the pore. In addition, they cast some doubt on whether the pore uses a hydrophobic gating mechanism in which ions are prevented from passing through a non-occluded pore as described previously, as this would not be necessary if the closed state of the pore contains a physical occlusion.

As noted previously, mesoscopic simulations indicate that only minimal widening of the pore is required to greatly increase the channel conductance (Sotomayor et al 2006; Vora et al 2006). Similarly, if the hydrophobic model of channel gating is to be believed, then evidence from other pores would suggest only minor conformational changes are required to open the pore. Such conformational changes have been observed in a number of MD simulations. When a large tension (20 dyn/cm, i.e. 20 mN/m) was applied to the membrane, the pore was seen to widen and remain fully hydrated (Sotomayor and Schulten, 2004). More surprisingly, when a large electric field (~1V) was applied across the membrane in the absence of any constraints, the pore was seen to widen significantly (Spronk et al 2006;

Sotomayor et al 2007), presumably related to the ability of water and ions to enter the hydrophobic region of the pore. The widening seen in the most recent study (Sotomayor et al 2007) has a direct impact on the channel conductance determined directly from MD simulations, yielding a value of 900–1100 ps in agreement with measured open channel values (Martinac et al., 1987; Sukharev et al., 1993). Thus, the widened pore witnessed in this study may be representative of the open state of the channel.

The conformation changes involved in gating MscS appear to be much smaller than those required to open MscL, which is not entirely surprising given the difference in currents carried by the two pores. Simulations in which the pore becomes physically constricted suggest a possible location of an occlusion gate. The pore lining TM3 helix in one of the seven subunits buckles slightly, moving Leu105 toward the center of the pore (Sotomayor and Schulten, 2004; Spronk et al., 2006). Salt bridges have also been noted between Asp62 in the TM1-TM2 linker and Arg128 at the cytoplasmic end of the TM3 helix on the adjacent subunit (Sotomayor and Schulten, 2004). The widening of the pore obtained under an applied field proceeded with a straightening of the TM3 helix to remove a kink midway through the membrane and establish contact between this helix and TM2 (Sotomayor et al 2007). Interactions between the lipid headgroups with a number of charged and polar residues in the TM1 and TM2 helices have been noted (Sotomayor and Schulten, 2004) that may provide a mechanism of linking membrane tension to channel opening. It has been suggested that this link between the lipid contacting TM1-TM2, cytoplasmic domain and pore lining TM3 may be important for gating or stability although no clear conclusions have been drawn.

The crystallized structure of MscS contains a very large cytoplasmic domain (Fig 4.1B) whose function remains unclear. It extends a long way from the membrane and contains a large central chamber accessible through seven side openings and one distal pore. This domain has been suggested to play a role in channel gating (Koprowski and Kubalski, 2003), desensitization and stability (Miller et al., 2003; Schuman et al., 2004) as well as in ion transport and selectivity (Edwards et al., 2004). Although the seven side entries to the chamber are surrounded by a number of basic amino acid residues that could play a role in valence selectivity, both K^+ and Cl^- ions have been seen to pass through them and enter the cytoplasmic chamber in MD (Sotomayor and Schulten, 2004; Sotomayor et al., 2007) and mesoscopic simulations (Sotomayor et al., 2006; Vora et al., 2006), but no ions pass through the distal opening. The distribution of ions within the chamber, however, is such that more Cl^- than K^+ ions reside near the transmembrane pore mouth, thus it is possible that this domain does play an important role in ionic selectivity. However, anion selectivity is also seen in simulations of only the transmembrane portion of the protein, and is largely controlled by the charged residue Lys169 (Vora et al., 2006; Sotomayor et al., 2007). It is possible that this residue is not normally charged, which may explain the extra anion selectivity seen in simulations compared with experiments. The conformation of the cytoplasmic domain has also been shown to influence the conductivity of the pore (Sotomayor et al., 2007). It has also been suggested that this domain may act as a molecular sieve (Koprowski and Kubalski, 2003), preventing large solutes from passing through the channel where they may either block it, or be expelled undesirably from the cell. Further molecular simulations including solutes may shed further light on this issue.

4.4 Conclusions

This review provides a brief summary on theoretical approaches used to model molecular dynamics and conduction properties of MscL and MscS, the two most extensively studied MS channels. Computational modeling of the two channels outlined here has significantly contributed to our understanding of basic principles of mechanosensory transduction in living cells. In particular it has helped to elucidate the nature of the imaged structures of the protein, the interactions between the protein and the lipid, the conformational changes involved in channel gating, and the specific roles played by various protein domains. We may expect further exciting developments of this research area in the future.

References

Ajouz, B., Berrier, C., Besnard, M., Martinac, B. and Ghazi, A. (2000) Contributions of the different extramembraneous domains of the mechanosensitive ion channel MscL to its response to membrane tension. J. Biol. Chem. 275: 1015–1022.

Aksimentiev, A., Schulten, K. (2005) Imaging alpha-hemolysin with molecular dynamics: Ionic conductance, osmotic permeability and the electrostatic potential map. Biophys. J. 88: 3745–3761.

Anishkin, A., V. Gendel, N. A. Sharifi, C. S. Chiang, L. Shirinian, H. R. Guy and S. I. Sukharev (2003) On the conformation of the COOH-terminal domain of the large mechanosensitive channel MscL. J. Gen. Physiol. 121: 227–244.

Anishkin, A., Sukharev, S. (2004) Water dynamics and dewetting transitions in the small mechanosensitive channel MscS, Biophys. J. 86: 2883–2895.

Bass, R.B., Strop, P., Barclay, M., Rees, D. (2002) Crystal structure of Escherichia coli MscS, a voltage-modulated and mechanosensitive channel. Science 298: 1582–1587.

Beckstein, O., Biggin, P.C., Sansom, M.S.P. (2001) A hydrophobic gating mechanism for nanopores, J. Phys. Chem. B. 105: 12902–12905.

Beckstein, O., Sansom, M.S.P. (2003) Liquid-vapor oscillations of water in hydrophobic nanopores, Proc. Natl. Acad. Sci. USA 100: 7063–7068.

Beckstein, O., Sansom, M.S.P. (2004) The influence of geometry, surface character, and flexibility on the permeation of ions and water through biological pores, Phys. Biol. 1: 42–52.

Beckstein, O., Sansom, M.S.P. (2006) A hydrophobic gate in an ion channel: the closed state of the nicotinic acetylcholine receptor, Phys. Biol. 3: 147–159.

Berrier, C., Besnard, M., Ajouz, B., Coulombe, A., Ghazi, A. (1996) Multiple mechano-sensitive ion channels from Escherichia coli, activated at different thresholds of applied pressure. J. Memb. Biol. 151: 175–187.

Betanzos, M., Chiang, C.-S., Guy, H.R., Sukharev, S. (2002) A large iris-like expansion of a mechanosensitive channel protein induced by membrane tension. Nature Struct. Biol. 9: 704–710.

Bilston, L.E., Mylvaganam, K. (2002) Molecular simulations of the large mechanosensitive channel (MscL) under mechanical loading, FEBS Lett., 512: 185–190.

Blount, P., Li, Y., Moe, P. C., and Iscla, I. (2007) Mechanosensitive channels gated by membrane tension: Bacteria and beyond. In: Mechanosensitive ion channels (a volume in the Mechanosensitivity in Cells and Tissues, Moscow Academia series), A. Kamkin, and I. Kiseleva, eds. (New York, Springer Press). (in press)

Chang, G., Spencer, R., Lee, A., Barclay, M., Rees, D. (1998) Structure of the MscL homologue from Mycobacterium tuberculosis: a gated mechanosensitive ion channel. Science 282: 2220–2226.

Colombo, G., Marrink, S.J., Mark, A.E. (2003) Simulation of MscL gating in a bilayer under stress, Biophys. J. 84: 2331–2337.

Corry, B., Rigby, P., Liu, Z.-W., Martinac, B. (2005) Conformational changes involved in MscL channel gating measured using FRET spectroscopy. Biophys. J. 89: L49-L51.

Corry, B. (2006) An energy-efficient gating mechanism in the acetylcholine receptor suggested by molecular and Brownian dynamics, Biophys. J. 90: 799–810.

Cruickshank, C.C., Minchin, R., Le Dain, A., Martinac, B. (1997) Estimation of the pore size of the large-conductance mechanosensitive ion channel of Escherichia coli. Biophys J 73: 1925–1931.

Dzubiella, J., Allen, R.J., Hansen, J.-P. (2004). Electric field-controlled water permeation coupled to ion transport through a nanopore. J. Chem. Phys. 120: 5001–5004.

Edwards, M.D., Booth, I.R., Miller, S. (2004) Gating the bacterial mechanosensitive channel: MscS a new paradigm? Curr., Opin. Microbiol. 7: 163–167.

Elmore, D.E., Dougherty, D.A. (2001) Molecular dynamics simulations of wild-type and mutant forms of the Mycobacterium tuberculosis MscL channel, Biophys. J. 81: 1345–1359.

Elmore, D.E., Dougherty, D.A. (2003) Investigating lipid composition effects on the mechanosensitive channel of large conductance (MscL) using molecular dynamics simulations, Biophys. J. 85: 1512–11524.

Gullingsrud, J., Kosztin, D., Schulten, K. (2001) Structural determinants of MscL gating studied by molecular dynamics simulations. Biophys. J. 80: 2074–2081.

Gullingsrud, J., Schulten, K. (2003) Gating of MscL studied by steered molecular dynamics, Biophys. J. 85: 2087–2099.

Gullingsrud, J., Schulten, K. (2004) Lipid bilayer pressure profiles and mechanosensitive channel gating, Biophys. J. 86: 3496–3509.

Hamill, O.P., Marty, A., Neher, E., Sackmann, B., Sigworth, F.J. (1981) Improved patch-clamp techniques for high-resolution current recording from cells and cell-free membrane patches. Pflügers Arch. Eur. J. Physiol. 391: 85–100.

Hummer, G., Rasaiah, J.C., Noworyta, J.P. (2001) Water conduction through the hydrophobic channel of a carbon nanotube, Nature 414: 188–190.

Kong, Y.F., Shen, Y.F., Warth, T.E., Ma, J.P. (2002) Conformational pathways in the gating of Escherichia coli mechanosensitive channel, Proc. Natl. Acad. Sci. USA 99: 5999–6004.

Koprowski, P., Kubalski, A. (2003) C-termini of the Escherichia coli mechanosensitive ion channel (MscS) move apart upon the channel opening, J. Biol. Chem. 278: 11237–11245.

Levina, N., Totemeyer, S., Stokes, N.R., Louis, P., Jones, M.A., Booth, I.R. (1999) Protection of Escherichia coli cells against extreme turgor by activation of MscS and MscL mechanosensitive channels: identification of genes required for MscS activity. EMBO J. 18: 1730–1737.

Martinac, B.: Mechanosensitive channels in prokaryotes. Cellular Physiol. Biochem. 11(2): 61–76, 2001.

Martinac, B. (2005) Structural plasticity in MS channels. Nat. Struct. Mol. Biol. 12: 104–105.

Martinac, B. (2006) Mechanosensitive channels. In: Biological Membrane Ion Channels: Dynamics, Structure, and Applications (eds. S. H. Chung, O. S. Andersen and V. Krishnamurthy). Springer, New York. Chapter 10, pp. 369–398.

Martinac, B., Buechner, M., Delcour, A.H., Adler, J., Kung, C. (1987) Pressure-sensitive ion channel in Escherichia coli. Proc. Natl. Acad. Sci. USA, 84: 2297–2301.

Martinac, B., Kloda, A. (2003) Evolutionary origins of mechanosensitive ion channels. Progress Biophys. Mol. Biol. 82: 11–24.

Meyer, G.R., Gullingsrud, J., Schulten, K., Martinac, B. (2006) Molecular dynamics study of MscL interactions with a curved lipid bilayer, Biophys. J. 91: 1630–1637.

Miller, S., Bartlett, W., Chandrasekaran, S., Simpson, S., Edwards, M., Booth, I.R. (2003) Domain organisation of the MscS mechanosensitive channel of Escherichia coli, EMBO J. 22: 36–46.

Oakley, A.J., Martinac, B., Wilce, M.C.J. (1999) Structure and function of the bacterial mechanosensitive channel of large conductance. Protein Sci. 8: 1915–1921.

Park, K.H., Berrier, C., Martinac, B., Ghazi, A. (2004) Purification and Functional Reconstitution of N-and C-Halves of the MscL Channel Biophys. J. 86: 2129–2136.

Perozo, E., Kloda, A. Cortes, D.M. and Martinac, B. (2001) Site-directed spin-labeling analysis of reconstituted MscL in the closed state. J. Gen. Physiol. 118: 193–206.

Perozo, E., Kloda, A., Cortes, D.M., Martinac, B. (2002a) Physical principles underlying the transduction of bilayer deformation forces during mechanosensitive channel gating. Nature Struct. Biol. 9: 696–703.

Perozo, E., Cortes, D.M., Sompornpisut, P., Kloda, A., Martinac, B. (2002b) Structure of MscL in the open state and the molecular mechanism of gating in mechanosensitive channels. Nature 418: 942–948.

Pivetti, C. D., Yen, M. R., Miller, S., Busch, W., Tseng, Y., Booth, I.R., & Saier M.H.J. (2003) Two families of mechanosensitive channel proteins. Microbiol. Mol. Biol. Rev., 67, 66–85.

Ruthe, H.J., Adler, J. (1985) Fusion of bacterial spheroplasts by electric fields. Biochim Biophys Acta. 819: 105–113.

Saint, N., Lacapere, J.J., Gu, L-Q., Ghazi, A., Martinac, B. and Rigaud, J.L. (1998) A hexameric transmembrane pore revealed by two-dimensional crystallization of the large mechanosensitive ion channel (MscL) of Escherichia coli. J. Biol. Chem., 273: 14667–14670.

Schumann, U., Edwards, M.D., Li, C., Booth, I.R. (2004) The conserved carboxy-terminus of the MscS mechanosensitive channel is not essential but increases stability and activity, FEBS Lett. 572: 233–237.

Sotomayor, M., Schulten, K., (2004) Molecular dynamics study of gating in the mechanosensitive channel of small conductance MscS, Biophys. J. 87: 3050–3065.

Sotomayor, M., van der Straaten, T.A., Ravaioli, U., Schulten, K. (2006) Electrostatic properties of the mechanosensitive channel of small conductance MscS, Biophys. J. 90: 3496–3510.

Sotomayor, M. Vásquez, V., Perozo, E., Schulten, K. (2007) Ion conduction though MscS as determined by electrophysiology and simulation. Biophys. J. 92: 886–902.

Spronk, S.A., Elmore, D.E., Dougherty, D.A. (2006) Voltage-dependent hydration and conduction properties of the hydrophobic pore of the mechanosensitive channel of small conductance, Biophys. J. 90: 3555–3569.

Sukharev, S.I., Blount, P., Martinac, B., Blattner, F.R., Kung, C. (1994) A large mechanosensitive channel in E. coli encoded by mscL alone. Nature 368: 265–268.

Sukharev, S.I., Martinac, B., Arshavsky, V.Y., and Kung, C. (1993) Two types of mechanosensitive channels in the E. coli cell envelope: solubilization and functional reconstitution. Biophys. J. 65: 177–183.

Tang, Y.Y., Cao, G.X., Chen, X., Yoo, J., Yethiraj, A., Cui, A. (2006) A finite element framework for studying the mechanical response of macromolecules: Application to the gating of the mechanosensitive channel MscL, Biophys. J. 91: 1248–1263.

Tsai, I-J., Zhen-Wei Liu, Z-W., Rayment, J, Norman, C., McKinley, A. and Martinac, B. (2005) The role of the periplasmic loop residue glutamine 65 for MscL mechanosensitivity. Eur. Biophys. J. 34: 403–413.

Vora, T., Corry, B., Chung, S.H. (2006) Brownian dynamics investigations into the conductance state of the MscS channel crystal structure, Biochim. Biophys. Acta 1758: 730–737.

Valadie, H., Lacapcre, J.J., Sanejouand, Y.H., Etchebest, C. (2003) Dynamical properties of the MscL of Escherichia coli: A normal mode analysis, J. Mol. Biol. 332: 657–674.

Wan, R., Li, J., Lu, H., Fang, H. (2005) Controllable water channel gating of nanometer dimensions, J. Am. Chem. Soc. 127: 7166–7170.

Wiggins, P., Phillips, R. (2004) Analytic models for mechanotransduction: Gating a mechanosensitive channel, Proc. Natl. Acad. Sci. USA, 101: 4071–4076.

Chapter 5
Mechanosensory Transduction in the Nematode *Caenorhabditis elegans*

Nikos Kourtis and Nektarios Tavernarakis

Abstract Mechanotransduction, the process of converting a mechanical stimulus into a biological signal, appeared very early in the evolution and underlies a plethora of fundamental biological processes such as osmosensation, touch, hearing, balance and proprioception. Mechanosensory transduction has been studied extensively in simple animal models such as the nematode *Caenorhabditis elegans* and the fruit fly *Drosophila melanogaster*. Genetic and physiological studies have revealed that specialized macromolecular complexes, encompassing mechanically gated ion channels, play a critical role in the conversion of mechanical energy into cellular response. Members of two large ion channel families, the degenerin/epithelial sodium channels (DEG/ENaC) and the transient receptor potential ion channels (TRP), have emerged as candidate mechanosensitive channels. Several auxiliary proteins associate with the core mechanosensitive channels to form the mechanotransducing apparatus in specialized mechanosensory cells. *C. elegans* displays a variety of mechanosensory behaviours. In this chapter, we survey the mechanisms of mechanosensory transduction in *C. elegans*. The exceptional amenability of this simple metazoan to genetic and molecular manipulations has facilitated the dissection of the mechanotransduction process to unprecedented detail.

Key words: Degenerin · Ion channels · Proprioception · Touch receptor neurons · TRP channels

5.1 Introduction

Animals receive and process information about their surroundings through specialized sensory cells. Ubiquitous mechanical stimuli permeate the environment of every living cell and every organism. The process by which cells convert mechanical energy into electrical or chemical signals is called *mechanotransduction*. Because mechanical force is everywhere, mechanosensation probably represents one

A. Kamkin and I. Kiseleva (eds.), *Mechanosensitive Ion Channels.*

of the oldest sensory transduction pathways that evolved in living organisms, from bacteria to humans (Blount and Moe, 1999; French, 1992; Gillespie and Walker, 2001; Sackin, 1995) The capacity to respond and adjust to mechanical inputs plays a pivotal role in numerous fundamental physiological phenomena such as the perception of sound and gravity, which underlie our senses of hearing and balance (Garcia-Anoveros and Corey, 1997; Hackney and Furness, 1995; Kellenberger and Schild, 2002). Touch sensation and proprioception (the coordinated movement of our body parts) are additional manifestations of responsiveness to mechanical stimulation (Garcia-Anoveros and Corey, 1996; Tavernarakis and Driscoll, 1997; Tavernarakis et al., 1997; Welsh et al., 2002). Moreover, mechanotransduction is equally critical for the stretch-activated reflexes of vascular epithelia and smooth muscle, and in the regulation of systemic fluid homeostasis and blood pressure (baroreception; Garcia-Anoveros and Corey, 1996; Lee and Huang, 2000; Tavernarakis and Driscoll, 2001a; Tavi et al., 2001; Welsh et al., 2002). Mechanotransduction is also important for the prevention of polyspermy during fertilization, cell volume and shape regulation, cell locomotion and tissue development and morphogenesis (Ingber, 1997; Ko and McCulloch, 2001; Rossier et al., 1994). In plants, mechanotransduction is the basis of gravitaxis and turgor control (Lynch et al., 1998; Pickard and Ding, 1993). In protists (*Paramecium, Stentor*) mechanotransduction underlies gravikinesis (the swimming against the gravity vector in order to avoid sedimentation; Block et al., 1999; Gebauer et al., 1999; Hemmersbach et al., 2001; Marino et al., 2001).

All living organisms have developed highly specialized structures that are receptive to mechanical forces originating either from the surrounding environment or from within the organism itself. Mechanotransducers are among the most elaborate and efficient, such structures, which are responsible for sensory awareness, for example, those facilitating touch, balance proprioception and hearing (Garcia-Anoveros and Corey, 1997; Gillespie and Walker, 2001; Hackney and Furness, 1995; Tavernarakis and Driscoll, 2001a). The mechanisms underlying the capability of living cells to receive and act in response to mechanical inputs are among the most anciently, implemented during evolution. Proteins with mechanosensitive properties are ubiquitously present in eubacteria, archaea and eukarya, and are postulated to have been an essential part of the physiology of the Last Universal Ancestor (Kloda and Martinac, 2001; Koch, 1994; Koprowski and Kubalski, 2001; Martinac, 2001). The first mechanosensitive processes may have evolved as backup mechanisms for cell protection, e.g. to reduce intracellular pressure and membrane tension during osmotic swelling. Subsequent organismal diversification and specialization resulted in variable requirements for mechanotransduction in different organisms (Norris et al., 1996). Hence, evolutionary pressure has shaped a large repertoire of mechanotransducers, optimized for a great assortment of tasks that range from maintenance of intracellular osmotic balance and pressure, to our impressive ability of hearing and discriminating sounds, and reading Braille code with our fingertips (Gillespie and Walker, 2001; Hamill and Martinac, 2001).

In this chapter, we describe *C. elegans* mechanosensory behaviours and survey the genes implicated in mechanosensory perception. We also discuss the relevant mechanisms underlying mechanotransduction in the nematode.

5.2 *C. elegans*: Background Information

Caenorhabditis elegans is a small (about 1mm length), hermaphroditic, soil-dwelling nematode. The size and simple dietary demands permit easy and cheap cultivation of the animal in the lab. The worm completes a reproductive life cycle in 2.5 days at 25°C progressing from a fertilized embryo through four larva stages, to become an egg-laying adult which lives for about 2–3 weeks (Brenner, 1974). Under non-favourable conditions such as starvation, high temperature or overcrowding, larvae may enter an alternative life stage, called dauer larva, a very resistant larval form that survives for months (Golden and Riddle, 1984). The simple body plan, the transparent egg and cuticle, and the nearly invariant developmental plan of this nematode has facilitated exceptionally detailed developmental and anatomical characterization of the animal (Ward et al., 1975; White et al., 1986; www.wormatlas.org). The complete sequence of somatic cell divisions from the fertilized egg to the 959-cell adult hermaphrodite has been described (Sulston and Horvitz, 1977). The *C. elegans* nervous system consists of 302 neurons of 118 types that interconnect in a stable and reproducible manner to create a variety of neural circuits (White et al., 1976; White et al., 1986). Individual neurons can be ablated through laser microsurgery and genetic manipulation, and neuronal pathways responsible for different behaviours have been characterized.

C. elegans is especially amenable to both forward and reverse genetic analysis. Mutagenized parents segregate homozygous F2 mutant progeny without any requirement for genetic crossing. Self-fertilization of heterozygotes leads to genetically homogeneous populations, while crossing with males that appear at a small percentage in the population facilitates genetic manipulations. Rapid and precise genetic mapping can be achieved, by taking advantage of a dense single nucleotide polymorphism map (Koch et al., 2000; Wicks et al., 2001). A physical map of the *C. elegans* genome, consisting of overlapping cosmid and YAC clones covering most of the six chromosomes, has been constructed to facilitate cloning of genes that have been positioned on the genetic map (Coulson et al., 1988; Waterston and Sulston, 1995). Double-stranded RNA mediated interference (dsRNAi) enables probable loss-of-function phenotypes to be readily evaluated (Fire et al., 1998; Tavernarakis et al., 2000). Transgenic animals carrying the gene of interest can be easily constructed by microinjecting the transgene into the gonad of hermaphrodite animals (Mello and Fire, 1995).

C. elegans displays a variety of behaviours, including mechanosensitive behaviours. The broad range of genetic and molecular tools available in the worm facilitates thorough and multifaceted investigation of the pathways that govern these behaviours.

5.3 *C. elegans* Mechanosensory Behaviours

Many behaviours displayed by *C. elegans* are direct manifestations of mechanosensitivity, making it exceptionally attractive for investigating mechanotransduction (Bargmann and Kaplan, 1998; Baumeister and Ge, 2002; Chalfie, 1993; Chalfie and Sulston, 1981; Chalfie et al., 1985; Driscoll and Tavernarakis, 1997; Herman, 1996; Kaplan and Horvitz, 1993; Mah and Rankin, 1992; O'Hagan and Chalfie, 2006; Rankin, 1991; Syntichaki and Tavernarakis, 2004; Tavernarakis and Driscoll, 1997; Wicks and Rankin, 1995; Wolinsky and Way, 1990). The best characterized such behaviour is the response to a gentle mechanical stimulus delivered transversely along the body of the animal, typically by means of an eyelash hair attached onto a toothpick (the 'gentle body touch response'; (Chalfie and Sulston, 1981; Chalfie et al., 1985; Herman, 1996; Tavernarakis and Driscoll, 1997)). We discuss studies elucidating the molecular mechanisms of this touch response in the following section. Other mechanosensory responses are the generation and maintenance of the characteristic coordinated sinusoidal pattern of locomotion (analogous to proprioception; (Li et al., 2006; Tavernarakis and Driscoll, 1997; Tavernarakis et al., 1997); see below), and the nose touch response, which can be further categorized into the head-on collision response and the head withdrawal response (Driscoll and Kaplan, 1996; Kaplan and Horvitz, 1993).

When animals collide with an obstacle in a nose-on fashion during the course of normal locomotion they respond by reversing their direction of movement (Bargmann and Kaplan, 1998; Colbert et al., 1997). This response is independent of touch receptor neurons, needed for gentle touch (Chalfie and Sulston, 1981). Three classes of mechanosensory neurons, ASH, FLP, and OLQ, mediate this avoidance response (Bargmann and Kaplan, 1998; Herman, 1996; Kaplan and Horvitz, 1993; Wicks and Rankin, 1995). Each of these sensory neurons accounts for a part of the normal response, which is quantitative with normal animals responding about 90% of the time. Laser ablation and genetic studies have demonstrated that each sensory neuron contributes to the overall responsiveness as follows: ASH, 45%; FLP, 29%; and OLQ, 5%. The remaining 10% of the responses are mediated by the ALM and AVM neurons, which sense anterior body touch (Driscoll and Kaplan, 1996; Kaplan and Horvitz, 1993). It is unclear what distinguishes the function of the three nose touch neurons. One attractive possibility is that these cells differ in their sensitivities and that the intensities of nose touch stimuli vary according to the violence of the collision. If this were the case, it would be expected that the most sensitive neuron (ASH) would account for the majority of responses while less sensitive neurons (FLP and OLQ) would account for the remainder. In addition to their mechanosensory properties, the ASH neurons are part of a chemosensory organ, the amphid sensilla, with their sensory endings exposed to the external environment (Perkins et al., 1986; Ward et al., 1975). The ASH neurons serve chemosensory and osmosensory functions, mediating avoidance of osmotic repellents (Hart et al., 1999; Hart et al., 1995). Several classes of chemosensory neurons respond to multiple chemical stimuli in *C. elegans*. However, ASH is unique among them in responding to such divergent stimuli. In this respect, ASH neurons are similar to vertebrate neurons that sense painful stimuli, which are called nociceptors. For their multi-sensory capabilities, the ASH neurons have been categorized as polymodal sensory neurons (Driscoll and Kaplan, 1996; Kaplan and Horvitz, 1993).

In addition to DEG/ENaC proteins, another major family of channel proteins implicated in sensory mechanotransduction is the transient receptor potential (TRP) ion channels. TRPs are non-specific cation-permeable channels that are present in diverse species ranging from yeast, flies, and worms to humans (Kahn-Kirby and Bargmann, 2006). All TRP channels appear to form tetrameric assemblies and include six predicted transmembrane domains and a variable number of ankyrin motifs, which are suggested to mediate protein-protein interactions. Members of individual subfamilies may bear several other domains, such as coiled-coil motifs, protein kinase domains, transmembrane segments, and TRP domains (Montell, 2005).

In *C. elegans*, the TRPV (vallinoid TRP) subfamily genes *osm-9* and *ocr-2* are required for the aversive responses of ASH neurons to various noxious stimuli, including high osmomolarity and noxious chemicals (Bargmann et al., 1990; Hilliard et al., 2005; Troemel et al., 1995). The OSM-9 and OCR-2 proteins localize to the sensory cilia of ASH, suggesting a direct role in sensory transduction (Colbert et al., 1997). Additionally, OSM9::GFP is expressed in FLP, OLQ and PVD mechanosensory neurons. Several genetic studies suggest that the function of the putative OSM-9/OCR-2 ion channel is regulated by G protein signalling and specific polyunsaturated fatty acids (PUFAs), which act upstream of OSM-9/OCR-2 to modulate nociceptive responses in ASH neurons, including the mechanosensory nose touch avoidance behaviour (Kahn-Kirby et al., 2004; Roayaie et al., 1998). Expression of a mammalian TRPV4 protein in ASH can rescue defects in osmotic and nose-touch avoidance in *osm-9* mutants (Liedtke et al., 2003).

C. elegans also distinguishes textural differences in its substrate. When worms enter a bacterial lawn they slow their movement, a behaviour known as basal slowing. This response is indeed mechanosensory since worms entering a lawn of Sephadex beads instead of bacteria, slow similarly (Sawin et al., 2000). This behaviour depends on the CEP, ADE and PDE dopaminergic neurons which have ciliated sensory endings, embedded in the cuticle (Perkins et al., 1986; Ward et al., 1975; White et al., 1986). Laser ablation experiments confirmed that these neurons are required for response to small particles (Sawin et al., 2000). In support of this model, dopaminergic neurons transduce an inhibitory signal to the motor circuit (White et al., 1986).

The complicated male-mating behaviour of *C. elegans* is probably based on chemosensory and mechanosensory cues (Liu and Sternberg, 1995). Males have 87 additional neurons, comparing with the hermaphrodite many of which are ciliated and considered to be sensory (Sulston et al., 1980). Two genes, *lov*-1 and *pkd*-2, needed for male mating have been implicated in mechanical signalling. *lov-1* and *pkd-2*. LOV-1 and PDK-2 are the nematode homologs of mammalian PDK-1 and PDK-2 TRPP ion channels respectively (Corey, 2003). Interestingly, mutations in the mammalian PDK-1 and PDK-2 cause autosomal dominant polycystic kidney disease (ADPKD). PDK-1 and PDK-2 form a Ca^{2+}-permeable ion channel which is mechanically activated by fluid flow in certain epithelial cells (Nauli et al., 2003).

An additional mechanosensitive behaviour in *C. elegans* is the tap withdrawal reflex, where animals retreat in response to a tap on the culture plate (Chiba and Rankin, 1990; Rankin et al., 2000; Wicks and Rankin, 1997). Worms respond to a diffuse mechanical stimulus (a tap to the side of the dish they are resting on) by either accelerating forward movement or by initiating backward movement (Chiba and Rankin, 1990; Rankin et al., 2000). Given that the stimulus is not spatially coherent and that the animal's response is variable, it was proposed that the tap response reflects the simultaneous activation of the anterior and posterior touch cells. The behavioural outcome is likely determined by the integration of these two antagonistic circuits.

Mechanotransduction appears to also play a regulatory role in processes such as egg laying, feeding, defecation, and maintenance of the pseudocoelomic body cavity pressure (Avery, 1993; Du and Chalfie, 2001; Liu and Thomas, 1994; Liu and Sternberg, 1995; Thomas et al., 1990; Wolinsky and Way, 1990). These behaviours add to the large repertoire of mechanosensitive phenomena, amenable to genetic and molecular dissection in the nematode (Bargmann and Kaplan, 1998; Syntichaki and Tavernarakis, 2004; Tavernarakis and Driscoll, 2001b).

5.4 The Gentle Body Touch Transduction System

In its natural habitat, the soil, *C. elegans* encounters a large number of mechanical stimuli. While crawling on surfaces of soil particles, the worm receives external forces generated by bumping on soil materials and other animals. The laboratory assay for the gentle body touch response involves a mild stroke of the animal with an eyelash hair attached to a toothpick, transversely to the anterior-posterior body axis (Chalfie and Sulston, 1981; Chalfie et al., 1985; Syntichaki and Tavernarakis, 2004). When no response is observed, animals are prodded with a thin platinum wire to confirm that they are touch insensitive rather than paralyzed (gentle-touch insensitive animals typically still respond to a strong stimulus-the harsh touch response) (Chalfie et al., 1985; Chalfie and Wolinsky, 1990; Driscoll and Kaplan, 1996; Wolinsky and Way, 1990). Depending on the part of the body touched, animals will either accelerate or initiate forward movement (when stimulated at the posterior or the tail), or reverse and move backwards (when stimulated at the anterior part of the body). Hermaphrodite, male, juvenile (except L1), and dauer animals respond identically to touch. The response is adaptive: repetitive stimulation leads to short periods of insensitivity (Mah and Rankin, 1992; Rankin et al., 2000; Wicks and Rankin, 1995).

5.4.1 The Sensory Cells

The touch reflex of the mature animal involves six touch receptor neurons, 5 pairs of interneurons and 69 motorneurons (Chalfie, 1993; Chalfie et al., 1985). The six touch receptor neurons share common ultrastructural features and express many of the same genes (Chalfie and Sulston, 1981; Chalfie and Thomson, 1979; Chalfie and Thomson, 1982). The six touch receptor neurons were originally designated as the microtubule cells because of distinctive bundles of 15-protofilament (pf; tubulin dimmer filaments) microtubules that fill their processes (ALML/R: anterior lateral microtubule cell, left/right; AVM: anterior ventral microtubule cell; PLML/R; posterior lateral microtubule cell, left/right; and PVM: posterior ventral microtubule cell; (Chalfie, 1993; Chalfie et al., 1986; Chalfie et al., 1985; Chalfie and Thomson, 1979; Chalfie and Thomson, 1982)). Two fields of touch sensitivity, anterior and posterior are defined by the arrangement of the six touch receptor neurons along the body axis (Fig. 5.1; Chalfie et al., 1985; Tavernarakis and Driscoll, 1997). All six cells have anteriorly directed processes and, except for PVM, an anterior branch. The processes of touch neurons are localized in the hypodermis, just beneath the cuticle, an ideal position for sensing external stimuli and vibrations. All six cells are dispensable for the viability of the organism. Apart from insensitivity to gentle body touch, laser ablation of all six neurons does not result in any additional adverse effects (Chalfie, 1995; Chalfie et al., 1985; Tavernarakis and Driscoll, 1997).

Laser microsurgery established that PLML and PLMR are required for response to a touch to the tail. If either is present, tail touch sensitivity is observed. When both are ablated, animals are completely insensitive to gentle touch stimuli administered to the posterior (Chalfie et al., 1985; Chalfie et al., 1983; Kitamura et al., 2001). Either ALML or ALMR can mediate a response to a mechanical stimulus delivered to the anterior part of the body. AVM, which is added into the touch circuitry postembryonically, can mediate a weak response to some stimuli but not all. In animals in which both ALM cells are killed, partial touch sensitivity returns 35–40 hours after hatching, which is attributable to AVM being generated. PVM alone does not produce a touch response, but its synaptic

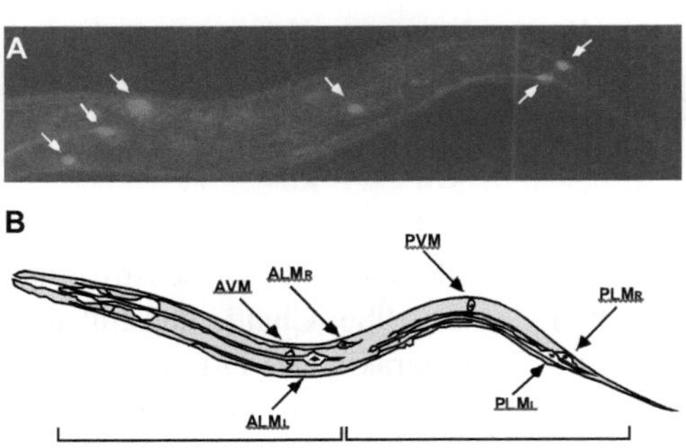

Fig. 5.1 The *C. elegans* touch receptor neurons. (A) Visualization of touch receptors. Worms are expressing the Green Fluorescence Protein (GFP) under the control of the *mec-4* promoter, which is active only in the six-touch receptor neurons. Arrows indicate the cell bodies of the neurons. (B) Schematic diagram , showing the position of the six touch receptor neurons in the body of the adult worm. The two fields of touch sensitivity are defined by the arrangement of touch receptor neurons along the body axis. The ALMs and AVM mediate the response to touch over the anterior field whereas PLMs mediate the response to touch over the posterior field. PVM does not mediate touch response by itself (Chalfie, 1995; Chalfie, 1997; Chalfie et al., 1985; Tavernarakis and Driscoll, 1997). Reproduced from (Voglis and Tavernarakis, 2005) with copyright permission of the Academia Publishing House Ltd

pattern implicates it in the touch behaviour (Chalfie, 1993; Chalfie and Au, 1989; Chalfie et al., 1985).

Bundles of darkly staining large diameter microtubules distinguish the touch receptor neurons (Chalfie and Thomson, 1979; White et al., 1976). Cross bridges between microtubules of a bundle are observed in micrographs obtained with electron microscopy, and may increase the structural integrity of the bundle. These microtubules are unique to the nematode touch receptor neurons and contain 15-protofilament microtubules, a unique feature of these six cells (Chalfie and Thomson, 1982; Tavernarakis and Driscoll, 1997). In most eukaryotic cells, α- and β-tubulin co-assemble into 13-protofilament microtubules, whereas the vast majority of microtubules in *C. elegans* cells have 11-protofilament (Chalfie and Thomson, 1982; White et al., 1986). In normal touch receptors, 11-protofilament microtubules typical of most other cells in this nematode are occasionally observed. If the 15-protofilament microtubules are eliminated by mutation, the number of 11-protofilament microtubules in the touch cell processes increases (Chalfie, 1993; Chalfie et al., 1986; Chalfie and Thomson, 1979; Fukushige et al., 1999; Savage et al., 1989). Individual microtubules are 10–20 μm, but overlap and create bundles, filling in that way the process of the neuron which is 400–500 μm long (Chalfie and Thomson, 1979). The distal ends of the microtubules are apposed to the plasma membrane and are associated with structures of a diameter up to twice that of the microtubules (Chalfie and Thomson, 1979).

Touch receptors also are uniquely surrounded by an osmiophillic extracellular material referred to as the mantle. The amount of mantle varies along the length of the process (Chalfie and Sulston, 1981). The mantle is needed for the attachment of the touch receptor process to the body wall, bringing it to an ideal position for detection of external mechanical stimuli.

5.4.2 The C. elegans Mechanosensory Apparatus

To identify molecules dedicated to touch transduction, Martin Chalfie and colleagues mounted a forward genetics approach to isolate gentle body touch-insensitive nematode mutants (Chalfie and Au, 1989; Chalfie et al., 1986; Chalfie and Sulston, 1981; Du and Chalfie, 2001; Gu et al., 1996; Tavernarakis and Driscoll, 1997). Briefly, populations of wild type, touch sensitive animals were mutagenized and touch insensitive individuals were sought among their descendants by stroking with an eyelash hair and prodding with a platinum wire (Chalfie, 1997). During the course of this very tedious screening process, over 417 mutations in 17 different genes, randomly distributed in all six chromosomes of *C. elegans* were isolated (Driscoll and Kaplan, 1996; Tavernarakis and Driscoll, 1997). By design, the screen yields mutations in genes that are fairly specific for normal gentle body touch perception. For example, gene mutations with pleotropic effects that result in lethality or uncoordinated and paralyzed phenotypes would have been missed. In addition to being touch insensitive *mec* mutants tend to be lethargic when grown normally in the presence of amble food (Driscoll and Kaplan, 1996). Reduced spontaneous movement is probably due to their inability to sense micro vibrations in their environment, interaction with external objects or stretch produced by the locomotory movements themselves. However, when starved or during mating they move as well as wild type. The 17 genes isolated are designated as the *mec* genes for their '*mec*hanosensory abnormal' phenotype. Corroborating the high specificity of the screen, while most of the alleles generated cause complete touch insensitivity, only few other abnormalities accompany the mutants (Driscoll and Kaplan, 1996). Depending on their role and point of action, *mec* genes can be loosely classified into three main categories. First, the regulatory/specification genes which control the expression touch receptor neuron specific genes or modify the activity of the mechanotransducer complex; second, the *mec* genes encoding core structural components of the mechanosensitive ion channel; and third the genes encoding peripheral, associated proteins.

5.4.2.1 Genes Needed for the Development of Touch Receptor Neurons

The UNC-86 and MEC-3 transcription factors are essential for proper development and differentiation of the six touch receptor neurons (Chalfie and Sulston, 1981; Duggan et al., 1998; Way and Chalfie, 1988; Way and Chalfie, 1989; Xue et al., 1992). UNC-86 is a POU domain protein which is required in several distinct neuroblast lineages for daughter cells to become different from their mothers (Finney and Ruvkun, 1990; Finney et al., 1988). UNC-86 activates the expression of *mec-3* which encodes a LIM-type homeodomain protein needed for the differentiation of the six touch receptor cells (Way and Chalfie, 1989). In mutants lacking *mec-3* activity, the touch receptors express none of their unique differentiated features and appear to be transformed to other types of neurons (Way and Chalfie, 1988). UNC-86 and MEC-3 bind cooperatively as a heterodimer to the *mec-3* promoter as well as to the promoters of other genes required for the function of touch cells (Duggan et al., 1998; Way and Chalfie, 1988; Way and Chalfie, 1989). *unc-86* and *mec-3* genes are also expressed in the PVD and FLP neurons which also function as mechanoreceptors but they do not express the same genes as the touch receptor neurons (Way and Chalfie, 1989).

5.4.2.2 Genes Needed for Function of Touch Receptor Neurons

Four *mec* genes can be classified in the category of core structural components of the putative mechanosensory ion channel in touch receptor neurons, *mec-2*, *mec-4*, *mec-6* and *mec-10*. MEC-4 and MEC-10 form the core ion channel, while MEC-2 and MEC-6 physically interact with the channel subunits to shape and modulate their gating properties. Animals bearing loss-of-function mutations in *mec-4* or *mec-10* are touch-insensitive despite the fact that in these mutant backgrounds the touch receptor neurons develop normally and exhibit no apparent defects in ultra structure (Driscoll and Chalfie, 1991; Huang and Chalfie, 1994; Tavernarakis and Driscoll, 1997). *mec-4* and *mec-10* encode homologous proteins related to subunits of the multimeric amiloride sensitive Na^+ channel which mediates Na^+ re-absorption in vertebrate kidney, intestine and lung epithelia (the ENaC channel; (Kellenberger and Schild, 2002; Rossier et al., 1994)). In addition to being involved in mechanotransduction, MEC-4, MEC-10 and several other related nematode proteins have a second, unusual property: specific amino acid substitutions result in aberrant channels that induce the swelling and subsequent necrotic death of the cells in which they are expressed (Driscoll, 1996; Driscoll and Chalfie, 1991; Hall et al., 1997; Harbinder et al., 1997). This pathological property is the reason that this family of proteins was originally called degenerins (Chalfie et al., 1993; Chalfie and Wolinsky, 1990; Tavernarakis and Driscoll, 2001a; Tavi et al., 2001). *C. elegans* degenerins, together with their mammalian relatives, the ENaCs, comprise the large DEG/ENaC family of ion channels (Fig. 5.2). The relationship of these channel subunits to subunits of an amiloride-sensitive ENaCs is intriguing because amiloride is a general inhibitor of mechanosensitive ion channels (Alvarez de la Rosa et al., 2000; Hamill et al., 1992; Hamill and McBride, 1996; Hoger et al., 1997; Lane et al., 1991; Rossier et al., 1994; Voilley et al., 1997).

mec-4 is expressed solely in the six touch receptor neurons, while *mec-10* in addition to the six touch receptor neurons, is expressed in two other neuron pairs that may mediate stretch-sensitive responses (FLPs and PVDs; (Driscoll and Chalfie, 1991; Driscoll and Tavernarakis, 1997; Huang and Chalfie, 1994; Tavernarakis and Driscoll, 1997)). Interestingly, a MEC-4::GFP fusion localizes in distinct puncta along the processes of the touch receptor neurons (Fig. 5.3). Such punctuate localization may reflect the distribution of mechanotransducing complexes on the axon of the touch receptor neuron. MEC-4 and MEC-10 co-localize exclusively in mechanosensitive neurons, where they may co-assemble into a mechanically-gated ion channel. Optical imaging studies using genetically encoded Ca^{2+} indicators (cameleons), which monitor intracellular Ca^{2+} changes in response to gentle body touch support this hypothesis (Suzuki et al., 2003). More definitive proof is provided by recent electrophysiological, whole-cell voltage-clamp recordings of mechanoreceptor currents (MRCs) from the PLM touch receptors *in vivo* (O'Hagan et al., 2005). The mechanoreceptor currents are extremely rapid; they turned on within a millisecond of force application and quickly decreased in amplitude, characteristic of adaptation. MRCs recorded from PLM neurons are probably representing transduction by direct activation of the channel and not through second messengers. MRCs are absent in MEC-4, MEC-2 and MEC-6 null mutants, whereas mechanoreceptor current was only partially decreased when testing mutations in non-conserved regions of the proteins (O'Hagan et al., 2005). In addition, MRCs are reduced but not eliminated in worms that carry a null mutation in *mec-7* and *mec-12* genes, affecting α and β-tubulin which forms the 15-protofilament microtubule bundles. These results question the absolute requirement of microtubules for touch transduction.

MEC-2 is an additional component of the channel complex. *mec-2* encodes a 481-amino acid protein and is expressed in the touch receptor neurons and in a few additional neurons in the nerve ring region (Du and Chalfie, 2001; Gu et al., 1996; Huang et al., 1995). MEC-2 features three candidate protein-protein interaction domains (Fig. 5.4). First, part of the amino-terminal domain (situated in part between AA 42–118) is needed for the proper localization of a *mec-2/lacZ* fusion

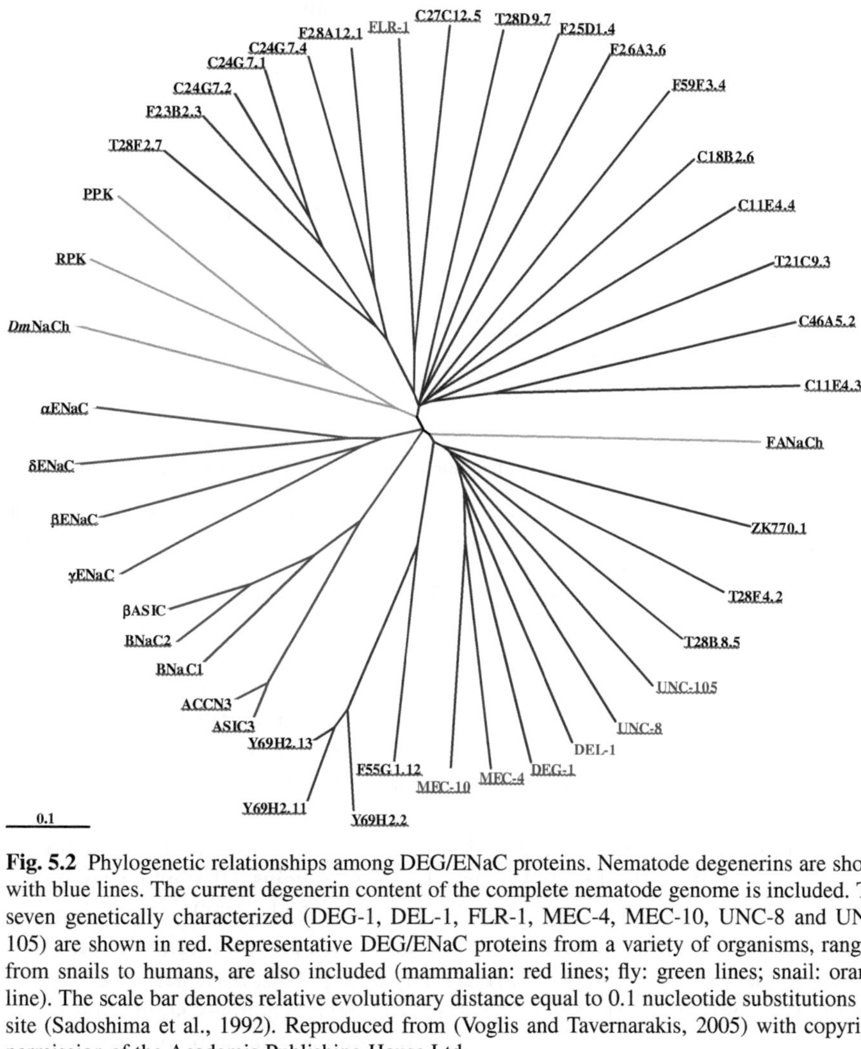

Fig. 5.2 Phylogenetic relationships among DEG/ENaC proteins. Nematode degenerins are shown with blue lines. The current degenerin content of the complete nematode genome is included. The seven genetically characterized (DEG-1, DEL-1, FLR-1, MEC-4, MEC-10, UNC-8 and UNC-105) are shown in red. Representative DEG/ENaC proteins from a variety of organisms, ranging from snails to humans, are also included (mammalian: red lines; fly: green lines; snail: orange line). The scale bar denotes relative evolutionary distance equal to 0.1 nucleotide substitutions per site (Sadoshima et al., 1992). Reproduced from (Voglis and Tavernarakis, 2005) with copyright permission of the Academia Publishing House Ltd

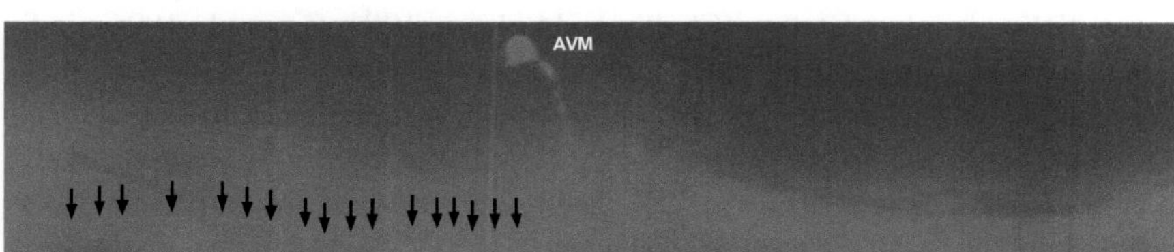

Fig. 5.3 Punctuate localization of a putative mechanosensitive ion channel subunit. Image of an AVM touch receptor neuron expressing a GFP-tagged MEC-4 protein. Fluorescence is unevenly distributed along the process of the neuron in distinct puncta, which may represent the location of the mechanotransducing apparatus. Reproduced from (Voglis and Tavernarakis, 2005) with copyright permission of the Academia Publishing House Ltd

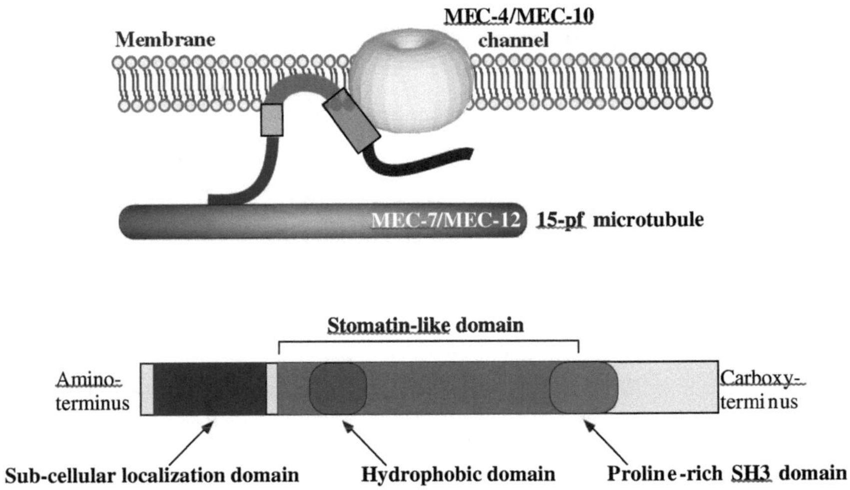

Fig. 5.4 Schematic representation and topology of the MEC-2 protein. Conserved domains as well as hydrophobic regions are highlighted. Putative interactions with the degenerin channel and the cytoskeleton are indicated (Goodman et al., 2002). Reproduced from (Voglis and Tavernarakis, 2005) with copyright permission of the Academia Publishing House Ltd

protein to the touch receptor process. Second, the carboxy-terminal domain includes a proline-rich region that is similar to SH3-binding domains. Third, the central region (AA 114–363) encompasses an SPFH domain with a membrane-associated hydrophobic part (AA 114–141) and a cytoplasmic hydrophilic part that together exhibit 65% identity to the human red blood cell protein stomatin (Huang et al., 1995; Tavernarakis et al., 1999). The SPFH domain is the common denominator of stomatins, prohibitins, flotilins and bacterial *Hfl*K/C proteins, all of which are membrane associated regulators (Fig. 5.5; Tavernarakis et al., 1999). Stomatin, also known as band 7.2b protein, is a membrane-associated protein originally identified as a component of human red cells (Delaunay et al., 1999; Sedensky et al., 2001; Snyers et al., 1998; Stewart, 1997; Stewart et al., 1993). In humans, stomatin is missing from erythrocyte membranes in autosomal dominant hemolytic disease overhydrated hereditary stomatocytosis, despite an apparent normal stomatin gene. Many of the 54 mutant *mec-2* alleles have dominant effects and exhibit a complex pattern of inter-allelic complementation (Chalfie and Sulston, 1981; Gu et al., 1996), indicating that MEC-2 protein molecules form higher order complexes. However, there is also genetic data suggesting that MEC-2 interacts with the specialized touch cell microtubules encoded by *mec-7* and *mec-12* (α-tubulin and β-tubulin respectively; (Gu et al., 1996; Huang et al., 1995)). Normally, a *mec-2/lacZ* fusion protein is distributed along the touch receptor axon (Huang et al., 1995). The axonal distribution of a MEC-2::lacZ fusion protein is mildly disrupted in a *mec-7* null or *mec-12* strong loss-of-function background, implying that the 15-protofilament microtubules are not essential for the localization of MEC-2 to the neuronal process. However, two specific *mec-12* missense alleles interfere dramatically with localization of MEC-2 fusion proteins, restricting the fusion proteins to the cell body (Huang et al., 1995).

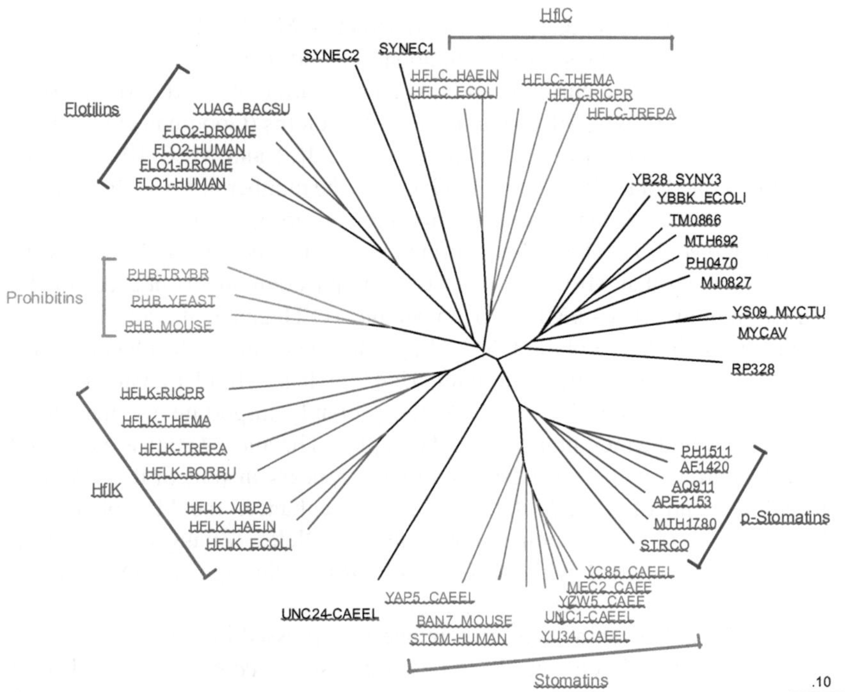

Fig. 5.5 Phylogenetic relations among SPFH domain proteins. A dendrogram showing distance relationships among most of the stomatin protein super-family members (the complete ClustalW generated alignment on which the dendrogram was based is available at http://www.imbb.forth.gr/worms/worms/alignment.gif). The dendrogram was constucted with the neighbor-joining method based on pairwise distance estimates of the expected number of amino acid replacements per site (0.10 in the scale bar), and visualized by TreeTool. Protein sub-families are denoted in different colours (Tavernarakis et al., 1999)

MEC-2 colocalizes with MEC-4 in the six touch receptor neurons and is distributed along neuronal processes in punctuate pattern (Zhang et al., 2004). This is consistent with the co-immunoprecipitation of the two proteins in Xenopus oocytes (Goodman et al., 2002). The stomatin-like domain of MEC-2 interacts specifically with the N-terminus cytoplasmic region of MEC-4 (Zhang et al., 2004). Punctuate expression of MEC-2 is disrupted in the *mec-4(u253)*, *mec-6(u450)* and *mec-10(u20)* loss-of-function mutants indicating that the MEC-2 subcellular localization depends on the other partners of the mechanosensory complex (Zhang et al., 2004). These genetic studies, which do not by themselves prove a direct interaction, have recently been complemented by elegant heterologous expression experiments in *Xenopus* oocytes that support physical interaction between MEC-2 and the channel subunits MEC-4 and MEC-10 (Goodman et al., 2002). Reconstitution of channel activity in *Xenopus* oocytes revealed that MEC-2 regulates the activity of the MEC-4/MEC-10 channel, providing the first direct support for the hypothesis that stomatin-like proteins interact with and regulate ion channels (Goodman et al., 2002; Stewart et al., 1993). This interaction appears to dramatically potentiate the conductivity of the channel in oocytes. Co-expression of MEC-2 with the hyperactive MEC-4(d) and MEC-10(d) derivatives in *Xenopus* oocytes resulted in about 40-fold increase in the amplitude of amiloride-sensitive ionic currents, and this amplification allowed currents to be detected even with wild-type MEC-4 and MEC-10 proteins (Goodman et al., 2002; Stewart et al., 1993). Visualization of tagged MEC-4(d) and MEC-10(d) in live oocytes demonstrated that MEC-2 does not increase the number of MEC-4(d)/MEC-10(d) channels that reach the plasma membrane, and probably acts by regulating their activity. In *mec-2(u37)* loss-of-function mutants the touch-evoked currents are abolished confirming that MEC-2 is one of the major components needed for the proper function of the MEC-4/MEC-10 ion channel (O'Hagan et al., 2005).

A second stomatin-like protein, UNC-24 appears to be part of the channel complex. In addition to the stomatin domain, UNC-24 has a lipid transfer domain (Barnes et al., 1996; Sedensky et al., 2001), probably important for the membrane localization of the membrane channel complex. The *unc-24* gene is expressed in the touch neurons, while the UNC-24 protein appears in puncta that colocalize with MEC-4 and MEC-2. Mutation of *unc-24* enhances the Mec phenotype caused by *mec-4* and *mec-6* temperature sensitive alleles (Zhang et al., 2004). UNC-24 appears to interact with MEC-2 and MEC-4 through its stomatin-like domain.

mec-6 encodes a 377-amino acid protein and is expressed in muscle cells, neurons and other tissues (Chelur et al., 2002). Recessive *mec-6* mutations disrupt touch sensitivity but do not cause detectable changes in touch cell ultrastructure (Chalfie and Sulston, 1981; Tavernarakis and Driscoll, 1997). *mec-6* alleles have the interesting property that they completely block *mec-4(d)* and *mec-4*(A673V)-induced touch cell degeneration (Harbinder et al., 1997; Huang and Chalfie, 1994; Tavernarakis and Driscoll, 1997). MEC-6 encodes a protein with limited similarity to Paraoxonases/Arylesterases that physically interacts with MEC-4 and MEC-10 (Chelur et al., 2002). MEC-6 has a short cytoplasmic N-terminus, a single transmembrane domain, and a large extracellular C-terminus. How exactly MEC-6 acts to influence MEC-4/MEC-10 channel activity is unknown. Nevertheless, it appears that *mec-6* mutations do not affect *mec-4* transcription, although they do cause full-length MEC-4::LacZ or MEC-4::GFP reporter fusion chimeras to be rapidly degraded (N. T. unpublished observations; (Chelur et al., 2002)). Thus, working hypotheses concerning the function of MEC-6 focus on two possibilities. First, MEC-6 is another subunit needed for channel function or assembly, or second, it mediates localization or post-translational modification essential for MEC-4 and MEC-10 activity/stability. It should be noted that MEC-6 function is not exclusively related to the MEC-4/MEC-10 touch receptor channel. *mec-6* mutations also suppress the deleterious consequences of neurodegeneration-inducing mutations in other *C. elegans* degenerins including *deg-1*, *unc-8* and partly *unc-105* ((Chalfie and Wolinsky, 1990; Liu et al., 1996; Shreffler et al., 1995; Tavernarakis et al., 1997); N.T unpublished observations). *mec-6* loss-of-function mutations affect localization of the MEC-4 channel and disrupt touch evoked membrane currents (Chelur et al., 2002).

Although the exact stoichiometry of the components of the mechanotransducer channel complex is not known, genetic data suggest that several proteins are present in multiple copies and in various combinations. Two subgroups of *mec* genes encoding peripheral components required for mechanotransduction in the touch receptor neurons can be defined, those encoding intracellular (*mec-7*, *mec-12*) and those encoding extracellular (*mec-1*, *mec-5*, *mec-9*) proteins (Driscoll and Kaplan, 1996; Driscoll and Tavernarakis, 1997; Tavernarakis and Driscoll, 1997). As described previously, the touch receptor processes are filled with bundled 15-protofilament microtubules. Mutations in two genes, *mec-7* and *mec-12*, disrupt the formation of these microtubules (Chalfie and Au, 1989; Chalfie and Sulston, 1981; Fukushige et al., 1999; Hamelin et al., 1992; Savage et al., 1989; Savage et al., 1994). *mec-7* encodes a β-tubulin expressed at high levels in the touch receptor neurons (Hamelin et al., 1992; Savage et al., 1989; Savage et al., 1994). MEC-7 is highly conserved–apart from the carboxy-terminal domain that is characteristically highly variable; only 7 amino acids differ from other β-tubulins. *mec-12* encodes an α-tubulin expressed at high levels in the touch receptor neurons but also expressed in several other neurons that do not assemble 15-protofilament microtubules (Fukushige et al., 1999). Thus, the presence of the MEC-12 tubulin is not sufficient to nucleate assembly of the touch-cell specific microtubules. As is the case for *mec-7*, many *mec-12* mutations are semi-dominant or dominant and are likely to disrupt subunit interactions or protofilament assembly (Gu et al., 1996). Recent data suggest that tethering of the channel complex to the microtubules is not essential for transduction, as mechanoreceptor currents are reduced but not eliminated in *mec-7* β-tubulin mutants (O'Hagan et al., 2005) and *mec-7* and *mec-12* null mutations do not prevent the formation of channel puncta (Emtage et al., 2004; Zhang et al., 2004).

In *mec-1* mutants, touch cells generally lack the mantle and associated periodic specializations of the overlying cuticle (Chalfie and Sulston, 1981; Gu et al., 1996; Savage et al., 1994). *mec-1* is expressed in touch receptor neurons, other lateral neurons and intestinal muscles. It encodes a likely secreted protein with multiple Kunitz-type serine protease inhibitor and EGF domains. The Kunitz and EGF domains are likely to be protein interaction domains. The C terminus of MEC-1 is needed for touch sensitivity, while the N terminus mediates the attachment of the touch neuron processes to the hypodermis (Emtage et al., 2004). MEC-1 is localized along the touch receptor processes in a punctuate manner and colocalizes with MEC-5 and the MEC-4/MEC-10 mechanosensory channel complex (Emtage et al., 2004).

mec-5 mutations disrupt the extracellular matrix in a subtle manner; the mantle in a wild-type animal can be stained with peanut lectin, whereas the mantle in *mec-5* mutants cannot (Chalfie and Sulston, 1981; Du et al., 1996; Gillespie and Walker, 2001). *mec-5* encodes a novel collagen type that is secreted by hypodermal cells (Du et al., 1996). The central portion of the *mec-5* protein is made up of Pro-rich Gly-X-Y repeats. *mec-5* mutations, many of which are temperature-sensitive, cluster toward the carboxy terminus of the protein and affect these repeats. Genetic interactions suggest that *mec-5* influences MEC-4/MEC-10 channel function (for example, *mec-4* and *mec-10* mutations can enhance the *mec-5(ts)* mutant phenotype; (Gu et al., 1996)). Thus, a specialized collagen could interact with the touch receptor channel, perhaps acting to provide gating tension.

mec-9 mutations do not alter mantle ultrastructure in a detectable manner, despite the fact that *mec-9* encodes a protein that appears to be secreted from the touch receptor neurons (Chalfie and Sulston, 1981; Du et al., 1996). The *mec-9* gene generates two transcripts, the larger of which encodes an 834 amino acid protein (MEC-9L) that is expressed only by the touch receptors (Chalfie and Sulston, 1981; Du et al., 1996). Akin to MEC-1, the predicted MEC-9L protein contains several domains related to the Kunitz-type serine protease inhibitor domain, the Ca^{2+}-binding EGF repeat, the non-Ca^{2+}-binding EGF repeat and a glutamic acid-rich domain (Chalfie and Sulston, 1981; Du et al., 1996). Single amino acid substitutions that disrupt MEC-9 function affect the two Ca^{2+}-binding EGF repeats, the sixth EGF repeat and the third Kunitz-type domain, thus implicating these regions as important in MEC-9 function (Chalfie and Sulston, 1981; Du et al., 1996). How MEC-9 is needed for touch cell activity is not clear, but it is interesting that MEC-9 appears specialized for protein interactions and that agrin, a protein that acts to localize acetylcholine receptors, has a domain structure that appears similarly specialized (agrin features multiple EGF and Kazal-type serine protease inhibitor repeats; (Rupp et al., 1992a; Rupp et al., 1992b; Rupp et al., 1991)). *mec-9* mutations are dominant enhancers of a *mec-5(ts)* allele, suggesting that these proteins might interact in the unique mantle extracellular matrix outside the touch receptor neuron (Du et al., 1996; Gu et al., 1996).

5.5 Proprioception

C. elegans senses forces arising within the body itself during movement, a phenomenon called proprioception. Animal locomotion results from alternate contraction and relaxation of dorsal and ventral body wall muscles, which generates a canonical sinusoidal pattern of movement (White et al., 1986; Wolinsky and Way, 1990). The arrangement of the body wall muscles and their synaptic inputs restricts locomotion to dorsal and ventral turns of the body. The body wall muscles are organized into two dorsal and two ventral rows. Each row consists of 23 or 24 diploid mononucleate muscle cells arranged in an interleaved pattern (Francis and Waterston, 1991; Moerman et al., 1996; Waterston et al., 1980; Williams and Waterston, 1994). Distinct classes of motorneurons control dorsal and ventral body muscles. To generate the sinusoidal pattern of movement, the contraction of the dorsal and ventral body muscles must be out of phase. For example, to turn the body dorsally, the dorsal muscles contract, while the opposing ventral muscles relax. Interactions between excitatory and inhibitory motorneurons produce a pattern of alternating dorsal and ventral contractions (Francis and Waterston, 1991; Hresko et al., 1994; Tavernarakis and Driscoll, 1997). Relatively little is known about how the sinusoidal wave is propagated along the body axis. Adjacent muscle cells are electrically coupled via gap junctions, which could couple excitation of adjacent body muscles. Alternatively, ventral cord motorneurons could promote wave propagation since gap junctions connect adjacent motorneurons of a given class (Chalfie et al., 1985; White et al., 1976; White et al., 1986). A third possibility is that motorneurons could themselves act as stretch receptors so that contraction of body muscles could regulate adjacent motorneuron activities, thereby propagating the wave (Tavernarakis and Driscoll, 1997; Tavernarakis and Driscoll, 2001b; Tavernarakis et al., 1997). Numerous mutations disrupt normal sinusoidal locomotion in *C. elegans*, resulting in animals with movement defects ranging from total paralysis, to severe uncoordination, to subtle and almost imperceptible irregularities in movement (Tavernarakis and Driscoll, 1997; Tavernarakis et al., 1997). Unusual, semi-dominant (sd), gain-of-function mutations in the gene *unc-8* induce transient neuronal swelling of embryonically derived motorneurons as well as some neurons in the head and tail ganglia, and severe uncoordination (Park and Horvitz, 1986b; Shreffler et al., 1995; Tavernarakis et al., 1997). Swelling is absent at hatching and peaks in severity late in L1 and L2. unc-8 encodes a degenerin expressed in several motor neuron classes, in some interneurons and in nose touch sensory neurons. Interestingly, semi-dominant unc-8 alleles alter an amino acid in the region hypothesized to be an extracellular channel-closing domain defined in studies of deg-1 and *mec-4* degenerins. The genetics of unc-8 are further similar to those of *mec-4* and *mec-10*; specific *unc-8* alleles can suppress or enhance *unc-8(sd)* mutations in trans, suggesting that UNC-8::UNC-8 interactions occur (Shreffler et al., 1995; Tavernarakis et al., 1997). Another degenerin family member, *del-1* (degenerin-like) is co-expressed in a subset of neurons that express *unc-8* (the VA and VB motor neurons) and is likely to assemble into a channel complex with UNC-8 in these cells (Tavernarakis et al., 1997). The UNC-8 and DEL-1 proteins include all domains characteristic of degenerin family members and are likely to adopt similar transmembrane topologies (amino and carboxy termini situated inside the cell and a large extracellular domain that includes three cysteine-rich regions). Neither degenerin has any primary sequence features that are markedly different from other *C. elegans* family members although one somewhat atypical feature of UNC-8 is that it has a relatively long C-terminal domain that shares some primary sequence homology with the extended C-terminus of another degenerin implicated in locomotion, UNC-105 (Liu et al., 1996; Park and Horvitz, 1986a).

The exact function of the UNC-8 degenerin channel in motorneurons was elucidated through genetic approaches. *unc-8* null mutants have a subtle locomotion defect; they inscribe a path in an E. coli lawn that is markedly reduced in both wavelength and amplitude as compared to wild type (Tavernarakis et al., 1997). This phenotype indicates that the UNC-8 degenerin channel functions to modulate the locomotory trajectory of the animal. How does the UNC-8 motor neuron channel influence locomotion? As mentioned earlier, one highly interesting morphological feature of some motorneurons (in particular, the VA and VB motorneurons that co-express unc-8 and del-1) is that their processes include extended regions that do not participate in neuromuscular junctions or neuronal synapses. These "undifferentiated" process regions have been hypothesized to be stretch-sensitive (White et al., 1986). Given the morphological features of certain motor neurons and the sequence similarity of UNC-8 and DEL-1 to the candidate mechanically-gated channels MEC-4 and MEC-10, we have proposed that these subunits co-assemble into a stretch-sensitive channel that might be localized to the undifferentiated regions of the motor neuron process (Tavernarakis et al., 1997). When activated by the localized body stretch that occurs during locomotion, this motor neuron channel potentiates signaling at the neuromuscular junction, which is situated at a distance from the site of stretch stimulus (Fig. 5.6). The stretch signal enhances motorneuron excitation of muscle, increasing the strength and duration of the pending muscle contraction and directing a full size body turn. In the absence of the stretch activation, the body wave and locomotion still occur, but with significantly reduced amplitude because the potentiating stretch signal is not transmitted. This model bears similarity to the chain reflex mechanism of movement pattern generation. However it does not exclude a central oscillator that would be responsible for the rhythmic locomotion. Instead, we suggest that the output of such an oscillator is further enhanced and modulated by stretch sensitive motorneurons (Tavernarakis and Driscoll, 1997; Tavernarakis et al., 1997). One important corollary of the *unc-8* mutant studies is that the UNC-8 channel does not appear to be essential for motor neuron function. If this were the case, animals lacking the *unc-8* gene would be severely paralyzed. This observation strengthens the argument that degenerin channels function directly in mechanotransduction rather than merely serving to maintain the osmotic environment so that other channels can function.

Muscle cells may also play part in the coordination of locomotion by sensing their own extent of stretch. Mutations in the muscle degenerin *unc-105* cause muscle hypercontraction (Garcia-Anoveros et al., 1998). The muscle hyper-contraction phenotype caused by dominant *unc-105* mutations can be suppressed by mutations near the carboxy-terminus of *let-2*, a gene that encodes the α2 chain of type-IV collagen found in the basement membrane between muscle cells and the hypodermis (Liu et al., 1996). It is tempting to speculate that LET-2 normally carries gating tension to the UNC-105 channel, when the muscle is stretched, thus providing regulatory feedback for muscle contraction. However, results from *in vivo* electrophysiogy suggest that UNC-105 is not involved in the formation of a muscle stretch receptor complex (Jospin et al., 2004).

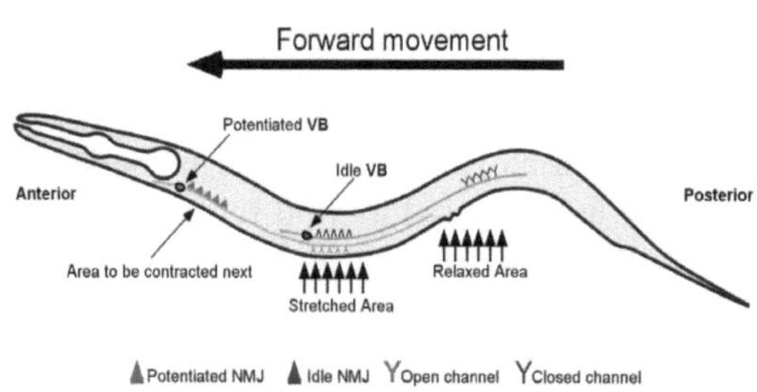

Fig. 5.6 A model for UNC-8 involvement in stretch-regulated control of locomotion. Schematic diagram of potentiated and inactive VB class motor neurons. Neuro-muscular junctions (signified by triangles) are made near the cell body (Tavernarakis et al., 1997; White et al., 1986). Mechanically-activated channels postulated to include UNC-8 (and, possibly in VB motor neurons, DEL-1) subunits (signified by Y figures) are hypothesized to be concentrated at the synapse-free, undifferentiated ends of the VB neuron. Mechanically-gated channels could potentiate local excitation of muscle. Body stretch is postulated to activate mechanically-gated channels which potentiate the motor neuron signal that excites a specific muscle field. Sequential activation of motor neurons that are distributed along the ventral nerve cord and signal non-overlapping groups of muscles, amplifies and propagates the sinusoidal body wave (NMJ: neuromuscular junction). Reproduced from (Voglis and Tavernarakis, 2005) with copyright permission of the Academia Publishing House Ltd

In addition to DEG/ENaC ion channels, a transient receptor potential (TRP) channel has recently been found to be critical for proprioception in *C. elegans*. TRP cation channels are present in all eukaryotes, from yeast to mammals and can be divided into seven subfamilies based on sequence similarities (Montell, 2005). TRP channels are linked to many physiological processes ranging from temperature sensation to mechanosensation and osmosensation (Caterina et al., 1999; Caterina et al., 1997; Colbert et al., 1997; Peier et al., 2002; Tobin et al., 2002). The *trp-4* gene is orthologous to the *Drosophila* nompC channel which is critical for hair cell mechanotransduction (Walker et al., 2000) and to the zebrafish nompC which is required for the function of auditory hair cells (Sidi et al., 2003). *C. elegans trp-4* mutants generate abnormal locomotion waves, with exaggerated body bands and larger than normal wave amplitudes (Li et al., 2006). TRP-4 acts in a single interneuron, called DVA and TRP-4 channels may act as stretch receptors in these neurons to provide sensory feedback to the locomotor control circuit. Thus, DVA appears to be a primary proprioreceptor neuron which modulates the shape of the locomotion sinusoidal wave.

5.6 Mechanotransduction in Other Organisms

Genetic studies of sensory mechanotransduction, which were initiated in *C. elegans* and are now also being carried out in *Drosophila* and in mammals, have converged to reveal a limited set of underlying mechanisms (Eberl, 1999; Gillespie and Walker, 2001; Harteneck et al., 2000; Kellenberger and Schild, 2002). For example, the model proposed for mechanotransduction in the touch receptor neurons (Fig. 5.7) and motorneurons of *C. elegans* shares the same underlying principle and features of the proposed gating mechanism of mechanosensory ion channels in Drosophila sensory bristles and the channels that respond to auditory stimuli in the hair cells of the vertebrate inner ear (Fettiplace and Fuchs, 1999; Gillespie and Walker, 2001; Hamill and McBride, 1996; Hudspeth, 1989; Hudspeth et al., 2000; Jaramillo and Hudspeth, 1991; Pickles and Corey, 1992; Pickles et al., 1991; Tavernarakis and Driscoll, 1997; Weinbaum et al., 2001). Hair cells have bundles of a few hundred stereocilia on their apical surface, which mediate sensory transduction Stereocilia are connected at their distal ends to neighboring stereocilia by filaments called tip links. The integrity of the tip links is essential for channel opening and the mechanosensitive channels appear to be situated at the ends of the stereocilia, near the connecting tip links. Directional deflection of the stereocilia relative to each other introduces tension on the tip links, which is proposed to open the mechanosensitive hair cell channels directly. This remarkable convergence of independent studies in distant species, strongly suggests that different mechanotransducers in different systems have evolved to strictly adhere to the same set of principles. Besides the DEG/ENaC family of ion channels, members of the TRP family are involved in mechanosensation in different organisms. (Alvarez de la Rosa et al., 2000; Duggan et al., 1998; Minke and Cook, 2002; Montell, 2001; Tavernarakis and Driscoll, 2001a; Welsh et al., 2002). Experiments in *Drosophila* revealed the involvement of TRP-like channel genes in the function of

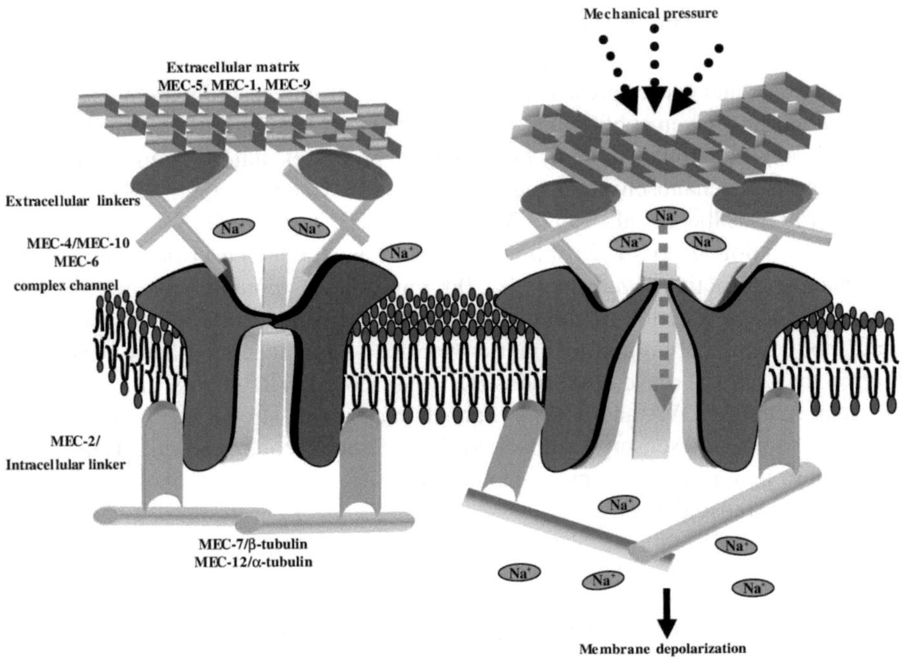

Fig. 5.7 A model for a mechanotransducing complex in *C. elegans* touch receptor neurons. The extracellular matrix contains MEC-5, MEC-9 and MEC-1; the sensory transduction channel is formed by MEC-4, MEC-10 and possibly MEC-6; MEC-2 enhances channel activity and tethers the channel to specialized microtubules containing MEC-7/β-tubulin and MEC-12/α-tubulin. Recent findings suggest that mechanotransduction can occur even when the 15-protofilament microtubules are missing in *mec-7* mutants, questioning in that way the necessity of tethering of the complex to microtubules (O'Hagan et al., 2005). In the absence of mechanical stimulation, the channel is closed and the sensory neuron is idle. Application of mechanical force to the body of the animal results in distortion of a network of interacting molecules that opens the degenerins channel. Na^+influx depolarizes the neuron, initiating a cascade that leads to the integration of the stimulus

mechanosensoty bristles (*nompC*; (Walker et al., 2000)), auditory receptors (*iav* and *nan*; (Gong et al., 2004)), and nociceptors (*painless*; (Tracey et al., 2003)). Another member of the TRP protein family, the TRPA1 channel has been identified as a candidate mechanotransducing channel in the mouse (Corey et al., 2004). *In situ* hybridization revealed that the TRPA1 channel is expressed in the cochlea organ of Corti, which contains the auditory hair cells. Additional colocalization experiments link TRPA1 to mechanosensation: TRPA1 is expressed together with two accessory proteins of the mechanosensory apparatus, myosin 1c and cadherin 23, at the tips of stereocilia throughout the kinocilium and in the pericuticular zone. Whole-cell patch clamp recording of inner hair cells in mice show that the transduction current produced is significantly reduced in the absence of TRPA1, indicating that this channel is a component of the mechanosensitive transduction channel of vertebrate hair cells (Corey et al., 2004). Additionally, two TRP proteins, a NompC-like protein and TRPA1 are required for hair cell function in zebrafish (Corey et al., 2004; Sidi et al., 2003).

Despite enormous progress on the illumination of vertebrate mechanosensory cell biology achieved in recent years, there is still a striking gap between the biophysical information that has accumulated and our understanding of the molecular aspects of mechanosensation. Sophisticated experiments in mice and humans revealed many genes involved in the development and function of the mammalian cochlea and have cumulated in the formulation of the gating-spring model for hair cell mechanotransduction (Gillespie, 1995; Gillespie and Walker, 2001). However, many pieces of the mechanotransducing apparatus puzzle are still missing. Work in lower vertebrates such as birds, amphibians and fish has also contributed significantly in complementing and extending the studies with mammals. In these animals mechanosensory structures are often much easier to access, follow and monitor providing large potential for investigating the molecular basis of auditory transduction (Ashmore, 1998; Smotherman and Narins, 2000). An increasing amount of evidence suggests that some mammalian DEG/ENaC proteins may play a role in mechanosensation similarly to their nematode counterparts. In mammals, there are strong indications that ENaC subunits may be components of the baroreceptor mechanotransducer, one of the most potent regulators of anterial pressure and neurohumoral control of the circulation (Drummond et al., 1998; Drummond et al., 2001). Members of the ASIC (acid sensing ion channel) subgroup of the DEG/ENaC family have been implicated in mechanotransduction in mammals. BNC1 (brain Na^+ channel; also known as MDEG, BNaC1, ASIC2; (Garcia-Anoveros and Corey, 1997; Price et al., 1996; Waldmann et al., 1996; Waldmann and Lazdunski, 1998)) has emerged as promising candidate for a mechanosensitive channel. In BNC1 null mice touch receptor neurons of the skin produce fewer action potentials than in wild type animals over a comparable range of stimuli (Price et al., 2000).

5.7 Conclusions

Studies in *C. elegans* have contributed critical insights into the cellular and molecular mechanisms of mechanotransduction (Fig. 5.7; Chalfie, 1997; Syntichaki and Tavernarakis, 2004; Tavernarakis and Driscoll, 1997; Tavernarakis and Driscoll, 2001b). Recently developed powerful methodologies, such as direct electrophysiological recordings from neurons and imaging of genetically encoded calcium sensors provide the unique opportunity to investigate the properties of the mechanotransduction apparatus in the context of live, behaving animals.

Although our understanding of metazoan mechanosensation has been advanced significantly, open questions still remain. While specific DEG/ENaC and TRP ion channels have been directly implicated in the process of mechanotransduction the *C. elegans* genome encodes many more members of these ion channel proteins. The role of these proteins in mechanotransduction remains to be elucidated.

An additional major question that remains to be addressed is whether the mammalian counterparts of the *C. elegans* degenerins play specialized roles in mechanical signalling in humans. A significant step toward addressing this question has been accomplished with the demonstration that BNC1 is involved in mechanosensory signalling in the skin as we have described above. Even though the candidacy of BNC1 for being in the core of a mechanotransducing complex was greatly boosted by these results, a demanding critic would argue that albeit very strong, it still remains just a candidacy. The potential role of BNC1 as part of the core mechanotransducing channel can still only be inferred from these experiments and is not directly proven. It is still possible that BNC1 forms or participates in an auxiliary channel that facilitates the function of the actual mechanotransducing channel. A BNC1 knockout does not completely eliminate the responses to mechanical stimuli (Price et al., 2000). The incomplete nature of the BNC1 deficiency effects indicates that even if BNC1 is indeed part of the core mechanosensory channel, it most likely is not the only critical one. Alternatively, there might be more than one, different mechanotransducing complexes within one neuron, with different properties and composition. The above arguments however, are by no means confined to BNC1. On the same basis, MEC-4/MEC-10 and UNC-8/DEL-1 in *C. elegans* as well as PPK in *Drosophila* might not be parts of the real mechanotransducer but only auxiliary ion channels. The recent identification of another strong candidate mechanosensory channel, the *Drosophila* NompC, adds to the list of candidate mechanosensitive ion channels (Walker et al., 2000). Evidence implicating NompC in mechanotransduction is especially convincing given the supporting electrophysiological analysis that is feasible in this system, and the availability of mutants with altered properties and intermediate effects (Walker et al., 2000). Therefore, NompC homologues in other organisms, including humans, emerge putative mechanosensitive ion channels. Even in this case however, there are caveats; the absence of NompC does not completely eliminate mechanosensitive currents in *Drosophila* hair bristles. Furthermore, the identities and properties of force-generating tethers of NompC in mechanotransducing complexes will need to be determined. Another issue that needs to be addressed is the potential interplay between DEG/ENaC and NompC channels in mechanosensory cells before a clear understanding of mechanotransduction can be achieved.

Acknowledgment We gratefully acknowledge the contributions of numerous investigators that we did not include in this review. Work in the authors' laboratory is funded by grants from EMBO and the EU 6th Framework Programme to N.T. N.T. is an EMBO Young Investigator.

References

Alvarez de la Rosa, D., C. M. Canessa, G. K. Fyfe, and P. Zhang. 2000. Structure and regulation of amiloride-sensitive sodium channels. Annu Rev Physiol 62:573–94.

Ashmore, J. 1998. Mechanosensation: swimming round in circles. Curr Biol 8:425–7.

Avery, L. 1993. The genetics of feeding in Caenorhabditis elegans. Genetics 133:897–917.

Bargmann, C. I., and J. M. Kaplan. 1998. Signal transduction in the Caenorhabditis elegans nervous system. Annu Rev Neurosci 21:279–308.

Bargmann, C. I., J. H. Thomas, and H. R. Horvitz. 1990. Chemosensory cell function in the behavior and development of Caenorhabditis elegans. Cold Spring Harb Symp Quant Biol 55:529–38.

Barnes, T. M., Y. Jin, H. R. Horvitz, G. Ruvkun, and S. Hekimi. 1996. The Caenorhabditis elegans behavioral gene unc-24 encodes a novel bipartite protein similar to both erythrocyte band 7.2 (stomatin) and nonspecific lipid transfer protein. J Neurochem 67:46–57.

Baumeister, R., and L. Ge. 2002. The worm in us - Caenorhabditis elegans as a model of human disease. Trends Biotechnol 20:147–8.

Block, I., N. Freiberger, O. Gavrilova, and R. Hemmersbach. 1999. Putative graviperception mechanisms of protists. Adv Space Res 24:877–82.

Blount, P., and P. C. Moe. 1999. Bacterial mechanosensitive channels: integrating physiology, structure and function. Trends Microbiol 7:420–4.

Brenner, S. 1974. The genetics of Caenorhabditis elegans. Genetics 77:71–94.

Caterina, M. J., T. A. Rosen, M. Tominaga, A. J. Brake, and D. Julius. 1999. A capsaicin-receptor homologue with a high threshold for noxious heat. Nature 398:436–41.

Caterina, M. J., M. A. Schumacher, M. Tominaga, T. A. Rosen, J. D. Levine, and D. Julius. 1997. The capsaicin receptor: a heat-activated ion channel in the pain pathway. Nature 389:816–24.

Chalfie, M. 1995. The differentiation and function of the touch receptor neurons of Caenorhabditis elegans. Prog Brain Res 105:179–82.

Chalfie, M. 1997. A molecular model for mechanosensation in Caenorhabditis elegans. Biol Bull 192:125.

Chalfie, M. 1993. Touch receptor development and function in Caenorhabditis elegans. J Neurobiol 24:1433–41.

Chalfie, M., and M. Au. 1989. Genetic control of differentiation of the Caenorhabditis elegans touch receptor neurons. Science 243:1027–33.

Chalfie, M., E. Dean, E. Reilly, K. Buck, and J. N. Thomson. 1986. Mutations affecting microtubule structure in Caenorhabditis elegans. J Cell Sci Suppl 5:257–71.

Chalfie, M., M. Driscoll, and M. Huang. 1993. Degenerin similarities. Nature 361:504.

Chalfie, M., and J. Sulston. 1981. Developmental genetics of the mechanosensory neurons of Caenorhabditis elegans. Dev Biol 82:358–70.

Chalfie, M., J. E. Sulston, J. G. White, E. Southgate, J. N. Thomson, and S. Brenner. 1985. The neural circuit for touch sensitivity in Caenorhabditis elegans. J Neurosci 5:956–64.

Chalfie, M., and J. N. Thomson. 1979. Organization of neuronal microtubules in the nematode Caenorhabditis elegans. J Cell Biol 82:278–89.

Chalfie, M., and J. N. Thomson. 1982. Structural and functional diversity in the neuronal microtubules of Caenorhabditis elegans. J Cell Biol 93:15–23.

Chalfie, M., J. N. Thomson, and J. E. Sulston. 1983. Induction of neuronal branching in Caenorhabditis elegans. Science 221:61–3.

Chalfie, M., and E. Wolinsky. 1990. The identification and suppression of inherited neurodegeneration in Caenorhabditis elegans. Nature 345:410–6.

Chelur, D. S., G. G. Ernstrom, M. B. Goodman, C. A. Yao, A. F. Chen, R. O'Hagan, and M. Chalfie. 2002. The mechanosensory protein MEC-6 is a subunit of the C. elegans touch-cell degenerin channel. Nature 420:669–73.

Chiba, C. M., and C. H. Rankin. 1990. A developmental analysis of spontaneous and reflexive reversals in the nematode Caenorhabditis elegans. J Neurobiol 21:543–54.

Colbert, H. A., T. L. Smith, and C. I. Bargmann. 1997. OSM-9, a novel protein with structural similarity to channels, is required for olfaction, mechanosensation, and olfactory adaptation in Caenorhabditis elegans. J Neurosci 17:8259–69.

Corey, D. P. 2003. New TRP channels in hearing and mechanosensation. Neuron 39:585–8.

Corey, D. P., J. Garcia-Anoveros, J. R. Holt, K. Y. Kwan, S. Y. Lin, M. A. Vollrath, A. Amalfitano, E. L. Cheung, B. H. Derfler, A. Duggan, G. S. Geleoc, P. A. Gray, M. P. Hoffman, H. L. Rehm, Tamasauskas, D., and D. S. Zhang. 2004. TRPA1 is a candidate for the mechanosensitive transduction channel of vertebrate hair cells. Nature 432:723–30.

Coulson, A., R. Waterston, J. Kiff, J. Sulston, and Y. Kohara. 1988. Genome linking with yeast artificial chromosomes. Nature 335:184–6.

Delaunay, J., G. Stewart, and A. Iolascon. 1999. Hereditary dehydrated and overhydrated stomatocytosis: recent advances. Curr Opin Hematol 6:110–4.

Driscoll, M. 1996. Cell death in C. elegans: molecular insights into mechanisms conserved between nematodes and mammals. Brain Pathol 6:411–25.

Driscoll, M., and M. Chalfie. 1991. The mec-4 gene is a member of a family of Caenorhabditis elegans genes that can mutate to induce neuronal degeneration. Nature 349:588–93.

Driscoll, M., and J. M. Kaplan. 1996. Mechanotransduction, p. 645–677. *In* D. L. Riddle, T. Blumenthal, B. J. Meyer, and J. R. Pries (ed.), The Nematode C. elegans, II. Cold Spring Harbor Laboratory Press, Cold Spring Harbor, NY.

Driscoll, M., and N. Tavernarakis. 1997. Molecules that mediate touch transduction in the nematode Caenorhabditis elegans. Gravit Space Biol Bull 10:33–42.

Drummond, H. A., M. P. Price, M. J. Welsh, and F. M. Abboud. 1998. A molecular component of the arterial baroreceptor mechanotransducer. Neuron 21:1435–41.

Drummond, H. A., M. J. Welsh, and F. M. Abboud. 2001. ENaC subunits are molecular components of the arterial baroreceptor complex. Ann N Y Acad Sci 940:42–7.

Du, H., and M. Chalfie. 2001. Genes regulating touch cell development in Caenorhabditis elegans. Genetics 158:197–207.

Du, H., G. Gu, C. M. William, and M. Chalfie. 1996. Extracellular proteins needed for C. elegans mechanosensation. Neuron 16:183–94.

Duggan, A., C. Ma, and M. Chalfie. 1998. Regulation of touch receptor differentiation by the Caenorhabditis elegans mec-3 and unc-86 genes. Development 125:4107–19.

Eberl, D. F. 1999. Feeling the vibes: chordotonal mechanisms in insect hearing. Curr Opin Neurobiol 9:389–93.

Emtage, L., G. Gu, E. Hartwieg, and M. Chalfie. 2004. Extracellular proteins organize the mechanosensory channel complex in C. elegans touch receptor neurons. Neuron 44:795–807.

Fettiplace, R., and P. A. Fuchs. 1999. Mechanisms of hair cell tuning. Annu Rev Physiol 61:809–34.

Finney, M., and G. Ruvkun. 1990. The unc-86 gene product couples cell lineage and cell identity in C. elegans. Cell 63:895–905.

Finney, M., G. Ruvkun, and H. R. Horvitz. 1988. The C. elegans cell lineage and differentiation gene unc-86 encodes a protein with a homeodomain and extended similarity to transcription factors. Cell 55:757–69.

Fire, A., S. Xu, M. K. Montgomery, S. A. Kostas, S. E. Driver, and C. C. Mello. 1998. Potent and specific genetic interference by double-stranded RNA in Caenorhabditis elegans. Nature 391:806–11.

Francis, R., and R. H. Waterston. 1991. Muscle cell attachment in Caenorhabditis elegans. J Cell Biol 114:465–79.

French, A. S. 1992. Mechanotransduction. Annu Rev Physiol 54:135–52.

Fukushige, T., Z. K. Siddiqui, M. Chou, J. G. Culotti, C. B. Gogonea, S. S. Siddiqui, and M. Hamelin. 1999. MEC-12, an alpha-tubulin required for touch sensitivity in C. elegans. J Cell Sci 112:395–403.

Garcia-Anoveros, J., and D. P. Corey. 1997. The molecules of mechanosensation. Annu Rev Neurosci 20:567–94.

Garcia-Anoveros, J., and D. P. Corey. 1996. Touch at the molecular level. Mechanosensation. Curr Biol 6:541–3.

Garcia-Anoveros, J., J. A. Garcia, J. D. Liu, and D. P. Corey. 1998. The nematode degenerin UNC-105 forms ion channels that are activated by degeneration- or hypercontraction-causing mutations. Neuron 20:1231–41.

Gebauer, M., D. Watzke, and H. Machemer. 1999. The gravikinetic response of Paramecium is based on orientation-dependent mechanotransduction. Naturwissenschaften 86:352–6.

Gillespie, P. G. 1995. Molecular machinery of auditory and vestibular transduction. Curr Opin Neurobiol 5:449–55.

Gillespie, P. G., and R. G. Walker. 2001. Molecular basis of mechanosensory transduction. Nature 413:194–202.

Golden, J. W., and D. L. Riddle. 1984. The Caenorhabditis elegans dauer larva: developmental effects of pheromone, food, and temperature. Dev Biol 102:368–78.

Gong, Z., W. Son, Y. D. Chung, J. Kim, D. W. Shin, C. A. McClung, Y. Lee, H. W. Lee, D. J. Chang, B. K. Kaang, H. Cho, U. Oh, J. Hirsh, M. J. Kernan, and C. Kim. 2004. Two interdependent TRPV channel subunits, inactive and Nanchung, mediate hearing in Drosophila. J Neurosci 24:9059–66.

Goodman, M. B., G. G. Ernstrom, D. S. Chelur, R. O'Hagan, C. A. Yao, and M. Chalfie. 2002. MEC-2 regulates C. elegans DEG/ENaC channels needed for mechanosensation. Nature 415:1039–42.

Gu, G., G. Caldwell, and M. Chalfie. 1996. Genetic interactions affecting touch sensitivity in Caenorhabditis elegans. Proc Natl Acad Sci U S A 93:6577 - 82.

Hackney, C. M., and D. N. Furness. 1995. Mechanotransduction in vertebrate hair cells: structure and function of the stereociliary bundle. Am J Physiol 268:C1–13.

Hall, D. H., G. Gu, J. Garcia-Anoveros, L. Gong, M. Chalfie, and M. Driscoll. 1997. Neuropathology of degenerative cell death in Caenorhabditis elegans. J Neurosci 17:1033–45.

Hamelin, M., I. M. Scott, J. C. Way, and J. G. Culotti. 1992. The mec-7 beta-tubulin gene of Caenorhabditis elegans is expressed primarily in the touch receptor neurons. Embo J 11:2885–93.

Hamill, O. P., J. W. Lane, and D. W. McBride, Jr. 1992. Amiloride: a molecular probe for mechanosensitive channels. Trends Pharmacol Sci 13:373–6.

Hamill, O. P., and B. Martinac. 2001. Molecular basis of mechanotransduction in living cells. Physiol Rev 81:685–740.

Hamill, O. P., and D. W. McBride, Jr. 1996. The pharmacology of mechanogated membrane ion channels. Pharmacol Rev 48:231–52.

Harbinder, S., N. Tavernarakis, L. A. Herndon, M. Kinnell, S. Q. Xu, A. Fire, and M. Driscoll. 1997. Genetically targeted cell disruption in Caenorhabditis elegans. Proc Natl Acad Sci U S A 94:13128–33.

Hart, A. C., J. Kass, J. E. Shapiro, and J. M. Kaplan. 1999. Distinct signaling pathways mediate touch and osmosensory responses in a polymodal sensory neuron. J Neurosci 19:1952–8.

Hart, A. C., S. Sims, and J. M. Kaplan. 1995. Synaptic code for sensory modalities revealed by C. elegans GLR-1 glutamate receptor. Nature 378:82–5.

Harteneck, C., T. D. Plant, and G. Schultz. 2000. From worm to man: three subfamilies of TRP channels. Trends Neurosci 23:159–66.

Hemmersbach, R., B. Bromeis, I. Block, R. Braucker, M. Krause, N. Freiberger, C. Stieber, and M. Wilczek. 2001. Paramecium–a model system for studying cellular graviperception. Adv Space Res 27:893–8.

Herman, R. K. 1996. Touch sensation in Caenorhabditis elegans. Bioessays 18:199–206.

Hilliard, M. A., A. J. Apicella, R. Kerr, H. Suzuki, P. Bazzicalupo, and W. R. Schafer. 2005. In vivo imaging of C. elegans ASH neurons: cellular response and adaptation to chemical repellents. Embo J 24:63–72.

Hoger, U., P. H. Torkkeli, E. A. Seyfarth, and A. S. French. 1997. Ionic selectivity of mechanically activated channels in spider mechanoreceptor neurons. J Neurophysiol 78:2079–85.

Hresko, M. C., B. D. Williams, and R. H. Waterston. 1994. Assembly of body wall muscle and muscle cell attachment structures in Caenorhabditis elegans. J Cell Biol 124: 491–506.

Huang, M., and M. Chalfie. 1994. Gene interactions affecting mechanosensory transduction in Caenorhabditis elegans. Nature 367:467–70.

Huang, M., G. Gu, E. L. Ferguson, and M. Chalfie. 1995. A stomatin-like protein necessary for mechanosensation in C. elegans. Nature 378:292–5.

Hudspeth, A. J. 1989. How the ear's works work. Nature 341:397–404.

Hudspeth, A. J., Y. Choe, A. D. Mehta, and P. Martin. 2000. Putting ion channels to work: mechanoelectrical transduction, adaptation, and amplification by hair cells. Proc Natl Acad Sci U S A 97:11765–72.

Ingber, D. E. 1997. Tensegrity: the architectural basis of cellular mechanotransduction. Annu Rev Physiol 59:575–99.

Jaramillo, F., and A. J. Hudspeth. 1991. Localization of the hair cell's transduction channels at the hair bundle's top by iontophoretic application of a channel blocker. Neuron 7:409–20.

Jospin, M., M. C. Mariol, L. Segalat, and B. Allard. 2004. Patch clamp study of the UNC-105 degenerin and its interaction with the LET-2 collagen in Caenorhabditis elegans muscle. J Physiol 557:379–88.
Kahn-Kirby, A. H., and C. I. Bargmann. 2006. TRP channels in C. elegans. Annu Rev Physiol 68:719–36.
Kahn-Kirby, A. H., J. L. Dantzker, A. J. Apicella, W. R. Schafer, J. Browse, C. I. Bargmann, and J. L. Watts. 2004. Specific polyunsaturated fatty acids drive TRPV-dependent sensory signaling in vivo. Cell 119:889–900.
Kaplan, J. M., and H. R. Horvitz. 1993. A dual mechanosensory and chemosensory neuron in Caenorhabditis elegans. Proc Natl Acad Sci U S A 90:2227–31.
Kellenberger, S., and L. Schild. 2002. Epithelial sodium channel/degenerin family of ion channels: a variety of functions for a shared structure. Physiol Rev 82:735–67.
Kitamura, K. I., S. Amano, and R. Hosono. 2001. Contribution of neurons to habituation to mechanical stimulation in Caenorhabditis elegans. J Neurobiol 46:29–40.
Kloda, A., and B. Martinac. 2001. Molecular identification of a mechanosensitive channel in archaea. Biophys J 80:229–40.
Ko, K. S., and C. A. McCulloch. 2001. Intercellular mechanotransduction: cellular circuits that coordinate tissue responses to mechanical loading. Biochem Biophys Res Commun 285:1077–83.
Koch, A. L. 1994. Development and diversification of the Last Universal Ancestor. J Theor Biol 168:269–80.
Koch, R., H. G. van Luenen, M. van der Horst, K. L. Thijssen, and R. H. Plasterk. 2000. Single nucleotide polymorphisms in wild isolates of Caenorhabditis elegans. Genome Res 10: 1690–6.
Koprowski, P., and A. Kubalski. 2001. Bacterial ion channels and their eukaryotic homologues. Bioessays 23:1148–58.
Lane, J. W., D. W. McBride, Jr., and O. P. Hamill. 1991. Amiloride block of the mechanosensitive cation channel in Xenopus oocytes. J Physiol 441:347–66.
Lee, R. T., and H. Huang. 2000. Mechanotransduction and arterial smooth muscle cells: new insight into hypertension and atherosclerosis. Ann Med 32:233–5.
Li, W., Z. Feng, P. W. Sternberg, and X. Z. Xu. 2006. A C. elegans stretch receptor neuron revealed by a mechanosensitive TRP channel homologue. Nature 440:684–7.
Liedtke, W., D. M. Tobin, C. I. Bargmann, and J. M. Friedman. 2003. Mammalian TRPV4 (VR-OAC) directs behavioral responses to osmotic and mechanical stimuli in Caenorhabditis elegans. Proc Natl Acad Sci U S A 100 Suppl 2:14531–6.
Liu, D. W., and J. H. Thomas. 1994. Regulation of a periodic motor program in C. elegans. J Neurosci 14:1953–62.
Liu, J., B. Schrank, and R. H. Waterston. 1996. Interaction between a putative mechanosensory membrane channel and a collagen. Science 273:361–4.
Liu, K. S., and P. W. Sternberg. 1995. Sensory regulation of male mating behavior in Caenorhabditis elegans. Neuron 14:79–89.
Lynch, T. M., P. M. Lintilhac, and D. Domozych. 1998. Mechanotransduction molecules in the plant gravisensory response: amyloplast/statolith membranes contain a beta 1 integrin-like protein. Protoplasma 201:92–100.
Mah, K. B., and C. H. Rankin. 1992. An analysis of behavioral plasticity in male Caenorhabditis elegans. Behav Neural Biol 58:211–21.
Marino, M. J., T. G. Sherman, and D. C. Wood. 2001. Partial cloning of putative G-proteins modulating mechanotransduction in the ciliate stentor. J Eukaryot Microbiol 48:527–36.
Martinac, B. 2001. Mechanosensitive channels in prokaryotes. Cell Physiol Biochem 11:61–76.
Mello, C., and A. Fire. 1995. DNA transformation. Methods Cell Biol 48:451–82.
Minke, B., and B. Cook. 2002. TRP channel proteins and signal transduction. Physiol Rev 82: 429–72.
Moerman, D. G., H. Hutter, G. P. Mullen, and R. Schnabel. 1996. Cell autonomous expression of perlecan and plasticity of cell shape in embryonic muscle of Caenorhabditis elegans. Dev Biol 173:228–42.

Montell, C. 2001. Physiology, phylogeny, and functions of the TRP superfamily of cation channels. Sci STKE 2001.

Montell, C. 2005. The TRP superfamily of cation channels. Sci STKE 2005.

Nauli, S. M., F. J. Alenghat, Y. Luo, E. Williams, P. Vassilev, X. Li, A. E. Elia, W. Lu, E. M. Brown, S. J. Quinn, D. E. Ingber, and J. Zhou. 2003. Polycystins 1 and 2 mediate mechanosensation in the primary cilium of kidney cells. Nat Genet 33:129–37.

Norris, V., M. S. Madsen, and P. Freestone. 1996. Elements of a unifying theory of biology. Acta Biotheor 44:209–18.

O'Hagan, R., and M. Chalfie. 2006. Mechanosensation in Caenorhabditis elegans. Int Rev Neurobiol 69:169–203.

O'Hagan, R., M. Chalfie, and M. B. Goodman. 2005. The MEC-4 DEG/ENaC channel of Caenorhabditis elegans touch receptor neurons transduces mechanical signals. Nat Neurosci 8:43–50.

Park, E. C., and H. R. Horvitz. 1986. C. elegans unc-105 mutations affect muscle and are suppressed by other mutations that affect muscle. Genetics 113:853–67.

Park, E. C., and H. R. Horvitz. 1986. Mutations with dominant effects on the behavior and morphology of the nematode Caenorhabditis elegans. Genetics 113:821–52.

Peier, A. M., A. J. Reeve, D. A. Andersson, A. Moqrich, T. J. Earley, A. C. Hergarden, G. M. Story, S. Colley, J. B. Hogenesch, P. McIntyre, S. Bevan, and A. Patapoutian. 2002. A heat-sensitive TRP channel expressed in keratinocytes. Science 296:2046–9.

Perkins, L. A., E. M. Hedgecock, J. N. Thomson, and J. G. Culotti. 1986. Mutant sensory cilia in the nematode Caenorhabditis elegans. Dev Biol 117:456–87.

Pickard, B. G., and J. P. Ding. 1993. The mechanosensory calcium-selective ion channel: key component of a plasmalemmal control centre? Aust J Plant Physiol 20:439–59.

Pickles, J. O., and D. P. Corey. 1992. Mechanoelectrical transduction by hair cells. Trends Neurosci 15:254–9.

Pickles, J. O., G. W. Rouse, and M. von Perger. 1991. Morphological correlates of mechanotransduction in acousticolateral hair cells. Scanning Microsc 5:1115–24.

Price, M. P., G. R. Lewin, S. L. McIlwrath, C. Cheng, J. Xie, P. A. Heppenstall, C. L. Stucky, A. G. Mannsfeldt, T. J. Brennan, H. A. Drummond, J. Qiao, C. J. Benson, D. E. Tarr, R. F. Hrstka, B. Yang, R. A. Williamson, and M. J. Welsh. 2000. The mammalian sodium channel BNC1 is required for normal touch sensation. Nature 407:1007–11.

Price, M. P., P. M. Snyder, and M. J. Welsh. 1996. Cloning and expression of a novel human brain Na+ channel. J Biol Chem 271:7879–82.

Rankin, C. H. 1991. Interactions between two antagonistic reflexes in the nematode Caenorhabditis elegans. J Comp Physiol [A] 169:59–67.

Rankin, C. H., T. Gannon, and S. R. Wicks. 2000. Developmental analysis of habituation in the Nematode C. elegans. Dev Psychobiol 36:261–70.

Roayaie, K., J. G. Crump, A. Sagasti, and C. I. Bargmann. 1998. The G alpha protein ODR-3 mediates olfactory and nociceptive function and controls cilium morphogenesis in C. elegans olfactory neurons. Neuron 20:55–67.

Rossier, B. C., C. M. Canessa, L. Schild, and J. D. Horisberger. 1994. Epithelial sodium channels. Curr Opin Nephrol Hypertens 3:487–96.

Rupp, F., W. Hoch, J. T. Campanelli, T. Kreiner, and R. H. Scheller. 1992. Agrin and the organization of the neuromuscular junction. Curr Opin Neurobiol 2:88–93.

Rupp, F., T. Ozcelik, M. Linial, K. Peterson, U. Francke, and R. Scheller. 1992. Structure and chromosomal localization of the mammalian agrin gene. J Neurosci 12:3535–44.

Rupp, F., D. G. Payan, C. Magill-Solc, D. M. Cowan, and R. H. Scheller. 1991. Structure and expression of a rat agrin. Neuron 6:811–23.

Sackin, H. 1995. Mechanosensitive channels. Annu Rev Physiol 57:333–53.

Sadoshima, J., T. Takahashi, L. Jahn, and S. Izumo. 1992. Roles of mechano-sensitive ion channels, cytoskeleton, and contractile activity in stretch-induced immediate-early gene expression and hypertrophy of cardiac myocytes. Proc Natl Acad Sci USA 89:9905–9.

Savage, C., M. Hamelin, J. G. Culotti, A. Coulson, D. G. Albertson, and M. Chalfie. 1989. mec-7 is a beta-tubulin gene required for the production of 15-protofilament microtubules in Caenorhabditis elegans. Genes Dev 3:870–81.

Savage, C., Y. Xue, S. Mitani, D. Hall, R. Zakhary, and M. Chalfie. 1994. Mutations in the Caenorhabditis elegans beta-tubulin gene mec-7: effects on microtubule assembly and stability and on tubulin autoregulation. J Cell Sci 107:2165–75.

Sawin, E. R., R. Ranganathan, and H. R. Horvitz. 2000. C. elegans locomotory rate is modulated by the environment through a dopaminergic pathway and by experience through a serotonergic pathway. Neuron 26:619–31.

Sedensky, M. M., J. M. Siefker, and P. G. Morgan. 2001. Model organisms: new insights into ion channel and transporter function. Stomatin homologues interact in Caenorhabditis elegans. Am J Physiol Cell Physiol 280:1340–8.

Shreffler, W., T. Magardino, K. Shekdar, and E. Wolinsky. 1995. The unc-8 and sup-40 genes regulate ion channel function in Caenorhabditis elegans motorneurons. Genetics 139:1261–72.

Sidi, S., R. W. Friedrich, and T. Nicolson. 2003. NompC TRP channel required for vertebrate sensory hair cell mechanotransduction. Science 301:96–9.

Smotherman, M. S., and P. M. Narins. 2000. Hair cells, hearing and hopping: a field guide to hair cell physiology in the frog. J Exp Biol 203:2237–46.

Snyers, L., E. Umlauf, and R. Prohaska. 1998. Oligomeric nature of the integral membrane protein stomatin. J Biol Chem 273:17221–6.

Stewart, G. W. 1997. Stomatin. Int J Biochem Cell Biol 29:271–4.

Stewart, G. W., A. C. Argent, and B. C. Dash. 1993. Stomatin: a putative cation transport regulator in the red cell membrane. Biochim Biophys Acta 1225:15–25.

Sulston, J. E., D. G. Albertson, and J. N. Thomson. 1980. The Caenorhabditis elegans male: postembryonic development of nongonadal structures. Dev Biol 78:542–76.

Sulston, J. E., and H. R. Horvitz. 1977. Post embriyonic cell lineages of the nematode *Caenorhabditis elegans*. Dev Biol 56:110–156.

Suzuki, H., R. Kerr, L. Bianchi, C. Frokjaer-Jensen, D. Slone, J. Xue, B. Gerstbrein, M. Driscoll, and W. R. Schafer. 2003. In vivo imaging of C. elegans mechanosensory neurons demonstrates a specific role for the MEC-4 channel in the process of gentle touch sensation. Neuron 39:1005–17.

Syntichaki, P., and N. Tavernarakis. 2004. Genetic models of mechanotransduction: the nematode Caenorhabditis elegans. Physiol Rev 84:1097–153.

Tavernarakis, N., and M. Driscoll. 2001. Degenerins. At the core of the metazoan mechanotransducer? Ann N Y Acad Sci 940:28–41.

Tavernarakis, N., and M. Driscoll. 2001. Mechanotransduction in Caenorhabditis elegans: the role of DEG/ENaC ion channels. Cell Biochem Biophys 35:1–18.

Tavernarakis, N., and M. Driscoll. 1997. Molecular modeling of mechanotransduction in the nematode Caenorhabditis elegans. Annu Rev Physiol 59:659–89.

Tavernarakis, N., M. Driscoll, and N. C. Kyrpides. 1999. The SPFH domain: implicated in regulating targeted protein turnover in stomatins and other membrane-associated proteins. Trends Biochem Sci 24:425–7.

Tavernarakis, N., W. Shreffler, S. Wang, and M. Driscoll. 1997. unc-8, a DEG/ENaC family member, encodes a subunit of a candidate mechanically gated channel that modulates C. elegans locomotion. Neuron 18:107–19.

Tavernarakis, N., S. L. Wang, M. Dorovkov, A. Ryazanov, and M. Driscoll. 2000. Heritable and inducible genetic interference by double-stranded RNA encoded by transgenes. Nat Genet 24:180–3.

Tavi, P., M. Laine, M. Weckstrom, and H. Ruskoaho. 2001. Cardiac mechanotransduction: from sensing to disease and treatment. Trends Pharmacol Sci 22:254–60.

Thomas, J. H., M. J. Stern, and H. R. Horvitz. 1990. Cell interactions coordinate the development of the C. elegans egg-laying system. Cell 62:1041–52.

Tobin, D., D. Madsen, A. Kahn-Kirby, E. Peckol, G. Moulder, R. Barstead, A. Maricq, and C. Bargmann. 2002. Combinatorial expression of TRPV channel proteins defines their sensory functions and subcellular localization in C. elegans neurons. Neuron 35:307–18.

Tracey, W. D., Jr., R. I. Wilson, G. Laurent, and S. Benzer. 2003. painless, a Drosophila gene essential for nociception. Cell 113:261–73.

Troemel, E. R., J. H. Chou, N. D. Dwyer, H. A. Colbert, and C. I. Bargmann. 1995. Divergent seven transmembrane receptors are candidate chemosensory receptors in C. elegans. Cell 83: 207–18.

Voglis and Tavernarakis. 2005. Transduction in the nematode *Caenorhabditis elegans*. In: Mechanosensitivity in Cells and Tissues. Ed. Kamkin, Andre and Kiseleva, Irina. Academia Publishing House Ltd, Moscow, p. 23–56.

Voilley, N., A. Galibert, F. Bassilana, S. Renard, E. Lingueglia, S. Coscoy, G. Champigny, P. Hofman, M. Lazdunski, and P. Barbry. 1997. The amiloride-sensitive Na+ channel: from primary structure to function. Comp Biochem Physiol A Physiol 118:193–200.

Waldmann, R., G. Champigny, N. Voilley, I. Lauritzen, and M. Lazdunski. 1996. The mammalian degenerin MDEG, an amiloride-sensitive cation channel activated by mutations causing neurodegeneration in Caenorhabditis elegans. J Biol Chem 271:10433–6.

Waldmann, R., and M. Lazdunski. 1998. H(+)-gated cation channels: neuronal acid sensors in the NaC/DEG family of ion channels. Curr Opin Neurobiol 8:418–24.

Walker, R. G., A. T. Willingham, and C. S. Zuker. 2000. A Drosophila mechanosensory transduction channel. Science 287:2229–34.

Ward, S., N. Thomson, J. G. White, and S. Brenner. 1975. Electron microscopical reconstruction of the anterior sensory anatomy of the nematode Caenorhabditis elegans. J Comp Neurol 160:313–37.

Waterston, R., and J. Sulston. 1995. The genome of *Caenorhabditis elegans*. Proc. Natl. Acad. Sci. USA 92:10836–10840.

Waterston, R. H., J. N. Thomson, and S. Brenner. 1980. Mutants with altered muscle structure of Caenorhabditis elegans. Dev Biol 77:271–302.

Way, J. C., and M. Chalfie. 1989. The mec-3 gene of Caenorhabditis elegans requires its own product for maintained expression and is expressed in three neuronal cell types. Genes Dev 3:1823–33.

Way, J. C., and M. Chalfie. 1988. mec-3, a homeobox-containing gene that specifies differentiation of the touch receptor neurons in C. elegans. Cell 54:5–16.

Weinbaum, S., P. Guo, and L. You. 2001. A new view of mechanotransduction and strain amplification in cells with microvilli and cell processes. Biorheology 38:119–42.

Welsh, M. J., M. P. Price, and J. Xie. 2002. Biochemical basis of touch perception: mechanosensory function of degenerin/epithelial Na+ channels. J Biol Chem 277:2369–72.

White, J. G., E. Southgate, J. N. Thomson, and S. Brenner. 1986. The structure of the nervous system of Caenorhabditis elegans. Philos Trans R Soc Lond B Biol Sci 314:1–340.

White, J. G., E. Southgate, J. N. Thomson, and S. Brenner. 1976. The structure of the ventral nerve cord of Caenorhabditis elegans. Philos Trans R Soc Lond B Biol Sci 275:327–48.

Wicks, S. R., and C. H. Rankin. 1997. Effects of tap withdrawal response habituation on other withdrawal behaviors: the localization of habituation in the nematode Caenorhabditis elegans. Behav Neurosci 111:342–53.

Wicks, S. R., and C. H. Rankin. 1995. Integration of mechanosensory stimuli in Caenorhabditis elegans. J Neurosci 15:2434–44.

Wicks, S. R., R. T. Yeh, W. R. Gish, R. H. Waterston, and R. H. Plasterk. 2001. Rapid gene mapping in Caenorhabditis elegans using a high density polymorphism map. Nat Genet 28:160–4.

Williams, B. D., and R. H. Waterston. 1994. Genes critical for muscle development and function in Caenorhabditis elegans identified through lethal mutations. J Cell Biol 124:475–90.

Wolinsky, E., and J. Way. 1990. The behavioral genetics of Caenorhabditis elegans. Behav Genet 20:169–89.

Xue, D., M. Finney, G. Ruvkun, and M. Chalfie. 1992. Regulation of the mec-3 gene by the C.elegans homeoproteins UNC-86 and MEC-3. Embo J 11:4969–79.

Zhang, S., J. Arnadottir, C. Keller, G. A. Caldwell, C. A. Yao, and M. Chalfie. 2004. MEC-2 is recruited to the putative mechanosensory complex in C. elegans touch receptor neurons through its stomatin-like domain. Curr Biol 14:1888–96.

Chapter 6
Mechanosensitive Ion Channels in Odontoblasts

Henry Magloire, Bruno Allard, Marie-Lise Couble, Jean-Christophe Maurin, and Françoise Bleicher

Abstract Odontoblasts are post-mitotic cells involved in the dentine formation throughout the life of the tooth and suspected to play a role in tooth pain transmission. They are organized as a single layer of specialized cells along the interface between dental pulp and calcified dentinal tubules into which run a cellular extension (odontoblast process) bathed in a liquid phase. Dense sensory unmyelinated nerve fibres surrounded the odontoblast bodies, coiled around the cell processes and give to this complex (nerve/odontoblast) a fundamental role as a barrier regulating molecules, fluid flow, ion transferts between dentine and pulp following external stimuli (mechanical thermal, electrical, osmotic shock...). Thus, this unique spatial situation of odontoblasts closely related with nerve endings and fluid movements suggest that odontoblasts could convert pain-evoking fluid displacement within dentinal tubules into electrical signals *via* at least mechanosensitive ion channels. Along this line, two kinds of mechanosensitive K^+ channels have been identified in human odontoblasts: I- TREK-1 channels belonging to the two-pore-domain potassium channel family and expressed in the plasma membrane of coronal odontoblasts; II- high-conductance Ca^{2+} -activated potassium channels (K_{Ca}) activated by stretch of the membrane as well as osmotic shock. These findings strengthened by the recent evidence for excitable properties of odontoblasts, concentration of mechanosensitive channels in the borderline between cell extension and bodies and clustering of key molecules at the site of odontoblast-nerve contact strongly suggest that odontoblasts may operate as sensor cells.

Key words: Teeth · Odontoblast · Mechanosensitive potassium channel · TREK-1 channel · K_{Ca} channel · Voltage-gated sodium channel · Tooth pain

6.1 Introduction

Three different calcified tissues constitute the structure of mammals teeth including enamel recovering the crown, cementum deposited on the root surface and dentine forming the bulk of the tooth. During tooth development, odontoblasts originating from the neural crest are organized as a layer of palisade cells along the interface

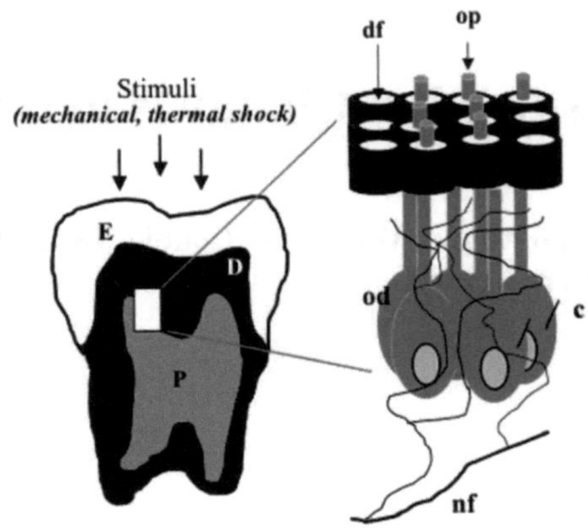

Fig. 6.1 Schematic representation of the spatial situation of odontoblasts (od) in tooth. External tooth stimuli in contact with enamel (E) cause continuous movements of dentine fluid (df) within dentinal tubules where odontoblast processes (op) extend no more than the inner third of dentine (D). Cell bodies are surrounded by a dense network of unmyelinated nerve fibres (nf) in close association with plasma membrane of odontoblasts. These latter exhibit a primary cilium (C) in the vicinity of the Golgi apparatus which could participate to the signal transduction

between the dental pulp (soft connective tissue) and dentine (Fig. 6.1). These post-mitotic cells play a central role during the formation of dentine in that they synthesize the organic matrix macromolecules (pre-dentine) and actively participate in the transportation and accumulation of calcium at the mineralization front (Ruch et al., 1995; Linde and Lundgren, 1995).

As dentinogenesis progresses in a pulp ward direction, there is continued deposition of pre-dentine, which further mineralizes forming dentinal tubules containing the odontoblast extension. This dynamic process gives to these cells a unique spatial situation (Fig. 6.2a) with cell bodies included in the soft pulp tissue, processes running in the calcified tubules and bathed in a liquid phase (dentinal

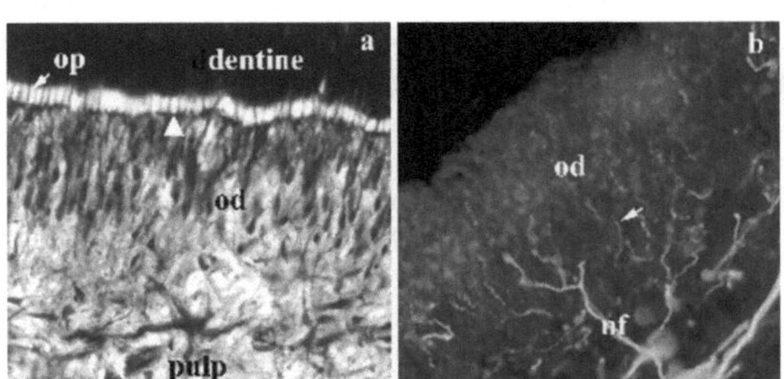

Fig. 6.2 (a) pulpal dentinal border of a human tooth showing the odontoblast layer (od) with cell processes (op) extending in the dentinal tubules. The spatial organization of junctional complexes corresponding to the terminal web (arrow head) represents the borderline between cell processes and bodies (Masson's trichrome staining). (b) Distribution of nerve fibres (nf) in the odontoblast layer of the crown. Thin nerve endings (arrow) extend into the layer (Immunodetection of neurofilaments with specific antibodies)

fluid). A dense network of sensory axons from neurons in the trigeminal ganglion branches extensively in the odontoblast region of the crown forming a marginal plexus at the pulpal-dentinal border (Fig. 6.2b). Interestingly, afferent unmyelinated nerve endings coiled around the cell bodies and processes of odontoblasts within dentinal canalicules with a 20 nm gap space (Byers, 1984, Ibuki et al., 1996). In addition, we have recently shown that reelin, a large extracellular matrix glycoprotein elaborated by odontoblasts could promote adhesion between nerves and cells (Maurin et al., 2004).

This close association suggested that odontoblasts and nerve terminals may directly interact although no synaptic structures or any junction could be detected between them and this event has been presupposed as the earliest step of tooth pain transmission. Indeed, the spatial situation of odontoblasts, nerves endings and fluid movements in dentinal tubules postulated that nociceptive responses might result in mechanical stimulation of odontoblasts cell membrane following external tooth stimuli (high pressure, osmotic, chemical or thermal shock). Thus, emerged the hypothesis that a transductive mechanism for somatic sensation could exist via mechanosensitive ion channels in odontoblasts.

6.2 Introducing Odontoblasts (Fig. 6.1, 6.2, 6.3)

At the morphological level mature odontoblasts are strongly polarized cells, the nuclei occupying the proximal part of the cell body, the rest of the organelles having a supra nuclear location. Interestingly, a primary cilium in the vicinity of the Golgi apparatus has been regularly described at the ultrastructural level and recently antibodies directed against detyrosinated α tubulin specifically identified this structure in human odontoblasts (Magloire et al., 2004). The role of this primary cilium remains unknown but it was suggested that it could constitute a critical link between the transfer of fluid, molecules or ions from dentinal tubules to pulp tissue and odontoblast response to stimuli.

At the distal end of the cell body, a monopolar extension, the odontoblast process contains numerous cytoskeleton structures and does not extend beyond the inner third of dentine in human. At the borderline between cell processes and bodies (named the terminal web), junctional complexes represent both a cell- to-cell communication at the molecular level and a selective barrier between dentine and pulp at the physiological level (Fig. 6.3).

The special location of odontoblasts at the dentine-pulp interface prevents sharp investigations at the electrophysiological or biological level. To overcome these difficulties, in vitro cultures of dental pulp tissue have been performed and mature odontoblasts were successfully obtained from human and mammals specimen. However, very few models of cultured cells exhibit typical features of odontoblasts at both the morphological and functional level (eccentric position of the nucleus, cellular extension, dentine sialophophoprotein, type I collagen, enamelysin, osteoadherin) (Buchaille et al., 2000; Couble et al., 2000; Shibukawa and Suzuki, 2003; Okumura et al., 2005) and represent the most convenient *in vitro* model for the study of cellular mechanisms underlying the signal transduction in odontoblast.

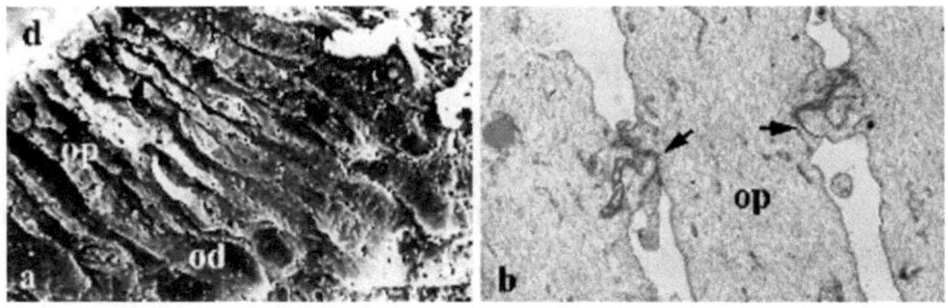

Fig. 6.3 (a) Scanning electron microscope observation of human odontoblasts (od) bodies and processes (op) entering the dentine tubules (D). (b) Transmission electron microscope picture of junctional complexes (arrows) between odontoblast processes at the pulpal-dentinal border referred to as terminal web

6.3 Ion Channels in Odontoblasts Plasma Membrane

In contrast to what is known about the factors that control dentine formation and repair, there is little information on the direct mechano-transduction response of dentine-forming cells to stimuli that contributes to the laying down of dentine throughout the life of the tooth. Cell membrane properties have been described in *in vitro* cultures of pulp cells, in freshly isolated odontoblasts from pulp cells and in surviving odontoblasts from pulp thick slices preparation. L-type channels ($Ca_v1.2$) and Ca^{2+} currents involving N-type (Cav2.2) have been respectively immuno-detected (Seux et al., 1994) and recorded using the patch clamp method in odontoblasts (Lundgren & Linde, 1997; Davidson & Guo, 2000). Inhibition of L-type Ca channels *in vivo* was shown to strongly impair the uptake of calcium ions into the dentine mineral phase during dentinogenesis (Lundgren & Linde, 1998). Recent findings provide evidence of their down regulation after dentine injury (Westenbroek et al., 2004) demonstrating the central role played by calcium channels in odontoblast behaviour (Shibukawa & Suzuki, 2003). Besides these Ca^{2+} channels, potassium and chloride-selective channels have also been described in the odontoblast membrane (Davidson, 1993; Guo & Davidson, 2000; Allard et al., 2000; Shibukawa & Suzuki, 2001, Magloire et al. 2003). Their role could probably be related to secretion, osmoregulation, pain sensation or formation of dentine similarly to osteoblasts during mineralization (Ypey et al., 1992; Allard et al., 2000). Finally, a special attention should be paid to voltage-gated sodium channels which have been very recently identified in human odontoblasts at the electrophysiological, gene and molecular level.

6.4 Mechanosensitive Ion Channels in the Odontoblast Membrane

Mechanosensitive ion channels (Ms channels) are involved in the transduction of mechanical stimuli into electrical cell signals. They are generally activated in response to membrane stretch in invertebrates and vertebrates (review: Martinac, 2004) but mechanisms underlying their activation are not completely understood. Two basic mechanisms have been suggested for gating (Sachs and Morris, 1998;

Syntichaki and Tavernarakis, 2003; Sukharev and Anishkin, 2004): I- the tension of the cell membrane alone is able to drive the channel open (this bilayer model is more appropriate to prokaryotic cells). II- The gating tension is transduced by both cytoskeletal and extracellular matrix to which the ion channel embedded in the cell membrane is linked *via* anchoring proteins (this tethered channel model is more relevant for eukaryotic cells).

Several types of Ms channels are highly selective for potassium and two kinds of Ms K^+ channels have been identified in odontoblasts cell membrane (Allard et al, 2000; Magloire et al., 2003): K_{Ca} (high conductance Ca^{2+}-activated potassium channel) and TREK-1 (TWIK-related K^+ channel 1).

6.4.1 TREK-1 Channel

TREK-1 belongs to the family of potassium channel subunits with two pore domains and 4 transmembrane segments named K_{2P} channels (Lesage and Lazdunski, 2000). This distinct class of K^+ channels has been identified by DNA database mining (Lesage et al., 1996) and TREK-1 was the first characterized at the molecular level in mammals (Patel et al., 1998). TREK channels are opened at resting membrane potentials in physiological conditions and gated by a variety of chemical and physical stimuli including stretch, cell swelling, intracellular acidosis, heat, polyunsaturated fatty acids and volatile general anaesthetic (Patel and Honoré, 2001).

TREK-1 is regulated by neurotransmitters (Lesage and Lazdunski, 2000; Lesage, 2003) that positioned it as target of mediator of pain (Maingret et al., 2000; Murbatian et al., 2005). It shares many pharmacological and electrophysiological properties with the Aplysia neuronal *S* channel (Patel et al., 1998), a presynaptic background K^+ channel involved in learning and memory. TREK-1 channel is present in the peripheral sensory system, particularly in small dorsal root ganglion neurons assumed as C-fibres nociceptors (Maingret et al., 2000). Interestingly, this Ms channel has been recently shown to co-localize with TRPV1 (vanilloid receptor 1), the capsaicin-activated non-selective ion channel (Alloui et al., 2006) and was shown to be sensitive to heat (30–45°C). It is now considered as thermo-sensor, which could play a role in polymodal pain perception.

In the pulpal-dentinal border, TREK-1 is strongly expressed in the membrane of the coronal odontoblasts and absent in the root. Interestingly, this pattern of expression could be related to the strong decreasing gradient of nerve fibres network from the cusp to the root region. In this respect, a close correlation could exist between nerve endings and the regulation of expression of TREK-1 channels in odontoblasts. Given the dense network of unmyelinated sensory axons (A delta and C-fibers), the unique nerve—odontoblasts relationships (Byers et al., 2003; Ibuki et al., 1996; Maurin et al., 2004; Allard et al., 2006) and the expression of TRPV1 in odontoblasts (Okumura et al., 2005), it could be speculated that TREK-1 channels, when stretched or temperature activated, could generate a signal to afferent nerve endings.

6.4.2 High Conductance Calcium-Activated K^+ Channels (K_{Ca})

Some voltage or ligand-gated channels such as N-type calcium (Calabrese et al., 2002), NMDA receptor channel (Casado & Asher, 1998) or high conductance K_{Ca} (Mallouk & Allard, 2000) have been shown to exhibit mechanosensitivity in different cell types. Among three potassium selective channels initially identified at the single channel level in rat isolated odontoblasts (Guo and Davidson, 1998), Allard et al. (2000) demonstrated that K_{Ca} channels displayed mechanosensitivity in cultured odontoblasts. An increase in the open probability of the channel was induced by application of a negative pressure or an osmotic shock (saccharose) to the cells. *In vivo* experiments confirmed that odontoblasts express *HSLO* gene encoding the pore-forming α subunit. Thus, odontoblasts may control *via* K_{Ca} channels a variety of metabolic processes including dentine formation, reparative dentine production and dentine sensitivity. Indeed, The K_{Ca} channel proteins were co-detected with L-type calcium channels at the apical pole of the cells, a specific physiological barrier that actively participates in the directional transportation of calcium to the mineralization front of the dentine (Lundgren and Linde, 1988, 1997). The Ca^{2+}entry pathway involves at least L-type Ca^{2+}channels (Seux et al., 1994; Lundgren and Linde, 1998; Allard et al., 2000; Shibukawa and Suzuki, 2003) and the co-localization of these latter with K_{Ca} channels suggests that stretch activation of K_{Ca} together with increase in intracellular Ca^{2+}could exert a negative control of Ca^{2+} entry by hyperpolarizing odontoblasts and closing Ca^{2+} channels. A similar process controlling the laying down of reparative dentine could occur under dentine injury. These K_{Ca} channels could also be involved in tooth pain sensation. In response to mechanical stimuli, the combination of increased intracellular Ca^{2+} plus membrane stretch could cause K_{Ca} channel opening in odontoblasts. The resulting elevation in the extracellular K^+ concentration in the restricted cleft delimited by the neuronal and odontoblast membranes may depolarize the nerve endings (or odontoblasts, see below) and lower threshold for firing in the sensory tract (or in odontoblasts). This could explain why K^+ -containing agents placed into deep dentinal cavities induce short tooth pain sensations.

6.5 Conclusion

The view that odontoblasts could have a sensory receptor function raises the question of excitable properties of the cells. Very recently, we brought evidence that voltage-gated tetrodotoxin-sensitive Na^+ channels are functional in odontoblasts cell membrane. We also demonstrated that these cells are excitable and produce all or none spikes in response to depolarizing currents. This finding has relevant physiological consequence: odontoblasts might be able to transduce, integrate diverse somatosensory signals known to elicit nociceptive responses, and initiate bursts of regenerative voltage responses as nerve cells were originally exclusively thought to be responsible for.

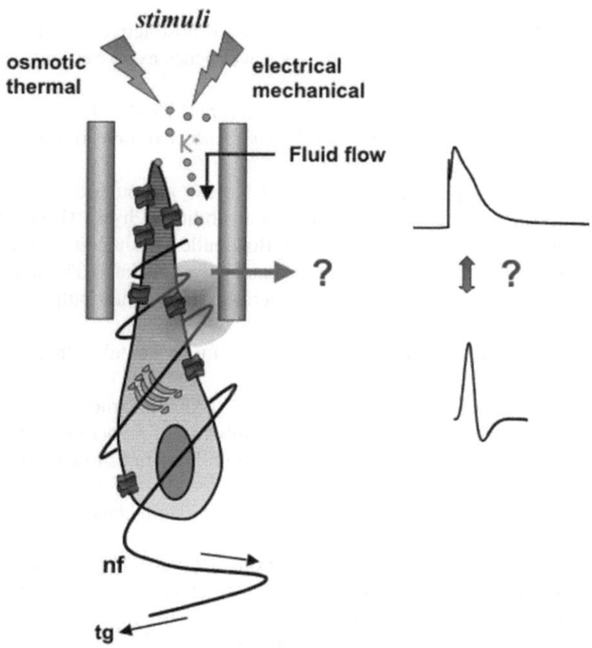

Fig. 6.4 Schematic representation of putative mechanisms underlying the Ms( ) channels signal transduction. Odontoblasts may operate as excitable sensor cells whose excitation is transmitted to nerve fibres (nf) and conducted to the trigeminal ganglion (tg). The question marks refer to the remaining open question of the type of transmission of excitation from odontoblasts to nerve endings

In this line, the excitable properties of odontoblasts, the concentration of Ms ion channels preferentially in the borderline between cell processes and bodies (terminal web) and cilium membrane, and the clustering of key molecules (a2/b2 sub-units, ankyrin $_G$) at the site of odontoblast-nerve close contact bring a new role for odontoblasts as sensor cells (Fig. 6.4). How the firing of odontoblasts is transmitted to neighbouring nerve cells remains the main open question.

References

Allard B, Couble ML, Magloire H, Bleicher F (2000) Characterization and gene expression of high conductance calcium-activated potassium channels displaying mechanosensitivity in human odontoblasts. J Biol Chem 275:25556–25561.

Allard B, Magloire H, Couble ML, Maurin JC, Bleicher F (2006) Voltage-gated Sodium Channels Confer Excitability to Human Odontoblasts: possible role in tooth pain transmission. J Biol Chem. 281:29002–29010.

Alloui A, Zimmermann K, Mamet J, Duprat F, Noel J, Chemin J, Guy N, Blondeau N, Voilley N, Rubat-Coudert C, Borsotto M, Romey G, Heurteaux C, Reeh P, Eschalier A, Lazdunski M (2006) TREK-1, a K^+ channel involved in polymodal pain perception. EMBO J 25: 2368–2376.

Buchaille R, Couble ML, Magloire H, Bleicher F (2000) a substractive PCR-based cDNA library from human odontoblast cells: identification of novel genes expressed in tooth forming cells. Matrix Biol 19:421–430.

Byers MR (1984) Dental sensory receptors. Int Rev Neurobiol 25:39–94.

Calabrese B, Tabarean IV, Juranka P, Morris CE (2002) Mechanosensitivity of N-type calcium channel currents. Biophys J 83:2560–2574.

Casado M, Ascher P (1998) Opposite modulation of NMDA receptors by lyophospholipids and arachidonic acid, common features with mechanosensibility. J Physiol (London) 513:317–330.

Couble ML, Farges JC, Bleicher F, Perrat-Mabillon, Boudeulle M, Magloire H (2000) Odontoblast differentiation of human dental pulp cells in explant cultures. Calcif Tissue Int 66:129–138.

Davidson RM (1993) Potassium currents in cells derived from dental pulp. Arch Oral Biol 38: 613–620.

Davidson RM, Guo L (2000) Calcium channel current in rat dental pulp cells. J membr Biol 178:21–30.

Ibuki T, Kido MA, Kiyoshima T, Terada Y, Tanaka T (1996) An ultrastructural study of the relationship between sensory trigeminal nerves and odontoblasts in rat/dentin pulp as demonstrated by the anterograde transport of wheat germ agglutinin-horseradish peroxidase (WGA-HRP). J Dent Res 75:1963–1970.

Lesage F (2003) Pharmacology of neuronal background potassium channels. Neuropharmacology 44:1–7.

Lesage F, Guillemare E, Fink M, Duprat F, Lazdunski M (1996) TWIK-1, a ubiquitous human weakly inward rectifying K^+ channel with a novel structure. EMBO J 15:1004–1011.

Lesage F, Lazdunski M (2000) Molecular and functional properties of two pore-domains potassium channels. Am J Physiol Renal Physiol 279:F793-F801.

Linde A, Lundgren T (1995) From serum to mineral phase. The role of the odontoblast in calcium transport and mineral formation. Int J Dev Biol 39:213–222.

Lundgren T, Linde A (1988) Na^+/Ca^{2+} antiports in membranes of rat incisor odontoblasts. J Oral Pathol 17:560–563.

Lundgren T, Linde A (1997) Voltage-gated calcium channels and non voltage-gated calcium uptake pathways in the rat incisor odontoblast plasma membrane. Calcif Tissue Int 60:79–85.

Lundgren T, Linde A (1998) Modulation of rat incisor odontoblast plasma membrane-associated Ca^{2+} with nifedipine. Biochem Biophys Acta 1373:341–346.

Magloire H, Couble ML, Romeas A, Bleicher F (2004) Odontoblast primary cilium: facts and hypotheses. Cell Biol Int 28:93–99.

Magloire H, Lesage F, Couble ML, Lazdunski M, Bleicher F (2003) Expression and localization of TREK-1 K^+ channels in human odontoblasts. J Dent Res 82:542–545.

Maingret F, Lauritzen I, Patel AJ, Heurteaux C, Reyes R, Lesage F, Lazdunski M (2000) TREK-1 is a heat-activated background K^+ channel. EMBO J 19:2483–2491.

Mallouk N, Allard B (2000) Stretch-induced activation of Ca(2+)-activated K(+) channels in mouse skeletal muscle fibers. Am J Physiol Cell Physiol. 278:C473–479.

Martinac B (2004) Mechanosensitive ion channels: molecules of mechanotransduction. J Cell Sci 117:2449–2460.

Maurin JC, Couble ML, Didier-Bazès M, Brisson C, Magloire H, Bleicher F (2004) Expression and localization of reelin in human odontoblasts. Matrix Biol 23:277–285.

Murbatian J, Lei Q, Sando JJ, Bayliss DA (2005) Sequential phosphorylation mediates receptor and kinase-induced inhibition of TREK-1 background potassium channels. J Biol Chem 280:30175–30184.

Okumura R, Shima K, Muramatsu T, Nakagawa KI, Shimono M, Suzuki T, Magloire H, Shibukawa Y (2005) The odontoblast as a sensory receptor cell? The expression of TRPV1 (VR-1) channels. Arch Histol Cytol 68:251–257.

Patel AJ, Honoré E (2001) Properties and modulation of mammalian 2P domain K^+ channels. Trends Neurosci 24:339–346.

Patel AJ, Honoré E, Maingret F, Lesage F, Fink M, Duprat F, Lazdunski M (1998) A mammalian two pore domains mechano-gated S-like K^+ channel. EMBO J 17:4283–4290.

Ruch JV, Lesot H, Bègue-Kirn C (1995) Odontoblast differentiation. Int J Dev Biol 39:51–68.

Sachs F, Morris CE (1998) Mechanosensitive ion channels in non specialized cells. Rev Physiol Biochem Pharmacol 132:1–77.

Seux D, Joffre A, Fosset M, Magloire H (1994) Immunohistochemical localization of L-type calcium channels in the developing first molar of the rat during odontoblast differentiation. Arch Oral Biol 39:167–170.

Shibukawa Y, Suzuki T (2001) A voltage-dependent transient K+ current in rat dental pulp cells. Jpn J Physiol 51:345–353.

Shibukawa Y, Suzuki T (2003) Ca^{2+} signalling mediated by IP3-dependent Ca^{2+} releasing and store-operated) Ca^{2+} channels in rat odontoblasts. J Bone Miner Res 18:30–38.

Sukharev S, Anishkin A (2004) Mechanosensitive channels: what can we learn from "simple" model systems? Trends Neurosci 27:345–351.

Syntichaki P, Tavernarakis N (2003) Genetic models of mechanotransduction: the nematode Caenorhabditis elegans. Physiol Rev 84:1097–1153.

Westenbroek RE, Anderson NL, Byers MR (2004) Altered localization of $Ca_V1.2$ (L-type) calcium channels in nerve fibers, Schwann cells, odontoblasts, and fibroblasts of tooth pulp after tooth injury. J Neurosci Res 75:371–383.

Ypey DL, Weidema AF, Hold KM, Van der Laarse A, Ravesloot JH, Van Der Plas A, Nijweide PJ (1992) Voltage, calcium, and stretch activated ionic channels and intracellular calcium in bone cells. J Bone Miner Res 7:S377–387.

Chapter 7
Potassium Ion Channels in Articular Chondrocytes

Putative Roles in Mechanotransduction, Metabolic Regulation and Cell Proliferation

Ali Mobasheri, Caroline Dart, and Richard Barrett-Jolley

Abstract Potassium ion channels belong to a large superfamily of integral membrane proteins that selectively transport K^+ across the plasma membrane. They are present in almost all mammalian cells and play a wide variety of physiological roles in both excitable and non-excitable cells. Despite sharing similar architectural and structural designs, the phenotypic diversity required to accomplish their diverse functional roles is created by subtle differences in conductance, time-course, mechanisms of gating and the interaction with a variety of ligands and accessory proteins. For example, the activities of members of the potassium channel superfamily are associated with the control of neuronal excitability, neurotransmitter release, cardiac and smooth muscle contraction, heart rate, endocrine secretion, epithelial electrolyte transport, cell proliferation, apoptosis and tumour progression. A number of different potassium channels have been identified in articular chondrocytes. Ongoing studies are aimed at deciphering the putative functions of potassium channels in these cells and determining the consequences of their pharmacological activation and inactivation on the unique chondrocyte phenotype. The behaviour of chondrocytes has been shown to be influenced by modulation of ion channel activity. In this review we will focus on recent experimental studies on the roles of potassium ion channels in chondrocytes within articular cartilage and discuss research which has implicated these proteins in metabolic regulation, mechanotransduction, cell volume regulation and cell proliferation. A better understanding of ion channel function may help elucidate the intricate processes involved in mechanotransduction, metabolic regulation and proliferation in chondrocytes.

Key words: Cartilage · Chondrocyte · Potassium channel · Membrane potential · Stretch-activation · Mechanotransduction · Metabolic regulation · Cell proliferation · Osteoarthritis

A. Kamkin and I. Kiseleva (eds.), *Mechanosensitive Ion Channels.*

7.1 Introduction

Physical, gravitational and mechanical forces are fundamental regulators of animal tissue development (Ingber, 1998). In order to influence morphogenesis, mechanical forces must alter growth and tissue function (Chicurel et al., 1998). Mechanical forces have long been known to regulate the growth, development and biological function of skeletal muscle (Goldspink et al., 1992; Goldspink, 1996, 1999). Mechanical forces are also important for the development, maintenance and remodeling of load bearing articular cartilage in diarthrodial joints of the musculoskeletal system (Urban, 1994; Grodzinsky et al., 2000; Lane Smith et al., 2000). Articular cartilage is a unique connective tissue that experiences a variety of stresses, strains and loading pressures that result from normal activities of daily life and physical activity (Lane Smith et al., 2000). In normal cartilage, the extracellular matrix exists as a highly organized composite of specialized macromolecules that distributes loads at the bony ends (Fig. 7.1A) (Lane Smith et al., 2000). Chondrocytes are highly specialized mesenchymal derived cells of articular cartilage (Fig. 7.1B) (Archer and Francis-West, 2003) that can detect and respond to applied mechanical loads by altering their metabolic state through a process known as mechanotransduction (Urban, 1994). Cellular mechanotransduction is defined as a series of dynamic processes that allow living cells to convert biomechanical stimuli into biochemical activity. Articular cartilage is loaded under normal gravity

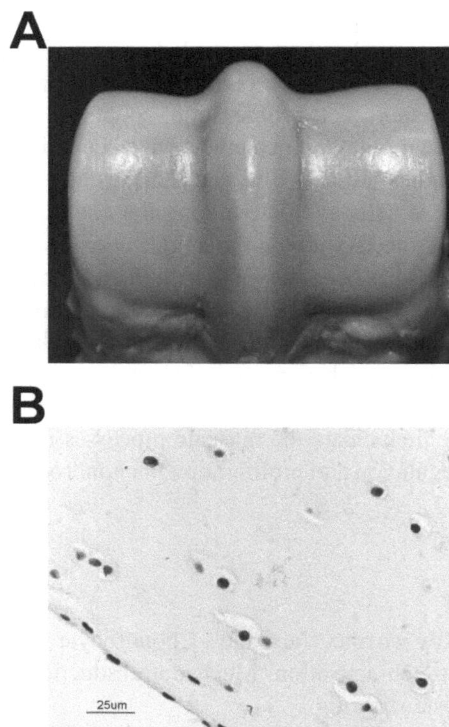

Fig. 7.1 A. Macroscopic appearance of a normal equine metacarpal joint. B. Articular chondrocytes from the superficial zone of equine articular cartilage

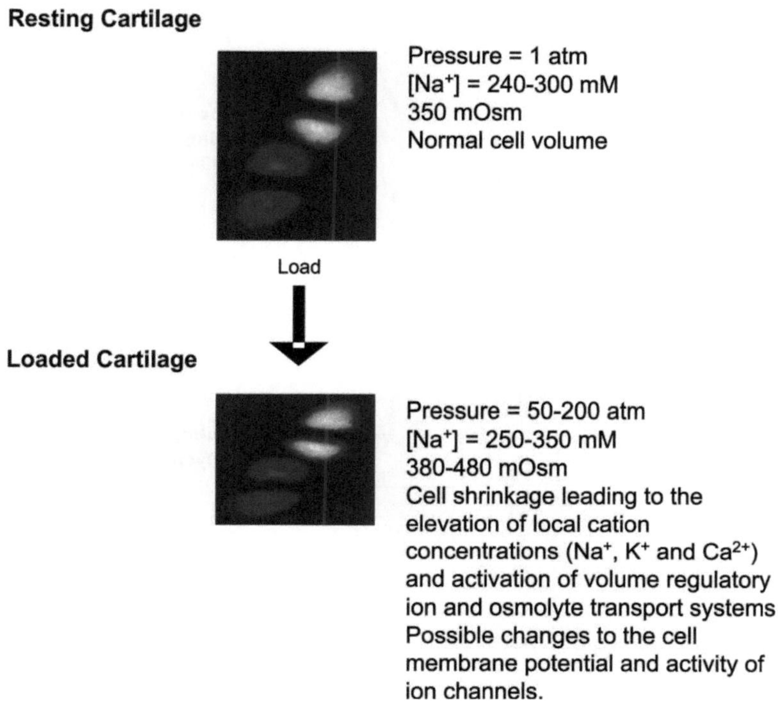

Fig. 7.2 The effects of biomechanical load on the physical environment of human articular chondrocytes (from (Mobasheri et al., 2002a))

(static loading) and frequently loaded during physical activity (dynamic loading; Fig. 7.2). Mechanotransduction in articular cartilage refers to the many cellular and extracellular matrix mechanisms by which chondrocytes quantitatively modulate the rates of matrix synthesis and degradation and alter the composition of the extracellular matrix. The extracellular matrix in cartilage consists primarily of type II collagen and aggregating proteoglycans which give cartilage the ability to resist tensile stress and physical load respectively (Fig. 7.3) (Benjamin et al., 1994).

Mechano-electrochemical responses of chondrocytes under mechanical load involve changes in osmotic pressure and the electrical membrane potential difference across the chondrocyte plasma membrane (Fig. 7.4). Chondrocytes are therefore excellent sensors of biomechanical, ionic, osmotic and electrical signals and respond to these varied signals in coordination with other environmental, hormonal and genetic factors to regulate metabolic activity (Mobasheri et al., 2002a). Despite this realization, very little is known about the molecular details of mechanotransduction in chondrocytes. We have previously reviewed the role of mechanoreceptors, β1-integrins, stretch activated cation channels and intracellular messengers (i.e. cAMP, intracellular Ca^{2+}) and their involvement in chondrocyte mechanotransduction (Mobasheri et al., 2002a). There is a distinct paucity of information with regard to the involvement of the plasma membrane and the role(s) of membrane proteins, particularly ion channels in mechanotransduction, metabolic regulation

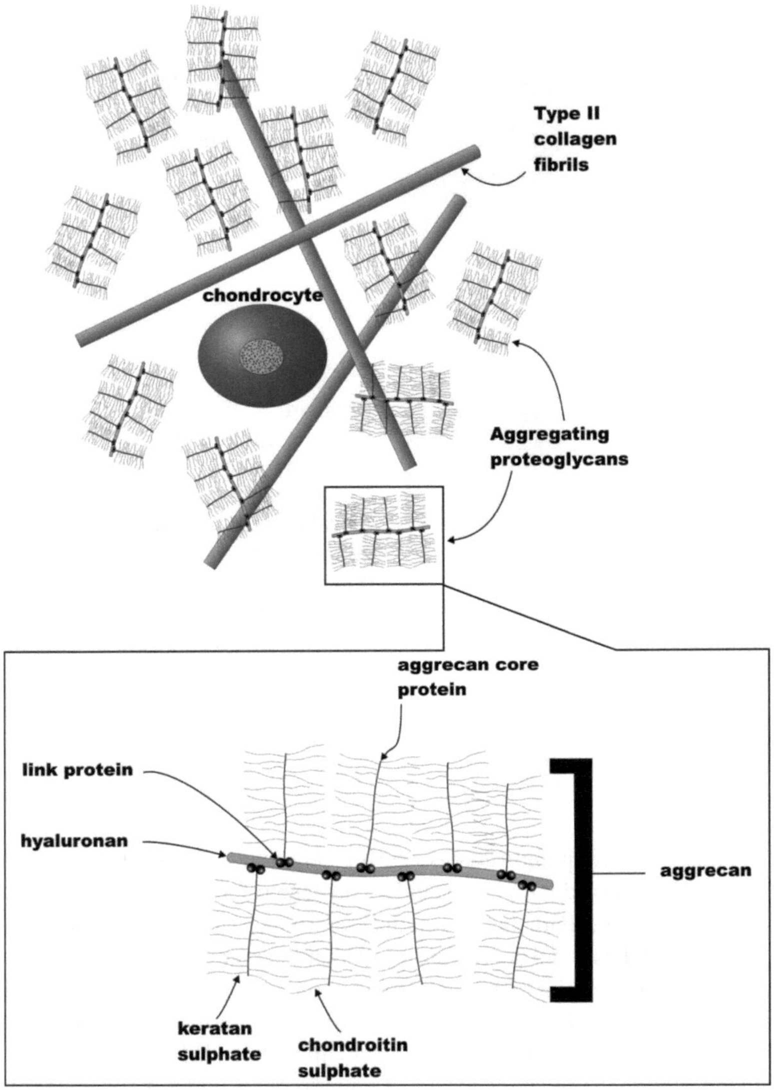

Fig. 7.3 Schematic of the extracellular matrix of articular cartilage which is composed primarily of type II collagen and aggregating proteoglycans

and proliferation in chondrocytes (Fig. 7.5). This review will focus on the role of potassium ion channels in these fundamental processes.

7.1.1 Mechanisms Involved in Chondrocyte Mechanotransduction

Mechanical load is an important regulator of metabolic and biosynthetic activity in articular chondrocytes (Urban, 1994). Mechanically induced cell membrane deformation is one of a number of possible mechanotransduction pathways by

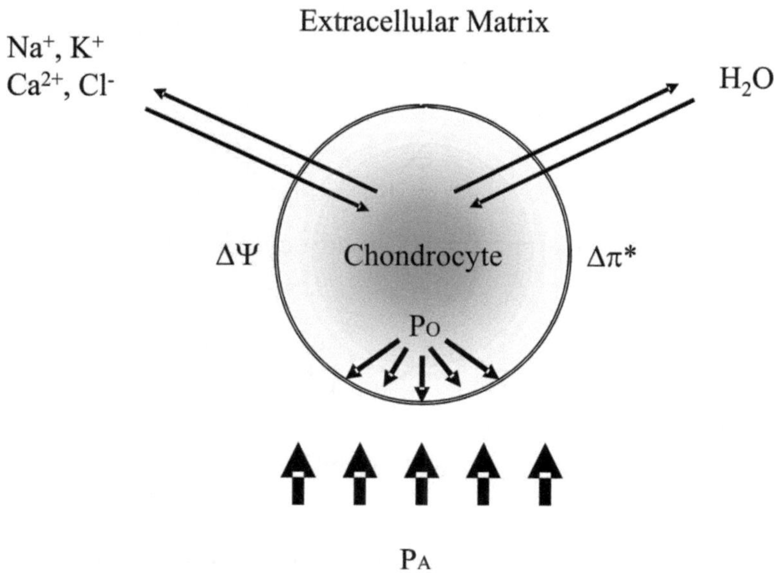

Fig. 7.4 Schematic illustration of mechano-electrochemical responses in chondrocytes under mechanical load and the interaction between the extracellular matrix and chondrocytes. $\Delta\pi^*$ represents changes in osmotic pressure and $\Delta\Psi$ is the electrical membrane potential difference across the chondrocyte plasma membrane (adapted from Mow et al., (Mow et al., 1999))

which chondrocytes sense and respond to changes in their mechanical environment (Guilak, 1995; Guilak et al., 1995; Knight et al., 1998). Changes in ionic and osmotic pressure, ion transport, fluid flow and electrical current across the chondrocyte membrane are important mechano-electrochemical phenomena in loaded articular cartilage (Fig. 7.6). In many cell types mechanical stimulation induces increases of the cytosolic free Ca^{2+} concentration that propagates from cell to cell as an intercellular Ca^{2+} wave (Cornell-Bell and Finkbeiner, 1991; Amundson and Clapham, 1993; Berridge and Dupont, 1994). The generation of intracellular calcium waves within chondrocytes and intercellular Ca^{2+}signals may provide mechanisms to co-ordinate tissue responses in cartilage physiology and biomechanical function (D'Andrea et al., 2000). Changes in amplitude or frequency of load have significant effects on the production of matrix macromolecules and of pro-inflammatory agents leading to cartilage breakdown (Urban, 1994). With inappropriate mechanical loading of the joint, as occurs with traumatic injury, ligament instability, bony malalignment or excessive weight bearing, cartilage exhibits manifestations characteristic of osteoarthritis (OA). The composition of cartilage reflects the net response of the chondrocytes to the prevailing loading pattern, with cartilage proteoglycan content highest in heavily loaded regions and removal of load leading to cartilage thinning and proteoglycan loss (Urban, 1994). Breakdown of cartilage matrix in OA involves degradation of the extracellular matrix macromolecules and decreased expression of chondrocyte matrix proteins necessary for normal joint function. OA cartilage often exhibits increased amounts of type I collagen and synthesis of proteoglycans characteristic of immature cartilage. The focus of work on chondrocyte

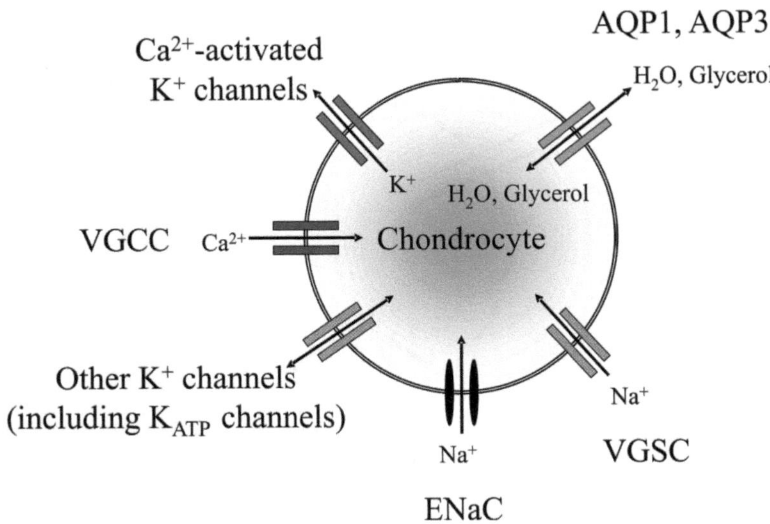

Fig. 7.5 Schematic illustration of plasma membrane ion and water channels potentially involved in mechano-electrochemical responses in chondrocytes under mechanical load. This model incorporates ion channels recently identified in chondrocytes and chondrocyte-like cells. Aquaporin 1 (AQP1) and aquaporin 3 (AQP3) mediating water and glycerol transport respectively (Mobasheri and Marples, 2004; Mobasheri et al., 2004; Trujillo et al., 2004), voltage-gated calcium channels (VGCC) responsible for calcium influx in response to changes in membrane potential (Wang et al., 2000; Shakibaei and Mobasheri, 2003), voltage-gated sodium channels (VGSC) mediating sodium influx following potential changes (Sugimoto et al., 1996), voltage gated potassium channels (VGPC) regulating the membrane potential (Wilson et al., 2004; Mobasheri et al., 2005c; Ponce, 2006), epithelial sodium channels (ENaC) mediating sodium and/or cation influx (Mobasheri and Martin-Vasallo, 1999; Trujillo et al., 1999; Schulze-Tanzil et al., 2004) and calcium-activated potassium channels (Wright et al., 1996). Recent studies by our group have also provided evidence for the presence of adenosine 5′-triphosphate-sensitive potassium channels (K_{ATP} channels) in chondrocytes (Mobasheri et al., 2006)

mechanotransduction has shifted from the biochemical responses of the extracellular matrix to the chondrocyte and its plasma membrane. We need to gain a better understanding of the signaling and regulatory pathways activated during mechanical signal transduction in normal articular cartilage. This knowledge may be important for formulating therapeutic strategies for the rational design of pharmaceutical compounds capable of modulating the metabolic and biosynthetic activities of chondrocytes. Immunohistochemical investigations of chondrocytes have shown the expression of aquaporin water channels (Mobasheri and Marples, 2004; Mobasheri et al., 2004; Trujillo et al., 2004) as well as; ENaC (Trujillo et al., 1999), N-methyl-D-aspartate (NMDA) (Salter et al., 2004; Orazizadeh et al., 2006; Shimazaki et al., 2006), calcium (Guilak et al., 1999; Yellowley et al., 2002; Shakibaei and Mobasheri, 2003; Wang et al., 2003), chloride (Sugimoto et al., 1996; Tsuga et al., 2002), and sodium (Kizer et al., 1997; Mobasheri et al., 2002a) ion channels. The most widely reported ion channels of chondrocytes are, however the potassium ion channels. This review will therefore focus on investigations that are progressing into how potassium ion channels fit into this general picture of chondrocyte regulation.

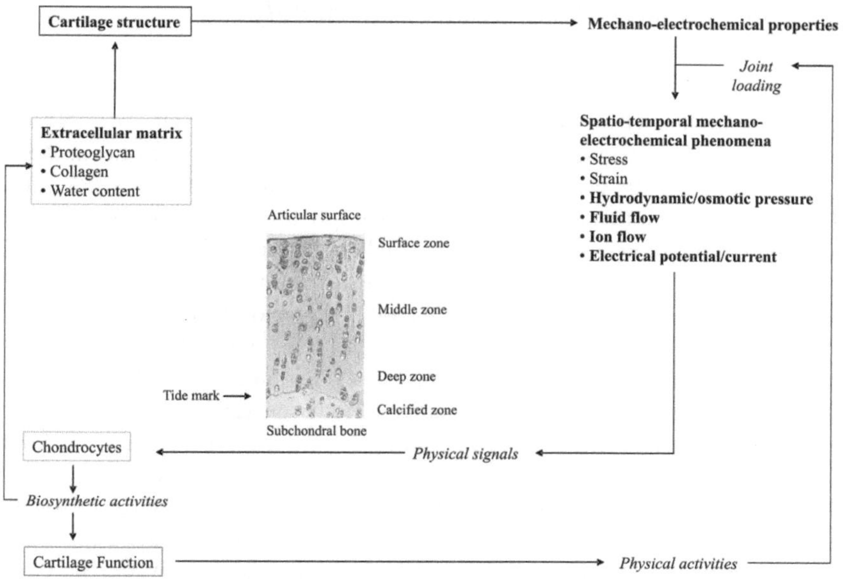

Fig. 7.6 Schematic illustration of the interrelationships between cartilage matrix composition, mechano-electrochemical signals and biosynthetic activities of articular chondrocytes in loaded cartilage (adapted from Mow et al., (Mow et al., 1999))

7.2 A Brief Introduction to Potassium Channels

Potassium channels represent a large and diverse family of integral membrane proteins that permit the passive flow of K^+across cellular membranes. In most cells they play an essential role in maintaining and stabilizing the resting membrane potential. In nerve and muscle cells, their ability to 'repolarize' the membrane helps them control action potential frequency and duration, while other functions include regulation of neurotransmitter release, hormone secretion, epithelial electrolyte transport, cell proliferation, apoptosis, tumour progression and potassium homeostasis (Jan and Jan, 1997; Nichols and Lopatin, 1997; Coetzee et al., 1999; Lesage and Lazdunski, 2000; Yellen, 2002).

Given their fundamental importance to cellular physiology, it is unsurprising that disruption of genes encoding K^+ channel subunits is linked to a number of human diseases, including hyperinsulinemia, disturbances of the heart rhythm (Long QT Syndrome) and certain forms of epilepsy (Sanguinetti and Spector, 1997; Ashcroft, 2006). K^+ channels also form the target for drugs used currently to treat conditions such diabetes, angina and intractable hypertension. Such therapies are likely to be of increasing clinical relevance given the putative role of K^+ channels in regulating cell proliferation and promoting tumour progression (Pardo et al., 1999; Garcia-Ferreiro et al., 2004; Pardo et al., 2005; Stuhmer et al., 2006).

7.2.1 The Potassium Ion Channel Superfamily

K^+ channels are made up of pore-forming α subunits that often coassemble with accessory proteins. They are by far the largest, most diverse (and possibly most ancient) of the ion channel families, with over 70 different genes in the human genome coding for the principal subunits (Coetzee et al., 1999; Alexander et al., 2006). Diversity is further assured by alternative splicing of α subunit genes, and by both homomeric and hetero-tetrameric assembly of the different α subunits. Despite their variety, K^+ channels can be divided into subgroups based on their structure. All K^+ channel α subunits possess a pore loop, a region of the protein that lines the narrowest part of the channel pore and forms the 'selectivity filter' that allows K^+ channels to discriminate between different ions (Doyle et al., 1998). The number of transmembrane domains in each α subunit, however, varies and can be used to divide the K^+ channel superfamily into 3 distinct groups (Alexander et al., 2006). The first major group is made up of K^+ channels that have six transmembrane (6TM) domains in each of their principal subunits. Four such subunits come together to form functional voltage- and/or Ca^{2+}-activated K^+ channels - these include the Kv family as well as members of the KCNQ, EAG, the Ca^{2+}-activated Slo (which actually have 7TM domains) and the Ca^{2+}-activated SK subfamilies (Bauer and Schwarz, 2001; Robbins, 2001; Yellen, 2002; Stocker, 2004). The second group is made up of K^+ channels that have four transmembrane (4TM) domains per subunit, the 'leak' channels TWIK, TREK, TASK, TALK, THIK and TRESK. These contain two pore domains in each α subunit and the functional channel probably forms as a dimer (Lesage and Lazdunski, 2000; Patel and Honore, 2001). The final group comprise the 2TM domain inward rectifier family, which include the ATP-sensitive K^+ (K_{ATP}) channels (Nichols and Lopatin, 1997; Ashcroft and Gribble, 1998; Stanfield et al., 2002).

7.3 Potassium Channels in Chondrocytes

A number of different potassium channels have been identified in articular chondrocytes including members of the voltage-gated Kv, Ca^{2+}-activated Slo and inward rectifier families. In the following sections we will focus on recent experimental studies on the roles of K^+ channels in chondrocytes within articular cartilage and discuss research which has implicated these proteins in metabolic regulation, mechanotransduction, cell volume regulation and cell proliferation.

7.3.1 Voltage-Gated (Kv) Potassium Channels

A number of authors, including our own group, have identified functional voltage-gated (Kv) channels in chondrocytes (Walsh et al., 1992; Sugimoto et al., 1996;

Wilson et al., 2004; Mobasheri et al., 2005; Ponce, 2006) (Figs. 7.7 & 7.8). Since chondrocytes are non-excitable cells, the role of these channels is speculative, although some authors report the resting membrane potential of dissociated chondrocytes to range from 15.3 +/– 0.24 mV to –21.1 +/– 0.28 mV (Wright et al., 1996), a potential at which *non-inactivating* Kv channels could be relatively active.

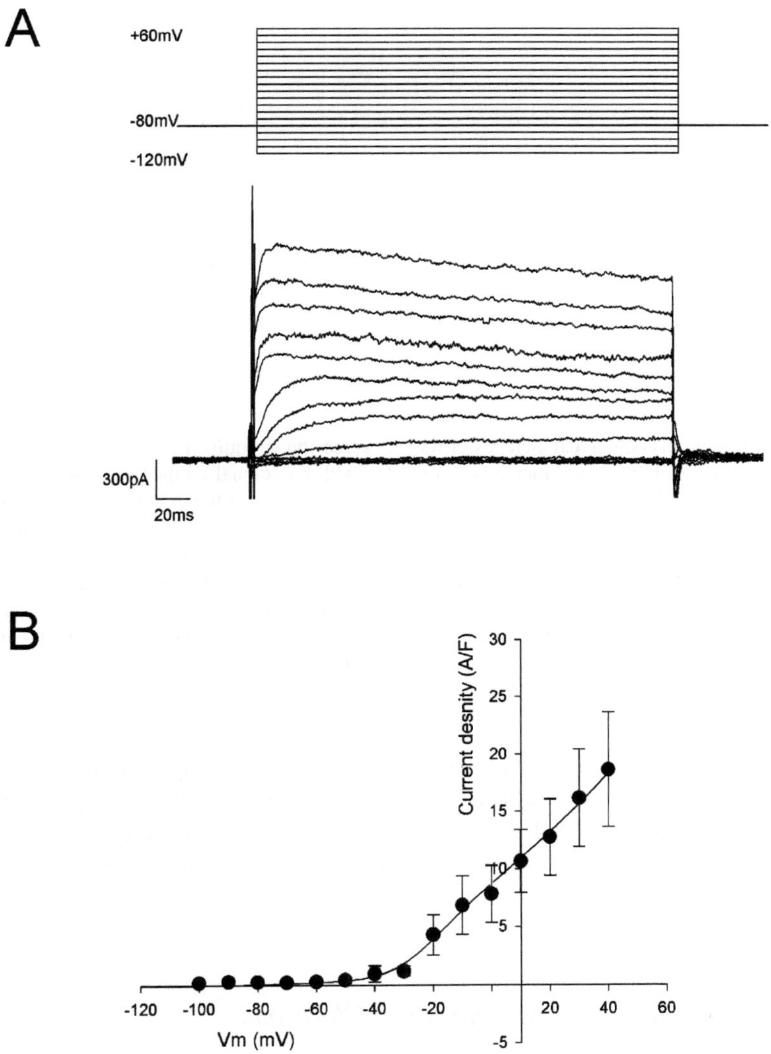

Fig. 7.7 Current-voltage curves of equine chondrocytes. A: a representative family of (leak subtracted) whole cell outward currents recorded from equine chondrocytes. Top: voltage protocol. From a holding potential of –80 mV, cells were stepped to command voltages in the range of –120 to +60 mV (in 10-mV increments). Bottom: leak-subtracted currents. B: mean peak current voltage data, calculated as current density by dividing peak current in each cell by the whole cell capacitance. V_m, membrane potential. Reproduced from (Mobasheri et al., 2005c) with copyright permission of the American Physiological Society

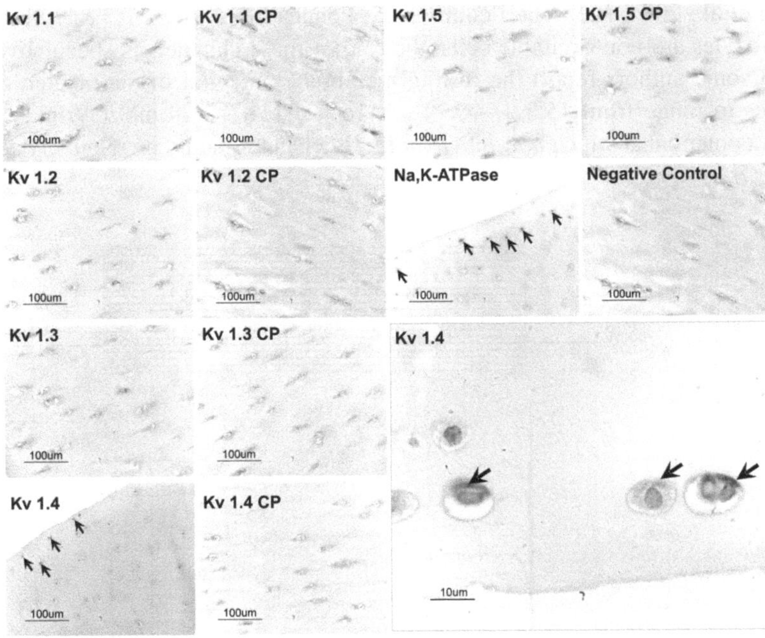

Fig. 7.8 Immunohistochemical analysis of Kv1.x expression in equine articular cartilage. Incubation of equine cartilage tissue microarrays (TMAs) with polyclonal antibodies to Kv1.1, Kv1.2, Kv1.3, and Kv1.4 did not produce any immunostaining with use of the extremely sensitive DakoCytomation EnVision+ Dual Link System-HRP (DAB+) kit. Positive staining was seen with polyclonal antibodies to Kv1.4. Representative images from TMA slides preincubated with primary antibody and control antigen [competing peptide (CP)] also are shown along with a negative control (primary antibody omitted) and a positive control (the a subunit of Na, K-ATPase). Arrows indicate positively stained chondrocytes. Bars = 100 μm (except for the enlarged Kv1.4 image, where the bar = 10 μm). Reproduced from (Mobasheri et al., 2005c) with copyright permission of the American Physiological Society

Interestingly, blockers of these channels, including tetraethyl-ammonium (TEA) and 4-aminopyridine (4AP), have been shown to influence the resting membrane potential (Wilson et al., 2004; Mobasheri et al., 2005d), although these non-selective compounds may also inhibit aquaporin water channels of canine chondrocytes (May et al., 2007).

7.3.2 Calcium-Activated Potassium Channels

Large conductance Ca^{2+}-activated Slo channels (also sometimes referred to as "Maxi-K" or BK) have also been identified in chondrocytes (Lee et al., 2000; Millward-Sadler et al., 2004; Perkins et al., 2005). Our own recent work has identified the BK channel at high density in the equine chondrocyte. While these channels are clearly activated by low levels of intracellular calcium and inhibited by low concentrations of TEA (indicative of BK) they are only weakly inhibited by the selective BK inhibitor toxin iberiotoxin (Womack et al., unpublished observations). This is sometimes a result of the presence of an accessory β -subunit (Lippiat et al., 2003). In further experiments we have located immunostaining for both the α and β subunit of BK channels (Mobasheri et al., unpublished observations), particularly in the superficial zone of cartilage (Mobasheri et al., unpublished observations). Interestingly, although BK channels are clearly identifiable in "normal" cartilage, their expression appears to be up-regulated in OA cartilage, suggesting a possible involvement with disease progression (Mobasheri et al., unpublished observations).

BK channels often demonstrate oxygen sensitivity in cell-free membrane patches suggesting that a significant component of the oxygen-sensing machinery must be closely associated with the channel protein complex (Kemp et al., 2006). Recent proteomic studies have identified the constitutive form of haem oxygenase, haem oxygenase 2 (HO-2), as a BK α-subunit protein partner. This enzyme-ion channel complex has been suggested to be directly involved in hypoxic inhibition of BK channel activity (Kemp et al., 2006). It is therefore possible that the chondrocyte BK channel may also be involved in oxygen sensing as it has been suggested to be in other cell systems (Kemp, 2006; Prabhakar, 2006).

7.3.3 Inward Rectifier Potassium Channels

Inward rectifiers, as their name implies, allow potassium ions to move easily into the cell at membrane potentials negative to the potassium equilibrium potential (E_K), but restrict potassium outflow at potentials positive to E_K. This asymmetry in the current-voltage relation results from the channel's susceptibility to voltage- dependent block by intracellular polyamines and magnesium ions (Matsuda et al., 1987; Lopatin et al., 1994) and ensures that, while being extremely active around E_K, they pass little or no current at membrane potentials positive to –40 mV. Thus, Kir channels maintain a tight control on the resting membrane potential but close in the face of significant membrane depolarization (such as that generated by the cardiac or neuronal action potential) to protect the cell from excessive K^+ loss. To date, only one member of the inwardly rectifying potassium channel family has been identified in chondrocytes; the K_{ATP} channel (Mobasheri et al., 2006).

K_{ATP} channels are closed by the binding of intracellular ATP and thus couple changes in cellular metabolism to membrane excitability (Nichols, 2006). They are expressed in pancreatic β-cells, certain types of neurones, cardiac, skeletal and smooth muscle and are important in regulating secretory processes, cardioprotection and muscle tone (Quayle et al., 1997; Ashcroft and Gribble, 1998; Yokoshiki et al., 1998). Their properties vary considerably from tissue to tissue, reflecting heterogeneity in channel structure. K_{ATP} channels form as 4+4 octamers of Kir6 pore-forming subunits and sulphonylurea receptor proteins (SURs) (Aguilar-Bryan et al., 1998). Two Kir6 subunits, Kir6.1 and 6.2, have been identified, and two SUR genes are known, SUR1 and SUR2, the latter giving rise to SUR2A and SUR2B by alternative splicing. β-cell and cardiac K_{ATP} channels comprise Kir6.2/SUR1 and

Kir6.2/SUR2A respectively, and it is likely that the dominant channel in most vascular smooth muscle comprises Kir6.1/SUR2B (Quayle et al., 1995; Babenko et al., 1998). In our own experiments, we demonstrated the presence of single channel activity which was inhibited by both intracellular ATP and by the sulphonylurea compound glibenclamide (Fig. 7.10). This was strongly suggestive of the presence of the full K_{ATP} complex, the SUR protein together with either Kir6.1 or Kir6.2. In subsequent immunohistochemical studies we have located both Kir 6.1 (Fig. 7.11) and Kir 6.2 (unpublished observations). As yet, we have not identified the specific SUR expressed in chondrocytes, however, by comparisons of glibenclamide sensitivities in other tissues SUR2 seems probable (Quayle et al., 1995; Babenko et al., 1998; Barrett-Jolley and McPherson, 1998).

The discovery of K_{ATP} channels in chondrocytes has quite striking implications. Articular cartilage is an avascular connective tissue in which the availability of oxygen and glucose is significantly lower than synovial fluid and plasma. Articular chondrocytes are capable of existing under hypoxic conditions (Fig. 7.9). In fact chondrocytes need such conditions for survival, chondrogenesis and matrix synthesis (Schipani et al., 2001; Coimbra et al., 2004; Schipani, 2005). Therefore, the chondrocyte requires sophisticated mechanisms to sense the quantities of available oxygen, glucose and ATP levels as well as the concentrations of other important metabolites (Mobasheri et al., 2005a). The presence of K_{ATP} channels in chondrocytes suggests that this channel may be involved in coupling metabolic and electrical activities in chondrocytes through sensing of extracellular glucose and intracellular adenosine triphosphate (ATP) levels. This raises the distinct possibility that the nutritional state of joints and synovial fluids may influence the functioning of chondrocytes and thus the health of cartilage; secondly, if this turns out to be the case, it also provides a possible therapeutic target. Our work also highlights the possibility that altered K_{ATP} channel expression in OA chondrocytes may result in impaired intracellular ATP sensing and sub-optimal metabolic regulation. It is also

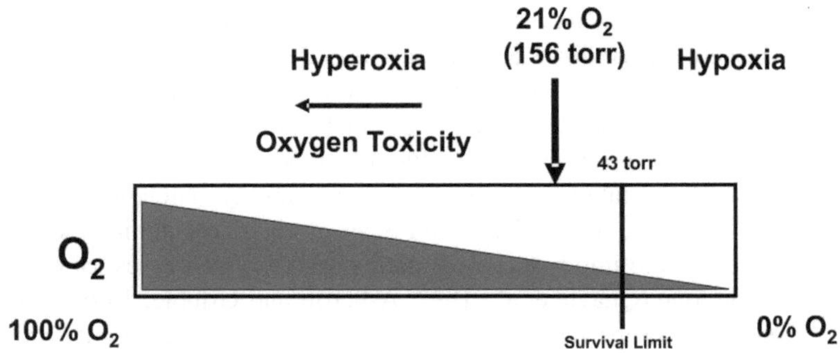

Fig. 7.9 Spectrum of tissue oxygenation. Hyperoxia and anoxia are lethal to all cells whereas most cells have the varied capacity to adapt to low oxygen tension (hypoxia). Chondrocytes are able to survive under hypoxic conditions (Mobasheri et al., 2005b). Hypoxia and HIF-1 alpha may play a crucial role in cartilage development and chondrocyte differentiation (Schipani et al., 2001; Schipani, 2005)

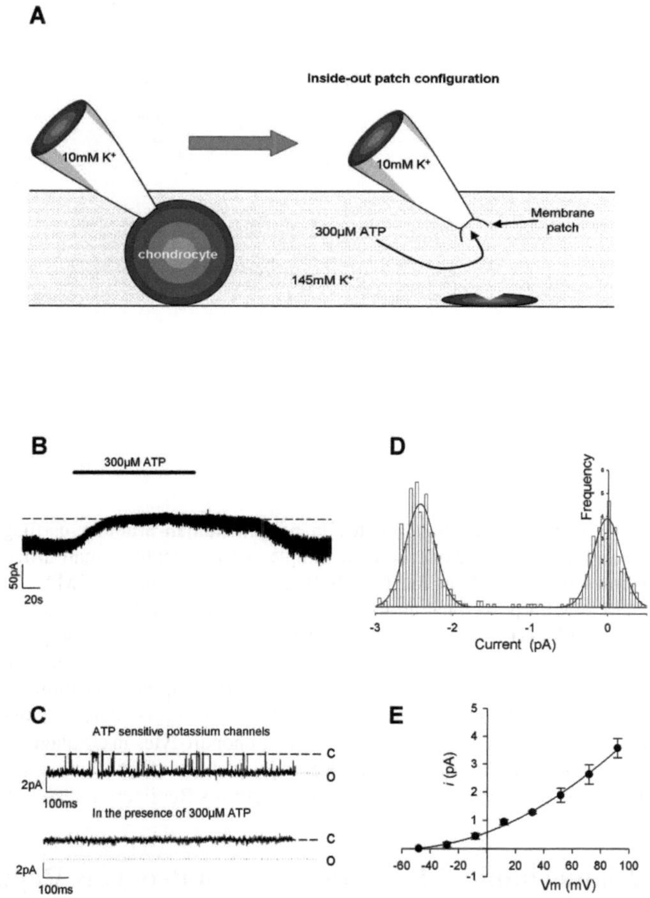

Fig. 7.10 Chondrocytes express functional ATP-sensitive potassium channels. (A) Outline of experimental design, see text. (B) Addition of 300 μM ATP (added by bath perfusion) inhibits currents in inside-out maxi-patches of equine articular chondrocytes. Dashed line indicates the zero current level, holding potential -60 mV. (C) Inside-out single-channel patch in the absence (upper panel) and presence (lower panel) of 300 μM ATP. The dashed line represents the zero current level, where the channel is closed ("C") and the dotted line represents the unitary current level, where the channel is open ("O"). Holding potential -60 mV. (D) All points amplitude histogram from the patch shown in (C). (E) Current–voltage curve for ATP-sensitive potassium channels recorded in a number of experiments similar to (C) and (D). Reproduced from (Mobasheri et al., 2006) with copyright permission of Elsevier Science and the OsteoArthritis Research Society International

possible that potassium channel function is affected by inflammatory mediators in arthritic diseases and the dysfunction of potassium channels may have long-lasting consequences for chondrocyte function and thus contribute to the progression of such diseases. Evidence from other studies suggests that this already occurs in other tissues and organ systems.

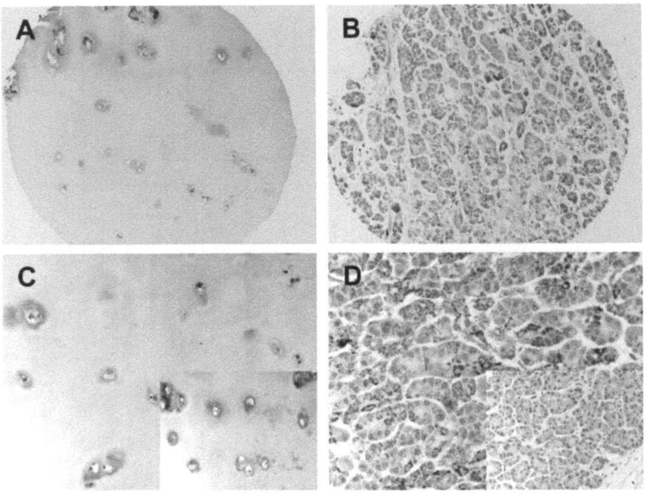

Fig. 7.11 Kir6.1 is expressed in chondrocytes in normal human articular cartilage. Immunohistochemical analysis of Kir6.1 expression in samples of full-depth human articular cartilage and human pancreas represented on the CHTN2002N1 multiple human TMAs. Incubation of CHTN2002N1 TMAs with polyclonal antibodies to Kir6.1 followed by horseradish peroxidase-labeled rabbit anti-goat IgG (DakoCytomation) produced positive immunostaining in chondrocytes in human knee cartilage (low magnification shown in panel A, high magnification shown in panel C) and human pancreas (low magnification shown in panel B, high magnification shown in panel D). Omission of primary antibody from the immunohistochemical procedure resulted in complete abrogation of specific immunostaining of human articular chondrocytes in cartilage samples (inset, panel C) and pancreatic cells (inset, panel D). Reproduced from (Mobasheri et al., 2006) with copyright permission of Elsevier Science and the OsteoArthritis Research Society International

7.4 Is the Biomechanical Function of Chondrocytes Regulated by Potassium Ion Channels?

The notion of having mechanosensitive ion channels has been around for decades. Studies in the yeast *Saccharomyces cerevisiae* have revealed voltage-dependent channels that are activated by, and adapted to, mechanical stretching of the plasma membrane (Gustin et al., 1988). Because these mechanosensitive channels pass both cations and anions, they may play a role in turgor regulation in this walled organism. Subsequent studies in snail neurones found evidence for stretch-activated K^+ channels (Sigurdson and Morris, 1989). Interestingly, early studies of potassium channels in human and avian fibroblasts revealed very clear mechanical sensitivity (French and Stockbridge, 1988). Patch-clamp electrophysiological recording of osteoblast-like cells (related to mesenchyme derived chondrocytes) have exposed non-selective cation channels (Davidson et al., 1990). These non-selective mechanosensitive cation channels have been proposed to be involved in the response of bone to mechanical loading (Davidson et al., 1990; Duncan and Hruska, 1994; Duncan, 1995). The activity of these channels is thought to raise intracellular Ca^{2+} in combination with inositol trisphosphate (IP3, the product of phospholipase C). Ca^{2+} and protein kinase C (PKC) are then thought to stimulate phospholipase A2 activity, arachidonic acid production, and ultimately PGE_2 release (Ajubi et al., 1999). Integrins are believed to be closely integrated in this process of mechanotransduction and form mechanoreceptor complexes with mechanosensitive ion channels (Mobasheri et al., 2002a). The mechanosensitive ion channel-dependent tyrosine phosphorylation of focal adhesion proteins and associated proteins are also involved in the responses to mechanical stimulation (Lee et al., 2000). Parathyroid hormone (PTH) and vitamin D are also implicated in this process as PTH significantly enhances the Ca^{2+} response to mechanical stress via modulation of mechanosensitive cation channels and voltage-gated calcium channels (Ryder and Duncan, 2001). The osteoactive steroid hormones also exert rapid non-genomic responses on bone cells by modulating calcium ion channel activities on the cell membrane of these cells (Zanello and Norman, 2003). Further evidence for the involvement of Ca^{2+} responses in bone cell mechanotransduction comes from studies of annexin V as a Ca^{2+} channel in the osteoblast (Haut Donahue et al., 2004). In summary, the osteoblastic mechanosensitive channel is thought to be a non-selective cation channel of small conductance. Recent studies in human tenocytes (tendon cells, also related to osteoblasts and chondrocytes) advocate a role for tandem pore domain potassium channels (TREK-1) and voltage operated calcium channels in tenocyte mechanotransduction (Magra et al., 2006).

In many cell types, the key signal for secretion is membrane depolarisation and, in turn, an increase of extracellular calcium (Penner and Neher, 1988). Although not fully understood, this also appears to be the case for chondrocytes, since ion channel blockers alter the chondrocyte membrane potential and reduce both proteoglycan secretion and calcium waves (Guilak et al., 1999).

In the majority of mammalian cells, the resting membrane potential is largely dependent upon potassium ion distribution and the activity of potassium ion. This too appears to be the case with chondrocytes; pressure changes, physical or hypertonic activate potassium channels and hyperpolarize the membrane of chondrocytes (Sanchez et al., 2003). These changes in membrane potential are accompanied by changes of intracellular calcium (Sanchez and Wilkins, 2004). Both the secretion and the change in membrane potential are reduced by classical potassium ion channel blockers (Sanchez and Wilkins, 2004; Wohlrab et al., 2004). The important and unaddressed question is, however, does the mechanical stretch of the membrane close potassium channels causing an influx of calcium and a calcium wave, or does stretch induce a calcium wave which *then* activates calcium activated potassium channels. This question needs to be answered with some urgency.

7.5 Ion Channels and Chondrocyte Proliferation

An increasing number of ion channels are being found to be causally involved in human and animal diseases, giving rise to "channelopathies". Several potassium ion channels have been linked to cell proliferation and to tumour progression (Pardo et al., 2005; Stuhmer et al., 2006). Original studies by Wohlrab and co-workers addressed whether the proliferation of human chondrocytes is regulated by ion channels. The authors used a number of ion channel modulators including TEA,

4-AP, 4′, 4′ diisothiocyanato-stilbene-2,2′-disulfonic acid (DIDS), 4-acetamido-4′-isothiocyano-2,2′-disulfonic acid stilbene (SITS), verapamil and lidocaine in their study and measured the membrane potential and the proliferation of human chondrocytes using flow cytometry and ^{3}H-thymidine incorporation (Wohlrab et al., 2002). They found that many ion channel modulators decrease cell proliferation and result in necrotic cell death (Wohlrab et al., 2004). Although these observations are not surprising they do suggest that ion channels are important for chondrocyte division and proliferation. They suggest that these results may serve as a basis for further investigations to include ion channel modulators in the development and formulation of therapeutic strategies to treat arthritis.

7.6 Conclusion

In the last decade we have witnessed a considerable proliferation of literature on cartilage mechanotransduction and metabolism. A very small number of these papers deal with the important issue of ion channels involved in mechanotransduction and metabolic regulation. Significantly, there are increasing reports of cation channels responsive to membrane tension in mesenchyme derived cells such as osteoblasts and tenocytes (Duncan and Hruska, 1994; Duncan, 1995; Haut Donahue et al., 2004; Magra et al., 2006). A number of research groups including ours have used multidisciplinary approaches to identify the transporters and channels responsible for ion and nutrient transport in articular and growth plate chondrocytes (Rajpurohit et al., 2002; Mobasheri et al., 2005b). There is considerable evidence accumulating to suggest that ion channels residing in the plasma membrane of chondrocytes and osteoblasts are involved in the transduction of mechanical signals. Several candidate ion channel systems have been proposed to participate in chondrocytes; these include voltage-gated sodium, potassium and calcium channels and the epithelial sodium channel. Further details about the physiological and pharmacological properties of these candidate electro-mechanical transducer channels may be found in one of our earlier reviews (Mobasheri et al., 2002a). Our recent studies suggest that a number of putative glucose (i.e. Kir6.1) and oxygen sensitive (i.e. BK) potassium ion channels are also present in chondrocytes. We have already identified a number of other nutrient (i.e. glucose) transporters regulated by the oxygen sensitive transcription factor hypoxia inducible factor 1 (HIF-1) in articular chondrocytes (Fig. 7.12) (Mobasheri et al., 2002b; Mobasheri et al., 2002c; Rajpurohit et al., 2002; Pfander et al., 2003; Richardson et al., 2003; Pfander et al., 2005; Phillips et al., 2005). Initially it might appear that many of these studies are unrelated and disparate. However, it is becoming more and more apparent that these studies are linked. It is highly likely that ion channels in chondrocytes are indeed multifunctional and serve a number of different physiological purposes. Therefore, the processes of mechanotransduction, metabolic regulation and chondrocyte proliferation may share a number of ion channels as common denominators. Further studies are required to shed light on these possible interrelationships. Potassium channels offer a rich source of targets for therapeutic development

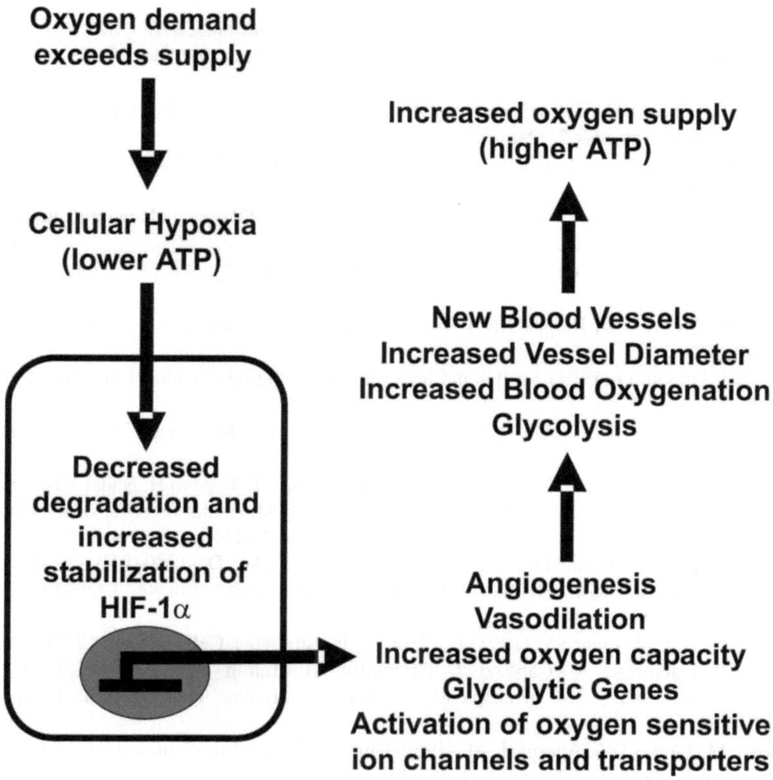

Fig. 7.12 Molecular responses to hypoxia; when oxygen demand exceeds supply the HIF-1 oxygen -sensing system results in activation of key genes involved in angiogenesis, vasodilatation, oxygen delivery and glycolysis in order to increase tissue oxygenation. Reproduced from (Mobasheri et al., 2005a) with copyright permission of Jiménez-Godoy, S.A

(Garcia and Kaczorowski, 2005). We anticipate that these channels may become viable targets for intervention in osteoarthritis and related arthritic conditions.

Acknowledgment Our work has been supported by grants from the Wellcome Trust, BBSRC, The British Heart Foundation, Novartis and The Waltham Foundation. The authors wish to thank Mr. A.F. Brandwood and Mr. S. Williams for excellent histology support.

References

Aguilar-Bryan L, Clement JP, Gonzalez G, Kunjilwar K, Babenko A, Bryan J (1998) Toward understanding the assembly and structure of K-ATP channels. Physiological Reviews 78:227–245.

Ajubi NE, Klein-Nulend J, Alblas MJ, Burger EH, Nijweide PJ (1999) Signal transduction pathways involved in fluid flow-induced PGE2 production by cultured osteocytes. Am J Physiol 276:E171–178.

Alexander SP, Mathie A, Peters JA (2006) Guide to receptors and channels, 2nd edition. Br J Pharmacol 147 Suppl 3:S1–168.

Amundson J, Clapham D (1993) Calcium waves. Curr Opin Neurobiol 3:375–382.

Archer CW, Francis-West P (2003) The chondrocyte. Int J Biochem Cell Biol 35:401–404.

Ashcroft FM (2006) From molecule to malady. Nature 440:440–447.

Ashcroft FM, Gribble FM (1998) Correlating structure and function in ATP-sensitive K^+ channels. Trends in Neurosciences 21:288–294.

Babenko AP, Aguilar-Bryan L, Bryan J (1998) A view of sur/KIR6.X, KATP channels. Annu Rev Physiol 60:667–687.

Barrett-Jolley R, McPherson GA (1998) Characterization of K(ATP) channels in intact mammalian skeletal muscle fibres. Br J Pharmacol 123:1103–1110.

Bauer CK, Schwarz JR (2001) Physiology of EAG K^+ channels. Journal Of Membrane Biology 182:1–15.

Benjamin M, Archer CW, Ralphs JR (1994) Cytoskeleton of cartilage cells. Microsc Res Tech 28:372–377.

Berridge MJ, Dupont G (1994) Spatial and temporal signalling by calcium. Curr Opin Cell Biol 6:267–274.

Chicurel ME, Chen CS, Ingber DE (1998) Cellular control lies in the balance of forces. Curr Opin Cell Biol 10:232–239.

Coetzee WA, Amarillo Y, Chiu J, Chow A, Lau D, McCormack T, Moreno H, Nadal MS, Ozaita A, Pountney D, Saganich M, De Miera EVS, Rudy B (1999) Molecular diversity of K^+ channels. In: Molecular And Functional Diversity Of Ion Channels And Receptors, pp 233–285.

Coimbra IB, Jimenez SA, Hawkins DF, Piera-Velazquez S, Stokes DG (2004) Hypoxia inducible factor-1 alpha expression in human normal and osteoarthritic chondrocytes. Osteoarthritis Cartilage 12:336–345.

Cornell-Bell AH, Finkbeiner SM (1991) Ca^{2+} waves in astrocytes. Cell Calcium 12:185–204.

D'Andrea P, Calabrese A, Capozzi I, Grandolfo M, Tonon R, Vittur F (2000) Intercellular Ca^{2+} waves in mechanically stimulated articular chondrocytes. Biorheology 37:75–83.

Davidson RM, Tatakis DW, Auerbach AL (1990) Multiple forms of mechanosensitive ion channels in osteoblast-like cells. Pflugers Arch 416:646–651.

Doyle DA, Cabral JM, Pfuetzner RA, Kuo AL, Gulbis JM, Cohen SL, Chait BT, MacKinnon R (1998) The structure of the potassium channel: Molecular basis of K+ conduction and selectivity. Science 280:69–77.

Duncan RL (1995) Transduction of mechanical strain in bone. ASGSB Bull 8:49–62.

Duncan RL, Hruska KA (1994) Chronic, intermittent loading alters mechanosensitive channel characteristics in osteoblast-like cells. Am J Physiol 267:F909–916.

French AS, Stockbridge LL (1988) Potassium channels in human and avian fibroblasts. Proc R Soc Lond B Biol Sci 232:395–412.

Garcia-Ferreiro RE, Kerschensteiner D, Major F, Monje F, Stuhmer W, Pardo LA (2004) Mechanism of block of hEag1 K+ channels by imipramine and astemizole. J Gen Physiol 124: 301–317.

Garcia ML, Kaczorowski GJ (2005) Potassium channels as targets for therapeutic intervention. Sci STKE 2005:pe46.

Goldspink G (1996) Muscle growth and muscle function: a molecular biological perspective. Res Vet Sci 60:193–204.

Goldspink G (1999) Changes in muscle mass and phenotype and the expression of autocrine and systemic growth factors by muscle in response to stretch and overload. J Anat 194 (Pt 3):323–334.

Goldspink G, Scutt A, Loughna PT, Wells DJ, Jaenicke T, Gerlach GF (1992) Gene expression in skeletal muscle in response to stretch and force generation. Am J Physiol 262:R356–363.

Grodzinsky AJ, Levenston ME, Jin M, Frank EH (2000) Cartilage tissue remodeling in response to mechanical forces. Annu Rev Biomed Eng 2:691–713.

Guilak F (1995) Compression-induced changes in the shape and volume of the chondrocyte nucleus. J Biomech 28:1529–1541.

Guilak F, Ratcliffe A, Mow VC (1995) Chondrocyte deformation and local tissue strain in articular cartilage: a confocal microscopy study. J Orthop Res 13:410–421.

Guilak F, Zell RA, Erickson GR, Grande DA, Rubin CT, McLeod KJ, Donahue HJ (1999) Mechanically induced calcium waves in articular chondrocytes are inhibited by gadolinium and amiloride. J Orthop Res 17:421–429.

Gustin MC, Zhou XL, Martinac B, Kung C (1988) A mechanosensitive ion channel in the yeast plasma membrane. Science 242:762–765.

Haut Donahue TL, Genetos DC, Jacobs CR, Donahue HJ, Yellowley CE (2004) Annexin V disruption impairs mechanically induced calcium signaling in osteoblastic cells. Bone 35:656–663.

Ingber DE (1998) Cellular basis of mechanotransduction. Biol Bull 194:323–325; discussion 325–327.

Jan LY, Jan YN (1997) Annual Review Prize Lecture - Voltage-gated and inwardly rectifying potassium channels. Journal Of Physiology-London 505:267–282.

Kemp PJ (2006) Detecting acute changes in oxygen: will the real sensor please stand up? Exp Physiol 91:829–834.

Kemp PJ, Williams SE, Mason HS, Wootton P, Iles DE, Riccardi D, Peers C (2006) Functional proteomics of BK potassium channels: defining the acute oxygen sensor. Novartis Found Symp 272:141–151; discussion 151–146, 214–147.

Kizer N, Guo XL, Hruska K (1997) Reconstitution of stretch-activated cation channels by expression of the alpha-subunit of the epithelial sodium channel cloned from osteoblasts. Proc Natl Acad Sci U S A 94:1013–1018.

Knight MM, Ghori SA, Lee DA, Bader DL (1998) Measurement of the deformation of isolated chondrocytes in agarose subjected to cyclic compression. Med Eng Phys 20:684–688.

Lane Smith R, Trindade MC, Ikenoue T, Mohtai M, Das P, Carter DR, Goodman SB, Schurman DJ (2000) Effects of shear stress on articular chondrocyte metabolism. Biorheology 37:95–107.

Lee HS, Millward-Sadler SJ, Wright MO, Nuki G, Salter DM (2000) Integrin and mechanosensitive ion channel-dependent tyrosine phosphorylation of focal adhesion proteins and beta-catenin in human articular chondrocytes after mechanical stimulation. J Bone Miner Res 15:1501–1509.

Lesage F, Lazdunski M (2000) Molecular and functional properties of two-pore-domain potassium channels. American Journal Of Physiology-Renal Physiology 279: F793–F801.

Lippiat JD, Standen NB, Harrow ID, Phillips SC, Davies NW (2003) Properties of BK(Ca) channels formed by bicistronic expression of hSloalpha and beta1–4 subunits in HEK293 cells. J Membr Biol 192:141–148.

Lopatin AN, Makhina EN, Nichols CG (1994) Potassium Channel Block By Cytoplasmic Polyamines As The Mechanism Of Intrinsic Rectification. Nature 372:366–369.

Magra M, Hughes S, El Haj AJ, Maffulli N (2006) VOCCs and TREK-1 ion channel expression in human tenocytes. Am J Physiol Cell Physiol.

Matsuda H, Saigusa A, Irisawa H (1987) Ohmic Conductance Through The Inwardly Rectifying K-Channel And Blocking By Internal Mg-2+. Nature 325:156–159.

May H, Mobasheri A, Womack M, Barrett-Jolley R (2007) Functional expression of aquaporins in canine chondrocytes. Biophys J 117A–117A Suppl. S.

Millward-Sadler SJ, Wright MO, Flatman PW, Salter DM (2004) ATP in the mechanotransduction pathway of normal human chondrocytes. Biorheology 41:567–575.

Mobasheri A, Martin-Vasallo P (1999) Epithelial sodium channels in skeletal cells; a role in mechanotransduction? Cell Biol Int 23:237–240.

Mobasheri A, Marples D (2004) Expression of the AQP-1 water channel in normal human tissues: a semiquantitative study using tissue microarray technology. Am J Physiol Cell Physiol 286:C529–537.

Mobasheri A, Carter SD, Martin-Vasallo P, Shakibaei M (2002a) Integrins and stretch activated ion channels; putative components of functional cell surface mechanoreceptors in articular chondrocytes. Cell Biol Int 26:1–18.

Mobasheri A, Neama G, Bell S, Richardson S, Carter SD (2002b) Human articular chondrocytes express three facilitative glucose transporter isoforms: GLUT1, GLUT3 and GLUT9. Cell Biol Int 26:297–300.

Mobasheri A, Richardson S, Mobasheri R, Shakibaei M, Hoyland JA (2005a) Hypoxia inducible factor-1 and facilitative glucose transporters GLUT1 and GLUT3: Putative molecular

components of the oxygen and glucose sensing apparatus in articular chondrocytes. Histology and Histopathology 20:1327–1338.
Mobasheri A, Richardson S, Mobasheri R, Shakibaei M, Hoyland JA (2005b) Hypoxia inducible factor-1 and facilitative glucose transporters GLUT1 and GLUT3: putative molecular components of the oxygen and glucose sensing apparatus in articular chondrocytes. Histol Histopathol 20:1327–1338.
Mobasheri A, Marples D, Young I, Moskaluk C, Frigeri A (2005c) Tissue distribution of the AQP-4 water channel: A study using normal human Tissue MicroArrays. Journal of General Physiology 126:76a–77a.
Mobasheri A, Gent TC, Womack MD, Carter SD, Clegg PD, Barrett-Jolley R (2005d) Quantitative analysis of voltage-gated potassium currents from primary equine (Equus caballus) and elephant (Loxodonta africana) articular chondrocytes. Am J Physiol Regul Integr Comp Physiol 289:R172–180.
Mobasheri A, Gent TC, Nash AI, Womack MD, Moskaluk CA, Barrett-Jolley R (2006) Evidence for functional ATP-sensitive (K(ATP)) potassium channels in human and equine articular chondrocytes. Osteoarthritis Cartilage.
Mobasheri A, Trujillo E, Bell S, Carter SD, Clegg PD, Martin-Vasallo P, Marples D (2004) Aquaporin water channels AQP1 and AQP3, are expressed in equine articular chondrocytes. Vet J 168:143–150.
Mobasheri A, Vannucci SJ, Bondy CA, Carter SD, Innes JF, Arteaga MF, Trujillo E, Ferraz I, Shakibaei M, Martin-Vasallo P (2002c) Glucose transport and metabolism in chondrocytes: a key to understanding chondrogenesis, skeletal development and cartilage degradation in osteoarthritis. Histol Histopathol 17:1239–1267.
Mow VC, Wang CC, Hung CT (1999) The extracellular matrix, interstitial fluid and ions as a mechanical signal transducer in articular cartilage. Osteoarthritis Cartilage 7:41–58.
Nichols CG (2006) K-ATP channels as molecular sensors of cellular metabolism. Nature 440:470–476.
Nichols CG, Lopatin AN (1997) Inward rectifier potassium channels. Annual Review Of Physiology 59:171–191.
Orazizadeh M, Cartlidge C, Wright MO, Millward-Sadler SJ, Nieman J, Halliday BP, Lee HS, Salter DM (2006) Mechanical responses and integrin associated protein expression by human ankle chondrocytes. Biorheology 43:249–258.
Pardo LA, Contreras-Jurado C, Zientkowska M, Alves F, Stuhmer W (2005) Role of voltage-gated potassium channels in cancer. J Membr Biol 205:115–124.
Pardo LA, del Camino D, Sanchez A, Alves F, Bruggemann A, Beckh S, Stuhmer W (1999) Oncogenic potential of EAG K(+) channels. Embo J 18:5540–5547.
Patel AJ, Honore E (2001) Properties and modulation of mammalian 2P domain K+ channels. Trends In Neurosciences 24:339–346.
Penner R, Neher E (1988) The role of calcium in stimulus-secretion coupling in excitable and non-excitable cells. J Exp Biol 139:329–345.
Perkins GL, Derfoul A, Ast A, Hall DJ (2005) An inhibitor of the stretch-activated cation receptor exerts a potent effect on chondrocyte phenotype. Differentiation 73:199–211.
Pfander D, Cramer T, Swoboda B (2005) Hypoxia and HIF-1alpha in osteoarthritis. Int Orthop 29:6–9.
Pfander D, Cramer T, Schipani E, Johnson RS (2003) HIF-1alpha controls extracellular matrix synthesis by epiphyseal chondrocytes. J Cell Sci 116:1819–1826.
Phillips T, Ferraz I, Bell S, Clegg PD, Carter SD, Mobasheri A (2005) Differential regulation of the GLUT1 and GLUT3 glucose transporters by growth factors and pro-inflammatory cytokines in equine articular chondrocytes. Vet J 169:216–222.
Ponce A (2006) Expression of voltage dependent potassium currents in freshly dissociated rat articular chondrocytes. Cell Physiol Biochem 18:35–46.
Prabhakar NR (2006) O2 sensing at the mammalian carotid body: why multiple O2 sensors and multiple transmitters? Exp Physiol 91:17–23.
Quayle JM, Nelson MT, Standen NB (1997) ATP-sensitive and inwardly rectifying potassium channels in smooth muscle. Physiological Reviews 77:1165–1232.

Quayle JM, Bonev AD, Brayden JE, Nelson MT (1995) Pharmacology of ATP-sensitive K+ currents in smooth muscle cells from rabbit mesenteric artery. Am J Physiol 269: C1112–1118.

Rajpurohit R, Risbud MV, Ducheyne P, Vresilovic EJ, Shapiro IM (2002) Phenotypic characteristics of the nucleus pulposus: expression of hypoxia inducing factor-1, glucose transporter-1 and MMP-2. Cell Tissue Res 308:401–407.

Richardson S, Neama G, Phillips T, Bell S, Carter SD, Moley KH, Moley JF, Vannucci SJ, Mobasheri A (2003) Molecular characterization and partial cDNA cloning of facilitative glucose transporters expressed in human articular chondrocytes; stimulation of 2-deoxyglucose uptake by IGF-I and elevated MMP-2 secretion by glucose deprivation. Osteoarthritis Cartilage 11:92–101.

Robbins J (2001) KCNQ potassium channels: physiology, pathophysiology, and pharmacology. Pharmacology & Therapeutics 90:1–19.

Ryder KD, Duncan RL (2001) Parathyroid hormone enhances fluid shear-induced [Ca^{2+}]i signaling in osteoblastic cells through activation of mechanosensitive and voltage-sensitive Ca^{2+} channels. J Bone Miner Res 16:240–248.

Salter DM, Wright MO, Millward-Sadler SJ (2004) NMDA receptor expression and roles in human articular chondrocyte mechanotransduction. Biorheology 41:273–281.

Sanchez JC, Wilkins RJ (2004) Changes in intracellular calcium concentration in response to hypertonicity in bovine articular chondrocytes. Comp Biochem Physiol A Mol Integr Physiol 137:173–182.

Sanchez JC, Danks TA, Wilkins RJ (2003) Mechanisms involved in the increase in intracellular calcium following hypotonic shock in bovine articular chondrocytes. Gen Physiol Biophys 22:487–500.

Sanguinetti MC, Spector PS (1997) Potassium channelopathies. Neuropharmacology 36:755–762.

Schipani E (2005) Hypoxia and HIF-1 alpha in chondrogenesis. Semin Cell Dev Biol 16:539–546.

Schipani E, Ryan HE, Didrickson S, Kobayashi T, Knight M, Johnson RS (2001) Hypoxia in cartilage: HIF-1alpha is essential for chondrocyte growth arrest and survival. Genes Dev 15:2865–2876.

Schulze-Tanzil G, Mobasheri A, de Souza P, John T, Shakibaei M (2004) Loss of chondrogenic potential in dedifferentiated chondrocytes correlates with deficient Shc-Erk interaction and apoptosis. Osteoarthritis Cartilage 12:448–458.

Shakibaei M, Mobasheri A (2003) Beta1-integrins co-localize with Na, K-ATPase, epithelial sodium channels (ENaC) and voltage activated calcium channels (VACC) in mechanoreceptor complexes of mouse limb-bud chondrocytes. Histol Histopathol 18:343–351.

Shimazaki A, Wright MO, Elliot K, Salter DM, Millward-Sadler SJ (2006) Calcium/calmodulin-dependent protein kinase II in human articular chondrocytes. Biorheology 43:223–233.

Sigurdson WJ, Morris CE (1989) Stretch-activated ion channels in growth cones of snail neurons. J Neurosci 9:2801–2808.

Stanfield PR, Nakajima S, Nakajima Y (2002) Constitutively active and G-protein coupled inward rectifier K+ channels: Kir2.0 and Kir3.0. In: Reviews Of Physiology Biochemistry And Pharmacology, Vol 145, pp 47–179.

Stocker M (2004) Ca^{2+}-activated K^{+} channels: Molecular determinants and function of the SK family. Nature Reviews Neuroscience 5:758–770.

Stuhmer W, Alves F, Hartung F, Zientkowska M, Pardo LA (2006) Potassium channels as tumour markers. FEBS Lett 580:2850–2852.

Sugimoto T, Yoshino M, Nagao M, Ishii S, Yabu H (1996) Voltage-gated ionic channels in cultured rabbit articular chondrocytes. Comp Biochem Physiol C Pharmacol Toxicol Endocrinol 115:223–232.

Trujillo E, Alvarez de la Rosa D, Mobasheri A, Gonzalez T, Canessa CM, Martin-Vasallo P (1999) Sodium transport systems in human chondrocytes. II. Expression of ENaC, $Na^{+}/K^{+}/2Cl^{-}$ cotransporter and Na+/H+ exchangers in healthy and arthritic chondrocytes. Histol Histopathol 14:1023–1031.

Trujillo E, Gonzalez T, Marin R, Martin-Vasallo P, Marples D, Mobasheri A (2004) Human articular chondrocytes, synoviocytes and synovial microvessels express aquaporin water channels; upregulation of AQP1 in rheumatoid arthritis. Histol Histopathol 19:435–444.

Tsuga K, Tohse N, Yoshino M, Sugimoto T, Yamashita T, Ishii S, Yabu H (2002) Chloride conductance determining membrane potential of rabbit articular chondrocytes. J Membr Biol 185:75–81.

Urban JP (1994) The chondrocyte: a cell under pressure. Br J Rheumatol 33:901–908.

Walsh KB, Cannon SD, Wuthier RE (1992) Characterization of a delayed rectifier potassium current in chicken growth plate chondrocytes. Am J Physiol 262:C1335–1340.

Wang W, Xu J, Kirsch T (2003) Annexin-mediated Ca^{2+}influx regulates growth plate chondrocyte maturation and apoptosis. J Biol Chem 278:3762–3769.

Wang XT, Nagaba S, Nagaba Y, Leung SW, Wang J, Qiu W, Zhao PL, Guggino SE (2000) Cardiac L-type calcium channel alpha 1-subunit is increased by cyclic adenosine monophosphate: messenger RNA and protein expression in intact bone. J Bone Miner Res 15:1275–1285.

Wilson JR, Duncan NA, Giles WR, Clark RB (2004) A voltage-dependent K+ current contributes to membrane potential of acutely isolated canine articular chondrocytes. J Physiol 557: 93–104.

Wohlrab D, Lebek S, Kruger T, Reichel H (2002) Influence of ion channels on the proliferation of human chondrocytes. Biorheology 39:55–61.

Wohlrab D, Vocke M, Klapperstuck T, Hein W (2004) Effects of potassium and anion channel blockers on the cellular response of human osteoarthritic chondrocytes. J Orthop Sci 9:364–371.

Wright M, Jobanputra P, Bavington C, Salter DM, Nuki G (1996) Effects of intermittent pressure-induced strain on the electrophysiology of cultured human chondrocytes: evidence for the presence of stretch-activated membrane ion channels. Clin Sci (Lond) 90:61–71.

Yellen G (2002) The voltage-gated potassium channels and their relatives. Nature 419:35–42.

Yellowley CE, Hancox JC, Donahue HJ (2002) Effects of cell swelling on intracellular calcium and membrane currents in bovine articular chondrocytes. J Cell Biochem 86:290–301.

Yokoshiki H, Sunagawa M, Seki T, Sperelakis N (1998) ATP-sensitive K^+ channels in pancreatic, cardiac, and vascular smooth muscle cells. American Journal of Physiology-Cell Physiology 43:C25–C37.

Zanello LP, Norman AW (2003) Multiple molecular mechanisms of 1 alpha,25(OH)2-vitamin D3 rapid modulation of three ion channel activities in osteoblasts. Bone 33:71–79.

Chapter 8
Osmotransduction Through Volume-Sensitive Cl^- Channels

Naomi Niisato and Yoshinori Marunaka

Abstract To sense environmental changes and various stresses is crucial for cells to survive by adjusting themselves to new environments. For this purpose, cells should have a mechanism to sense the environmental changes including osmotic stress and to respond immediately to them. Osmotic stress has been recognized to produce mechanical stress by membrane stretch and/or deformation through alteration of cell volume. A drastic cell volume change is observed by hypo-osmotic stress, namely an initial cell swelling followed by a regulatory volume decrease (RVD). Membrane stretch/deformation caused by the initial cell swelling might be the first step to transduce the stress into intracellular signaling cascades by modulating functions of membrane proteins including volume-sensitive ion channels and transporters. The conversion of physical forces produced by membrane stretch/deformation into biochemical signals has been discussed to clarify the regulatory mechanism of osmotransduction and to identify the osmosensitive signal molecules. Recent studies have indicated that stretch-activated ion channels, integrins, growth factor receptors, cytoskelton, extracellular matrix and other various molecules contribute to the osmotransduction. However, a little is known how these molecules sense the osmotic- and mechanical-stresses and function to transduce the stress. So far, RVD is considered to be required for recovering the original cell volume after the initial cell swelling through activation of volume-sensitive Cl^-/K^+ channels. Recently, we have indicated that a part of hypo-osmotic stress might be converted into the RVD-induced decrease in intracellular Cl^- concentration ($[Cl^-]_i$), which plays a key role in response to hypo-osmotic stress. In a renal epithelium, plasma hypo-osmolality stimulates Na^+ reabsorption to maintain normal plasma osmolality by inducing epithelial Na^+ channel (ENaC) gene expression through the RVD-induced decrease in $[Cl^-]_i$. This is a newly recognized physiological role of RVD in response to hypo-osmotic stress. In this review, we focus our discussion on understanding of regulatory mechanism of volume-sensitive Cl^- channel by hypo-osmotic stress and the cytosolic Cl^--dependent cellular responses through volume-sensitive Cl^- channels.

Key words: Cytosolic Cl^- · Volume-sensitive Cl^- Channel · RVD · ENaC · Swelling

8.1 Introduction

This review introduces the mechanisms how cells sense physical forces such as mechanical and osmotic stress. This sensing mechanism is essential for cell function on cellular growth, development and differentiation. Specially, this chapter provides researchers with evidences on involvement of ion channels/transporters in sensing physical forces, and perspective for future research on roles of ion channels/transporters as sensors detecting physical forces.

8.1.1 Cell Volume Regulation

The maintenance of a constant cell volume is an imperative for cellular homeostasis in most of cells. Even at constant extracellular osmolality, cell volume constancy is maintained by transport of osmotically active substances across cell membrane and formation or disappearance of cellular osmolality by metabolism. Thus, to keep cell volume constancy requires the continued operation of cell volume regulatory mechanisms, such as ion transport across the cell membrane as well as accumulation or disposal of organic osmolytes and metabolites. The cell volume is regulated by various cellular events including alteration of membrane potential, intracellular ionic composition, phosphorylation of diverse targets of proteins, hormonal stimuli and altered gene expression (Lang et al, 1998a). Further, the cell volume is also regulated by extracellular environments including alteration of fluid osmolality. Accordingly, alterations of cell volume and volume regulatory mechanisms participate in various cellular functions including epithelial transport, metabolism, excitation, migration, cell proliferation and cell death.

8.1.2 Cell Volume Regulation in Osmotic Stress

Generally, most of cells have high permeability to water, and the water flow across the plasma membrane is directed by osmotic pressure gradients and hydrostatic pressure. The reduction of extracellular osmolality leads to rapid cell swelling in a magnitude proportional to the change in osmolality by water movement across the plasma membrane. The swollen cell undergoes subsequent shrinkage known as regulatory volume decrease (RVD). Then, cell volume reaches a plateau level, which is higher than the initial one and depends on the osmolality of extracellular fluid (Fig. 8.1). Cell swelling-induced activation of K^+, Cl^- and organic osmolyte channels have been demonstrated in a variety of cells. Activation of a Cl^- efflux pathway by cell swelling is first demonstrated in Ehrlich ascites tumor cells (Hoffman et al, 1978) and in lymphocytes (Grinstein et al, 1982), and subsequently demonstrated in a variety of cells (Nilius et al, 1997; Okada, 1997). Electrophysiological studies of swelling-activated K^+ and Cl^- currents were performed in human intestine cell

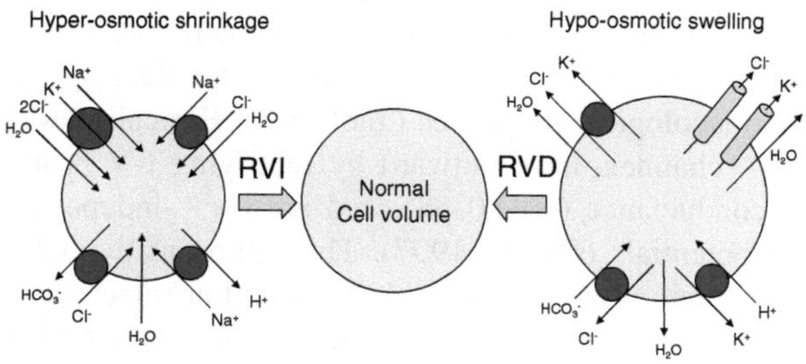

Fig. 8.1 Regulatory mechanisms of cell volume by hypo- or hyper-osmotic stress. In hyper-osmotic shrinkage, the shrunk cell incorporates ions such as Na^+ and Cl^- for regulatory volume increase (RVI) by activating tranporters including NKCC, NCC, NHE and Cl^-/HCO_3^- exchanger. In hypo-osmotic swelling, the swollen cell excretes ions such as K^+, Cl^- for regulatory volume decrease (RVD) by activating volume-sensitive K^+, Cl^- channels

line 407 cells (Lang et al, 1998b). Later, activation of separate conductive pathways for K^+ and Cl^- during RVD has been indicated in a variety of cells. Further, electroneutral K^+/Cl^- cotransporter (KCC) activated by hypo-osmotic swelling also contributes to extrusion of K^+/Cl^- during RVD in several cells. On the contrary, the increase in extracellular osmolality leads to a cell shrinkage followed by a recovery process of cell volume, termed regulatory volume increase (RVI). The shrunk cell incorporates solutes, ions and small organic osmolytes to recover the original cell volume by stimulating active-transporters such as the $Na^+/K^+/2Cl^-$ contranporter (NKCC) (Haas et al, 1994, Haas et al, 1995; Putney et al, 1999), the Na^+/H^+ antiporters (NHE), the Cl^-/HCO_3^- exchanger and the Na^+/Cl^- cotransporter (NCC) (Fig. 8.1), and produces osmolytes through induction of gene expression. Although the processes of RVD and RVI occurring at osmotic stress are well understood as described above, how these volume-sensitive ion channels and transporters sense the alteration of cell volume is still unclear.

8.1.3 Osmosensitive Cl$^-$ Channels

Cl^- channels have various physiological functions, including the transcellular transport of ions and fluid, cell volume regulation, controlling the intracellular Cl^- concentration ($[Cl^-]_i$) and stabilization of membrane potential. To date Cl^- channels activated by hypo-osmotic cell swelling have been reported to exist in various types of cells such as human epithelial cells (Kubo et al, 1992), rabbit renal cortical collecting duct cell line (Schwiebert et al, 1994), rabbit cardiac myocytes (Hagiwara et al, 1992) and rat osteoblast-like cells (Gosling et al, 1995). These osmosensitive (volume-sensitive) Cl^- channels have been indicated to be responsible for RVD and cell volume regulation in hypo-osmotic stress, because abolishment of the Cl^- current by Cl^- channel blockers has shown suppression of the RVD in various types of cells (Kubo et al, 1992; Nilius et al, 1995; Schwiebert et al, 1994).

Characterization of the osmosensitive Cl^- channels has been performed in various cells and indicates that the channels have similar electrophysiological, biophysical and pharmacological properties (Table 8.1). Basically swelling-activated osmosensitive Cl^- channels have outward by rectifying I-V profiles with intermediate unitary conductance (40–80 pS), and are Ca^{2+}-independent and inactivated at positive potentials (Okada, 1997). The osmosensitive Cl^- channels are highly selective for anion over cation with broad anion selectivity. Activation of some types of osmosensitive Cl^- channels requires intracellular adenosine triphosphate (ATP) without hydrolysis, while extracellular ATP has an inhibitory effect on the Cl^- channels (Tsumura et al, 1996). The Cl^- channel is sensitive to most of general Cl^- channel blockers; 5-nitro-2-(3-phenylpropylamino) benzoic (NPPB), diphenylamine-2-carboxylate (DPC), 4,4'-diisothiocyanatostilbene- 2,2'-disulphonic acid (DIDS), 4-acetamide-4'-isothiocyanostilbene-2,2'-disulphonic acid (SITS). Further, the volume-sensitive Cl^- current is known to be abolished by anthrancene-9-carboxylic acid, niflumic acid, tamoxifen, verapamil, and phloretin (Table 8.2). Historically, the ion channel/current elicited by cell swelling is described as an anion current and given many different names, e.g. VOSAC (volume-sensitive osmolyte and anion channel), $I_{Cl,\ swell}$ (swelling-dependent Cl^- current), $I_{Cl,\ vol}$ (volume-dependent Cl^- current), VRAC (volume-regulated anion channel), VSOR (volume-sensitive outwardly rectifying Cl^- channel). Therefore, in this review, we use a term of "volume-sensitive Cl^- channel or current" based on the meaning of the hypo-osmotic cell swelling activated Cl^- channel or current. Although volume-sensitive Cl^- currents induced by hypo-osmotic stress are observed in various cells, molecular identification of volume-sensitive Cl^- channels has still been under investigation.

8.1.3.1 CLC Family Cl^- Channels; CLC-2 and CLC-3

Two members of the CLC family of Cl^- channels, CLC-2 and CLC-3, are considered to be involved in RVD as volume-sensitive Cl^- channels. Previous studies have indicated that the CLC-2 Cl^- channel is activated by hypo-osmotic cell swelling and contributes to volume regulation in several types of cells (Bond et al, 1998;

Table 8.1 Characteristics of the volume-sensitive Cl^- channel

Ubiquitous expression in most of animal cells
Moderate outward rectifiers
Intermediate unitary conductance (20–100 pS)
Inactivation at positive potential
High selelctivity for anion over cation
Broad selectivity for monovalent anions
Anion permeability ($I^- > Br^- > Cl^-$)
Ca^{2+}-independent activation
Requirement of intracellular ATP without hydrolysis
Inhibition by extracellular ATP
Blocking by general Cl^- channel blockers and other transporter inhbitors

Table 8.2 Pharmacological properties of volume-sensitive Cl^- channel–
Blockers of volume-sensitive Cl^- channel -

Stilbene derivatives	
	4,4'-diisothiocyanatostilbene-2,2'-disulphonic acid (DIDS)
	4-acetamide-4'-isothiocyanostilbene-2,2'-disulphonic acid (SITS)
Carboxylate derivatives	
	5-nitro-2-(3-phenylpropylamino) benzoic (NPPB)
	diphenylamine-2-carboxylate (DPC)
	niflumic acid
	anthrancene-9-carboxylate
P-glycoprotein inhibitors	
	verapamil (L-type Ca^{2+} channel blocker)
	nifedipine (L-type Ca^{2+} channel blocker)
	tamoxifen (antagonist for estrogen)
	1,9-dideoxyforskolin
	quinidine (K^+ channel blocker)
Glucose transporter inhibitor	
	phloretin

Furukawa et al, 1998). The CLC-2 is activated upon hyperpolarization, is blocked by Cd^{2+}, and has inward rectification. As these characters are quite different from those generally reported for volume-sensitive Cl^- channels, the physiological role of CLC-2 as a volume-sensitive Cl^- channel is still open to argue. On the other hand, the role of CLC-3 in contribution to RVD is also controvertible. Although CLC-3 may be involved in volume regulation in several types of cells (Duan et al, 1997; Hermoso et al, 2002), others have reported results against the CLC-3 hypothesis (Li et al, 2000, Sardini et al, 2003). To determine the molecular entity of CLC-3 Cl^- channel in cardiomyocytes, the properties of volume-sensitive Cl^- currents in single ventricular myocytes isolated from CLC-3-deficient (Clcn3(−/−)) mice were compared with those in CLC-3-expressing wild-type (Clcn3(+/+)) and heterozygous (Clcn3(+/−)) mice (Gong et al, 2004). Volume-sensitive Cl^- channels at least in the plasma membrane of mouse cardiomyocytes is functionally expressed without the molecular expression of CLC-3. Thus, molecular identification of CLC-2 and CLC-3 as a volume-sensitive Cl^- channel has not been determined completely.

8.1.3.2 Nuclotide-Sensitive Cl^- Current (ICln)

Nuclotide-sensitive Cl^- current (ICln) was first cloned from Madin Darby canine kidney (MDCK) cells. ICln expressed in *Xenopus oocytes* shows nucleotide-sensitive, calcium-insensitive, slow inactivation at positive potentials, and an outwardly rectifying Cl^- current similar to biophysical properties of volume-sensitive Cl^- channels (Paulmichl et al, 1992). The intracellular distribution of ICln depends on the cell type and physiological status of cells. For example, in the kidney, ICln is located in the apical membrane of the proximal tubles and both in the intracellular membrane and the plasma membrane of distal tubles. In neonatal rat heart cells, hypo-osmotic challenge elicits translocation of ICln channel protein from the intracellular sites to the plasma membrane. Over-expression of ICln channel protein increases volume-sensitive Cl^- currents and accelerates activation of volume-sensitive Cl^- channel during hypo-osmotic challenge. Specific knock-down of ICln channel protein by using a specific antibodies in *Xenopus oocytes* (Krapivinsky et al, 1994) or anti-sense oligo-nucleotides in NIH 3T3 fibroblasts (Chen et al, 1999) significantly reduces volume-sensitive Cl^- currents. These findings indicate that ICln functionally contributes to RVD as a volume-sensitive Cl^- channel. On the other hand, Nilius and his colleagues have shown that ICln and volume-sensitive Cl^- current can be clearly discriminated by biophysical, pharmacological, and biological criteria in human ICln-expressed *Xenopus oocytes* (Voets et al, 1996), concluding that ICln and volume-sensitive Cl^- current are two different Cl^- currents. Therefore, molecular identification of ICln as a volume-sensitive Cl^- current is still controversial and the possibility of ICln as a volume-sensitive Cl^- channel can not be eliminated completely.

8.1.3.3 Voltage-Dependent Anion Channel (VDAC)

VDAC is discovered in the outer membrane of mitochondria in all eukaryotes where they control mitochondrial homeostasis by transporting metabolites, solutes and nucleotides. VDAC is also shown to be located in the plasma membrane. This is supported by the result of co-purification of VDAC with GABA receptor complex and with caveolae. Regarding VDAC function as a volume-sensitive Cl^- channel, the finding that preincubation of HaLa cells with monoclonal antibodies against VDAC blocks RVD (Thinnes et al, 2000) indicates that VDAC is involved in RVD. However, the single channel conductance (Coulombe et al, 1992), voltage dependence and ion selectivity of VDAC are seemed to be different from those of volume-sensitive Cl^- channel. These findings argue against that VDAC is a candidate for volume-sensitive Cl^- channel but do not exclude the possibility that VDAC is partly involved in cell volume regulation after hypo-osmotic stress.

8.1.3.4 ABC Transporter (P-Glycoprotein/MDR1)

P-glycoprotein, the MDR1 gene product, is an ATP-dependent transporter responsible for multi-drug resistance. P-glycoprotein has been characterized as a bifunctional protein; i.e., based on several experimental evidences a drug pump activated upon exposure to its substrates and a Cl^- channel activated upon osmotic swelling. Although Valverde et al. (Valverde et al, 1992) have reported that expression of P-glycoprotein generates volume-regulated, ATP-dependent, Cl^--selective channels, with properties similar to channels characterized previously as volume-sensitive Cl^- channels in epithelial cells, recent reports have suggested that endogenous p-glycoprotein is not a volume-sensitive Cl^- channel in human epithelial (intestine 407) cells (Okada, 1997; Tominaga et al, 1995). Functional suppression of P-glycoprotein by antisense oligonucleotides against MDR1 gene and by four monoclonal antibodies against P-glycoprotein failed to inhibit the activities of volume-sensitive Cl^- channels in intestine 407 cells. Further, in a cell line of the human epidermoid KB cell (KB-3–1) and the corresponding MDR1-transfected cell line (KB-G2), osmotic swelling activates Cl^- channels not only in P-glycoprotein-expressing (KB3–1) but also in P-glycoprotein -lacking (KB-G2) cells. However, there remains a possibility to discuss some functional correlation with each other, because inhibitors for drug pump function of P-glycoprotein such as phloretin suppressed volume-sensitive Cl^- current (Tominaga et al, 1995).

8.1.4 Osmosensitive K^+ Channels

Hypo-osmolality-induced RVD is accomplished by cellular loss of KCl followed by water efflux through parallel activation of K^+ and Cl^- channels. Osmotic swelling has been indicated to cause the activation of different types of volume-sensitive K^+ channels including Ca^{2+}-activated K^+ channel (Park et al, 1994; Ubl et al, 1988; Weiss et al, 1992), stretch activated K^+-channels (Reuss et al. 2000; Sackin, 1989), voltage-gated K^+ channels (Baraban et al, 1997; Deutsch et al, 1993; Schoenmakers et al, 1995), two-pore K^+ (TASK-2) channels (Niemeyer et al, 2001) and Min K^+ channels (Lock et al, 2000). The molecular identification of volume-sensitive K^+ channels has been still discussed and the cell or tissue type-dependent K^+ channels contribute to cell volume regulation during RVD.

8.1.4.1 Ca^{2+}-Activated K^+ Channels

Hypo-osmotic swelling elicits an increase of intracellular Ca^{2+} ($[Ca^{2+}]_i$) in most types of cells. Cellular loss of KCl followed by water efflux mainly contributes to the process of RVD in hypo-osmotic stress. Ca^{2+}-activated K^+ (K_{ca}) channels have been reported to play a role in the RVD process in several types of cells including osteoblasts (Weskamp et al, 2000), T cells (Khanna et al, 1999), human epithelial cells (Hazama et al, 1988; Pasantes-Morales et al, 2000) and tracheal epithelial cells (Vazquez et al, 2001). There are three groups of K_{ca} channels: big conductance (BK_{ca}), intermediate conductance (IK_{ca}) and small conductance (SK_{ca}) channels.

BK_{ca} channels are ubiquitously expressed, and their physiological roles are not fully understood. BK_{ca} channels have been indicated to be activated by hypo-osmotic stress and contribute to K^+ efflux during RVD. In clonal kidney cells, hypo-osmolality activates BK_{ca} channels via elevation of $[Ca^{2+}]_i$ through activation of purinergic receptor (Hafting et al, 2006). Further, hypo-osmotic swelling causes a significant increase in $[Ca^{2+}]_i$ in insulinoma cells and RVD in insulinoma cells is impaired by removal of extracellular Ca^{2+} and inhibited by the BK_{ca} channel blockers, tetraethylammonium (TEA) and iberiotoxin (Sheader et al, 2001). These observations have indicated that BK_{ca} channel is the major K^+ conductive pathway during RVD in several types of cells.

Previous studies have shown the evidence that IK_{ca} channel is also swelling-activated and contributes to cell volume regulation during RVD (Jorgensen et al, 2003; Vazquez et al, 2001). In human epithelial intestine 407 cells, ionomycin and hypo-osmotic stress elevate inwardly rectifying K^+ currents in whole cell recording that are inhibited by IK_{ca} channel blocker (Wang et al, 2003). Moreover, the RVD is suppressed by an IK_{ca} channel blocker but not by BK_{ca} and SK_{ca} channel blockers (Wang et al, 2003). On the other hand, in intermediate (hIK) and small (rSK3) conductance K^+ channels expressed in HEK293 cells, inhibitors of IK_{ca} and SK_{ca} channels significantly reduced the rate of RVD. The hIK channels are activated by cell swelling. The cell-swelling induced activation of hIK channel is significantly inhibited by an inhibitor of F-actin polymerization, cytochalasin D. This observation suggests a role of F-actin cytoskelton in the swelling-induced activation of hIK channels. In osteoblast-like cells, an elevation of $[Ca^{2+}]_i$ is observed during cell swelling, but is not required for activation of K^+ (BK_{ca}) channels (Weskamp et al, 2000). Although K_{ca} channel is basically activated by an increase in cytosolic Ca^{2+}, F-actin cytoskelton also contributes to cell swelling-induced activation of K_{ca} channels through a Ca^{2+}-independent mechanism. In neuronal cells, membrane repolarization following action potential is regulated by small-conductance, Ca^{2+}-activated K^+ (SK_{ca}) channels (Kohler et al, 1996). The SK_{ca} channels have a unitary conductance of 2–20 pS and their opening is relatively insensitive to the membrane voltage. Some SK_{ca} channels are inhibited selectively by bee venom toxin apamin (Ishii et al, 1997) which is useful for functional identification of SK_{ca} channels. In cholangiocarcinoma (Mz-ChA-1) cells and human HuH-7 hepatoma cells, exposure to a hypo-osmotic solution increases the K^+ current density, and apamin partially inhibits the K^+ current activation and cell volume recovery from swelling (Roman et al, 2002). The SK_{ca} channels also contribute to cell volume regulation during RVD as volume-sensitive K^+ channels in several cells.

8.1.4.2 Voltage-Dependent K^+ (Kv) Channels

Cell swelling has been indicated to activate stretch-sensitive ion channels. The voltage-gated K^+ current is also affected by osmotic stress-mediated mechanical changes. It has been suggested that exposure to a hypo-osmotic solution produces a mechanical stress in channel proteins which leads the channel pore to opening, causing a greater ion flux. Cell swelling evokes depolarization in various cells by rapidly activating swelling-sensitive Cl^- channels associated with the subsequent Cl^- efflux. This depolarization then activates voltage-gated K^+ (Kv) channels that contribute to cell volume regulation. As described here, Kv channels are involved in cell volume regulation of lymphocytes (Deutsch et al, 1993) and hippocampal pyramidal neurons (Baraban et al, 1997).

8.1.5 Osmosensitive K^+/Cl^- Cotransporter (KCC)

Activation of electoroneutral K^+/Cl^- cotransporter (KCC) occurs in RVD in most of cells. Three KCC isoforms (KCC1, KCC3 and KCC4) are known to be involved in RVD and activated by hyposmolality (Mercado et al, 2000), whereas KCC2 is insensitive to cell swelling. The KCC isoforms are activated through

Ser/Thr-directed dephosphorylation (Lauf et al, 2000), although a role of tyrosine phosphorylation has been poorly understood.

8.2 Osmotransduction

Mechanotransduction is a process of conversion of physical forces into biochemical signals through transmitter machineries in the plasma membrane and integrating these signals into cellular responses. Mechanical stress is caused by cell volume changes such as osmotic stress-induced cell swelling and cell shrinkage, fluid flow-induced shear stress and mechanical pressure and exercise.

A purpose in most of mechanical stress is stimulation of growth in the mechanosensitive tissues and cells including osteoblasts (Weyts et al, 2003), chondrocytes (Wu et al, 2000) in the bone, and fibroblasts and cardiomyocytes (Lammerding et al, 2004) in the heart. For instances, in gastrointestinal tract, shear stress following feeding increases proliferation of the gut mucosa (Basson et al, 2003). In the heart, a mechanical pressure leads to fibroblasts proliferation and cardiomyocyte hypertrophy (Ruwhof et al, 2000). Therefore, the lack of mechanical stress leads to tissue degradation and cell death in most of mechanosensitive tissues. Thus, mechanical stress is essential for cellular proliferation. Osmotic stress also produces mechanical stress by changing cell volume, i.e., cell swelling and cell shrinkage. Especially, cell swelling caused by hypo-osmotic stress produces membrane stretch and deformation that affect the functions of various membrane proteins and their associated proteins, i.e. cell surface receptors (Lezama et al, 2005), integrins, cytoskeltal proteins, ion channels and transporters. Therefore, membrane stretch/deformation caused by cell swelling might be the first signal in response to hypo-osmotic stress. In this process, hypo-osmotic stress as a physical force is converted into biochemical signals through functional regulation of membrane proteins. Basically functional regulation of protein depends upon its conformational property that can be altered by chemical modification or physical force. The chemical modification is mainly mediated through protein phosphorylation and dephosphorylation associated with conformational alteration in the intra-molecule. The physical force causes direct stretch of ion channels and clustering of surface receptors through alteration of membrane rigidity, fluidity and composition.

The change in cytosolic ion environments in response to osmotic stress and hormonal stimulation is considerable as a possible signal, whereas membrane stretch/deformation is a crucial signal. Hypo-osmotic cell swelling caused by water movement across plasma membrane due to the osmotic pressure gradients alters cytosolic ion concentrations and strengths. In particular, the intracellular Cl^- concentration ($[Cl^-]_i$) is altered during RVD in response to hypo-osmotic stress (Zhou et al, 2005). Although hypo-osmotic cell swelling causes reduction of $[Cl^-]_i$ by water influx in proportion to extracelllular osmolality, further reduction of $[Cl^-]_i$ occurs during RVD accompanied with K^+/Cl^- efflux. In renal A6 cells, additional reduction of $[Cl^-]_i$ during RVD in response to hypo-osmotic stress is observed and the reduction of $[Cl^-]_i$ might be a signal to regulate Na^+ reabsorption through epithelial Na^+ channel (ENaC) gene expression for recovering normal plasma osmolality

(Niisato et al, 2004). Although the detailed mechanism will be discussed later, this evidence leads us to becoming aware of the importance of cytosolic Cl^- as a signal molecule in hypo-osmotic stress.

8.2.1 Osmotransduction Through Membrane Stretch

As mentioned above, cell swelling-induced membrane stretch/deformation might be the first signal to sense the change in extracellular fluid hypo-osmolality. The physical force (i.e. membrane stretch/deformation by hypo-osmotic stress) affects the structures and functions of membrane proteins and their associated proteins including hormone receptors (G protein coupled- receptor; GPCR), growth factor receptors, integrins, cytoskeltal proteins, ion channels and ion transporters that transduce the signal into intracellular signaling cascades (Fig. 8.2).

One of major growth factor receptors affected by membrane stretch/deformation is epidermal growth factor receptor (EGFR). Generally, unstimulated EGFR tyroine kinase is a monomer consisting of an extracellular ligand binding domain, a single trans-membrane domain and a cytosolic kinase domain. With ligand binding, EGFR kinase dimerizes, autophosphorylating six specific tyrosine residues within the non-catalytic cytoplasmic tail. Especially, conformational alteration of EGFR by tyrosine phosphorylation produces high affinity binding sites for src homology 2 (SH2) domain of adaptor proteins. EGFR kinase then recruits and phosphorylates various adaptors and signaling molecules that bind phosphotyrosine motifs with

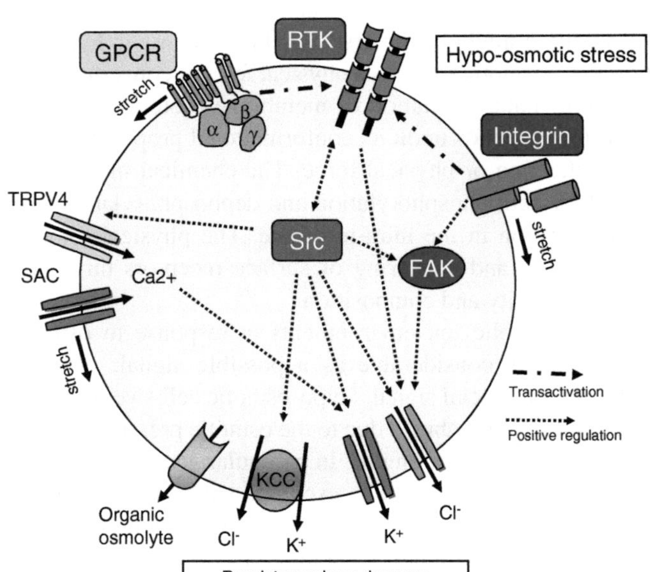

Fig. 8.2 Hypo-osmotic stress activates PTK-mediated signaling pathways through activation of various surface receptors by cell swelling-induced membrane stretch/deformation. The activated PTKs activate volume-sensitive ion channels and transporters involved in RVD

their SH2 domain and phospho-tyrosine binding (PTB) domain, and thereby EGFR kinase transduces the signal to its downstream signaling cascades. On the other hand, integrins are transmembrane heterodimeric receptors that transmit physical force from extracellular matrix into the cytoskelton and surface receptors, and activate multiple signaling pathways.

Rosette and Karin (Rosette et al, 1996) have discovered a novel activation mechanism of EGFR by osmotic stress and UV light, so-called transactivation of EGFR, plays a key role in cell proliferation and differetiation. Previous studies have shown that EGFR kinase undergoes transactivation upon membrane stretch without ligand binding (Lezama et al, 2005). Several mode of EGFR transactivation is mediated via integrins (Moro et al, 1998), intracellular Ca^{2+} (Eguchi et al, 1998) and G protein-coupled receptor (Daub et al, 1996). In human keratinocyte cell line (HaCaT), mechanical stretch causes transactivation of EGFR via pertussis toxin-sensitive G protein coupled receptor; i.e., angiotensin II type 1 (AT1) receptor (Daub et al, 1996; Kippenberger et a, 2004). Further, in rat vascular smooth muscle cells, angiotensin-II induces Ca^{2+}-dependent transactivation of EGFR that is mimicked by Ca^{2+} ionophore and blocked by an intracellular Ca^{2+} chelator (Eguchi et al, 1998). The ligand-independent transactivation mechanism is crucial for mechanotransduction through membrane stretch, and EGFR transactivation is a major signal in response to mechanical stress including osmotic stress. In contrast, in fetal lung alveolar epithelial cells, EGFR activated by mechanical stretch might be a mechanosensor during fetal lung development (Sanchez-Esteban et al, 2004). This observation suggests that EGFR directly senses mechanical stress for lung development. The differences in activation mechanism of EGFR by mechanical stress possibly depend upon types of cells and tissues. EGFR as a transmembrane protein is a crucial transmitter to sense mechanical stress directly or indirectly, and contributes to cell proliferation and differentiation depending upon cells, tissues and species (Fig. 8.2).

8.2.2 Osmotranduction Through Volume-Sensitive Ion Channels

In mechano-sensitive tissues or cells, the mechanical stress is essential for proliferation. Therefore, sensing mechanisms of mechanical stress or cell volume are essentially required and crucial for cell proliferation. One possible mechanism to sense mechanical stress for proliferation in mechanosensitive tissues is stretch-dependent activation of EGFR kinases located in the plasma membrane as described above. Another possible mechanism is functional regulation of volume-sensitive ion channels and transporters. Particularly, volume-sensitive Cl^- channels have been recently indicated to be associated with proliferation (Varela et al, 2004, Wondergem et al, 2001).

8.2.2.1 Functional Regulation of Volume-Sensitive Cl^- Channel

An important question has been raised as to whether volume-sensitive Cl^- channels really sense mechanical stress. In addition to mechanical effects, cell swelling reduces ionic strength, dilutes cytoplasmic ions and macromolecules, thereby activating signal molecules. These signal molecules, rather than mechanical stress, might be responsible to the activation of volume-sensitive Cl^- channel during cell swelling. In support of this idea, direct stretch generally activates cation current but not volume-sensitive Cl^- current (Hu et al, 1997; Kamkin et al, 2003; Niisato et al, 2001). These observations indicate the requirement of alternative activation mechanism of volume-sensitive Cl^- channel in addition to direct sense of mechanical stress.

It has been recognized that tyrosine phosphorylation steps might be involved in activation of volume-sensitive Cl^- channel (Browe et al, 2006, Niisato et al, 1999c, Sadoshima et al, 1996, Tilly et al, 1996). The idea has been supported by following evidences. The cell swelling activates PTKs in various cells (Sadoshima et al, 1996, Tilly et al, 1996) and various PTK inhibitors abolish swelling-induced volume-sensitive Cl^- currents in canine arterial cells (Sorota et al, 1995), human intestine (Shi et al, 2002), bovine endothelial (Voets et al, 1998), human prostate cancer cells (Shuba et al, 2000) and other various types of cells (Browe et al, 2003; Crepel et al, 1998; Lepple-Wienhues et al, 1998; Shen et al, 2001, Shuba et al, 2000, Tilly et al, 1993). Moreover, direct application of purified PTK activates volume-sensitive Cl^- currents (Lepple-Wienhues et al, 1998). In contrast, inhibitors of protein tyrosine phosphatase (PTP) potentiate swelling-induced activation of Cl^- currents (Shen et al, 2001). Conversely, in some tissues, tyrosine phosphorylation is associated with an inhibition of swelling-induced volume-sensitive Cl^- current (Du et al, 2004; Ren et al, 2005). Such diversity arises because the regulation of volume-sensitive Cl^- current by tyrosine phosphorylation may be tissues or species specific. Regulatory mechanisms for volume-sensitive Cl^- currents have not been firmly understood, however one of the major mechanisms is tyrosine phosphorylation through membrane stretch/deformation-induced activation of PTKs.

Next question is to identify PTK involved in volume-sensitive Cl^- channels. Previous studies have suggested that EGFR kinase activated by cell swelling or membrane stretch is involved in functional regulation of volume-sensitive Cl^- channels in several types of cells. In mouse C127 cells, EGF peptides significantly upregulates the outwardly rectifying Cl^- channels inhibitable by tyrphotin B46 (an inhibitor of EGFR kinase) (Abdullaev et al, 2003). In rabbit ventricular myocytes, stretch of beta1-integrin activates outwardly rectifying volume-sensitive Cl^- channels via EGFR transactivation (Browe et al, 2006). Thus, one of PTKs involved in regulation of volume-sensitive Cl^- channels is EGFR which can be transactivated by cell swelling and mechanical stress. On the other hand, the involvements of several tyrosine kinases including src kinase (Lepple-Wienhues et al, 1998) and forcal adhesion kinase (FAK) in regulation of volume-sensitive Cl^- channels have been also suggested (Moro et al, 2002). FAK is a nonreceptor tyrosine kinase localized in focal adhesion. FAK undergoes autophosphorylation at Tyr397 in response to cell adhesion, growth factor stimulation (Parsons, 2003). Phosphorylation of this site creates a high affinity binding site for SH2 domain of src kinases. The src kinases phosphorylate FAK at other tyrosine residues including Tyr576/577 (Calalb et al, 1995). After the phosphorylation, FAK and src kinase stimulate their downstream signaling molecules such as MAP kinases, paxillin, and cytoskeltal proteins.

A recent observation has indicated that FAK activated by mechanical stretch regulates an outwardly rectifying volume-sensitive Cl$^-$ channel in the heart (Browe et al, 2003). Furthermore, in neonatal rat myocytes, FAK and src kinases contribute to the regulation of cardiac volume-sensitive Cl$^-$ channels through a complex mechanism (Walsh et al, 2005). Src-like kinase also activates an outwardly rectifying volume-sensitive Cl$^-$ channel in lymphocytes (Lepple-Wienhues et a, 2001). Inhibition of src kinase by a specific inhibitor leads to elevation of volume-sensitive Cl$^-$ current, whereas a specific inhibitor of EGFR causes suppression of volume-sensitive Cl$^-$ current in human atrial myocytes (Du et al, 2004, Ren et al, 2005). In this observation, src kinase and EGFR kinase are involved in the antagonistic regulation of the volume-sensitive Cl$^-$ channel. Thus, the volume-sensitive Cl$^-$ channels are differentially regulated by various types of PTK, and the distinct and complex regulation of volume-sensitive Cl$^-$ channels by various PTKs occur in cells, tissues and species specific manners.

In addition to PTK-dependent mechanisms, cytoskeleton also may contribute to regulation of volume-sensitive Cl$^-$ channel by hypo-osmotic stress (Tilly et al, 1996). Mechanical force is once transmitted by integrins that are extracellular matrix (ECM) receptors and physically associated with ECM and cytoskeleton (Geiger et al, 2001; Ingber et al, 1989). The dynamic rearrangement of the cytoskeleton by hypo-osmotic stress has been shown to play a crucial role in cell volume regulation. Actin filaments are depolymerized during cell swelling in a variety of cells (Lang et al, 1998a), whereas subsequent polymerization of actin filaments may occur in the phase of volume recovery (Ziyadeh et al, 1992). An actin depolymerizing drug has varying effects on RVD and volume-sensitive Cl$^-$ channels in different types of cells, suggesting the cell type specific requirement of cytoskelton for cell volume regulation. Cytochalasin B, an inhibitor of F-actin polymerization, elevates volume-sensitive Cl$^-$ currents induced by hypo-osmotic stress, while phalloidin, a stabilizer of actin polymerization, abolishes the increase in the Cl$^-$ current elicited by hypo-osmotic cell swelling and subsequent RVD in human cervical cancer HT-3 cells (Shen et al, 1999). The observation has indicated that functional integrity of actin filaments and microtubles plays critical roles in maintaining the RVD responses and activation of the Cl$^-$ channels in human cervical cancer HT-3 cells. In renal collecting duct cells, hypo-osmotic swelling activates the volume-sensitive Cl$^-$ channel which contributes to RVD through a mechanism involving membrane stretch and disruption of F-actin (Schwiebert et al, 1994). Although the precise activation mechanism is not fully understood, activation of volume-sensitive Cl$^-$ channels by hypo-osmotic stress is mainly mediated through tyrosine phosphorylation that is regulated by directly or indirectly activated PTKs. Further, the cytoskelton-dependent activation mechanism alternatively contributes to regulation of volume-sensitive Cl$^-$ channels.

8.3 Role of Cell Volume Regulation in Cellular functions

So far a physiological role of volume-sensitive Cl^- channels in response to hypo-osmotic stress is considered to recover the original cell volume through RVD. Hypo-osmotic stress-induced cell swelling directly or indirectly activates volume-sensitive Cl^- channels and K^+ channels, and then cell volume returns toward the original one through K^+/Cl^- release causing water movement. Considering physiological roles of RVD through volume-sensitive Cl^- channels in response to hypo-osmotic stress in addition to cell volume recovering, a possible role is to control cytosolic Cl^- concentration ($[Cl^-]_i$) by modulating Cl^- movement. Principally, $[Cl^-]_i$ under the basal condition is lower compared with cytosolic K^+ concentration. Therefore, the same amounts of Cl^- and K^+ effluxes during RVD through volume-sensitive K^+ and Cl^- channels lead to a substantial decrease in $[Cl^-]_i$. Although hypo-osmotic cell swelling decreases $[Cl^-]_i$ by the osmotic pressure gradient-dependent water influx, additional reduction in $[Cl^-]_i$ occurs during RVD by the K^+/Cl^- efflux. The reduced $[Cl^-]_i$ has been proposed to be a novel signal molecule to regulate fundamental cellular functions, i.e., cell proliferation, cell death and gene expression (Shiozaki et al, 2006).

8.3.1 *Epithelial Transport; Cytosolic Cl^- Regulates Na^+ Reabsorption*

One of important physiological roles in the distal nephron is Na^+ reabsorption to maintain body Na^+ homeostasis. The plasma osmolality is finely controlled by the nervous-hormone system through the osmosensor in the hypothalamus. On the other hand, renal epithelial cells themselves can sense the changes in osmolality of extracellular fluid and respond to them. Basolateral exposure of monolayered A6 cell, a cortical collecting duct model cell-line, to a hypo-osmotic fluid stimulates Na^+ reabsorption to maintain the normal plasma osmolality. An acute response (1–2 h) to the hypo-osmotic stress is to stimulate ENaC translocation to the apical membrane (Niisato et al, 2000) and a chronic response (6–24 h) is to induce ENaC gene expression (Niisato et al, 2004). The question is how A6 cells sense the changes in extracellular osmolality and stimulate Na^+ reabsorption. The first signal responding to hypo-osmotic stress might be a cell swelling-induced membrane stretch/deformation to modulate functions of membrane proteins including membrane surface receptors, ion channels and transporters. Hypo-osmotic swelling is converted into tyrosine phosphorylation of multiple proteins through membrane stretch/deformation. In A6 cells, a tyrosine phosphorylation-dependent signal is involved in ENaC translocation from cytosolic site to the apical membrane, resulting in stimulation of Na^+ reabsorption. A cell-attached single channel recording by patch-clamp technique has indicated that hypo-osmotic stress increases the number of functional ENaC in the apical membrane (Niisato et al, 2004). Pretreatment with tyrphostin A23 and genistein, non-specific PTK inhibitors, abolished the hytpo-osmotic stress-stimulated Na^+ reabsorption (Niisato et al, 2000). A possible regulatory mechanism in hypo-osmotic stimulation of Na^+ reabsorption is the cell swelling-induced EGFR kinase / c-jun N-terminal kinase (JNK)-dependent pathway (Taruno A et al., 2007). This mechanism contributes to the acute stimulation of Na^+ reabsorption. On the other hand, induction of ENaC gene expression as a chronic response depends upon the RVD-induced reduction of $[Cl^-]_i$. If RVD is inhibited by a Cl^- channel blocker, the hypo-osmotic stress-induced ENaC gene expression is suppressed and the stimulation of Na^+ reabsorption is abolished (Niisato et al, 2004). Previously, several flavones have been shown to stimulate Cl^- secretion by activating $Na^+/K^+/2Cl^-$ cotransporter (NKCC1) in the basolateral membrane of A6 cells (Niisato et al, 1999a). Quercetin, a flavone which activates NKCC1 but not the apical Cl^- channels, significantly suppresses the hypo-osmotic stress-induced ENaC gene expression via a mechanism of the quercetin-induced increase in $[Cl^-]_i$ (Fujimoto et al, 2005). These experimental observations suggest that the reduced $[Cl^-]_i$ is a possible signal to regulate ENaC gene expression.

In renal epithelial A6 cells, hypo-osmotic signals are considered to be divided into two pathways; one is a PTK-dependent regulation of ENaC translocation in the acute phase, the other is a $[Cl^-]_i$ -dependent regulation of ENaC gene expression in the chronic phase (Fig. 8.3). The PTK-dependent signal is caused by activation of EGFR kinase through membrane stretch/deformation. The $[Cl^-]_i$-dependent signal is caused K^+/Cl^- efflux during RVD. The reduced $[Cl^-]_i$ might activate transcriptional factors for ENaC gene expression.

Furthermore, it has been reported that cytosolic Cl^- plays a crucial role in regulation of ion channels and transporters in rat fetal lung alveolar epithelial type II cells. Terbutaline (a specific beta$_2$-adrenergic agonist) increases the open probability (Po) of an amiloride-sensitive Na^+-permeable non-selective cation (NSC) channel in rat fetal lung alveolar epithelial type II cells. The NSC channels contribute to lung fluid

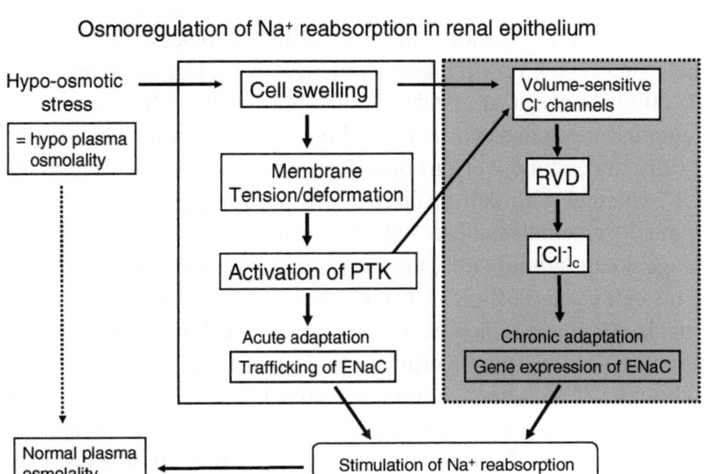

Fig. 8.3 A hypothesis for osmoregulation of Na^+ reabsorption through cell swelling and RVD in renal epithelial A6 cells

clearance in response to catecholamine in delivery. Terbutaline stimulation causes a transient elevation of $[Ca^{2+}]_i$ and a reduction of $[Cl^-]_i$ palalleled to cell shrinkage through K^+/Cl^- release under an isotonic condition (Marunaka et al, 1999). Terbutaline activates the NSC channel, which is activated by $Ca^{2+}$and inhibited by Cl^-, in fetal rat alveolar epithelium in two ways: first, through an increase in Ca^{2+} sensitivity, and second, through a reduction in the effect of cytosolic Cl^- to promote channel closing (Marunaka et al, 1999). These observations have suggested that cytosoic Cl^- has an ability to regulate function of ion channel protein likely Ca^{2+}/ calmodulin (Marunaka et al, 1999). On the other hand, Cl^- conductance also plays an important role in regulation of the Na^+/K^+ ATPase in the basolateral membrane in hypo-osmotic stress. In monolayered A6 cells, hypo-osmotic stress increases the Na^+/K^+ ATPase capacity and Cl^- conductance partly caused by activation of volume-sensitive Cl^- channels. The observation has suggested that activation of volume-sensitive Cl^- channel may contribute to the increase in Cl^- conductance and the Na^+/K^+ ATPase capacity. Blockade of Cl^- conductance by a Cl^- channel blocker or a reduction of extracellular Cl^- drastically reduces the Na^+/K^+ ATPase capacity (Niisato et al, 1999b), suggesting the importance of Cl^- channel activity in regulation of the Na^+/K^+ ATPase in hypo-osmotic stress. These observations suggest that the signal caused by cell volume changes under iso-osmotic and hypo-osmotic conditions is converted to the reduction of $[Cl^-]_i$ which regulates functions of several proteins and enzymes.

8.3.2 Cell Proliferation and Death

Cl^- channel plays a key role in cell volume regulation and epithelial Cl^- transport, maintenance of intracellular Cl^-. Cl^- channels are expressed ubiquitously and their roles in proliferation and cell cycle have been recently suggested. It has been shown that activation of volume-sensitive Cl^- channels is associated with cell proliferation and cell cycle progression in various cells (Lang et al, 2000; Nilius et al, 2001; Varela et al, 2004; Wondergem et al, 2001). Blockade of volume-sensitive Cl^- channels suppresses proliferation in many types of cells (Jiang et al, 2004b; Shen et al, 2000; Voets et al, 1995; Wondergem et al, 2001; Xiao et al, 2002). Since the detailed mechanism of cell proliferation by volume-sensitive Cl^- channels is not fully understood, various observations indicating correlation of volume-sensitive Cl^- channel with cell proliferation are discrepant. Cell cycle-dependent expression of Cl^- channels is observed, similar to that of K^+ channels. In CNE-2Z (nasopharyngeal carcinoma) cells, the expression of volume-sensitive Cl^- current dependent on cell cycle is high in G1 phase and downregulated in S phase, but increased again M phase (Chen et al, 2002b). The volume-sensitive Cl^- current associated with RVD may play a crucial role in cell cycle progression (Chen et al, 2002b). Although ClC-5 is not a volume-sensitive Cl^- channel, ClC-5 expression is increased during S and G2/M, and is decreased in G0/G1 phase (Jiang et al, 2004a). Further, a volume-regulated anion conductance in differentiation of muscle cells was downregulated (Voets et al, 1997) and blockade of volume-sensitive Cl^- channels inhibits angiogenesis in rat endothelial cells (Manolopoulos et al, 2000). Thus, the volume-sensitive Cl^- channel is involved in cell proliferation, cell cycle regulation and cell differentiation.

Since cell volume changes alter cell structure, cell hydration, cell membrane integrity and cytosolic ion environments, in some cases these alterations eventually lead to cell damage and ultimately to cell death. Cell shrinkage under a normotonic condition termed apoptotic volume decrease (AVD) induces the programmed cell death (i.e., apoptosis). It is suggested that AVD is associated with activation of volume-sensitive Cl^- channel and blockade of the Cl^- channel leads to prevention of AVD and apoptosis. (Okada et al, 2001a, Okada et al, 2001b, Okada et al, 2006). The functional relationship between volume-sensitive Cl^- channels and cell proliferation, cell cycle progression and apoptosis are well recognized based upon various observations, however regulatory mechanisms of cell proliferation, cell cycle progression and apoptosis by the volume-sensitive Cl^- channels are poorly understood and still under investigation. Recently, we have reported that removal of extracellular Cl^- causes drastic reduction of $[Cl^-]_i$ and suppression of cell proliferation with G0/G1 arrest (Shiozaki et al, 2006). The observation suggests a possible mechanism to understand volume-sensitive Cl^- channel-dependent regulation of cell proliferation (i.e., controlling $[Cl^-]_i$). Further investigation is requied for complete understanding the mechanisms.

8.3.3 *Gene Transcription*

Decreased extracellular salt or Cl^- up-regulates the cortical thick ascending limb of Hnele (cTALH) COX-2 expression via p38-dependent pathway (Cheng et al, 2000; Yang et al, 2000). A report has more recently shown that p38 MAP kinase stimulates COX-2 expression in cTALH and macula densa by transcriptional regulation predominantly via an NF-kB-dependent pathway and by post transcriptional increases in mRNA stability (Cheng et al, 2002a). On the other hand, furosemide causes three-fold increases COX-2 expression by inhibiting NKCC2 which uptakes Na^+, K^+ and Cl^- into intracellular space (Mann et al, 2001). These observations are very interesting to consider a role of cytosolic Cl^- in gene expression, because it is expected that inhibition of NKCC and removal of extracellular Cl^- result in the reduction of $[Cl^-]_i$. The authors of the reports (Mann et al, 2001) do not refer to the contribution of cytosolic Cl^- to the gene expression, however cytosolic Cl^- might be a possible molecule to regulate COX-2 expression. Further, ENaC gene expression is also regulated in a cytosolic Cl^--dependent pathway as mentioned above. These results lead us to an idea that the cytosolic Cl^- plays an important role in control of cell function through regulation of cell volume and volume-sensitive Cl^- channel.

8.4 Conclusion

Cell volume regulation is important for fundamental physiological events including cell proliferation, cell differentiation and cell death. Cell volume homeostasis is regulated by functional and transcriptional regulation of various ion channels and transporters. Basically the volume-sensitive Cl^- channel is considered to contribute to maintenance of cell volume homeostasis especially responding to mechanical and osmotic stress. However, an alternative role of volume-sensitive Cl^- channels in addition to cell volume regulation is revealed based on various evidences; i.e., it is control of cytosolic Cl^- concentration. The $[Cl^-]_i$ is controlled by a fine balance between Cl^- uptake and Cl^- release. Therefore, the Cl^- uptake mechanism is also important in addition to the regulation of Cl^- release. Recently, $[Cl^-]_i$ has been suggested to play a crucial role in regulation of gene expression, cell cycle progression, ion channel and transporter activities. Ca^{2+} signaling is widely recognized to be involved in regulation of various physiological responses. We would like to propose a novel signaling mechanism via cytosolic Cl^- by regulating volume-sensitive Cl^- channels and cell volume. Hopefully, this review provides the investigators with an opportunity to investigate a possible role of cytosolic Cl^-.

References

Abdullaev IF, Sabirov RZ, Okada Y (2003) Upregulation of swelling-activated Cl^- channel sensitivity to cell volume by activation of EGF receptors in murine mammary cells. J Physiol 549:749–758

Baraban SC, Bellingham MC, Berger AJ, Schwartzkroin PA (1997) Osmolarity modulates K^+ channel function on rat hippocampal interneurons but not CA1 pyramidal neurons. J Physiol 498:679–89

Basson MD (2003) Paradigms for mechanical signal transduction in the intestinal epithelium. Category: molecular, cell, and developmental biology. Digestion 68:217–225

Bond TD, Ambikapathy S, Mohammad S, Valverde MA (1998) Osmosensitive $C1^-$ currents and their relevance to regulatory volume decrease in human intestinal T84 cells: outwardly vs. inwardly rectifying currents. J Physiol 511:45–54

Browe DM, Baumgarten CM (2003) Stretch of beta 1 integrin activates an outwardly rectifying chloride current via FAK and Src in rabbit ventricular myocytes. J Gen Physiol 122:689–702

Browe DM, Baumgarten CM (2006) EGFR kinase regulates volume-sensitive chloride current elicited by integrin stretch via PI-3K and NADPH oxidase in ventricular myocytes. J Gen Physiol 127:237–251

Calalb MB, Polte TR, Hanks SK (1995) Tyrosine phosphorylation of focal adhesion kinase at sites in the catalytic domain regulates kinase activity: a role for Src family kinases. Mol Cell Biol 15:954–663

Chen L, Wang L, Jacob TJ (1999) Association of intrinsic pICln with volume-activated Cl^- current and volume regulation in a native epithelial cell. Am J Physiol Cell Physiol 276:C182–C192

Cheng HF, Wang JL, Zhang MZ, McKanna JA, Harris RC (2000) Role of p38 in the regulation of renal cortical cyclooxygenase-2 expression by extracellular chloride. J Clin Invest 106: 681–688

Cheng HF, Harris RC (2002a) Cyclooxygenase-2 expression in cultured cortical thick ascending limb of Henle increases in response to decreased extracellular ionic content by both transcriptional and post-transcriptional mechanisms. Role of p38-mediated pathways. J Biol Chem 277:45638–45643

Chen L, Wang L, Zhu L, Nie S, Zhang J, Zhong P, Cai B, Luo H, Jacob TJ (2002b) Cell cycle-dependent expression of volume-activated chloride currents in nasopharyngeal carcinoma cells. Am J Physiol Cell Physiol 283:C1313–C1323

Coulombe A, Coraboeuf E (1992) Large-conductance chloride channels of new-born rat cardiac myocytes are activated by hypotonic media. Pflugers Arch. 422:143–150

Crepel V, Panenka W, Kelly ME, MacVicar BA (1998) Mitogen-activated protein and tyrosine kinases in the activation of astrocyte volume-activated chloride current. J Neurosci 18:1196–1206

Daub H, Weiss FU, Wallasch C, Ullrich A (1996) Role of transactivation of the EGF receptor in signalling by G-protein-coupled receptors. Nature 379:557–560

Deutsch C, Chen LQ (1993) Heterologous expression of specific K^+ channels in T lymphocytes: functional consequences for volume regulation. Proc Natl Acad Sci U S Abreak 90:10036–10040

Du XL, Gao Z, Lau CP, Chiu SW, Tse HF, Baumgarten CM, Li GR (2004) Differential effects of tyrosine kinase inhibitors on volume-sensitive chloride current in human atrial myocytes: evidence for dual regulation by Src and EGFR kinases. J Gen Physiol 123:427–439

Duan D, Winter C, Cowley S, Hume JR, Horowitz B (1997) Molecular identification of a volume-regulated chloride channel. Nature 390:417–421

Eguchi S, Numaguchi K, Iwasaki H, Matsumoto T, Yamakawa T, Utsunomiya H, Motley ED, Kawakatsu H, Owada KM, Hirata Y, Marumo F, Inagami T (1998) Calcium-dependent epidermal growth factor receptor transactivation mediates the angiotensin II-induced mitogen-activated protein kinase activation in vascular smooth muscle cells. J Biol Chem 273:8890–8896

Fujimoto S, Niisato N, Sugimoto T, Marunaka Y (2005) Quercetin and NPPB-induced diminution of aldosterone action on Na^+ absorption and ENaC expression in renal epithelium. Biochem Biophys Res Commun 336:401–407

Furukawa T, Ogura T, Katayama Y, Hiraoka M (1998) Characteristics of rabbit ClC-2 current expressed in Xenopus oocytes and its contribution to volume regulation. Am J Physiol Cell Physiol 274:C500-C512

Geiger B, Bershadsky A (2001) Assembly and mechanosensory function of focal contacts. Curr Opin Cell Biol 13:584–92

Gong W, Xu H, Shimizu T, Morishima S, Tanabe S, Tachibe T, Uchida S, Sasaki S, Okada Y (2004) ClC-3-independent, PKC-dependent activity of volume-sensitive Cl channel in mouse ventricular cardiomyocytes. Cell Physiol Biochem 14:213–224

Gosling M, Smith JW, Poyner DR (1995) Characterization of a volume-sensitive chloride current in rat osteoblast-like (ROS 17/2.8) cells. J Physiol 485:671–682

Grinstein S, Clarke CA, Rothstein A (1982) Increased anion permeability during volume regulation in human lymphocytes. Philos Trans R Soc Lond B Biol Sci 299:509–518

Haas M and McBrayer DG (1994) Na-K-Cl cotransport in nystatin-treated tracheal cells: regulation by isoproterenol, apical UTP, and $[Cl]_i$. Am J Physiol Cell Physiol 266: C1440-C1452

Haas M, McBrayer DG, and Lytle C (1995) [Cl]-dependent Phosphorylation of the Na-K-Cl Cotransport Protein of Dog Tracheal Epithelial Cells. J Biol Chem 270: 28955–28961

Hafting T, Haug TM, Ellefsen S, Sand O (2006) Hypotonic stress activates BK channels in clonal kidney cells via purinergic receptors, presumably of the P_2Y subtype. Acta Physiol (Oxf) 188:21–31

Hagiwara N, Masuda H, Shoda M, Irisawa H (1992) Stretch-activated anion currents of rabbit cardiac myocytes. J Physiol 456:285–302

Hazama A, Okada Y (1988) Ca^{2+} sensitivity of volume-regulatory K^+ and Cl^- channels in cultured human epithelial cells. J Physiol 402:687–702

Hermoso M, Satterwhite CM, Andrade YN, Hidalgo J, Wilson SM, Horowitz B, Hume JR (2002) ClC-3 is a fundamental molecular component of volume-sensitive outwardly rectifying Cl^- channels and volume regulation in HeLa cells and Xenopus laevis oocytes. J Biol Chem 277:40066–40074

Hoffman EK (1978) Regulation of cell volume by selective changes in the leak permeabilities of Ehrlich ascites tumor cells. In: Jorgensen CB, Strange K (eds) Alfred Benzon Symposium IX, Osmotic and volume regulation. Munksgaard, Kobenhavn, pp397–417

Hu H, Sachs F (1997) Stretch-activated ion channels in the heart. J Mol Cell Cardiol 29:1511–1523

Ingber DE, Folkman J (1989) Mechanochemical switching between growth and differentiation during fibroblast growth factor-stimulated angiogenesis in vitro: role of extracellular matrix. J Cell Biol 109:317–330

Ishii TM, Maylie J, Adelman JP (1997) Determinants of apamin and d-tubocurarine block in SK potassium channels. J Biol Chem 272:23195–200

Jiang B, Hattori N, Liu B, Nakayama Y, Kitagawa K, Sumita K, Inagaki C (2004a) Expression and roles of Cl^- channel ClC-5 in cell cycles of myeloid cells. Biochem Biophys Res Commun 317:192–197

Jiang B, Hattori N, Liu B, Nakayama Y, Kitagawa K, Inagaki C (2004b) Suppression of cell proliferation with induction of p21 by Cl^- channel blockers in human leukemic cells. Eur J Pharmacol 488:27–34

Jorgensen NK, Pedersen SF, Rasmussen HB, Grunnet M, Klaerke DA, Olesen SP (2003) Cell swelling activates cloned Ca^{2+}-activated K^+ channels: a role for the F-actin cytoskeleton. Biochim Biophys Acta 1615:115–125

Kamkin A, Kiseleva I, Isenberg G (2003) Ion selectivity of stretch-activated cation currents in mouse ventricular myocytes. Pflugers Arch 446:220–231

Kippenberger S, Loitsch S, Guschel M, Muller J, Knies Y, Kaufmann R, Bernd A (2005) Mechanical stretch stimulates protein kinase B/Akt phosphorylation in epidermal cells via angiotensin II type 1 receptor and epidermal growth factor receptor. J Biol Chem 280:3060–3067

Kohler M, Hirschberg B, Bond CT, Kinzie JM, Marrion NV, Maylie J, Adelman JP (1996) Small-conductance, calcium-activated potassium channels from mammalian brain. Science 273:1709–1714

Khanna R, Chang MC, Joiner WJ, Kaczmarek LK, Schlichter LC (1999) hSK4/hIK1, a calmodulin-binding KCa channel in human T lymphocytes. Roles in proliferation and volume regulation. J Biol Chem 274:14838–14849

Krapivinsky GB, Ackerman MJ, Gordon EA, Krapivinsky LD, Clapham DE (1994) Molecular characterization of a swelling-induced chloride conductance regulatory protein, pICln. Cell. 76:439–448

Kubo M, Okada Y (1992) Volume-regulatory Cl^- channel currents in cultured human epithelial cells. J Physiol 456:351–371

Lammerding J, Kamm RD, Lee RT (2004) Mechanotransduction in cardiac myocytes. Ann N Y Acad Sci 1015:53–70

Lang F, Busch GL, Ritter M, Volkl H, Waldegger S, Gulbins E, Haussinger D (1998a) Functional significance of cell volume regulatory mechanisms. Physiol Rev 78:247–306

Lang F, Busch GL, Volkl H (1998b) The diversity of volume regulatory mechanisms. Cell Physiol Biochem 8:1–45

Lang F, Ritter M, Gamper N, Huber S, Fillon S, Tanneur V, Lepple-Wienhues A, Szabo I, Gulbins E (2000) Cell volume in the regulation of cell proliferation and apoptotic cell death. Cell Physiol Biochem 10:417–428

Lauf PK, Adragna NC (2000) K-Cl cotransport: properties and molecular mechanism. Cell Physiol Biochem. 10:341–354

Lepple-Wienhues A, Szabo I, Laun T (1998) The tyrosine kinase p56lck mediates activation of swelling-induced chloride channels in lymphocytes. J Cell Biol 141:281–286

Lepple-Wienhues A, Wieland U, Laun T, Heil L, Stern M, Lang F (2001) A src-like kinase activates outwardly rectifying chloride channels in CFTR-defective lymphocytes. FASEB J 15:927–931

Lezama R, Diaz-Tellez A, Ramos-Mandujano G, Oropeza L, Pasantes-Morales H (2005) Epidermal growth factor receptor is a common element in the signaling pathways activated by cell volume changes in isosmotic, hyposmotic or hyperosmotic conditions. Neurochem Res 30:1589–9157

Li X, Shimada K, Showalter LA, Weinman SA (2000) Biophysical properties of ClC-3 differentiate it from swelling-activated chloride channels in Chinese hamster ovary-K1 cells. J Biol Chem 275:35994–35998

Lock H, Valverde MA (2000) Contribution of the IsK (MinK) potassium channel subunit to regulatory volume decrease in murine tracheal epithelial cells. J Biol Chem 275:34849–34852

Mann B, Hartner A, Jensen BL, Kammerl M, Kramer BK, Kurtz A (2001) Furosemide stimulates macula densa cyclooxygenase-2 expression in rats. Kidney Int 59:62–68

Manolopoulos VG, Liekens S, Koolwijk P, Voets T, Peters E, Droogmans G, Lelkes PI, De Clercq E, Nilius B (2000) Inhibition of angiogenesis by blockers of volume-regulated anion channels. Gen Pharmacol 34:107–116

Marunaka Y, Niisato N, O'Brodovich H, Eaton DC (1999) Regulation of an amiloride-sensitive Na^+-permeable channel by a beta2-adrenergic agonist, cytosolic Ca^{2+} and Cl^- in fetal rat alveolar epithelium. J Physiol 515:669–683

Mercado A, Song L, Vazquez N, Mount DB, Gamba G (2000) Functional comparison of the K^+-Cl^- cotransporters KCC1 and KCC4. J Biol Chem 275:30326–30334

Moro L, Venturino M, Bozzo C, Silengo L, Altruda F, Beguinot L, Tarone G, Defilippi P (1998) Integrins induce activation of EGF receptor: role in MAP kinase induction and adhesion-dependent cell survival. EMBO J 17:6622–3332

Moro L, Dolce L, Cabodi S, Bergatto E, Erba EB, Smeriglio M, Turco E, Retta SF, Giuffrida MG, M, Godovac-Zimmermann J, Conti A, Schaefer E, Beguinot L, Tacchetti C, Gaggini P, Silengo L, Tarone G, Defilippi P (2002) Integrin-induced epidermal growth factor (EGF) receptor activation requires c-Src and p130Cas and leads to phosphorylation of specific EGF receptor tyrosines. J Biol Chem 277:9405–9414

Niemeyer MI, Cid LP, Barros LF, Sepulveda FV (2001) Modulation of the two-pore domain acid-sensitive K^+ channel TASK-2 (KCNK5) by changes in cell volume. J Biol Chem 276:43166–43174

Niisato N, Ito Y, Marunaka Y (1999a) Activation of Cl^- channel and $Na^+/K^+/2Cl^-$ cotransporter in renal epithelial A6 cells by flavonoids: genistein, daidzein, and apigenin. Biochem Biophys Res Commun 254:368–371

Niisato N, Marunaka Y (1999b) Activation of the Na^+-K^+ pump by hyposmolality through tyrosine kinase-dependent Cl^- conductance in Xenopus renal epithelial A6 cells. J Physiol 518:417–432

Niisato N, Post M, Van Driessche W, Marunaka Y (1999c) Cell swelling activates stress-activated protein kinases, p38 MAP kinase and JNK, in renal epithelial A6 cells. Biochem Biophys Res Commun 266:547–550

Niisato N, Van Driessche W, Liu M, Marunaka Y (2000) Involvement of protein tyrosine kinase in osmoregulation of Na^+ transport and membrane capacitance in renal A6 cells. J Membr Biol 175:63–77

Niisato N, Marunaka Y (2001) Blocking action of cytochalasin D on protein kinase A stimulation of a stretch-activated cation channel in renal epithelial A6 cells. Biochem Pharmacol 61:761–765

Niisato N, Eaton DC, Marunaka Y (2004) Involvement of cytosolic Cl^- in osmoregulation of alpha-ENaC gene expression. Am J Physiol Renal Physiol 287:F932-F939

Nilius B, Sehrer J, De Smet P, Van Driessche W, Droogmans G (1995) Volume regulation in a toad epithelial cell line: role of coactivation of K^+ and Cl^- channels. J Physiol. 487:367–378

Nilius B, Eggermont J, Voets T, Buyse G, Manolopoulos V, Droogmans G (1997) Properties of volume-regulated anion channels in mammalian cells. Prog Biophys Mol Biol 68:69–119

Nilius B (2001) Chloride channels go cell cycling. J Physiol 532:581

Okada Y (1997) Volume expansion-sensing outward-rectifier Cl^- channel: fresh start to the molecular identity and volume sensor. Am J Physiol 273:C755–789

Okada Y, Maeno E (2001a) Apoptosis, cell volume regulation and volume-regulatory chloride channels. Comp Biochem Physiol A Mol Integr Physiol 130:377–383

Okada Y, Maeno E, Shimizu T, Dezaki K, Wang J, Morishima S (2001b) Receptor-mediated control of regulatory volume decrease (RVD) and apoptotic volume decrease (AVD). J Physiol 532:3–16

Okada Y, Shimizu T, Maeno E, Tanabe S, Wang X, Takahashi N (2006) Volume-sensitive chloride channels involved in apoptotic volume decrease and cell death. J Membr Biol 209:21–29

Park KP, Beck JS, Douglas IJ, Brown PD (1994) Ca^{2+}-activated K^+ channels are involved in regulatory volume decrease in acinar cells isolated from the rat lacrimal gland. J Membr Biol 141:193–201

Parsons JT (2003) Focal adhesion kinase: the first ten years. J Cell Sci 116:1409–1416

Pasantes-Morales H, Morales Mulia S (2000) Influence of calcium on regulatory volume decrease: role of potassium channels. Nephron 86:414–427

Paulmichl M, Li Y, Wickman K, Ackerman M Peralta E, Clapham D (1992) New mammalian chloride channel identified by expression cloning. Nature 356:238–241

Putney LK, Vibat CR, O'Donnell ME (1999) Intracellular Cl regulates Na-K-Cl cotransport activity in human trabecular meshwork cells. Am J Physiol Cell Physiol 277: C373–C383

Ren Z, Baumgarten CM (2005) Antagonistic regulation of swelling-activated Cl^- current in rabbit ventricle by Src and EGFR protein tyrosine kinases. Am J Physiol Heart Circ Physiol 288:H2628–H2636

Reuss L, Vanoye CA, Altenberg GA, Vergara L, Subramaniam M, Torres R (2000) Cell-volume changes and ion conductances in amphibian gallbladder epithelium. Cell Physiol Biochem 10:385–392

Roman R, Feranchak AP, Troetsch M, Dunkelberg JC, Kilic G, Schlenker T, Schaack J, Fitz JG (2002) Molecular characterization of volume-sensitive SK(Ca) channels in human liver cell lines. Am J Physiol Gastrointest Liver Physiol 282:G116–G122

Rosette C, Karin M (1996) Ultraviolet light and osmotic stress: activation of the JNK cascade through multiple growth factor and cytokine receptors. Science 274:1194–1197

Ruwhof C, van der Laarse A (2000) Mechanical stress-induced cardiac hypertrophy: mechanisms and signal transduction pathways. Cardiovasc Res 47:23–37

Sackin H (1989) A stretch-activated K^+ channel sensitive to cell volume. Proc Natl Acad Sci USA 86:1731–1735

Sadoshima J, Izumo S (1996) The heterotrimeric G q protein-coupled angiotensin II receptor activates p21 ras via the tyrosine kinase-Shc-Grb2-Sos pathway in cardiac myocytes. EMBO J 15:775–787

Sanchez-Esteban J, Wang Y, Gruppuso PA, Rubin LP (2004) Mechanical stretch induces fetal type II cell differentiation via an epidermal growth factor receptor-extracellular-regulated protein kinase signaling pathway. Am J Respir Cell Mol Biol 30:76–83

Sardini A, Amey JS, Weylandt KH, Nobles M, Valverde MA, Higgins CF (2003) Cell volume regulation and swelling-activated chloride channels. Biochim Biophys Acta 1618:153–162

Schoenmakers TJ, Vaudry H, Cazin L (1995) Osmo- and mechanosensitivity of the transient outward K^+ current in a mammalian neuronal cell line. J Physiol 489:419–430

Schwiebert EM, Mills JW, Stanton BA (1994) Actin-based cytoskeleton regulates a chloride channel and cell volume in a renal cortical collecting duct cell line. J Biol Chem 269:7081–7089

Sheader EA, Brown PD, Best L (2001) Swelling-induced changes in cytosolic $[Ca^{2+}]$ in insulin-secreting cells: a role in regulatory volume decrease? Mol Cell Endocrinol 181:179–187

Shen MR, Chou CY, Hsu KF, Hsu KS, Wu M (1999) Modulation of volume-sensitive Cl^- channels and cell volume by actin filaments and microtubules in human cervical cancer HT-3 cells. Acta Physiol Scand 167:215–225

Shen MR, Droogmans G, Eggermont J, Voets T, Ellory JC, Nilius B (2000) Differential expression of volume-regulated anion channels during cell cycle progression of human cervical cancer cells. J Physiol 529:385–394

Shen MR, Chou CY, Browning JA, Wilkins RJ, Ellory JC (2001) Human cervical cancer cells use Ca^{2+} signalling, protein tyrosine phosphorylation and MAP kinase in regulatory volume decrease. J Physiol 537:347–362

Shi C, Barnes S, Coca-Prados M, Kelly ME (2002) Protein tyrosine kinase and protein phosphatase signaling pathways regulate volume-sensitive chloride currents in a nonpigmented ciliary epithelial cell line. Invest Ophthalmol Vis Sci 43:1525–1532

Shiozaki A, Miyazaki H, Niisato N, Nakahari T, Iwasaki Y, Itoi H, Ueda Y, Yamagishi H, Marunaka Y (2006) Furosemide, a blocker of $Na^+/K^+/2Cl^-$ cotransporter, diminishes proliferation of poorly differentiated human gastric cancer cells by affecting G0/G1 state. J Physiol Sci 56: 401–406

Shuba YM, Prevarskaya N, Lemonnier L, Van Coppenolle F, Kostyuk PG, Mauroy B, Skryma R (2000) Volume-regulated chloride conductance in the LNCaP human prostate cancer cell line. Am J Physiol Cell Physiol 279:C1144–C1154

Sorota S (1995) Tyrosine protein kinase inhibitors prevent activation of cardiac swelling-induced chloride current. Pflugers Arch 431:178–185

Taruno A, Niisato N, Marunaka Y (2007) Hypertonicity stimulates renal epithelial sodium transport by activating JNK via receptor tyrosine kinase. Am J Physiol Renal Physiol 293:F128–F138.

Thinnes FP, Hellmann KP, Hellmann T, Merker R, Brockhaus-Pruchniewicz U, Schwarzer C, Walter G, Gotz H, Hilschmann N (2000) Studies on human porin XXII: cell membrane integrated human porin channels are involved in regulatory volume decrease (RVD) of HeLa cells. Mol Genet Metab 69:331–337

Tilly BC, van den Berghe N, Tertoolen LG, Edixhoven MJ, de Jonge HR (1993) Protein tyrosine phosphorylation is involved in osmoregulation of ionic conductances. J Biol Chem 268:19919–19922

Tilly BC, Edixhoven MJ, Tertoolen LG, Morii N, Saitoh Y, Narumiya S, de Jonge HR (1996) Activation of the osmo-sensitive chloride conductance involves P21rho and is accompanied by a transient reorganization of the F-actin cytoskeleton. Mol Biol Cell 7:1419–1427

Tominaga M, Tominaga T, Miwa A, Okada Y (1995) Volume-sensitive chloride channel activity does not depend on endogenous P-glycoprotein. J Biol Chem 270:27887–27893

Tsumura T, Oiki S, Ueda S, Okuma M, Okada Y (1996) Sensitivity of volume-sensitive Cl^- conductance in human epithelial cells to extracellular nucleotides. Am J Physiol Cell Physiol 271: C1872–C1878

Ubl J, Murer H, Kolb HA (1988) Hypotonic shock evokes opening of Ca^{2+}-activated K channels in opossum kidney cells. Pflugers Arch 412:551–553

Valverde MA, Diaz M, Sepulveda FV, Gill DR, Hyde SC, Higgins CF (1992) Volume-regulated chloride channels associated with the human multidrug-resistance P-glycoprotein. Nature 355:830–833

Varela D, Simon F, Riveros A, Jorgensen F, Stutzin A (2004) NAD(P)H oxidase-derived H_2O_2 signals chloride channel activation in cell volume regulation and cell proliferation. J Biol Chem 279:13301–13304

Vazquez E, Nobles M, Valverde MA (2001) Defective regulatory volume decrease in human cystic fibrosis tracheal cells because of altered regulation of intermediate conductance Ca^{2+}-dependent potassium channels. Proc Natl Acad Sci U S A 98:5329–5334

Voets T, Szucs G, Droogmans G, Nilius B (1995) Blockers of volume-activated Cl^- currents inhibit endothelial cell proliferation. Pflugers Arch 431:132–134

Voets T, Buyse G, Tytgat J, Droogmans G, Eggermont J, Nilius B (1996) The chloride current induced by expression of the protein pICln in Xenopus oocytes differs from the endogenous volume-sensitive chloride current. J Physiol 495:441–447

Voets T, Wei L, De Smet P, Van Driessche W, Eggermont J, Droogmans G, Nilius B. (1997) Downregulation of volume-activated Cl^- currents during muscle differentiation. Am J Physiol 272:C667–C674

Voets T, Manolopoulos V, Eggermont J, Ellory C, Droogmans G, Nilius B (1998) Regulation of a swelling-activated chloride current in bovine endothelium by protein tyrosine phosphorylation and G proteins. J Physiol 506:341–352

Walsh KB, Zhang J (2005) Regulation of cardiac volume-sensitive chloride channel by focal adhesion kinase and Src kinase. Am J Physiol Heart Circ Physiol 289:H2566–H2574

Wang J, Morishima S, Okada Y (2003) IK channels are involved in the regulatory volume decrease in human epithelial cells. Am J Physiol Cell Physiol 284:C77–C84

Weiss H, Lang F (1992) Ion channels activated by swelling of Madin Darby canine kidney (MDCK) cells. J Membr Biol 126:109–114

Weskamp M, Seidl W, Grissmer S (2000) Characterization of the increase in $[Ca^{2+}]_i$ during hypotonic shock and the involvement of Ca^{2+}-activated K^+ channels in the regulatory volume decrease in human osteoblast-like cells. J Membr Biol 178: 11–20

Weyts FA, Bosmans B, Niesing R, van Leeuwen JP, Weinans H (2003) Mechanical control of human osteoblast apoptosis and proliferation in relation to differentiation. Calcif Tissue Int 72:505–512

Wondergem R, Gong W, Monen SH, Dooley SN, Gonce JL, Conner TD, Houser M, Ecay TW, Ferslew KE (2001) Blocking swelling-activated chloride current inhibits mouse liver cell proliferation. J Physiol 532:661–672

Wu QQ, Chen Q (2000) Mechanoregulation of chondrocyte proliferation, maturation, and hypertrophy: ion-channel dependent transduction of matrix deformation signals. Exp Cell Res 256:383–391

Xiao GN, Guan YY, He H (2002) Effects of Cl^- channel blockers on endothelin-1-induced proliferation of rat vascular smooth muscle cells. Life Sci 70:2233–2241

Yang T, Park JM, Arend L, Huang Y, Topaloglu R, Pasumarthy A, Praetorius H, Spring K, Briggs JP, Schnermann J (2000) Low chloride stimulation of prostaglandin E_2 release and cyclooxygenase-2 expression in a mouse macula densa cell line. J Biol Chem 275: 37922–37929

Zhou JG, Ren JL, Qiu QY, He H, Guan YY (2005) Regulation of intracellular Cl^- concentration through volume-regulated ClC-3 chloride channels in A10 vascular smooth muscle cells. J Biol Chem 280:7301–7308

Ziyadeh FN, Mills JW, Kleinzeller A (1992) Hypotonicity and cell volume regulation in shark rectal gland: role of organic osmolytes and F-actin. Am J Physiol 262:F468–F479

Chapter 9
Mechanosensitive Channel TRPV4

A Micro-Machine Converting Physical Force into an Ion Flow

Makoto Suzuki

Abstract The transient receptor potential vanilloid 4 (TRPV4), a member of the TRP family that has a 6-transmembrane (TM) segment, was first identified as a hypoosmolality-activated Ca^{2+} permeable channel. The mechanism of opening by hypoosmolality is distinct from direct coupling of membrane tension with conformational changes, such as that which occurs with Msc or TREK, in which phospholipase A2 is sensitive to hypoosmolality and the resulting arachidonic acid metabolites open this channel. TRPV4 is, however, a "multi-micromachine" that senses more diverse physical stimuli and converts them to the Ca^{2+} signal in various mammalian tissues. Recent analyses suggest that TRPV4 and other molecules together constitute a physiological response, such as cell volume regulation, epithelial permeability, and vascular dilatation. A mouse lacking the TRPV4 gene appeared normal but had unexpected responses to various stimuli; this will provide a clue to elucidate a novel physiological, pathophysiological, and pharmacological significance in the sensation of physical factors.

Key words: Sensory system · Nociceptor · Cation channel · Osmolality · Pressure · Temperature

9.1 Introduction

Beginning with the identification of the gene responsive to the *Drosophila* mutant, which shows transient receptor potential (TRP) in comparison with the control, which shows a sustained receptor potential in the eye, several novel Ca^{2+} entry channels belonging to the large family of TRP cation channels have been discovered. TRP genes encoding membrane proteins with six transmembrane segments (TM1–TM6) and a putative pore region formed by a short hydrophobic stretch between TM5 and TM6 have been identified (Clapham 2001, Montell 2001, Montell et al 2002a; Owsianik et al, 2006). On the basis of their homology, mammalian TRP proteins are classified into subfamilies (Montell et al 2002b): TRPC (canonical), TRPV (vanilloid), TRPM (melastatin), TRPN, TRPP (polycystin), TRPA, and TRPML.

A. Kamkin and I. Kiseleva (eds.), *Mechanosensitive Ion Channels.*

TRPV1 (a capsaicin receptor) was independently isolated by Caterina et al. in 1997. This TRP-like receptor is the first channel that converts a physical input, such as hot temperature, into electrical stimuli (Caterina et al, 2000). The ion channel that converts the pressure or anisoosmomolarity (hyper- and hypoosmolality) into electrical stimuli was therefore supposed to be involved in this family. A clue to its identity came from the discovery of Osm-9 in a genetic screen of high-osmolality-insensitive *Caenorhabditis elegans* mutants (Colbert et al 1997). TRPV4 is a candidate of mammalian Osm-9 and was thus investigated.

TRPV4 (the transient receptor potential vanilloid 4, also formerly called OTRPC4, VR-OAC, VRL2, or TRP12), a member of the TRPV subfamily, was first identified as an osmo-sensing channel (Strotmann et al. 2000; Liedtke et al. 2000; Nilius et al 2001a). TRPV4 mRNA is widely and locally detected in the neuron, kidney, lung, skin, vascular endothelium, and other tissues (Strotmann et al. 2000; Liedtke et al. 2000). Growing numbers of experiments show consistent results that these cells express a swell-activated current by the electrophysiological technique or Ca^{2+} influx by Ca^{2+}-sensitive fluorescence measurements, while those using TRPV4-deficient (*Trp4*$^{-/-}$) mouse do not show any such results. The TRPV4 channel is activated by cell swelling, shear stress, acidic pH, moderate heat, arachidonic acid metabolites, and probably pressure in cultured cells (Watanabe et al. 2002a; Suzuki et al. 2003a; Vriens et al. 2003). The swelling of the cells activates phospholipase A2, which triggers the arachidonic acid cascade, leading to the phosphorylation of the channel. The thermal stimulus changes the lipid moiety around the channels, which directly induces the opening of the channel. Other mechanisms under the activation of TRPV4 involving the β -subunit may be considered. Our recent studies using *Trp4*$^{-/-}$mice have suggested that TRPV4 channels function as nociceptors for pressure (Suzuki et al. 2003a), warmth (Chung et al. 2004a, Lee et al. 2005), noxious compounds (Todaka et al. 2004; Alessandri-Haber2006), and noxious sounds (Tabuchi et al. 2005) as well as flow sensor (Taniguchi et al 2006). Physiological data of *Trp4*$^{-/-}$ mice further suggested that TRPV4 is necessary for normal osmotic regulation (Mizuno et al. 2003; Liedtke et al. 2003a). TRPV4 is also involved in the mechanism underlying asthma (Liedtke 2004) and, possibly, in its development. In view of these findings, the TRPV4 channel is a "multi-micromachine" that senses diverse physical stimuli and converts them into the Ca^{2+} signal (or possibly other signals) in various mammalian tissues.

9.2 TRPV4 Gene and Localization

9.2.1 TRPV4 Gene

A part of the TRPV4 nucleotide sequence was found in EST files when the prototype, TRPV1, a vanilloid receptor, was discovered in the course of excellent and novel research using expression cloning (Caterina et al, 1997). TRPV1 is the first TRP Ca^{2+} channel responsive to a physical factor, such as heat. Therefore, a family member, TRPV4, may be proposed to sense a physical factor, namely, the mechanical stimuli. The TRPV family is composed of 6 members. TRPV1–4 are located in the sensory systems from the skin to the brain, playing a direct role in the sensing of physical stimuli. On the other hand, TRPV5 and 6 are located in the intestine and the kidney, playing a role in Ca^{2+} transport in the body (Hoenderop et al, 2005; Nilius et al, 2004). Within the TRPV subfamily, TRPV4 displays significantly stronger homology with TRPV1–TRPV3 than with TRPV5 and TRPV6. Species differences for TRPV4 are minimal (human/mouse, 95.2/96.9%; human/rat, 94.8/97.0%, mouse/rat, 98.9/99.2% identity/similarity).

Trpv4 is encoded in 12q24.11, and any relationship to genetic diseases is unknown. A report has suggested that delayed-onset, progressive, high-frequency, non-syndromic sensorineural hearing loss (DFNA25) is linked with a 20-cM region of chromosome 12q21–24 between D12S327 and D12S84, with a maximum two-point LOD score of 6.82 (Greene et al, 2001). A single nucleotide polymorphism is found in the TRPV4 gene and TRPV4 exons. However, no functional relationship has been determined (http://www.ncbi.nlm.nih.gov/). The human TRPV4 variants, TRPV4-A to –E, have been identified (Liedtke et al, 2000; Arniges et al, 2006). All alternative spliced variants involved deletions in the cytoplasmic N-terminal region, affecting (except for TRPV4-D) the ankyrin domains. TRPV4 undergoes glycosylation and oligomerization in the endoplasmic reticulum followed by transfer to the Golgi apparatus. The ankyrin domains are necessary for the oligomerization of TRPV4, and the lack of TRPV4 oligomerization determines its accumulation in the endoplasmic reticulum. Thus, the amount of spliced variants may determine the quantity of the functional TRPV4 in the plasma membrane. TRP channels may constitute homomeric or heteromeric complexes. The multimerization was examined with fluorescence resonance energy transfer and coimmunoprecipitation in TRPV1–6. Except for TRPV5 and TRPV6, TRPV channel subunits preferentially assemble into homomeric pore complexes (Hellwig et al, 2005), although no combination with any other class of TRP channels has been examined.

9.2.2 TRPV4 mRNA

TRPV4 mRNA is mainly detected in the lung and the kidney, but RT-PCR allows its detection in a broad range of tissues. Distribution of TRPV4 is summarized in Fig. 9.1.

Northern blot analysis shows that TRPV4 transcripts are detected with the strongest levels present in the trachea, followed by the kidney, prostate, pancreas, placenta (Liedtke et al, 2000; Wissenbach et al, 2000), liver, and heart (Strotmann et al, 2000). Bronchial epithelial cells, smooth muscle cells, and renal distal tubules express the TRPV4 protein. In the kidney, the basolateral membrane from the proximal to the connecting tubules is positively stained using an anti-C-terminal antibody. The water-permeant cells of the macula densa are, however, unstained. Moderate TRPV4 expression is noted in all collecting duct portions and in the papillary epithelium, including intercalated cells (Tian et al 2004). However, luminal localization

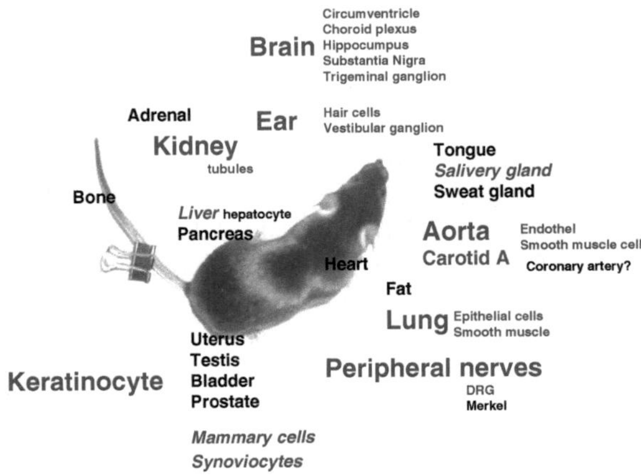

Fig. 9.1 Distribution of TRPV4 protein. TRPV4$^{-/-}$ mice are less sensitive to pressure on the tail. By using this mechano-insensitive knockout mouse, TRPV4 is proven to play a role in variable sensation and in physiological process (red letter). Although the mRNA of TRPV4 is detected in other tissues (black letter), the functional roles remain unclarified. TRPV4 may be detected more widely in other tissues. TRPV4 function: Osmo-sensation and thermo-regulation in the brain. Osmo-sensation in hair cells of the ear. Flow- and osmo- sensation in the kidney. Shear-stress (or pressure) sensation in the arteries. Osm- and mechano-sensation supposed in the lung. Pressure-, warmth- and inflammatory signals-sensation in peripheral nerves. Warmth- and osmo- sensation in keratinocytes

has also been demonstrated using an anti-N-terminal antibody (Cuajungco et al, 2006). In the brain, *in situ* hybridization revealed that the lamina terminalis (VOLT), the subfornical organ (SFO), and the median preoptic area (MnOP) in the hypothalamus express the TRPV4 transcript. The ependymal cells of the choroids plexus of the third ventricle and the lateral ventricle are also positive (Liedtke et al, 2000). All these portions are closely related to organs that regulate body fluid osmolality. The trigeminal ganglion is positively stained by *in situ* hybridization and the anti-C-terminal antibody (Liedtke et al, 2000; Suzuki et al, 2003). In the murine cochlea, TPV4 is expressed in most cells lining the endolymphatic duct of the mouse ear, including the hair cells and the marginal cells of the stria vascularis (Liedtke et al, 2000; Shen et al, 2006). Dorsal root ganglion (DRG) neurons and primary sensory neurons expressed TRPV4 (Alessandri-Haber et al, 2003) in murine but not in human (Delany et al, 2001). Interestingly, TRPV4 mRNA, but not the protein, could be detected in the soma of DRG neurons, suggesting that there might be a mechanism for the transport of the TRPV4 protein from the neuronal bodies to the sensory terminals (Guler et al, 2002). Human sympathetic ganglia have been reported to express TRPV4 immunoreactivity (Delany et al, 2001). Adrenal and sweat glands are other sites suggested to express TRPV4 mRNA, although their immunolocalization is not available. Vagal afferent nerves may also express TRPV4 (Zhang et al, 2006).

In the skin, keratinocytes in deep rather than surface areas express TRPV4. Merkel bodies (Liedtke et al, 2000) and other mechanosensitive terminals (Suzuki et al, 2003b) may express TRPV4. Both white and brown adipocytes in rodents strongly express TRPV4. Endothelial cells in mouse aortas show robust positive signals for TRPV4 by Northern blot and histochemical analyses, but those in small vessels, such as the lung, show negative signals. In contrast, human endothelial cells in the lung, but not in the aorta, show TRPV4 by RT-PCR. A species difference in the expression of TRPV4 in endothelial cells may exist. Vascular smooth muscle cells in the aorta and pulmonary arteries of rodents express TRPV4. The expression of endothelial or/and vascular smooth muscle cells of other small vessels remains obscure. Liver expresses TRPV4 mRNA by Northern blot analysis (Stroteman et al, 2000), and human hepatoblastoma cells show TRPV4 mRNA by RT-PCR, as well as a 4aPDD-induced Ca^{2+} increase (Vriens et al, 2004). Human synoviocytes also show mRNA as well as increases in Ca^{2+} (Kochukov, et al, 2006). It is not known whether heart muscles express TRPV4 function, although an early report suggested the presence of mRNA by northern blot (Stroteman et al, 2000). Coronary vessels may express TRPV4 (unpublished observation).

mRNA could be further detected by RT-PCR in chondrocytes, osteoblast, sperm, and urinary bladder, but the functional roles of TRPV4 remain obscure.

Attention should be given to the antibody used, since the resultant localization is dependent on the antibodies. For example, an anti-COOH-terminal peptide antibody detected TRPV4 in the basolateral membrane of renal tubules (Tian et al, 2004), while an anti-NH2-terminal antibody detected it in the luminal membrane of renal tubules (Cuajungco et al, 2006). Both antibodies specify the positive 98Kd signal with exogenously expressed cells by Western blot analysis. Thus, controversy remains regarding the precise immunohistological localization of TRPV4 *in situ*.

9.3 Characterization of the Channel

9.3.1 Structure and Channel Function(Fig. 9.2)

TRPV4 consists of 871 amino acids with at least three ankyrin repeats in the NH2 terminus. Putative domain structure of TRPV4 includes PKC (S88,S134,S528), PKA phosphorylation sites, ankyrin binding repeats (Blue circle), the 6 transmembrane regions (TM1–TM6), a glycosylation site close to the pore (N651), the pore region (P), a Ca^{2+}/calmodulin binding site and the tyrosine kinase phosphorylation site (Y253).

Exogenously expressed TRPV4 in Chinese hamster ovary (CHO) or Human epithelial kidney (HEK) cells shows a robust outward-rectified current when activated by hypotonic stress or an agonist, i.e., 4α-phorbol 12,13-didecanoate (4αPDD) (Watanabe et al, 2002a; Nilius 2003). Without any stimuli, the current is inactive when measured at room temperature. Gao et al. (2003) revealed that the basal level of Ca^{2+} at body temperature (37°C) is significantly higher than that at room temperature and that this range of warm temperature is essential in the activation of TRPV4 by phorbol ester (PMA), hypotonic stress, 4αPDD and shear stress. Thus, TRPV4 is constitutively active in *in situ* -expressed cells.

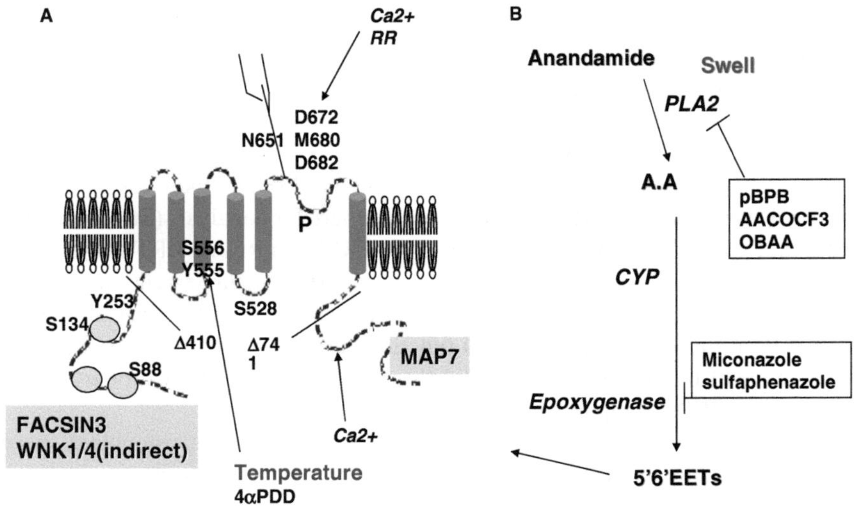

Fig. 9.2 Putative TRPV4 structure and the regulation. **A**. TRPV4 is supposed to have three ankyrin repeats (blue circle), and six putative transmembrane segments with pore (P), followed by Ca^{2+} (or Ca-calmodulin) binding C-terminal domain. N-terminal intracellular domain binds PACSIN3 and is modified by WNK1/4 kinase. Microtubule-associated protein (MAP) binds to C-terminal. Deletion before 410 or after 741 looses ability in response to mechanical stress. N-terminal deletion mutant does not move toward the membrane surface. Putative protein kinase C phosphorylation sites are S88, S134, and S528. D672, M680, and D682 are important in Ca^{2+} permeability and RR inhibition. N651 is the site of glycosylation. Warm temperature (33°C <), presumably though a release of lipids, and 4αPDD interact Y555/S556 motif. **B**. Swelling of the cells activate anandamide-dependent PLA2 cascade Arachidonic acid (AA) metabolites, cytochrome P 450, epoxigenase-catalyzed end-products such as 5'6'-EET, activate TRPV4 thought unknown the motif. Reagents in the right boxes are the inhibitors of this cascade and prevent the TRPV4 activation by swelling of the cell

9.3.2 Pore, Blockers, Permeation and Rectification (Fig. 9.3)

TRPV channels are activated by acid lower than pH 6.0 like as TRPV1 and the influx of Ca^{2+}, or the inward part of the current is blocked by ruthenium red (RR) at μM concentration. The TRPV4 channel permeation is non-selective for the cation. Permeability values relative to Na^{+} are 6–10 for Ca^{2+} and 2–3 for Mg^{2+} (Lidetke et al, 2000; Nilius et al, 2001b; Strotmann et al, 2000; Strotmann et al, 2003; Watanabe et al, 2002a). The magnitude of the current was diminished by external Ca^{2+}, and the removal of Ca^{2+} by EGTA or citrate potentiates the amplitude.

Residues of aspartate (D) play a cardinal role in this selectivity (Martinez et al, 2000; Nilius et al, 2001b, Nilius 2004; Liedtke et al, 2003b, Voets et al, 2002) (**Fig. 9.3**).

The mutation of the neutralization of D to alanine (A) between TM5 and TM6 reveals the pore information of the TRPV4 channel. The Ca^{2+}permeability is moderately influenced by the mutation of 672D without significantly altering the monovalent permeability. A further change in the permeability (involving monovalent permeability), rectification and sensitivity to RR is achieved in the mutation of

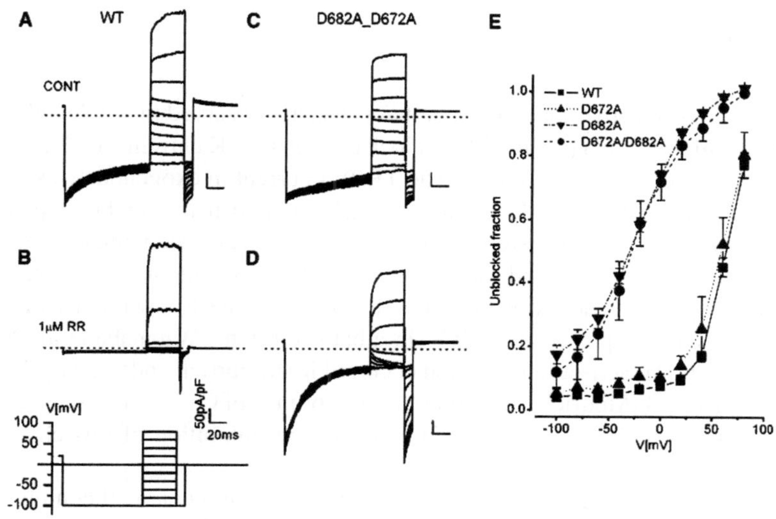

Fig. 9.3 Block of TRPV4 by RR. **A.** currents are shown through wild-type (WT) TRPV4. Channels were activated by 1 μM 4aPDD. Holding potential was 20 mV. The voltage protocol consisted of a hyperpolarizing prestep to 100 mV, followed by test steps from 100 to 80 mV spaced by 20 mV and a further step back to 100 mV ($[Na^+]$ = 150 mM, $[Ca^{2+}]$ =5 mM). The slow decay of the inward current is likely due to inhibition by Ca^{2+}. **B.** 1 μM RR completely abolished the inward current but did not affect the outward currents in TRPV4. This again indicates that the block of TRPV4 by RR is voltage dependent. **C.** The double mutant D672A-D682A currents are similar to those for the WT; however, the Ca^{2+}-dependent decay was delayed. **D.** RR had much less effect on the mutant channel than on the WT. Inward currents were still large and decayed slowly, probably due to the slower entrance of RR into the pore vestibule. **E.** Voltage dependence of the block by 1 μM RR for WT TRPV4 and the 3 mutants. The voltage at the abscissa is the test potential after the first step to 100 mV. The unblocked fraction in the presence of RR was obtained by measuring peak tail currents during the second step to 100 mV and normalizing them to the current in the absence of the blocker (Nilius et al, 2004 with permission)

682D/A. A mutation of 680M also reduces the whole-cell current amplitude and impairs Ca^{2+} permeation. Thus, these three amino acids (D672, M680, D682) in the pore are essential for TRPV4 permeation, selectivity, RR sensitivity, and outward rectification.

The TRPV4 current was slowly blocked by Gd^{3+} (Liedtke et al 2000, Strotmann et al, 2000) at μM concentration. Unlike RR, Gd^{3+} blocks both the inward and the outward current. The blockers RR and Gd^{3+} are, however, non-specific for the TRPV4 channel but, they may be specific for the TRP channels. The reagent, 2-aminoethoxydiphenyl borate activates TRPV1, 2 and 3 but not TRPV4 (Chung 2004b; Hu et al, 2004). Recently, bisandrographolide, an ingredient of plant, is found to activate TRPV4 channel (Smith et al, 2006). The discovery of an agonist will open TRPV4 as a target of pharmacological therapeutics (Krause et al, 2005).

9.3.3 Interaction of Ca^{2+}

TRPV is a non-selective cation current that is usually observed in whole-cell patch clamping with CsCl in a pipette fill (to block endogenous K current). In this setting, the reversal potential is nearly zero. The TRPV4 current in exogenously expressed cells is outward-rectified, and the current density at a positive membrane potential is about twice that at a negative membrane potential. When the voltage is held for example, at -40 mV, the inward current is induced by hypotonic stress. The repeated addition of hypotonic stress or that of 4αPDD causes the following activation to rapidly decay in the presence of Ca^{2+} in a bath solution. (Watanabe et al, 2002a, Nilius et al, 2004). Self-desensitization occurred in the current endowed by TRPV4 but not by TRPV3 (Moqrich et al, 2005). In the absence of Ca^{2+}, the desensitization is much slower indicating that the Ca^{2+-}dependent inhibition of this channel is involved in the mechanism of desensitization.

An increase in intracellular Ca^{2+} by ionomycin in the whole-cell configuration was shown to stimulate TRPV4, and TRPV4 currents stimulated by hypotonic solutions were strongly reduced in the absence of extracellular Ca^{2+}(Strotmann et al, 2003). The proposed calmodulin-binding site at 809–832 is responsible for this Ca^{2+}-dependent potentiation. V814 seems essential for the spontaneous opening of TRPV4 channels responsible for the observed elevated Ca^{2+}levels in non-stimulated TRPV4-expressing cells (Watanabe et al, 2003a). An increase in Ca^{2+} at higher level inhibits the TRPV4 current; a TRPV4 with a single mutation of E797 was constitutively open, suggesting that this site may interfere with Ca^{2+} binding at the neighboring calmodulin-binding motif.

The single TRPV4 channel is outward-rectified. Single channel conductance is 90–110 pS for outward currents and 50–70 pS for inward currents (Liedtke et al, 2000; Watanabe et al, 2002a). The effects of negative and positive pressure are applied to the patch pipette on channel activity in the cell-attached mode, but no significant influence on the appearance of TRPV4 is observed (Strotmann et al, 2000).

9.3.4 NH_2-Terminal Domain

The NH2-terminal intracellular domain of TRPV4 contains three or more ankyrin repeat domains and is important in the mechano- and thermo-sensation of the channel as well as the translocation of this protein within a cell. TRPV4 activation by mechanical stimuli is delayed if these ankyrin repeats are lacking (Liedtke et al, 2000). The exogenous expression of an NH2-terminal deletion form of TRPV4 into *C. elegans* also supports it *in vivo* because the mutant is less sensitive to touch. These repeats may anchor the channel to the cytoskeleton and form a mechanical link for gating.

The ankyrin repeats of human TRPV4 are also involved in the alterative splicing mechanism, and the TRPV4-B (Δ384–444), TRPV4-C (Δ237–284), and TRPV4-E

(Δ 237–284 and Δ 384–444) variants are not carried toward the plasma membrane (Arniges et al, 2004). On the other hand, TRPV4-D (Δ 27–61) variants, in which the defect is not included in the ankyrin repeats, are expressed in the plasma membrane, such as in the wild-type TRPV4. Thus, the NH2-terminal of TRPV4 is crucial in plasma membrane expression. Recently, PACSIN (protein kinase C and casein kinase substrate in neuron protein) 3, but not PACSINs 1 and 2, shifted the ratio of the plasma membrane versus cytosolic TRPV4 toward an increase of plasma membrane-associated plasma form in the mechanism involving dynamin-mediated processes (Cuajungco et al, 2006). PACSIN 3 has been identified as a binding protein of the NH2 terminal of TRPV4, and it specifically affects the endocytosis of TRPV4, thereby modulating its subcellular localization. In particular, co-immunostaining of PACSIN3 and TRPV4 reveals that they are a localized luminal membrane of the renal tubules. Thus, the interaction of the binding protein affects not only the membrane surface expression of TRPV4 but also the polarity of expression in epithelia.

The WNK family of kinases was described by Wilson et al. (2001) in a screen for novel mitogen-activated protein (MAP)/extracellular signal-regulated protein kinase (ERK) kinase (MEK) family members in the rat brain. Human hypertension (pseudohypoaldosteronism type II, PHAII) is occasionally found to be caused by a mutation in WNK kinases. Biochemically, WNK kinases are activated by a variety of stimuli, including hypertonicity and hypotonicity (Lenertz et al, 2004). Interestingly, WNK4 down-regulates TRPV4 membrane expression, although no direct coupling of WNK4 and TRPV4 has been demonstrated (Fu et al, 2006). The NH2-terminal deletion mutant (Δ2–147) is unable to interact with WNK4, suggesting that an unknown protein(s) regulates the membrane expression of TRPV4 under the influence of WNK kinases. Loss of the inhibitory role of WNK4 in PHAII suggests the involvement of TRPV4 into the pathogenesis of hypercalcemia and hypertension (Gamba 2006).

9.3.5 COOH-Terminal Domain

The COOH terminal is also important in the mechanosensation of the channel and modulation by Ca^{2+}. Liedtke et al, (2003b) have expressed TRPV4 in *C. elegans* sensory neurons and examined its ability to generate behavioral responses to sensory stimuli. The deletion mutant of the COOH terminal (Δ741-) as well as the NH2 terminal (Δ -410) showed reduced sensitivity to touch on the nose and avoidance to high osmolality. The currents evoked by hypotonic shock in HEK cells were poor in the NH2-terminal and COOH-terminal deletion mutants (personal observation), suggesting that both ends are important in the transduction of mechanical stimuli or the expression of TRPV4 on the plasma membrane.

Binding proteins to the C terminal were sought by yeast two-hybrid screening; candidate proteins involve the microtubule-associated protein 7 (MAP7=E-MAP-115), the heat shock protein and the EH domain containing protein 1. MAP7 is a coiled-coil protein that binds to cytoskeleton (Suzuki et al, 2003c) and can transmit the geometrical signal toward the channel's C terminal. However, the co-expression of MAP7 and TRPV4 did not make TRPV4 a more mechanically vulnerable channel. MAP7 contributes to the membrane expression of TRPV4. Other proteins that link TRPV4 to cytoskeleton might be required for better mechanosensitive complex.

9.3.6 Glycosylation Site

There is one report regarding the gylcosylation of TRPV4 (Xu et al, 2006). A single high-probability N-linked glycosylation site in TRPV4 that faces the extracellular milieu is phylogenetically conserved. From a structural perspective, this site (N651) is adjacent to the hydrophobic hairpin of the pore-forming loop. The mutation of this residue reveals loss of glycosylation and results in enhanced trafficking of the channel to the plasma membrane. Thus, glycosylation is closely related to the membrane trafficking of the TRPV4 protein and the functional expression of this channel.

9.3.7 Modulation by Phosphorylation

Protein kinase A did not modify the TRPV4 current. 4αPDD, an inactive analogue of phorbolester, is a potent activator of TRPV4 (Watanabe et al, 2002a). The classic protein kinase C (PKC) activator phorbol 12-myristate 13-acetate (PMA) a known activator of PKC, also activates TRPV4 (Gao et al, 2003; Xu et al, 2003; Wissenbach et al, 2000). The concentration of PMA is 10 or 50 times higher than that of 4αPDD at 37°C in the activation of TRPV4, which is reduced in the presence of the PKC inhibitors calphostin C and staurosporine (Gao et al, 2003), indicating that phorbols activate TRPV4 via PKC-independent and -dependent pathways. Putative PKC phosphorylation sites have been found; probably, S88, S134, and S528 are the most likely candidates for mediating functional effects. The activation by PMA at room temperature is not clear but becomes obvious at warm temperature (37°C) (Gao et al, 2003).

The phosphorylation of TRPV4 by a member of the Src family of tyrosine kinases is involved in the mechanism of activation (Xu et al, 2003). Whether this phosphorylation is essential for hypoosmolality-induced channel activation is however controversial. Xu et al., (2003) described that hypotonic stress resulted in the genistein-sensitive phosphorylation of TRPV4 at residue Y253 and that cells expressing the Y253F mutant no longer responded to hypotonic cell swelling. However, the result was not reproduced by another group (Vriens et al, 2003). On the other hand, taxol, irritable chemicals, hypotonic and hypertonic solution-induced hyperalgesia are TRPV4-mediated and dependent on integrin/Src tyrosine kinase signaling in sensory cells *in vitro* and in the rat *in vivo* (Alessandri-Haber et al, 2004; Alessandri-Haber et al, 2005; Alessandri-Haber et al, 2006), and this signaling is potentiated by the prostaglandin E pathway (Alessandri-Haber et al 2003). Therefore, tyrosine phosphorylation should be involved in the mechanism of TRPV4 activation, although the quality of stimuli such as irritants or anisoosmolality is not uniform in all cells expressing TRPV4.

9.4 Molecular Mechanism of the Conversion of Physical Stimuli into Channel Opening

9.4.1 Stimuli that Activate TRPV4

On the basis of the stimuli activating TRPV4 and/or Osm-9, anisosmolality, noxious odors, touch, and warmth were candidate stimuli as channel openers. In *in vivo* and *in vitro* experiments with TRPV4, anisoosmolality, algesic substances, pressure (shear stress), and warmth activated TRPV4 on possibly independent mechanisms. Extensive studies by Nilius's group have discovered the molecular mechanisms underlying the conversion of thermal and osmolar stimuli into channel opening.

9.4.2 Hypoosmolality

Basically, cell swelling by hypoosmolality can be considered to be different from the stretching of the membrane. Since swelling as a result of exposure to a hypotonic solution enabling TRPV4 opening does not induce a significant rise in the capacitance of the whole cell, the swelling signal involves biochemical cascades rather than a simple increase in physical force, such as that occurring in stretching. In TRPV4-expressing HEK cells, the frequency of the appearance of the stretch-activated (SA) channel is unchanged from that in non-transfected control cells, suggesting that the TRPV4 molecule itself does not constitute the SA channel (Strotmann et al 2000, O'Neil and Heller 2005), as it does in the case of the TRPC1 channel (Maroto et al, 2005).

Based on the signal transduction (such as an unidentified tyrosine kinase or phosphatase, the arachidonic acid metabolite 5,6-epoxyeicosatrienoic acid (5,6-EET), cGMP, 1,2-diacylglycerol (DAG), or metabolized arachidonic acid derivatives) observed in endothelial Ca^{2+} entry channels, researchers observed that TRPV4 was activated by arachidonic acid (AA) (Nilius et al, 2004). Extending this observation by electrophysiological studies led to the discovery of the molecular mechanism of the hypoosmolality-activation of TRPV4 (Fig. 9.2). Cells exposed to hypoosmolality activates TRPV4 by means of the PLA2-dependent formation of AA (Basavappa et al, 1998) and its subsequent metabolization to 5,6-EET through a cytochrome P450 epoxygenase-dependent pathway. Thus, the hypoosmolality-activated currents driven by TRPV4 are completely blocked by all PLA2 inhibitors; methylarachidonyl fluorophosphate (MAFP), arachidonyl trifluoromethyl ketone (AACOCF3), 3-[(4-octadecyl) benzoyl]acrylic acid (OBAA, 100 μM), *N*-(*p*-amylcinnamoyl) anthranilic acid, (ACA, 80 μM), and p-bromophenacyl bromide (BPB, 100 μM) in TRPV4-expressing HEK cells (Watanabe et al 2003b, Vriens et al, 2004). PLA2 is divided into two groups cytosolic PLA2 (cPLA2) in the cell interior and secretory PLA2 (sPLA2) in plasma. The effect of inhibitors on PLAs subclass is different; MAFP, AACOCF3, and OBAA are inhibitors of cPLA2, while BPB is a specific inhibitor of sPLA2 (Hernåndez et al, 1998). AA released from PLA2 is metabolized into three pathways (via cyclooxygenase, lipoxygenase and epoxygenase). Among these, the epoxygenase endoproduct, 5,6-EET has been proven to be a direct activator of TRPV4 in single-channel analysis (Watanabe et al, 2003b). Furthermore, the inhibition of the cytochrome P-450 epoxygenase pathway by 5,8,11,13- eicosatetraynoic acid, miconazole, 17-octadecynoic acid (all 10μM) (Vriens et al, 2004) or sulfaphenazole (for CYP2C9) (Vriens et al, 2005) blocks the hypoosmolality activation of TRPV4. The motif of TRPV4 amino acids in response to 5,6-EET remains unclear.

AA is derived from various fatty acids of the plasma membrane. Hydrolysis of the endocannabinoids, anandamide (AEA) and 2-arachidonoylglycerol (2-AG), which are endogenous ligands of the CB1 and CB2 metabotropic cannabinoid receptors, is one of these pathways resulting in the formation of AA. AEA and 2-AG activate TRPV4 in an AA-dependent pathway (Watanabe et al, 2003b). All mechanisms under the hypoosmolality activation and stimulation by endocannabinoid have been examined in TRPV4-expressing HEK cells and primary cultured endothelial cells. In these cells, the primary target of the swelling is the activation of PLA2, although the molecular mechanism of the direct coupling of the enzyme and the physical force has not been fully clairified. Interestingly, methanandamide, which blocks the conversion of anandamide to AA, completely blocks the hypotonic reduction of AQP5 in mouse lung epithelial cells through the activation of TRPV4 (Sidhaye et al, 2006). Therefore, PLA2 in particular which hydrolyzes anandamide into AA may be a primary target of cell swelling.

The hypoosmolality-activated or 4αPDD-activated Ca^{2+} influx that is blocked by RR is observed widely in TRPV4-expressing airway epithelial cells (Arniges et al, 2006), smooth muscles cells (Jia et al, 2004; Yang et al, 2006), aortic endothelial cells (Koehler et al, 2006; Yao and Garland 2005; Vriens et al, 2005), vascular smooth muscle cells (Earley et al 2005; Koehler et al, 2006; Dietrich et al, 2006), keratinocytes (Becker et al, 2005), hepatocyte cells (Vriens et al 2004), synoviocytes (Kochukov et al, 2006), renal tubular cells (unpublished observation), salivary gland cells (Liu et al, 2006), CA1 neurons (Lipski et al, 2006), and early chondrocytes (unpublished observation). It is not known whether methanandamide or other PLA2 blockers abolish the hypoosmolality-activated TRPV4 in all of these cells. Using the primary cultured cells from *TRPV4*$^{-/-}$ mice, TRPV4 has been proven to be a hypoosmolality-activated channel in keratinocytes (Chung et al, 2003), aortic endothelial cells (Vriens et al, 2004), hippocampus neurons, renal tubular cells, outer hair cells (Shen et al, 2006), and early chondrocytes.

Shear stress induced the release of arachidonic acids in lung epithelial cells or cultured human endothelial cells (Frangos et al, 1985). Hamster oviductal ciliated cells (Teilmann et al, 2005) express the TRPV4 currents, which is activated by 4αPDD and increased by the viscous load with a PLA2-dependent pathway (Andrade et al, 2005). In endothelial cells, shear stress and 4αPDD cause vasodilatation, which is blocked by a PLA2 blocker, AACOCF3 (Koehler et al, 2006). Therefore, the PLA2-dependent mechanism operates under the sensation of shear stress in these cells.

9.4.3 Temperature

In TRPV4-expressing model cells (Watanabe et al, 2002b) or primary cultured cells (Chen et al, 2003), warm temperature (33°C<) induces the characteristic outward-rectified current or a rise in $[Ca^{2+}]$i. As suggested by Gao et al (2003), temperature (37°C) amplifies the magnitude of other stimuli that activate TRPV4, such as shear stress, phorbols, and hypoosmolality and integrates them.

The mechanism of the thermo-sensation of the TRP family entails changes in temperature that lead to the production of a channel-activating ligand or temperature sensitivity that is largely dependent on their transmembrane voltage. Most thermo-TRPs (at least TRPV1 and TRPM8) are in the latter group, while TRPV4 is involved in the former (Voets et al, 2005). The effect of the temperature on TRPV4 requires an unknown ligand released into a cell interior, since the opening of TRPV4 in cell-attached patches by warming is abolished by detaching the membrane from TRPV4-expressing HEK cells (Watanabe et al, 2002) or from 308 keratinocytes (Chung et al, 2003).

The tyrosine-serine (YS) motif in the TM2-TM3 loop domain is important in the activation of TRPV1 by capsaicin (Jordt and Julius 2002). In an experiment involving analogy, Vriens et al. (2004) mutated a Y residue (Y-555) or YS (Y-555 and S556) in the N-terminal part of TM3 to alanine (A), which strongly impaired the activation of TRPV4 by 4αPDD and heat but had no effect on the activation by cell swelling or AA. Thus, the mechanism under the activation of TRPV4 by warmth would involve a ligand-dependent conformational change of the TM3 in TRPV4. This simple schematic mechanism may provoke questions. Under this mechanism, heat and 4αPDD share the same molecular mechanism on the activation and, thereby, should compete with the magnitude of activation in each other. However, warm temperatures not only activate TRPV4 independently but they also help or amplify other stimuli involving 4αPDD (Gao et al, 2003). Thus, the molecular mechanism of thermo-sensation might be more complex than it is in this model.

9.4.4 Other

The analysis of *TRPV4*$^{-/-}$ mice not only indicates the importance in thermo- and mechano-sensation *in vivo* but implies a possible contribution by other stimuli, such as an algegic substance and hyperosmolality, as has been suggested *in vivo* experiments. The molecular mechanism under these stimuli has not been clarified, since it is not reproduced in the model cells expressing TRPV4. Accordingly, the mechanisms may be cell-dependent: for example, if hyperosmolality activates PLA2 in some neurons, these neurons are hypertonicity-sensitive neurons that may regulate body fluid tonicity.

9.5 TRPV4 Knockout Mice and their Phenotype

9.5.1 General Features of the TRPV4$^{-/-}$Mice

A PKG-neo cassette was inserted into the fourth exon encoding the ankyrin-repeat domain, which is downstream of the splicing site (Mizuno et al 2003, Suzuki et al, 2003). Another group disrupted the gene at a more downstream site encoding 6TM (Liedtke et al, 2003a). Both strategies successfully knocked out the functional TRPV4 protein. TRPV4 null mouse (*TRPV4*$^{-/-}$), heterozygous (*TRPV4*$^{+/-}$), and wild-type (*TRPV4*$^{+/+}$) mice were produced in the expected 1:2:1 Mendelian ratio. The *TRPV4*$^{-/-}$appeared normal in growth, size, body temperature (anal), and fertility and showed no obvious behavioral (Shirpa's protocol, including drinking) abnormalities. Continuous careful observations revealed following subtle differences (unpublished); (1) a slightly shorter tail length than that in the wild-type mice, (2) slightly lower blood pressure (BP) at 11 weeks of age than that in the wild-type mice, and (3) cross maze test, dark-bright box test and tail suspension test may show *TRPV4*$^{-/-}$ relax. By maintaining the littermates for 2 years, no changes in life span and no abnormality in histological findings of heart, kidney, brain and fat at autopsy between brothers are observed.

The mice were isolated and maintained in a restricted cage for the measurement of their daily intake and excretion (Table 9.1). The resultant chemical data are summarized in Table 9.1 The urinary Na concentration, but not the total excretion of the Na and K/Na ratio, was lower in *TRPV4*$^{-/-}$mice. The urinary osmolality was not different, but total excretion of osmolality was lower in *TRPV4*$^{-/-}$mice. There was no significant difference in the plasma osmolality and Na concentration. However, when tap water was changed to 2% NaCl for a week, the plasma Na concentration was significantly lower in *TRPV4*$^{-/-}$mice. On the other hand, Liedtke et al (2003) revealed a different result as discussed below. All these results at least suggest that *TRPV4*$^{-/-}$mice have an abnormal metabolism of Na and osmolality.

Although not statistically significant, the blood pressure, fasting glucose, and serum calcium concentration were lower in *TRPV4*$^{-/-}$mice than in the matched controls. Although almost normal in their appearance at the steady state, *TRPV4*$^{-/-}$ mice revealed an unexpected response to several stimuli, suggesting a possible model of stress-induced abnormality.

Table 9.1 Electrolytes, blood pressure and water balance in TRPV4+/+ and TRPV4-/- mice Urinary chemical data and water intake are determined for the normal diet averaged triplicate over one week. At the end of the maintenance, blood was sampled. Tail blood pressure was measured duplicate at the given age. (Mizuno et al, 2003)

Plasma (n=10)

	Cr (mg/dl)	Na (mEq/l)	K (m Eq/l)	Ca(mg/dl)	Cl (mEq/l)	Osm (mOsm/Kg)
Trpv4+/+	0.10 ± 0.003	137.9 ± 1.0	5.4 ± 0.1	9.8 ± 0.4	115 ± 2.8	320 ± 4.0
Trpv4-/-	0.11 ± 0.02	140.7 ± 0.8	5.4 ± 0.4	8.8 ± 0.3	110 ± 1.8	320 ± 4.9

Urine (n=10)

	Cr (mg/dl)	Na (mEq/l)	K (mEq/l)	Ca(mg/dl)	Cl (mEq/l)	Osm	OsmxUV
Trpv4+/+	36.2 ± 3.8	164.9 ± 20.6	277.5± 29.2	20.9 ± 1.9	178 ±27.3	2402 ± 897	3147±1360
Trpv4-/-	32.5 ± 3.2	123.1±14.6*	217 ± 27.7	17.1 ± 1.5	136.9 ±18.6	2236 ± 930	4063±2160

*P<0.05

After 2% NaCl for a week (n = 10)

	P Na	U Na	Uosm	U Na/day (mEq)	U/P Na	U K/Na
Trpv4+/+	151 ± 10.8	646.5 ± 27.6	3308± 170	0.56± 0.07	4,1 ± 0.15	2.2 ± 0.2
Trpv4-/-	133 ± 7.1*	592.7 ± 25.8	3120± 180	0.50± 0.12	4.5±0.19	1.95±0.0.24

				Systolic	Diastolic
Blood pressure 11W (n=10)			TRPV4+/+	123.2 ± 2.7	71.2 ± 4.6 (mmHg)
Systolic	Diastolic		TRPV4-/-	120.1 ± 8.1	64.2 ± 5.3 (mmHg)
TRPV4+/+	116. 1± 5.2	69.8 ± 3.8	(mmHg)	Water Intake	Urine Output (ml/day) (n=15)
TRPV4-/-	104.8 ± 4.2	64.2 ± 5.4 (mmHg)	TRPV4+/+	4.0 ± 0.8	1.4 ± 0.4
Blood pressure 20W (n=10)			TRPV4-/-	4.5 ± 1.0	1.8 ± 0.5

9.5.2 Impaired Mechanosensation in TRPV4$^{-/-}$Mice (Fig. 9.5)

Impairment of the pressure sensation in *TRPV4*$^{-/-}$mice *in vivo* has been indicated (Suzuki et al, 2003; Liedtkee t al, 2003a), although at slightly different ranges of pressure. To test the mechanosensation, von-Fray hair is a useful tool to detect the touch sensation in animals and humans. Von-Fray hair is composed of variably sized plastic hairs providing variable pressure (gram). When a mouse senses the touch of the hair on its hind paw, it shakes the hair off. Liedtke et al. (2003a) indicated, using an automated von-Fray detector, that the elevation in the threshold of avoidance was observed in *TRPV4*$^{-/-}$mice. We performed the test manually and failed to find a significant difference. However, we used our own device to measure the pressure sensation on the tail and found that sensation to mechanical stimuli is impaired in *TRPV4*$^{-/-}$mice. An additional study, in which the direct neural activity of femoral nerves evoked by pressure on the paw was measured, confirmed the result. As reported above, Liedtke et al, (2003b) have expressed TRPV4 in *C. elegans* sensory neurons and demonstrated that avoidance of touch on nose is restored by exogenously expressed TRPV4. Thus, TRPV4 is a direct sensor of mechanical pressure *in vivo*.

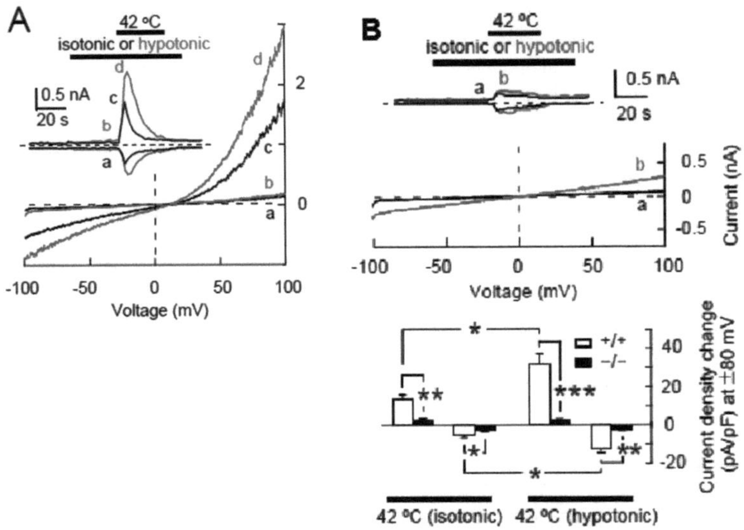

Fig. 9.4 Warmth- and hypotonic stress- induced current in keratinocytes in TRPV4$^{+/+}$and TRPV4$^{-/-}$mice. **A.** Representative I-V traces recorded at 24°C (a, b) and 42°C (c, d) under isotonic (a, c; black) or hypotonic (b, d; red) conditions. Data were derived from two different TRPV4$^{+/+}$cells in primary culture. Inset Superimposed current amplitude traces derived from the same two cells. **B.** Upper Superimposed current amplitude changes in two different TRPV4$^{-/-}$ keratinocytes during a heat stimulus under isotonic (black) or hypotonic (red) conditions. Middle I-V relationships obtained at points a and b in the upper panel. Lower Mean (± sem) heat-evoked current density change, obtained by dividing the heat-evoked current amplitude change (amplitude at 42°C minus amplitude at 24°C) by the membrane capacitance, in keratinocytes derived from TRPV4 +/+ (open bars) versus TRPV4$^{-/-}$ keratinocytes (filled bars) under isotonic (left) or hypotonic (right) conditions at +80 mV (upward bars) or –80 mV (downward bars). (n=6; *, p<0.05; **, p<0.005; ***, p<0.001). (Chung et al, 2004a)

The type of mammalian neurons possessing TRPV4 and sensing the mechanical stimuli has been investigated (Suzuki et al, 2003b). Double staining of ant-TRPV4 and peripherin (C-fibers) or NF200 (larger fibers) suggests that both types of neurons possess the TRPV4 protein. Furthermore, TRPV4 is located in the Merkel body, Meisner body, penicillate, and Ruffini endings, suggesting that at least larger-sized (Aγ) axons play a role in mechanical detection by TRPV4. On the other hand, TRPV4 corresponds to PGE2- and hypotonicity-induced firing of C fiber in rat peripheral nerves (Alessandri-Haber et al, 2003). Thus, both smaller sized fibers, Aγand C, may be the mechanosensitive fibers containing TRPV4. Impaired sensation of mechanical stimuli was also observed in mice lacking TRPA1 using the von-Fray hair (Kwan et al, 2006). A cell-type that corresponds to mechanosensation should be identified in the future in these knockout mice.

The magnitude of the pressure can be discussed in a recent finding (Taniguchi et al, 2006): isolated mouse renal tubules are perfused *in vitro*. The intraluminal hydrostatic pressure is changed by increasing the perfusate (from 5 to 15 cm), leading to a calculated pressure of 0.7 to 1.2 kPa. The manipulation on the renal collecting duct influences tubular Na reabsorption and K excretion. When the hydrostatic

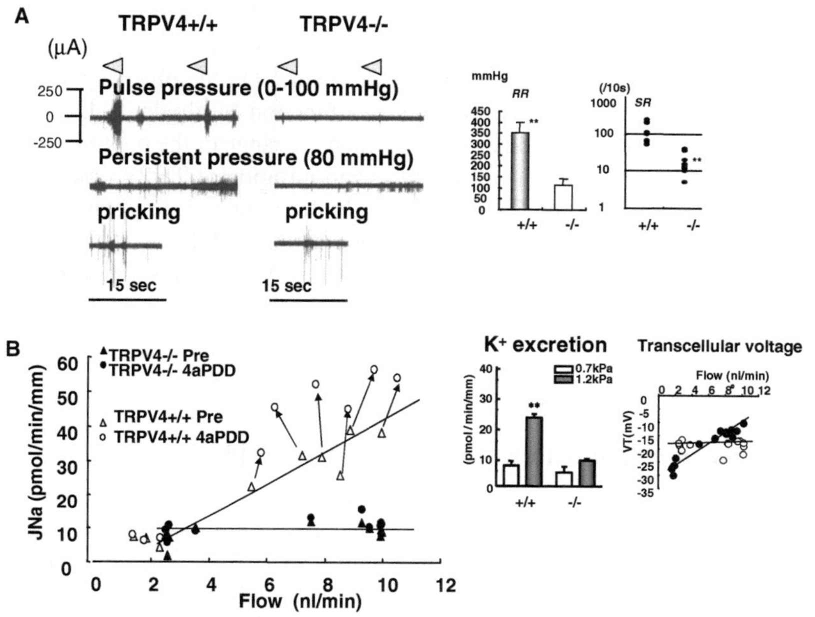

Fig. 9.5 Impaired response to pressure in TRPV4$^{-/-}$ mice. **A.** The neuronal activity (extracellular recording) was recorded in femoral nerves and pressure was applied on its hindpaw. Left; Representative current amplitudes are shown. Pulse pressure, the ressure rising to a crescendo, is marked as triangle and its response is observed in TRPV4$^{+/+}$and in TRPV4$^{-/-}$ (RR response, upper traces). When the pressure is maintained at 80 mmHg, lasting neural activity is recorded in TRPV4$^{+/+}$and in TRPV4$^{-/-}$ (SR response, middle traces). On the other hand, pricking the 23G needle (>300mmHg) induced the similar burst in TRPV4$^{+/+}$and in TRPV4$^{-/-}$ (bottom traces). Center; The threshold of RR response in 16 mice. Right; Frequency of discharges in SR response for 1 min are counted (** $p<0.01$). (Suzuki et al, 2003a) **B.** Renal cortical collecting tubules were perfused in vitro, and the hydrostatic pressure/flow rate was changed. Left; Na reabsorption in low and high flow rates was measured before (triangle) and after (circle) treatment with 4αPDD in TRPV4$^{+/+}$(filled) and TRPV4$^{-/-}$ (open). Center; K excretion was concomitantly measured and summarized. ** $p < 0.01$. Right; Transcellular voltage difference (VT), the lumen negative voltage compared to basolateral bath solution, was measured in TRPV4$^{+/+}$mice (filled circle), which was depolarized in response to high flow rate. Whereas, VT in TRPV4$^{-/-}$ (open circle) was significantly smaller ($p < 0.05$) and not changed in response to high flow rate. (Taniguchi et al, 2006)

pressure (or shear stress) is increased within this physiological range, influx of Ca^{2+} from the lumen into the rabbit tubular cells occurs. The rise in $[Ca^{2+}]$i leads to the activation of the Ca^{2+}-activated large conductance K channel (BKCa) channel to secrete K into the lumen, and hyperpolarization by increased K permeability elicits an electrical driving force of Na reabsorption. Using this method of microperfusion, the enhanced reabsorption of Na and excretion of K by the rise in hydrostatic pressure, which is entirely dependent on the presence of luminal Ca^{2+}, are completely abolished in *TRPV4*$^{-/-}$mice. TRPV4 was first reported to be located in the basolateral membrane but was later found in the luminal membrane. Thus, preferable luminal TRPV4 may act as a sensor of the flow rate, resulting in the initiation of the Ca^{2+} influx. The range of pressure that activates TRPV4 is thereby suggested to be around 1–10 kPa from the hydrostatic pressure obtained in microperfusion to direct protruding and calculated from touch pressure. On the other hand, TRPV4 has been implicated in the regulation of ciliary function (Andrade et al, 2005) and in endothelial-dependent vasodilatation, in which the range of the pressure, or shear stress, is around 1–100 mPa. Thus, the mechanosensation of TRPV4 may cover a wide range of physical force found in the body. It is otherwise hypothesized that the shear stress- and hypotonicity-induced activation of TRPV4 at the range of mPa is dependent of the release of arachidonic acid metabolites and independent of membrane tension, while another mechanism may operate in a higher range of pressure.

9.5.3 Impaired Osmosensation in TRPV4$^{-/-}$Mice

Impairment of osmotic sensation results in abnormality of vasopressin regulation and in that of thirst sensation. The abnormalities in *TRPV4*$^{-/-}$mice *in vivo* have been indicated (Mizuno et al, 2003; Liedtke et al, 2003), although the conclusive phenotype is controversial and complex (Liedtke 2005a). Mizuno et al (2003) have shown that *TRPV4*$^{-/-}$mice reveal an inappropriate secretion of an ADH-like phenotype, and other researchers have shown that *TRPV4*$^{-/-}$mice reveal a diabetes insipedes-like phenotype. Based on the localization of TRPV4 in VOLT and SFO reported in 2000, it was anticipated that an osmotic sensation is altered in *TRPV4*$^{-/-}$mice. Systemic osmolality was not different in one group (Mizuno et al, 2003), but it was significantly elevated in the other group (Liedtke et al, 2003). Hypertonic solution injected into the abdomen to reduce the volume of circulation resulted in some impaired responses in the serum osmolality of *TRPV4*$^{-/-}$mice. The response of ADH to hypertonicity was exaggerated in one group and diminished in the other. In addition to the fluctuation in vasopressin regulation, the thirst sensation is also involved in the role of TRPV4. Deprivation of water for three days showed that drinking behavior and water consumption were significantly reduced in *TRPV4*$^{-/-}$mice. We followed this method precisely and found a similar but slight difference in thirst sensation. In contrast, a direct injection of 4αPDD into a rat cerebral ventricle resulted in an angiotensin II-dependent decrease in thirst sensation (Tsushima and Mori 2006). If this is the result by stimulating TRPV4, the resultant phenotype is opposite to the findings in *TRPV4*$^{-/-}$mice. Thus, complex and opposite results are reported in variable situations. The conclusions with *TRPV4*$^{-/-}$mice have suggested that there is at least an abnormal osmotic regulation in *TRPV4*$^{-/-}$mice when they are exposed to a hypertonic sate. *TRPV4*$^{-/-}$mice are resistant to noxious irritants, including hyperosmotic as well as hypoosmotic solutions injected into the paw (Alessandri-Haber et al, 2005). Therefore, TRPV4 *de novo* in some neurons may be activated by hyperosmotic pressure rather than by a hypotonic solution.

One of hyperosmolality-activated channel was recently identified. A stretch-inhibitable channel (SIC), a hyperosmolality-activated channel, is an N-terminal truncated form of the TRPV channel (Suzuki et al, 1999). While this reported SIC was a PCR-based chimera of TRPV1 and TRPV4, the truncated form of TRPV1 was also supposed to encode a ubiquitously expressed cation channel in 2000 (Schumacher et al, 2000). Thus, the truncated form of the TRPV channel might be a functional channel in response to physical stimuli. Recently, this N-terminal variant of TRPV1 was shown to be the hyperosmolality-activated channel, and the mice lacking TRPV1 revealed a diabetes insipedes-like phenotype (Naeini et al, 2006). A recent hypothesis in the regulation of body fluid osmolality indicates that the concert of the hyperosmolality- and hypoosmolality-activated channel in the circumventricular nucleus is essential (Bourque and Oliet 1997; Chakfe and Bourque 2000; Morris and Sigurdson 1989). Thus, the TRPV1 variant and TRPV4 may be corresponding to these channels. The precise localization and co-operation of these channels will be clarified in the future.

9.5.4 Impaired Thermosensation in TRPV4$^{-/-}$Mice

Impairment of warmth but not a hot sensation in *TRPV4*$^{-/-}$mice *in vivo* is indicated by the direct measurement of neural activity (Todaka et al, 2004) and by a thermotaxis assay (Lee et al, 2005). *TRPV4*$^{-/-}$ mice display no change in escape latency from acute painful heat stimuli on the foot plate (>50°C) or upon radiant paw heating. Lee et al, (2005) conducted a detailed analysis of responses to warm temperature. On a thermal gradient, the *TRPV4*$^{-/-}$ mice selected warmer floor temperatures than *TRPV4*$^{+/+}$ mice. *TRPV4*$^{-/-}$ mice showed a strong preference for 34°C, whereas *TRPV*$^{+/+}$ failed to discriminate between 30°C and 34°C. These assays consistently showed that the absence of TRPV4 results in a tolerance of higher temperatures. *TRPV4*$^{-/-}$ mice unexpectedly showed longer escape latencies on the hot plate and lower neural activity in femoral nerves by warming of their hind paws than *TRPV4*$^{+/+}$ mice after subcutaneous injection of carageenan, suggesting a role for TRPV4 in thermal hyperalgesia, or after cutaneous injection of capsaicin, suggesting a role in inflammatory hyperalgesia. (Todaka et al 2004). Thus, the TRPV4 in the peripheral neurons play a role in sensing warmth.

TRPV4 exists in keratinocytes, and the keratinocytes cultured from *TRPV4*$^{-/-}$ mice do not respond to the warm temperature (Chung et al, 2004a). Thus, TRPV4 in keratinocytes also plays an essential role in sensing warmth *in vivo* and *in vitro*. On the other hand, TRPV3 is located in keratinocytes but not in neurons. The mice lacking TRPV3 show a heat-insensitive phenotype (Moqrich et al, 2005), indicating that both TRPV3 and TRPV4 cover the sensation of warmth on the skin. Details of thermo-sensation have been described in several reviews (Caterina 2006; Dhaka et al 2006; Voets et al, 2005. Benham et al, 2003).

In contrast to thermo-sensation, TRPV4 may participate in the regulation of body temperature in the brain. Changes in temperature of up to several degrees have been reported in different brain regions during various behaviors or in response to environmental stimuli (Hori et al, 1999; Boulant 2000; Wechselberger et al, 2006).

TRPV4 found in dopaminergic neurons of the rat substantia nigra pars compacta (Guatteo et al, 2005) or mouse hippocampus is constitutively active and supposed to play a role in body temperature regulation or in ischemia (Lipski et al, 2006). The response to fever is attenuated in mice lacking TRPV1 (Iida, et al, 2005). However, no significant difference in the maintenance of body temperature or the response to fever has been observed in *TRPV4*$^{-/-}$mice until now.

9.5.5 Impaired Pain Sensation in TRPV4$^{-/-}$Mice

Several irritants, including algesic substances, induce their own responses, which are absent in *TRPV4*$^{-/-}$mice. Acetic acid injected into the abdomen induced a characteristic writhing of the abdomen in rodents. This test was used in screening analgesics. In *TRPV4*$^{-/-}$mice, this response was completely absent (Suzuki et al, 2003). Capsaicin and taxol, which are chemical algegic substances, induced a lower response in neural activity in *TRPV4*$^{-/-}$mice than *TRPV4*$^{+/+}$mice or in the rat after the treatment with antisense oligonucleotides against TRPV4 in DRG than before (Alessandri-Haber et al, 2003). The contribution of TRPV4 to inflammatory pain has also been suggested in *TRPV4*$^{-/-}$ mice by Src kinase dependent manner (Alessandri-Haber et al, 2006). The release of arachidonic acids and activation of *Src* kinase are supposed to be a signal transmission, although more complex mechanisms are involved in the pain sensation. Accordingly, TRPV4 can be a target in analgesic pharmacology, and the discovery of antagonists is key in the future.

9.5.6 Hearing Impairment Found in TRPV4$^{-/-}$Mice

TRPV4 is located in the inner ear and is supposed to play a role in hearing sensation. Its localization in guinea pig has been investigated (Takumida et al, 2005); TRPV4 was found in hair cells and supporting cells of the organ of Corti, in marginal cells of the stria vascularis, spiral ganglion cells, sensory cells, transitional cells, dark cells in the vestibular end organs, vestibular ganglion cells and epithelial cells of the endolymphatic sac. Several efforts to detect the localization of TRPV4 have suggested that inner hair cells are not the predominant cells (Corey 2006). *In vitro*, a swell-activated rise in $[Ca^{2+}]i$ is abolished in freshly isolated outer hair cells from *TRPV4*$^{-/-}$mice. Outer hair cells are exposed to lymphatic fluid, and mechanical or osmotic shock might be achieved in Meniere's disease. *In vivo*, audiometrical analysis suggests that *TRPV4*$^{-/-}$mice exhibit normal hearing ability at 8 weeks of age. A slight but significant difference was observed in hearing ability at 20 weeks of age and the recovery after 128 db SPL (Tabuchi et al, 2005). Thus adult-onset or noise induced hearing loss is likely to be affected in the functional loss of TRPV4.

9.5.7 Sodium and Water Metabolism in the $TRPV4^{-/-}$ Kidney

Since TRPV4 mRNA is abundant in the kidney, the roles of TRPV4 in the regulation of the sodium/water metabolism have been proposed (Cohen et al, 2005). (Note: Please include the year of publication in this and other in-text citations.) As noted above, the TRPV4 protein is predominantly expressed in the basolateral membrane of distal tubules, but its luminal localization and presence in proximal tubules has also been suggested. By using *TRPV4*$^{-/-}$mice, we (Taniguchi et al, 2006) have shown that (1) TRPV4 senses the flow rate in the collecting duct and plays a key role in flow-dependent Na reabsorption and K excretion in isolated tubules. Thus, TRPV4 is a flow-sensor in the luminal membrane, which converts the flow rate to Ca^{2+} influx. (2) The urinary K excretion was significantly low in the *TRPV4*$^{-/-}$mice after treatment with aldosterone and furosemide. This may be because impaired flow-dependent K excretion becomes obvious with a high urine output or because the impaired response to the decreased interstitial osmality is induced by furosemide. The decrement of interstitial osmolality may activate the basolateral TRPV4. The following additional features of the *TRPV4*$^{-/-}$ kidney can be noted: (1) The urinary Na concentration may be different in the steady state. This might be because paracellular permeability at a low flow rate is reduced in the *TRPV4*$^{-/-}$-collecting tubule. (2) The plasma Ca^{2+} concentration might be low, and urinary Ca^{2+} might be higher in the *TRPV4*$^{-/-}$mouse. This suggests that TRPV4 may play a role in Ca^{2+} reabsorption under the influence of WNK4. There is one other possibility: Ca-sensing receptor (CaR) senses plasma Ca^{2+} concentration and regulates the Ca^{2+} reabsorption. The signal transducation of CaR in a thick ascending limb, where TRV4 is expressed, involves arachinonic acid and cytochrome P450 metabolite, that is known to inhibit the luminal K channel (Hebert et al, 1997). Thus, TRPV4 may mediate the signal cascade by CaR and play a role in Ca homeostasis. TRPV4 may play a role in the sodium and water metabolism, leading to the possible regulation of BP, and it probably plays a role in Ca metabolism as well; however, these mechanisms are intertwined and complex, and more studies using *TRPV4*$^{-/-}$mice will be required.

9.6 Functional Coupling of TRPV4 with Other Molecules

Various studies have been performed on the physiological roles of TRPV4 in a variety of tissues. The functional coupling of TRPV4 and other molecules is discovered, in which TRPV4 does not only function as the Ca^{2+} channel but also play as a signal transmitter. In the former example, TRPV4 couples with BKCa channel and thereby mechanical stress, through the influx of Ca^{2+}, results in membrane repolalization in airway and vascular smooth muscle, or in K excretion during the cell volume regulation. In the latter example, activation of TRPV4 may stimulate unknown cascade to regulate aquaporin expression, ryanodine receptor or tight junction proteins. Most studies have been performed from a pharmacological approach in which a

4αPDD-evoked current and RR-sensitive Ca^{2+} influx were the hallmark of functional TRPV4. Precise studies have added a molecular biological approach by using antisense oligonucleotides to knockdown TRPV4 and examining the functional difference. It is noteworthy that other functional TRP channels are co-expressed in the same cell; anandamide activates TRPV4/TRPV1, mechanical stress affects TRPV4/TRPV2/TRPA1/TRPC1, and heat activates TRPV1/TRPV2/TRPV3/TRPV4. Thus, a functional difference may not be restricted to the defect of TRPV4 alone. The result of these studies is interesting; TRPV4 is a sensor of physical stimuli, converting the signal to other molecules to create a "physiological or pathophysiological response."

9.6.1 TRPV4 and Cell Volume Regulation: (TRPV4 and AQP)

Cells swell in a hypotonic medium by the influx of water through an aquaporin channel (AQP). Most cells respond to changes in hypotonicity and recover their original volume (regulatory volume decrease) by the activation of cation and anion channels. An efflux of ions through this pathway reduces the tonicity of the cell interior, leading to restoration of the cell volume. Furthermore, AQP channels reduced their activity to enhance the regulatory volume changes. TRPV4, initially activated by swelling, enhances the influx of Ca^{2+} and then opens Ca^{2+} activated-K^{+} and -Cl^{-} channels. These monovalent ions play roles in cell volume restoration. Interestingly, TRPV4 directly reduces the abundance of AQP5 channels in lung (Sidhaye et al, 2006) and salivary epithelial cells (Liu, et al 2006) when exposed to hypotonic solution. Direct coupling of TRPV4 with an AQP5 molecule or the contribution of other proteins that regulate AQP5 membrane expression and are regulated by TRPV4 may be involved.

Bronchial epithelial cells cultured from cystic fibrosis patients does not show the swell-activated BKCa current, although the 4αPDD-induced current remains active (Arniges et al, 2006). The epithelial cell from a normal airway, however, exhibits the swell-activated BK_{Ca} current, which is abolished by anti-sense oligonucleotides against TRPV4. Thus, a cystic fibrosis transmembrane regulator may regulate TRPV4 expression (Arniges et al, 2004).

9.6.2 TRPV4 and Vascular Tone (TRPV4 and Other Channels)

TRPV4 is expressed widely in the vasculature involving endothelial cells and smooth muscle cells. In endothelial cells, hypotonic solution or anandamides activates TRPV4 to increase $[Ca^{2+}]I$ (Vriens et al, 2005). This may lead to NO production, and then NO-dependent vasodilation in smooth muscle occurs (Koehler et al, 2006). On the other hand, cytochrome P450 arachidonic acid metabolites, such as 11,12-EET are considered to be an endothelial-dependent hyperpolalization factor, (EDHF) and activate TRPV4. TRPV4 is coupled with a ryanodine receptor and BKCa. (Earley et al, 2005), and the activation of TRPV4 by EDHF leads these other channels on, as a result, the membrane potential is hyperpolalized, resulting in vasodilatation. Arguments and unsolved questions, however, remain in the involvement of TRPV4 in the mechanism of EDHF (Kotlikoff 2005). However, direct activation of TRPV4 by mechanical stress or by EDHF dilates arteries possessing endothelial cells or smooth muscle and expressing TRPV4.

In fact, 4aPDD induced vasodilation of carotid arteries and a smaller resistant vessel, *Arteria gracilis* by using pressure myograph *in vitro*. The vasodilatation is blocked by RR. The shear stress induced by the addition of 5% dextrose in the perfusate dilates the two small arteries that are blocked by reagents that inhibit NO-production, PLA2 activity, and by RR (Koehler et al, 2006). This direct evidence suggests that shear stress activates TRPV4 in endothelial cells by the activation of PLA2 and the influx of Ca^{2+} initiates the NO-dependent vasodilation in the aorta, carotid artery and smaller resistant vessels.

In contrast to the acute change in BP, *TRPV4*$^{-/-}$mice that are permanently deficient in functional TRPV4 exhibit lower rather than higher BP at a younger age than *TRPV4*$^{+/+}$ mice. BP was measured by the tail-cuff method or by direct measurement (mean BP 81 vs. 70 mmHg, n = 10, <0.05, unpublished observation). Experiments with baroreflex are required to elucidate the significance in shear-stress sensing but have not yet been reported. As noted above, WNK1 and 4 down regulate TRPV4 membrane expression (Fu et al, 2006). Loss of function in WNK1/4 was observed in the rare human disease PHAII (Wilson et al, 2001), in which hypertension and hypercalciuria are involved. *TRPV4*$^{-/-}$mice show a tendency of hypotension and hypocalciuria (hypercalcemia) that are matched with opposite signs in PHAII. The pathway involving WNK1/4 and TRPV4 may also be important in the regulation of BP.

9.6.3 TRPV4 and Epithelial Permeability (TRPV4 and Tight-Junction)

Activation of TRPV4 caused an increase in [Ca^{2+}]i through an influx of extracellular Ca^{2+}in the mouse mammary cell line HC11, triggering an unexpected event in which there was a slow increase in paracellular permeability for small solutes (Reiter et al 2006). Thereby TRPV4 is supposed to regulate the paracellular permeability by unknown mechanisms. The findings by Reiter et al indicated that the effect was accompanied by a downregulation of the tight junctional proteins claudin -1, -3, -4, -5, -7, and -8 and by dramatic changes in tight junction morphology, including frequent large breaks in the tight junction strands. TRPV4 in HC11 is predominantly localized in the basolateral membrane. This effect is independent of another event in which apical BKCa is activated. Thus, a specific pathway connecting basolateral TRPV4 with tight junction proteins should be considered. Although unknown the mechanisms, the relation of TRPV4 to cell-junction has been suggested in other tissues.

TRPV4 mRNA is rich in the lung in which airway epithelia and smooth muscle show functionally active TRPV4 channels. Swell-induced activation of TRPV4 is critically described in these airway cells. The airway cells are exposed to hypotonic milieu in patients suffering from sputum. Mechanical stress is also proposed in asthma patients. Therefore, TRPV4 may play a role in the pathophysiology of asthma (Liedtke 2005b). However, no studies have conducted experiments on the allergen challenge to *TRPV4*$^{-/-}$mice. Recently, TRPV4 is reported to play a crucial role in barrier function of alveolar network by using the agonists and *TRPV4*$^{-/-}$ (Alvarez et al, 2006)

In our experiment with the microperfusion of collecting tubules, transepithelial voltage (dependent on the paracellular permeability) was significantly lower in *TRPV4*$^{-/-}$mice (Taniguchi et al, 2006). This may be compatible with that observed in mammary cells, in which TRPV4 slowly upregulates the paracellular permeability to constitute its own transepithelial voltage.

Therefore, TRPV4 plays a role in barrier function of variable epithelia by regulating of cell-cell communication and tight junction proteins.

9.7 Future Problems in TRPV4 Expressing Tissues

TRPV4 is highly expressed in white and brown fat. However, its functional expression and physiologic roles have not yet been investigated. Temperature generation and volume in fat tissues may be regulated by TRPV4. TRPV4 is detected in the urinary bladder. The TRPV4 may be a sensor of bladder extension signals. TRPV4 is detected in sperm, but its functional role is unknown. TRPV4 is expressed in chondrocyte and osteocytes but not in osteoclasts. TRPV4 may play a role in sensing "gravity" to enhance bone formation.

References

Alessandri-Haber N, Yeh JJ, Boyd AE, Parada CA, Chen X, Reichling DB, Levine JD. (2003) Hypotonicity induces TRPV4-mediated nociception in rat. Neuron. 39:497–511.

Alessandri-Haber N, Dina OA, Yeh JJ, Parada CA, Reichling DB, Levine JD. (2004) Transient receptor potential vanilloid 4 is essential in chemotherapy-induced neuropathic pain in the rat. J Neurosci 24:4444–4452.

Alessandri-Haber N, Joseph E, Dina OA, Liedtke W, Levine JD. (2005) TRPV4 mediates pain-related behavior induced by mild hypertonic stimuli in the presence of inflammatory mediator. Pain 118:70–79.

Alessandri-Haber N, Dina OA, Joseph EK, Reichling D, Levine JD. (2006) A transient receptor potential vanilloid 4-dependent mechanism of hyperalgesia is engaged by concerted action of inflammatory mediators. J Neurosci 26:3864–3874.

Alvarez DF, King JA, Weber D, Addison E, Liedtke W, Townsley MI.(2006) Transient Receptor Potential Vanilloid 4-Mediated Disruption of the Alveolar Septal Barrier. A Novel Mechanism of Acute Lung Injury. Circ Res (in press).

Andrade YN, Fernandes J, Vazquez E, Fernandez-Fernandez JM, Arniges M, Sanchez TM, Villalon M, Valverde MA. (2005) TRPV4 channel is involved in the coupling of fluid viscosity changes to epithelial ciliary activity. J Cell Biol. 168:869–874.

Arniges M, Fernandez-Fernandez JM, Albrecht N, Schaefer M, Valverde MA (2006). Human TRPV4 channel splice variants revealed a key role of ankyrin domains in multimerization and trafficking. J Biol Chem 281:1580–1586.

Arniges M, Vazquez E, Fernandez-Fernandez JM, Valverde MA. (2004) Swelling-activated Ca2+ entry via TRPV4 channel is defective in cystic fibrosis airway epithelia. J Biol Chem 279:54062–54068.

Basavappa S, Pedersen SF, Jorgensen NK, Ellory JC, and Hoffmann EK. (1998) Swelling-induced arachidonic acid release via the 85-kDa cPLA2 in human neuroblastoma cells. J Neurophysiol 79: 1441–1449.

Becker D, Blase C, Bereiter-Hahn J, Jendrach M. (2005) TRPV4 exhibits a functional role in cell-volume regulation. J Cell Sci 118:2435–2440.

Benham CD, Gunthorpe MJ, Davis JB. (2003) TRPV channels as temperature sensors Cell Calcium 33:479–487.

Boulant JA. (2000) Role of the preoptic-anterior hypothalamus in thermoregulation and fever. Clin Infect Dis 31: S157–S161, 2000.

Bourque CW, Oliet SH. (1997) Osmoreceptors in the central nervous system. Annu Rev Physiol 59: 601–619.

Caterina MJ, Schumacher MA, Tominaga M, Rosen TA, Levine JD, Julius D. (1997) The capsaicin receptor: a heat-activated ion channel in the pain pathway. Nature. 389, 816–824.

Caterina MJ, Leffler A, Malmberg AB, Martin WJ, Trafton J, Petersen-Zeitz KR, Koltzenburg M, Basbaum AI, Julius D. (2000) Impaired nociception and pain sensation in mice lacking the capsaicin receptor. Science 288: 306–313.

Caterina MJ. (2006) Transient receptor potential ion channels as participants in thermosensation and thermoregulation. Am J Physiol Regul Integr Comp Physiol. (in press).

Chakfe Y, Bourque CW. (2000) Excitatory peptides and osmotic pressure modulate mechanosensitive cation channels in concert. Nat Neurosci 3:572–579.

Chung MK, Lee H, Caterina MJ. (2003) Warm temperatures activate TRPV4 in mouse 308 keratinocytes. J Biol Chem. 278:32037–32046.

Chung MK, Lee H, Mizuno A, Suzuki M, Caterina MJ (2004a) TRPV3 and TRPV4 mediate warmth-evoked currents in primary mouse keratinocytes. J Biol Chem. 279:21569–21575.

Chung MK, Lee H, Mizuno A, Suzuki M, Caterina MJ. (2004b) 2-aminoethoxydiphenyl borate activates and sensitizes the heat-gated ion channel TRPV3. J Neurosci. 24: 5177–5182.

Clapham DE, Runnels LW, Strubing C. (2001) The trp ion channel family. Nat Rev Neurosci 2: 387–396.

Cohen DM (2005) TRPV4 and the mammalian kidney. Pflugers Arch. 451:168–175.

Colbert HA, Smith TL, Bargmann CI. (1997) OSM-9, a novel protein with structural similarity to channels, is required for olfaction, mechanosensation, and olfactory adaptation in Caenorhabditis elegans. J Neurosci 17:8259–69.

Corey DP. (2006) What is the hair cell transduction channel? J Physiol. (in press).

Cuajungco MP, Grimm C, Oshima K, D'hoedt D, Nilius B, Mensenkamp AR, Bindels RJ, Plomann M, Heller S. (2006) PACSINs bind to the TRPV4 cation channel. PACSIN 3 modulates the subcellular localization of TRPV4. J Biol Chem 281:18753–18762.

Dhaka A, Viswanath V, Patapoutian A. (2006) Trp ion channels and temperature sensation. Annu Rev Neurosci. 29:135–161.

Delany NS, Hurle M, Facer P, Alnadaf T, Plumpton C, Kinghorn I, See CG, Costigan M, Anand P, Woolf CJ, Crowther D, Sanseau P, Tate SN. (2001) Identification and characterization of a novel human vanilloid receptor-like protein, VRL-2. Physiol Genomics. 19:165–167.

Dietrich A, Chubanov V, Kalwa H, Rost BR, Gudermann T. (2006) Cation channels of the transient receptor potential superfamily: Their role in physiological and pathophysiological processes of smooth muscle cells. Pharmacol Ther. 2006 (in press).

Earley S, Heppner TJ, Nelson MT, Brayden JE (2005) TRPV4 forms a novel Ca2+ signaling complex with ryanodine receptors and BKCa channels.Circ Res. 97: 1270–1279.

Frangos JA, Eskin SG, McIntire LV, Ives CL. (1985) Flow effects on prostacyclin production by cultured human endothelial cells. Science. 227:1477–1479.

Fu Y, Subramanya A, Rozansky D, Cohen DM. (2006) WNK kinases influence TRPV4 channel function and localization. Am J Physiol Renal Physiol. 290:F1305–1314.

Gamba G. (2006) TRPV4: a new target for the hypertension-related kinases WNK1 and WNK4.Am J Physiol Renal Physiol. 290:F1303–4.

Gao X, Wu L, O'Neil RG.J (2003) Temperature-modulated diversity of TRPV4 channel gating: activation by physical stresses and phorbol ester derivatives through protein kinase C-dependent and -independent pathways. Biol Chem. 278:27129–27137.

Greene CG, Pamella M. McMillan, SE. Barker, Purnima K, Margaret I. Burmeister LM, Lesperance LM. (2001) DFNA25, a Novel Locus for Dominant Nonsyndromic Hereditary Hearing Impairment, Maps to 12q21–24 Am. J Hum. Genet 68:254–260.

Guatteo E, Chung KK, Bowala TK, Bernardi G, Mercuri NB, Lipski J. (2005) Temperature sensitivity of dopaminergic neurons of the substantia nigra pars compacta: involvement of transient receptor potential channels.J Neurophysiol. 94:3069–80.

Gunthorpe MJ, Benham CD, Randall A, Davis JB. (2002) The diversity in the vanilloid (TRPV) receptor family of ion channels. Trends Pharmacol Sci 23: 183–191.

Guler AD, Lee H, Iida T, Shimizu I, Tominaga M, Caterina M.J (2002) Heat-evoked activation of the ion channel, TRPV4. Neurosci. 22:6408–6414.

Hebert SC, Brown EM, Harris HW. (1997) Role of the Ca(2+)-sensing receptor in divalent mineral ion homeostasis. J Exp Biol 200:295–302.

Hellwig N, Albrecht N, Harteneck C, Schultz G, Schaefer M. (2005) Homo- and heteromeric assembly of TRPV channel subunits J Cell Sci. 118:917–928.

Hoenderop JG, Nilius B, Bindels RJ. (2005) Calcium absorption across epithelia. Physiol Rev 85:373–422.

Hori A, Minato K, Kobayashi S. (1999) Warming-activated channels of warmsensitiveneurons in rat hypothalamic slices. Neurosci Lett 275: 93–96.

Hu HZ, Gu Q, Wang C, Colton CK, Tang J, Kinoshita-Kawada M, Lee LY, Wood JD, Zhu MX. (2004) 2-aminoethoxydiphenyl borate is a common activator of TRPV1, TRPV2, and TRPV3. J Biol Chem 279: 35741–35748.

Jia Y, Wang X, Varty L, Rizzo CA, Yang R, Correll CC, Phelps PT, Egan RW, Hey JA. (2004) Functional TRPV4 channels are expressed in human airway smooth muscle cells. Am J Physiol Lung Cell Mol Physiol. 287:L272–278.

Jordt SE, Julius D. Molecular basis for species-specific sensitivity to "hot" chili peppers. (2002) Cell 108: 421–430.

Kochukov MY, McNearney TA, Fu Y, Westlund KN. (2006) Thermosensitive TRP Ion Channels Mediate Cytosolic Calcium Response in Human Synoviocytes. Am J Physiol Cell Physiol. 291: C424–432.

Koehler R, Heyken WT, Heinau P, Schubert R, Si H, Kacik M, Busch C, Grgic I, Maier T, Hoyer J. (2006) Evidence for a Functional Role of Endothelial Transient Receptor Potential V4 in Shear Stress-Induced Vasodilatation. Arterioscler Thromb Vasc Biol. 26:1495–1502.

Kotlikoff MI (2005) EDHF redux: EETs, TRPV4, and Ca2+ sparks. Circ Res. 97:1209–1210.

Krause JE, Chenard BL, Cortright DN (2005) Transient receptor potential ion channels as targets for the discovery of pain therapeutics.Curr Opin Investig Drugs 6:48–57.

Kunert-Keil C, Bisping F, Kruger J, Brinkmeier H. (2006) Tissue-specific expression of TRP channel genes in the mouse and its variation in three different mouse strains. BMC Genomics. 7:159.

Kwan KY, Allchorne AJ, Vollrath MA, Christensen AP, Zhang DS, Woolf CJ, Corey DP. (2006) TRPA1 contributes to cold, mechanical, and chemical nociception but is not essential for hair-cell transduction. Neuron. 250:277–289.

Lee H, Iida T, Mizuno A, Suzuki M, Caterina MJ. (2005) Altered thermal selection behavior in mice lacking transient receptor potential vanilloid 4. J Neurosci. 25:1304–1310.

Lenertz LY, Lee BH, Min X, Xu BE, Wedin K, Earnest S, Goldsmith EJ, Cobb MH. (2005) Properties of WNK1 and implications for other family members. J Biol Chem 280: 26653–26658.

Liedtke W, Choe Y, Marti-Renom MA, Bell AM, Denis CS, Sali A, Hudspeth AJ, Friedman JM, Heller S. (2000) Vanilloid receptor-related osmotically activated channel (VR-OAC), a candidate vertebrate osmoreceptor. Cell. 103:525–535.

Liedtke W, Friedman JM. (2003a) Abnormal osmotic regulation in trpv4-/- mice. Proc Natl Acad Sci U S A. 100:13698–13703.

Liedtke W, Tobin DM, Bargmann CI, Friedman JM. (2003b) Mammalian TRPV4 (VR-OAC) directs behavioral responses to osmotic and mechanical stimuli in Caenorhabditis elegans. Proc Natl Acad Sci U S A. 100:14531–6.

Liedtke W, Simon SA. (2004) A possible role for TRPV4 receptors in asthma. Am J Physiol Lung Cell Mol Physiol. 287:L269–71.

Liedtke W. (2005a) TRPV4 as osmosensor: a transgenic approach. Pflugers Arch. 451:176–80.

Liedtke W. (2005b) TRPV4 plays an evolutionary conserved role in the transduction of osmotic and mechanical stimuli in live animals. J Physiol. 567:53–8.

Lipski J, Park TI, Li D, Lee SC, Trevarton AJ, Chung KK, Freestone PS, Bai JZ. (2006) Involvement of TRP-like channels in the acute ischemic response of hippocampal CA1 neurons in brain slices. Brain Res. 1077:187–199.

Liu X, Bandyopadhyay BB, Nakamoto T, Singh BB, Liedtke W, Melvin JE, Ambudkar IS. (2006) A role for AQP5 in activation of TRPV4 by hypotonicity: concerted involvement of AQP5 and TRPV4 in regulation of cell volume recovery. J Biol Chem 281: 15485–15495.

Hernández M, Burillo SL, Crespo MS, Nieto ML. (1998) Secretory phospholipase A2 activates the cascade of mitogen-activated protein kinases and cytosolic phospholipase A2 in the human astrocytoma cell line 1321N1. Biol. Chem 273: 606–612.

Iida T, Shimizu I, Nealen MI, Campbell A, Caterina M. (2005) Attenuated fever response in mice lacking TRPV1. Neuroscience let 378: 28–32.

Maroto R, Raso A, Wood TG, Kurosky A, Martinac B, Hamill OP. (2005) TRPC1 forms the stretch-activated cation channel in vertebrate cells. Nat Cell Biol. 7:179–185.

Martinez CG, Palao CM, Cases RP, Merino JM, Montiel F A. (2000) Identification of an aspartic residue in the P-loop of the vanilloid receptor that modulates pore properties. J Biol Chem 275: 32552–32558.

Mizuno A, Matsumoto N, Imai M, Suzuki M. (2003) Impaired osmotic sensation in mice lacking TRPV4. Am J Physiol Cell Physiol. 285: C96–101.

Montell C. (2001) Physiology, phylogeny, and functions of the TRP superfamily of cation channels. Science's STKE 10.

Montell C, Birnbaumer L, Flockerzi V. (2002a) The TRP channels, a remarkable functional family. Cell 108: 595–598.

Montell C, Birnbaumer L, Flockerzi V, Bindels RJ, Bruford EA, Caterina MJ, Clapham D, Harteneck C, Heller S, Julius D, Kojima I, Mori Y, Penner R, Prawitt D, Scharenberg AM, Schultz G, Shimizu S, Zhu MX. (2002b) A unified nomenclature for the superfamily of TRP cation channels. Mol Cell 9: 229–231.

Moqrich A, Hwang SW, Earley TJ, Petrus MJ, Murray AN, Spencer KS, Andahazy M, Story GM, Patapoutian A. (2005) Impaired thermosensation in mice lacking TRPV3, a heat and camphor sensor in the skin. Science. 307: 1468–1472.

Morris, CE, Sigurdson, WJ. (1989) Stretch-inactivated ion channels co-exist with stretch-activated ion channels. Science 243: 807–809.

Naeini SR, Witty MF, Seguela P, Bourque CW. (2006) An N-terminal variant of Trpv1 channel is required for osmosensory transduction. Nat Neurosci. 9:93–98.

Nilius B, Prenen J, Wissenbach U, Bodding M, Droogmans G. (2001a) Differential activation of the volume-sensitive cation channel TRP12 (OTRPC4) and volume-regulated anion currents in HEK-293 cells. Pflugers Arch 443:227–233.

Nilius B, Vennekens R, Prenen J, Hoenderop JG, Droogmans G, Bindels RJ. (2001b) The single pore residue Asp542 determines Ca2+ permeation and Mg2+ block of the epithelial Ca2+ channel. J Biol Chem 276: 1020–1025.

Nilius B, Watanabe H, Vriens J. (2003) The TRPV4 channel: structure-function relationship and promiscuous gating behaviour. Pflugers Arch. 446:298–303.

Nilius B, Vriens J, Prenen J, Droogmans G, Voets T. (2004) TRPV4 calcium entry channel: a paradigm for gating diversity. Am J Physiol Cell Physiol. 286:C195–205.

O'Neil RG, Heller S (2005) The mechanosensitive nature of TRPV channels. Pflugers Arch. 451:193–203.

Owsianik G, D'hoedt D, Voets T, Nilius B. (2006) Structure-function relationship of the TRP channel superfamily. Rev Physiol Biochem Pharmacol. 156:61–90.

Reiter B, Kraft R, Gunzel D, Zeissig S, Schulzke JD, Fromm M, Harteneck C.(2006) TRPV4-mediated regulation of epithelial permeability. FASEB J. 20:1802–1812.

Schumacher MA, Moff I, Sudanagunta SP, Levine JD. (2000) Molecular Cloning of an N-terminal Splice Variant of the Capsaicin Receptor. J. Biol. Chem 275: 2756 - 2762.

Shen, J., Harada S., N, Kubo N, Liu B, Mizuno A, Suzuki M, Yamashita T. (2006) Functional expression of transient receptor potential vanilloid 4 in the mouse cochlea. Neuroreport 17:135–139.

Sidhaye VK, Guler AD, Schweitzer KS, D'Alessio F, Caterina MJ, King LS. (2006) Transient receptor potential vanilloid 4 regulates aquaporin-5 abundance under hypotonic conditions. Proc Natl Acad Sci U S A. 103:4747–4752.

Smith PL, Maloney KN, Pothen RG, Clardy J, Clapham DE. (2006) Bisandrographolide from Andrographis Paniculata activates TRPV4 channels. J Biol Chem. (in press).

Strotmann R, Harteneck C, Nunnenmacher K, Schultz G, Plant TD. (2000) OTRPC4, a nonselective cation channel that confers sensitivity to extracellular osmolarity. Nat Cell Biol. 2:695–702.

Strotmann R, Schultz G, Plant TD. (2003) Ca2+-dependent potentiation of the nonselective cation channel TRPV4 is mediated by a carboxy terminal calmodulin binding site. J Biol Chem 278: 26541–26549.

Suzuki M, Sato J, Kutsuwada K, Ooki G, Imai M. (1999) Cloning of a stretch-inhibitable nonselective cation channel. J Biol Chem 274:6330–5335.

Suzuki M, Mizuno A, Kodaira K, Imai M. (2003a) Impaired pressure sensation in mice lacking TRPV4. J Biol Chem 278:22664–22668.

Suzuki M, Watanabe Y, Oyama Y, Mizuno A, Kusano E, Hirao A, Ookawara S. (2003b) Localization of mechanosensitive channel TRPV4 in mouse skin. Neurosci Lett. 353:189–192.

Suzuki M, Hirao A, Mizuno A. (2003c) Microtubule-associated [corrected] protein 7 increases the membrane expression of transient receptor potential vanilloid 4 (TRPV4).J Biol Chem 278:51448–51453.

Tabuchi K, Suzuki M, Mizuno A, Hara A. (2005) Hearing impairment in TRPV4 knockout mice. Neurosci Lett. 382:304–308.

Takumida M, Kubo N, Ohtani M, Suzuka Y, Anniko M. (2005) Transient receptor potential channels in the inner ear: presence of transient receptor potential channel subfamily 1 and 4 in the guinea pig inner ear. Acta Otolaryngol. 125:929–934.

Taniguchi J, Tsuruoka S, Mizuno A, Sato J, Fujimura A, Suzuki M. (2006) TRPV4 as a flow sensor in flow-dependent K^+ secretion from the cortical collecting duct. Am J Physiol Renal Physiol. (in press).

Teilmann SC, Byskov AG, Pedersen PA, Wheatley DN, Pazour GJ, Christensen ST. (2005) Localization of transient receptor potential ion channels in primary and motile cilia of the female murine reproductive organs. Mol Reprod Dev. 71:444–452.

Tian W, Salanova M, Xu H, Lindsley JN, Oyama TT, Anderson S, Bachmann S, Cohen DM. (2004) Renal expression of osmotically responsive cation channel TRPV4 is restricted to water-impermeant nephron segments.Am J Physiol Renal Physiol. 287: F17–24.

Todaka H, Taniguchi J, Satoh J, Mizuno A, Suzuki M. (2004) Warm temperature-sensitive transient receptor potential vanilloid 4 (TRPV4) plays an essential role in thermal hyperalgesia. J Biol Chem 279: 35133–35138.

Tsushima H, Mori M. (2006) Antidipsogenic effects of a TRPV4 agonist, 4{alpha}-phorbol 12, 13-didecanoate, injected into the cerebroventricle. Am J Physiol Regul Integr Comp Physiol. 290:R1736–1741.

Voets T, Prenen J, Vriens J, Watanabe H, Janssens A, Wissenbach U, Bodding M, Droogmans G, Nilius B. (2002) Molecular determinants of permeation through the cation channel TRPV4. J Biol Chem 277:33704–33710.

Voets T, Talavera K, Owsianik G, Nilius B. (2005) Sensing with TRP channels. Nat Chem Biol.1: 85–92.

Vriens J, Watanabe H, Janssens A, Droogmans G, Voets T, Nilius B. (2003) Cell swelling, heat, and chemical agonists use distinct pathways for the activation of the cation channel TRPV4. Proc Natl Acad Sci U S A. 101:396–401.

Vriens J, Janssens A, Prenen J, Nilius B, Wondergem R. (2004) TRPV channels and modulation by hepatocyte growth factor/scatter factor in human hepatoblastoma (HepG2) cells. Cell Calcium. 36:19–28.

Vriens J, Owsianik G, Fisslthaler B, Suzuki M, Janssens A, Voets T, Morisseau C, Hammock BD, Fleming I, Busse R, Nilius B. (2005) Modulation of the Ca2+ permeable cation channel TRPV4 by cytochrome P450 epoxygenases in vascular endothelium Circ Res. 28;:908–915.

Watanabe H, Davis JB, Smart D, Jerman JC, Smith GD, Hayes P, Vriens J, Cairns W, Wissenbach U, Prenen J, Flockerzi V, Droogmans G, Benham CD, Nilius B (2002a) Activation of TRPV4 channels (hVRL-2/mTRP12) by phorbol derivatives. J Biol Chem 277:13569–13577.

Watanabe H, Vriens J, Suh SH, Benham CD, Droogmans G, Nilius B. (2002b) Heat-evoked activation of TRPV4 channels in a HEK293 cell expression system and in native mouse aorta endothelial cells. J Biol Chem. 277: 47044–47051.

Watanabe H, Vriens J, Janssens A, Wondergem R, Droogmans G, Nilius B. (2003a) Modulation of TRPV4 gating by intra- and extracellular Ca2+. Cell Calcium. 33:489–495.

Watanabe H, Vriens J, Prenen J, Droogmans G, Voets T, Nilius B. (2003b) Anandamide and arachidonic acid use epoxyeicosatrienoic acids to activate TRPV4 channels. Nature 424: 434–438.

Wechselberger M, Wright CL, Bishop GA, Boulant JA (2006) Ionic channels and conductance-based models for hypothalamic neuronal thermosensitivity. Am J Physiol Regul Integr Comp Physiol 291: R518–29.

Wilson FH, Disse-Nicodeme S, Choate KA, Ishikawa K, Nelson-Williams C, Desitter I, Gunel M, Milford DV, Lipkin GW, Achard JM, Feely MP, Dussol B, Berland Y, Unwin RJ, Mayan H, Simon DB, Farfel Z, Jeunemaitre X, Lifton RP. (2001) Human hypertension caused by mutations in WNK kinases. Science 293: 1107–1112.

Wissenbach U, Bodding M, Freichel M, Flockerzi V. (2000) Trp12, a novel Trp related protein from kidney. FEBS Lett 485: 127–134.

Xu F, Satoh E, Iijima T. (2003) Protein kinase C-mediated Ca2+ entry in HEK 293 cells transiently expressing human TRPV4.Br J Pharmacol. 140:413–421.

Xu H, Zhao H, Tian W, Yoshida K, Roullet JB, Cohen DM (2003) Regulation of a transient receptor potential (TRP) channel by tyrosine phosphorylation. SRC family kinase-dependent tyrosine phosphorylation of TRPV4 on TYR-253 mediates its response to hypotonic stressJ Biol Chem 278:11520–11527.

Xu H, Fu Y, Tian W, Cohen DM. (2006) Glycosylation of the osmoresponsive transient receptor potential channel TRPV4 on Asn-651 influences membrane trafficking. Am J Physiol Renal Physiol. 2290: F1103–9.

Yao X, Garland CJ (2005) Recent developments in vascular endothelial cell transient receptor potential channels. Circ Res 97:853–863.

Yang XR, Lin MJ, McIntosh LS, Sham JS. (2006) Functional expression of transient receptor potential melastatin- and vanilloid-related channels in pulmonary arterial and aortic smooth muscle. Am J Physiol Lung Cell Mol Physiol 290:L1267–1276.

Zhang L, Jones S, Brody K, Costa M, Brookes SJ. (2004) Thermosensitive transient receptor potential channels in vagal afferent neurons of the mouse. Am J Physiol Gastrointest Liver Physiol. 286:G983–991.

Part II
Mechanosensitive Signalling Cascades

Chapter 10
Mechanosensitive Purinergic Calcium Signalling in Articular Chondrocytes

Belinda Pingguan-Murphy and Martin M. Knight

Abstract Mechanotransduction is the mechanism through which living cells sense and respond to mechanical stimuli. In many tissues, including muscle, blood vessel, bone, ligament and cartilage, mechanotransduction is essential for tissue health and function. Although mechanotransduction mechanisms have been identified in a number of tissues, the signalling pathways have not yet been elucidated in articular cartilage. This chapter explores the influence of physiological mechanical stimuli on intracellular Ca^{2+} signalling in articular chondrocytes and its potential involvement in cartilage mechanotransduction. The review focuses primarily on Ca^{2+} signalling activated by deformation of isolated primary articular chondrocytes cultured within the well-established agarose model system. The chapter discusses the involvement of a mechanosensitive purinergic pathway which trigger Ca^{2+} signalling through the release of ATP and the activation of purine receptors. The influence of different loading parameters on the Ca^{2+} signalling characteristics is discussed as a putative mechanism through which cells differentiate between different loading conditions. Finally the chapter examines the downstream cellular response to mechanically-activated purinergic Ca^{2+} signalling and its importance in tissue health and homeostasis.

Key words: Cartilage · Chondrocyte · Mechanotransduction · Calcium · ATP · Purine receptors

10.1 Introduction

Articular cartilage is the specialised connective tissue that covers the articulating surfaces in synovial joints. Its function is to provide a low friction, low wear bearing surface, whilst reducing the contact stresses to the underlying bone. This mechanical functionality is provided by the extracellular matrix, which makes up at least 90% of the tissue volume. The extracellular matrix is composed of a hydrated proteoglycan gel, whose swelling is restrained by a dense meshwork of collagen fibres arranged in characteristic 'Benninghoff arcades' of parallel sheets (Benninghoff, 1925; Boyde and Jones, 1983). The chondrocytes, which constitute the only cell type in articular cartilage, are responsible for the maintenance of the extracellular matrix.

A. Kamkin and I. Kiseleva (eds.), *Mechanosensitive Ion Channels.*

During normal activity, cartilage is subjected to dynamic loading which is essential for the development, health and homeostasis of the tissue (for review see Grodzinsky et al, 2000). *In vivo* studies on canine knee cartilage have shown that a period of immobilisation produced a reduction in proteoglycan content in the extracellular matrix (Kiviranta et al, 1987; Behrens et al, 1989; Palmoski and Brandt, 1984). Conversely, the increased weight-bearing through the contralateral joint resulted in an increased proteoglycan content (Behrens et al, 1989). In other studies it has been shown that areas of a joint which are exposed to high levels of stress, have a higher proteoglycan content than those normally exposed to lower stresses (Kiviranta et al, 1987; Kempson et al, 1971). These results are consistent with the well established concept of functional adaptation of cartilage to its mechanical environment.

Numerous *in vitro* studies, using cartilage explants, have demonstrated that mechanical loading, particularly compression, affects chondrocyte metabolism and the synthesis and degradation of extracellular matrix molecules. While static compression leads to matrix catabolism, dynamic compression generally up-regulates extracellular matrix synthesis in a complex dose dependent manner (Sah et al, 1989; Buschmann et al, 1999; Valhmu et al, 1998; Chowdhury et al, 2003; Lee and Bader, 1997; Fitzgerald et al, 2006). Thus it has been proposed that dynamic mechanical conditioning may be utilised *in vitro* to stimulate the production of mechanically functional tissue engineered cartilage for the repair of diseased or damaged joints (Hung et al, 2004; Mauck et al, 2000; Seidel et al, 2004; Li et al, 2004). Alternatively pharmaceutical manipulation of mechanotransduction may provide a mechanism for stimulating cartilage repair *in vivo*. However, the mechanotransduction pathways through which chondrocytes sense and respond to mechanical loading are, as yet,

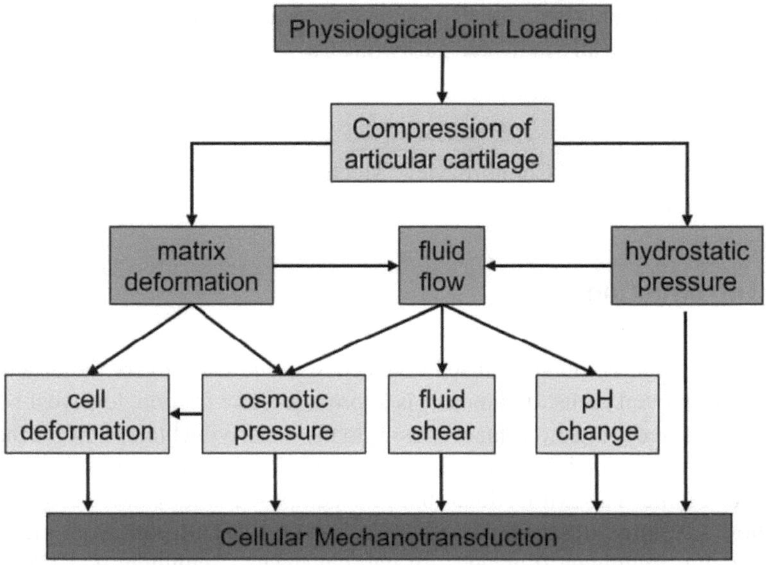

Fig. 10.1 Schemmatic flow chart indicating the putative primary cellular mechanotransduction events associated with physiological compression of articular cartilage

unclear. In order to fully optimise the therapeutic potential of mechanotransduction, it will be necessary to elucidate the fundamental signalling pathways.

Physiological compressive loading of articular cartilage induces a range of putative extracellular signalling events including cell deformation, increased hydrostatic pressure, fluid flow, electrical streaming potentials and changes in pH and osmolarity as summarised in Fig. 10.1 (for review see Urban, 1994). In particular, cell deformation has been widely implicated in mechanotransduction for chondrocytes, both in intact cartilage (Guilak et al, 1995) and in isolated cell culture models (Mauck et al, 2000; Buschmann et al, 1995; Lee et al, 2000). However, the subsequent intracellular mechanotransduction signalling pathways by which chondrocytes sense and respond to mechanical deformation, remain unclear. Never the less, an increasing number of studies now suggest that intracellular Ca^{2+} signalling may be involved in chondrocyte mechanotransduction.

10.2 Experimental Models for Investigating Cartilage Mechanotransduction

A variety *in vitro* experimental model systems have been employed in the study of chondrocyte mechanotransduction. These include cartilage explants, with and without subchondral bone, and various different isolated cell culture models as summarised in Table 10.1. This chapter concentrates on the authors previous studies examining Ca^{2+} signalling associated with compression of chondrocytes cultured within 3D agarose constructs. Numerous previous studies have employed this well-established agarose model system to investigate chondrocyte mechanotransduction. One advantage of this system, is that it enables the role of cell deformation in mechanotransduction to be examined in isolation from factors associated with compression of the charged extracellular matrix (Fig. 10.1). Furthermore, it also provides greater understanding of mechanotransduction events within cell seeded scaffold such as those proposed for cartilage tissue engineering. The chondrocyte-agarose model typically uses chondrocytes from the articular cartilage of the bovine metacarpal phalangeal joint (Fig. 10.2) although other cell types have also been used including porcine, equine and murine chondrocytes as well as mesenchymal stem cells. The primary chondrocytes are isolated from their extracellular matrix by sequential enzyme digestion and then seeded in molten agarose which is subsequently gelled in specially prepared moulds (Lee and Knight, 2004). In some cases, the agarose may be gelled between porous glass endplates (Fig. 10.2d) to enable the construct to be gripped securely for the application of cyclic compression (Pingguan-Murphy et al, 2005). Within the agarose gel, the chondrocytes adopt and maintain a rounded morphology associated with the maintenance of chondrocytic phenotype (Knight et al, 1996; Pingguan-Murphy et al, 2005). This phenotypic stability is demonstrated by the synthesis of type II collagen and the proteoglycan, aggrecan (Aydelotte and Keuttner, 1988; Hauselmann et al, 1994). At early time points, less than 72 hours post isolation, gross compression of the chondrocyte-agarose constructs results in deformation of the cells from a spherical to an oblate

Table 10.1 Summary of previous in vitro model system used for investigating the role of Ca^{2+} signalling in chondrocyte mechanotransduction

Chondrocyte Model System	Mechanical Loading	Reference
Cartilage Explant	Shear	(Fitzgerald et al, 2006)
	Hydrostatic pressure	(Bourret and Rodan, 1976)
	Compression	(Fitzgerald et al, 2004; Trickey et al, 2004; Fitzgerald et al, 2006)
Isolated Chondrocytes:		
Suspension	Hydrostatic Pressure	(Browning et al, 2004)
	Micropipette aspiration	(Ohashi et al, 2006)
Monolayer culture	Hydrostatic Pressure	(Mizuno, 2005; Zhang et al, 2005)
	Fluid shear	(Edlich et al, 2001; Yellowley et al, 1999; D'Andrea et al, 2000; Edlich et al, 2004)
	Substrate deformation	(Wright et al, 1996; Tanaka et al, 2005)
	Micropipette indentation	(Guilak et al, 1999; Kono et al, 2006)
Culture in 3D hydrogels (e.g. agarose)	Hydrostatic pressure	(Toyoda et al, 2003; Mizuno et al, 2002)
	Compression	(Pingguan-Murphy et al, 2005; Pingguan-Murphy et al, 2006; Roberts et al, 2001)

ellipsoid morphology (Lee et al, 2000). However, at later time points the presence of a freshly elaborated and relatively stiff pericellular matrix prevents cell deformation during compression (Knight et al, 1998). This ability to apply physiologically relevant levels of cell deformation in the absence of pericellular matrix, coupled with the translucent nature of the agarose gel, enables studies to visualise and quantify mechanically induced intracellular Ca^{2+} signalling using fluorescent calcium indicators and live cell microscopy (Pingguan-Murphy et al, 2005; Roberts et al, 2001).

10.3 Ca^{2+} Signalling in Unstrained Chondrocytes

Intracellular (Ca^{2+}) is a ubiquitous second messenger which controls many aspects of cell function including metabolism, proliferation, gene transcription and the response to external stimuli including mechanical deformation (Berridge et al, 1998; Berridge et al, 2000). The ability of Ca^{2+} ions to control such a diverse range of cellular processes arises from the fact that the associated signalling pathways are able to produce enormous versatility in Ca^{2+} signalling characteristics in space, time and amplitude (see (Bootman et al, 2001) for review). Transient increases in intracellular Ca^{2+} concentration are produced by the activation of Ca^{2+} channels in the cell membrane and/or release of Ca^{2+} from intracellular stores. Local changes in Ca^{2+} concentration may propagate throughout the cytoplasm triggering global rises in intracellular Ca^{2+}. The resulting Ca^{2+} signals are translated into specific

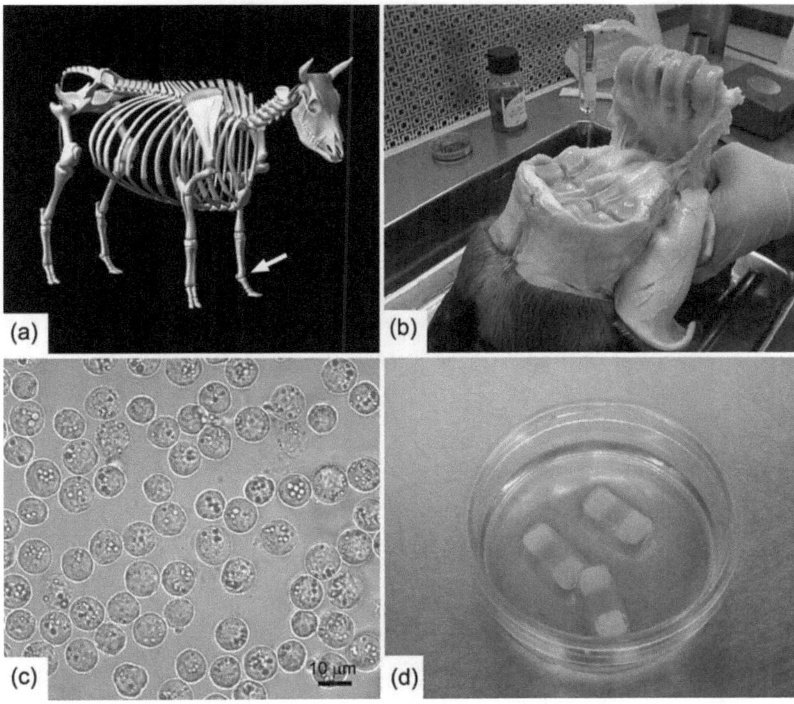

Fig. 10.2 The well-established chondrocyte-agarose model typically involves the removal of articular cartilage from the proximal surfaces of the bovine metacarpal phalangeal joint (a) and (b). The tissue is then minced and digested in pronase and collagenase to release a population of viable chondrocytes (c), which are then seeded into molten agarose gel. The cell-agarose suspension may be gelled between porous glass endplates (d). This enables the constructs to be securely gripped so that cyclic compression can be applied without damage

cellular responses through various Ca^{2+}-binding proteins that undergo conformational changes to activate downstream effectors, as well as other signalling pathways such as those involving cyclicAMP and MAP kinases.

Articular chondrocytes in unloaded cartilage explants exhibit spontaneous global Ca^{2+} transients (Kono et al, 2006). Similar Ca^{2+} transients have also been observed in isolated chondrocytes cultured in monolayer (Kono et al, 2006) and agarose constructs (Pingguan-Murphy et al, 2005; Pingguan-Murphy et al, 2006; Roberts et al, 2001). An example of these global Ca^{2+} transients is illustrated in Fig. 10.3 for isolated chondrocytes visualised in an unstrained agarose construct using confocal microscopy and the Ca^{2+} indicator, Fluo-4 AM. The Ca^{2+} transients have a characteristic shape with a rise time (T_R) which is shorter than the fall time (T_F) such that the rise time represents approximately 40% of the total transient time (Pingguan-Murphy et al, 2006). The percentage of cells exhibiting these Ca^{2+} transients and the frequency at which they occur are increased by the presence of serum. This is probably due to the action of growth factors contained within the serum, although only a sub-population of approximately 50% of bovine articular chondrocytes is responsive in this manner (Pingguan-Murphy et al, 2005).

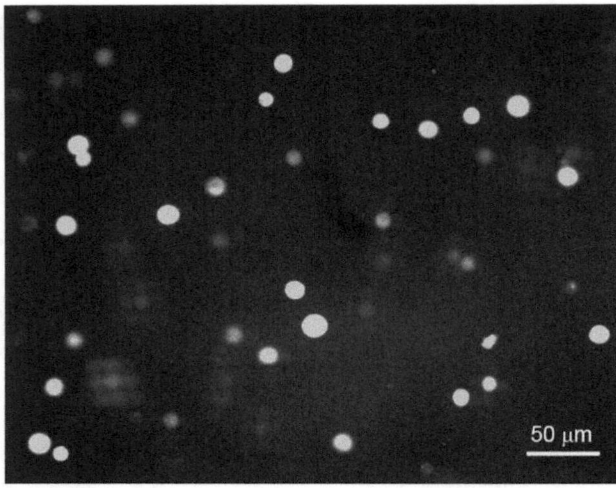

(a)

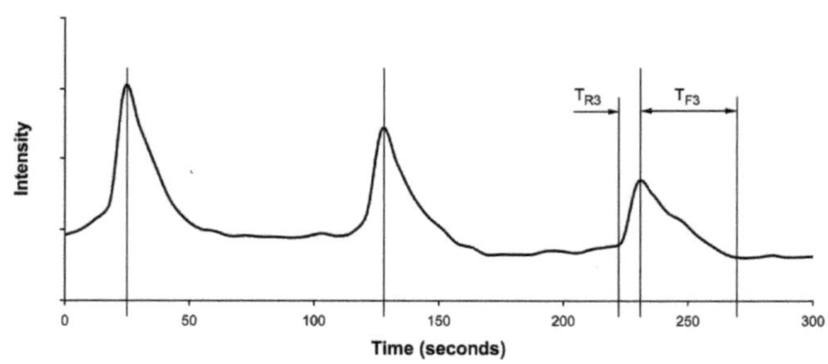

(b)

Fig. 10.3 Typical image from a confocal time-lapse series, showing isolated chondrocytes labelled with the Ca^{2+} indicator, Fluo-4, within an unstrained agarose construct (a). The cells were imaged every 4 seconds over a 5 minute period and the temporal changes in Ca^{2+} concentration for individual cells examined by quantifying the change in fluorescent intensity. A sub-population of cells exhibit characteristic Ca^{2+} transients (b) from which the transient rise and fall times could be calculated (T_R and T_F respectively). Adapted from(Pingguan-Murphy et al, 2005)

10.4 Mechanosensitive Ca^{2+} Signalling in Chondrocytes

Previous studies using articular chondrocytes, have reported activation of global Ca^{2+} signalling by mechanical stimuli in the form of fluid shear (Edlich et al, 2004; Edlich et al, 2001), micropipette indentation (Guilak et al, 1999; D'Andrea et al, 2000; Kono et al, 2006) and aspiration (Ohashi et al, 2006), hydrostatic pressure (Mizuno, 2005), osmotic challenge (Wilkins et al, 2003; Erickson et al, 2001) and uniaxial compression in agarose (Roberts et al, 2001; Pingguan-Murphy et al, 2005) (see Table 10.1 for summary). For chondrocytes in agarose constructs, on which this chapter is focussed, the percentage of cells exhibiting global Ca^{2+} transients is enhanced in the 5 minute period following 10 cycles of 0–10% compression at 1Hz and 100%.s^{-1} strain rate (Pingguan-Murphy et al, 2005). The up-regulation is maintained for several minutes before returning to levels observed in unstrained cells. This maintenance of signalling after the removal of the mechanical stimuli suggests the involvement of a purinergic signalling pathway such as that described for other cell types mediated by the mechanosensitive release of adenosine 5'-triphosphate (ATP) (for review see Bodin and Burnstock, 2001; Pingguan-Murphy et al, 2005). Chondrocytes are capable of releasing ATP into the extracellular milieu at sufficient local concentrations to activate downstream signalling pathways (Hatori et al, 1995). In deed, previous studies have reported that chondrocytes release ATP in response to mechanical stimuli in the form of compression (Graff et al, 2000) or tensile stretch (Millward-Sadler et al, 2004). Studies suggest that the ATP release and associated mechanotransduction requires the involvement of $\alpha5\beta1$ integrins and an intact actin cytoskeleton (Millward-Sadler and Salter, 2004; Millward-Sadler et al, 2004). However, the precise mechanism of ATP release in chondrocytes has not been established. In fact, in all cell types there is currently an active debate about the physiological transport mechanisms that facilitate ATP release of which there are three putative mechanisms: (i) anion channels; (ii) connexons or hemichannels; and (iii) exocytosis of ATP-filled vesicles.

10.5 Activation of Purinergic Ca^{2+} Signalling

Once released from the cell, ATP and its associated breakdown products, adenosine diphosphate (ADP) and adenosine monophosphate (AMP), activate specific purine receptors on the cell membrane. Purine receptors can be classified into two groups, P1 (adenosine receptors) and P2 (ATP receptors). The presence of P2 receptors on chondrocytes was first reported by Russell and colleagues (Caswell et al, 1991; Leong et al, 1994). Koolpe *et al.*, have subsequently reported the expression of both P1 and P2 purine receptor genes in human chondrocytes (Koolpe et al, 1999). The latter is divided in two major superfamilies: the intrinsic ATP-gated ion channel type receptor, P2X, and the G protein-coupled receptor, P2Y. Both receptor types can be further divided into subtypes dependent on molecular structure and pharmacological properties: $P2X_1$-$P2X_7$ and $P2Y_{1,2,4,6}$ and $_{11}$ (Ralevic and Burnstock, 1998; Surprenant et al, 1996). Immunohistochemistry has been used to demonstrated that chondrocytes express $P2X_2$ and $P2X_5$ receptors, while *in situ* hybridisation has identified the presence of $P2Y_1$ and $P2Y_2$ receptors (Hoebertz et al, 2000).

In chondrocytes, as in other cell types, it is well established that extracellular ATP activates purine receptors triggering global Ca^{2+} transients, thereby confirming the functional presence of P2 receptors on the chondrocyte membrane (Kono et al, 2006; Koolpe and Benton, 1997; Elfervig et al, 2001; Caswell et al, 1991). However, the characteristics of the Ca^{2+} response are influenced by the characteristics of the specific receptors involved. Activation of P2Y receptors triggers Ca^{2+} mobilisation through the classical IP3 mediated release of intracellular Ca^{2+} stores. Alternatively, activation of P2X channel-type receptors may facilitate extracellular Ca^{2+} entry, either directly or via Na^+ influx triggering membrane depolarisation and activation of voltage operated Ca^{2+} channels (VOCC).

Studies by Millward-Sadler *et al.* report the presence of mRNA for P2Y2 purine receptors in both normal and osteoarthritic human chondrocytes (Millward-Sadler et al, 2004). However further studies by the same authors appear to implicate the P2X pathway, in that deformation of chondrocytes in monolayer activates membrane depolarisation and the opening of VOCC (Millward-Sadler et al, 2004). It should be noted that the P2X receptor shows remarkable similarity to the epithelial sodium channel (ENaC) which has also been implicated in chondrocyte mechanotransduction (Mobasheri and Martin-Vasallo, 1999). However for chondrocytes in agarose, cyclic compression appears to activate Ca^{2+} signalling through both the P2Y and P2X pathways (Pingguan-Murphy et al, 2006).

Interestingly, the up-regulation of Ca^{2+} signalling following cyclic compression in agarose is not present in the entire cell population despite the fact that the homogenous nature of the agarose constructs ensures that all cells received identical mechanical stimuli. This heterogeneity in purinergic signalling may be caused by differences in the expression of purine receptors, possibly associated with the original position of the cells within the cartilage. Indeed, studies have reported that chondrocyte isolated separately from surface and deep zones of articular cartilage exhibit different Ca^{2+} signalling in response to hydrostatic pressure (Mizuno, 2005). However, the authors have failed to detect any differences in response to static compression in agarose. An alternatively explanation is that the heterogeneity in mechanotransduction response may reflect heterogeneity in the intracellular mechanical environment, as demonstrated in a recent study of chondrocyte deformation in agarose (Knight et al, 2006).

Despite the expression of purine receptors by articular chondrocytes and their functional characterisation, their involvement in mechanically induced Ca^{2+} signalling has not been extensively studied. To further investigate this purinergic pathway, previous studies have employed various inhibitors. For example, Ca^{2+} signalling associated with micropipette indentation (Kono et al, 2006) or compression in agarose (Pingguan-Murphy et al, 2006) can be completely blocked by pre-treatment with apyrase, which catalyses the hydrolysis of ATP. Similarly treatment with P2 receptor blockers, suramin and basilen blue, significantly reduced the percentage of cells exhibiting Ca^{2+} transients following cyclic compression, compared to untreated cells (Pingguan-Murphy et al, 2006). Furthermore, treatment with gadolinium, an inhibitor of mechanosensitive ATP release (Boudreault and Grygorczyk, 2002), prevented mechanically induced Ca^{2+} signalling in chondrocytes subjected to compression in agarose (Pingguan-Murphy et al, 2006; Roberts et al, 2001), micropipette indentation (Guilak et al, 1999), hydrostatic pressure (Mizuno, 2005) or fluid shear (Yellowley et al, 2002). However, gadolinium is also a putative inhibitor of stretch-activated Ca^{2+} channels and hence this alternative Ca^{2+} pathway may also be involved. Never-the-less, taken together, these results strongly support the hypothesis that mechanical deformation activates chondrocytes Ca^{2+} signalling through a purinergic pathway involving the mechanosensitive release of ATP, as illustrated schematically in Fig. 10.4

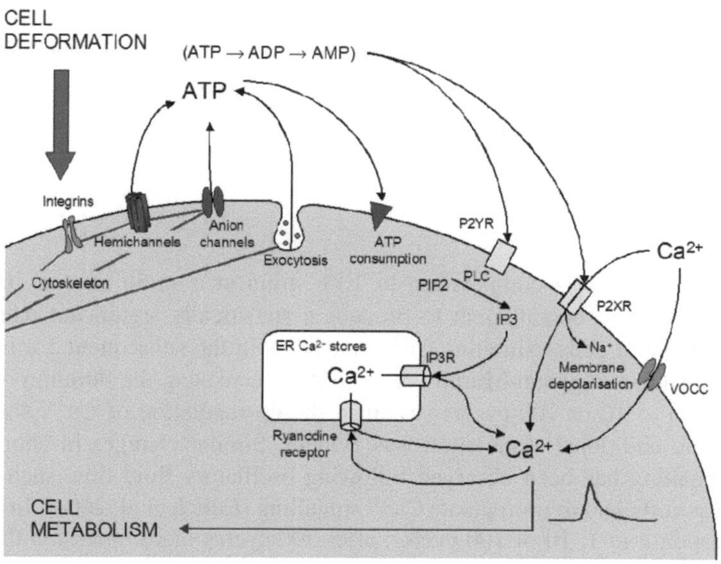

Fig. 10.4 Schemmatic diagram showing the putative mechanosensitive purinergic pathways leading to activation of Ca^{2+} transients within articular chondrocytes. Three possible ATP release mechanisms are illustrated, namely: hemichannels, anion channels and exocytosis. Once release into the extracellular milieu, ATP and its breakdown products may activate P2 receptors thereby initiating intracellular Ca^{2+} signalling. Adapted from (Pingguan-Murphy et al, 2006)

10.6 Mechanical Modulation of Chondrocyte Ca^{2+} Signalling

It is well known that the chondrocyte response to mechanical loading is sensitive to the precise nature of the loading, such as the magnitude, rate, frequency and duty cycle of the applied strain (Fitzgerald et al, 2006; Chowdhury et al, 2003). It is also established that the characteristics of Ca^{2+} signalling and, in particular, the amplitude and frequency of the Ca^{2+} transients, enables cells to distinguish between different stimuli (Berridge et al, 2000). However it is unclear whether this Ca^{2+} signalling modulation enables chondrocytes to differentiate between different loading regimes. Previous studies by the authors have applied a range of loading conditions to chondrocytes in agarose constructs in order to examine whether mechanical modulation of cell function is associated with differences in Ca^{2+} signalling. In particular, studies have quantified the influence of compression on the following:

1) The percentage of cells exhibiting Ca^{2+} transients, so called 'population modulation'.
2) The rise and fall times of these transients (T_R and T_F, Fig. 10.3), termed 'Ca^{2+} transient shape modulation'.
3) The frequency at which Ca^{2+} transients occur in a given cell, know as 'frequency modulation'.

Unfortunately previous studies have often used non ratiometric Ca^{2+} indicators, such as Fluo-4, which has prevented the analysis of Ca^{2+} transient 'amplitude modulation'.

10.6.1 Number of Cycles

A single cycle of 1Hz compression to 10% strain at a strain rate of 100%.s^{-1} has been shown to be sufficient to produce a statistically significant increase in the percentage of cells exhibiting Ca^{2+} transients in the subsequent 5 minute period (Fig. 10.5) (Pingguan-Murphy et al, 2005). Increasing the duration of cyclic compression to 10 or 100 cycles sustained the up-regulation of Ca^{2+} signalling, although no additional stimulation was evident. Similar changes in chondrocyte Ca^{2+} signalling has been observed following oscillatory fluid flow such that 11 cycles was sufficient to up-regulate Ca^{2+} signalling (Edlich et al, 2004). In contrast to the response to 1, 10 or 100 cycles, after 300 cycles of compression there was no statistically significant change in Ca^{2+} signalling compared to unloaded controls (Pingguan-Murphy et al, 2005). This lack of response following 300 cycles was due to the increased time from the start of compression. This can be demonstrated by plotting the percentage of cells exhibiting Ca^{2+} transients in one minute intervals against the time following the start of compression (Fig. 10.5). It can be seem that

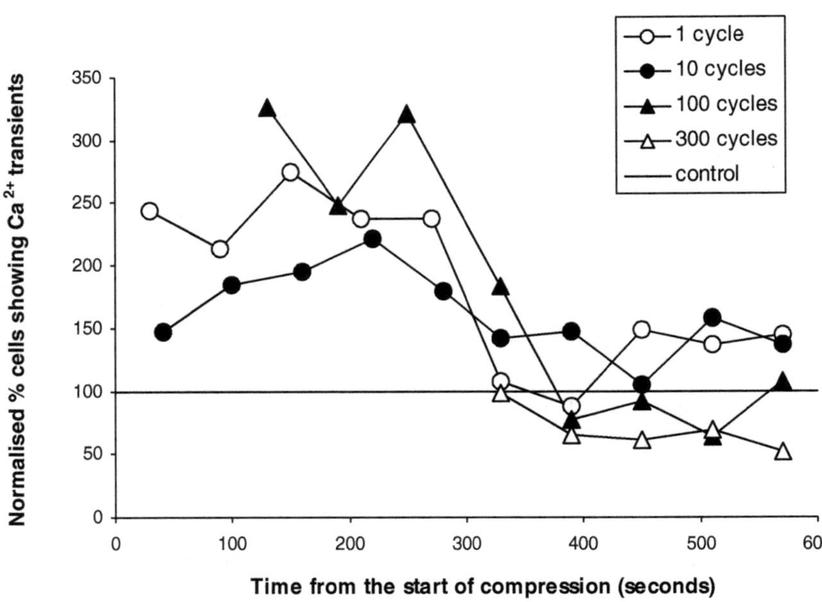

Fig. 10.5 Influence of time following the start of cyclic compression on the percentage of cells exhibiting Ca^{2+} transients. Chondrocyte-agarose constructs were subjected to cyclic compression between 0% and 10% strain at 1 Hz and 100%.s^{-1} strain rate for 1, 10, 100 or 300 cycles prior to Ca^{2+} imaging. Adapted from (Pingguan-Murphy et al, 2005)

the temporal change in the percentage of cells responding follows a similar trend for 1, 10, 100 or 300 cycles. Therefore, Ca^{2+}signalling is activated by the first cycle of compression with subsequent cycles have little additional affect indicating a redundancy of mechanical stimuli. For example, the last cycle of a 300 cycle loading regime does not have the same stimulatory effect as the first cycle although the levels of cell deformation are identical. This may be associated with desensitization of the purine receptors or fatigue of the ATP release channels (Ralevic and Burnstock, 1998). Alternatively it may be associated with diffusion or degradation of extracellular ATP. Further studies are required to determine the exact mechanism and the length of the refractory period, before cells will again respond to cycles of compression. This may be useful in determining an optimum duty cycle for intermittent compressive loading regimes used in cartilage tissue engineering bioreactors (Hung et al, 2004; Mauck et al, 2000; Seidel et al, 2004; Li et al, 2004).

In addition to population modulation, cyclic compression in agarose for 1, 10 or 100 cycles, not only increased the percentage of cells exhibiting Ca^{2+} transients, but also increased the transient frequency. This was reflected by a statistically significant increase in the number of transients per responding cell, compared to unstrained controls (Pingguan-Murphy et al, 2005). Furthermore, 0–10% cyclic compression of chondrocytes in agarose at 1Hz and a strain rate of $100\%.s^{-1}$, also altered the shape of the Ca^{2+} transients as indicated by statistically significant reductions in the rise and fall times compared to unstrained controls (Pingguan-Murphy et al, 2006).

10.6.2 Strain Rate and Frequency

Variation in the frequency of applied gross compression have little effect on Ca^{2+} signalling, providing the strain rate is kept constant by the use of a trapezoidal waveform (Pingguan-Murphy et al, 2006). Compression at both 0.3 Hz and 1 Hz between 0% and 10% strain, induced a similar population modulation in the subsequent 5 minute period (Fig. 10.6). However, there were differences in the precise nature of the response between the two frequencies. At 1 Hz, the up-regulation of Ca^{2+} signalling following 10 cycles of compression was maintained throughout the subsequent 5 minute imaging period (Fig. 10.5). By contrast, at 0.3 Hz there was a large increase in Ca^{2+} transients in the first minute post loading, but thereafter the level of Ca^{2+} signalling quickly reduced to that in unloaded controls.

The authors have also examined the influence of strain rate on Ca^{2+} signalling characteristics (Pingguan-Murphy et al, 2006). After 10 cycles of compression between 0% and 10% strain at a frequency of 1 Hz, Ca^{2+} signalling population modulation occurred at a strain rate of $100\ \%.s^{-1}$, but not at slower strain rates of 25 or $50\ \%.s^{-1}$. The mechanism underlying this strain rate-dependent behaviour is unclear, although previous studies have reported similar findings in a variety of other cell types (McKnight and Frangos, 2003). One possibility is that changes in strain rate are likely to be associated with changes in fluid flow which influences chondrocyte Ca^{2+} signalling (Edlich et al, 2004) and metabolic response to cyclic compression in agarose constructs (Buschmann et al, 1999). However, the activation of

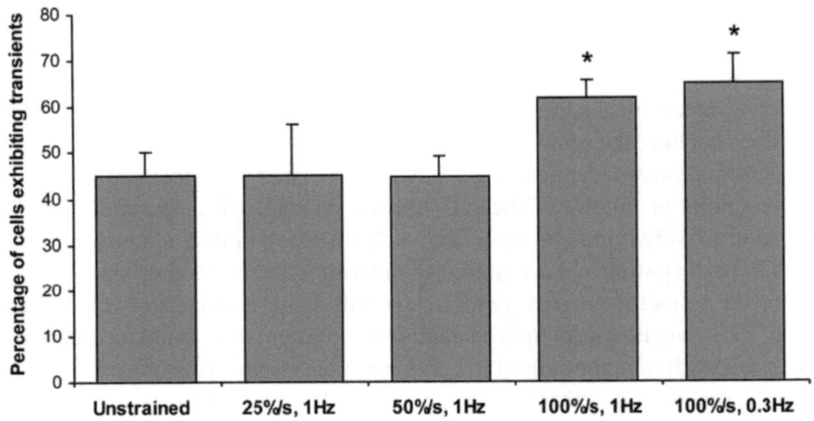

Fig. 10.6 Influence of strain rate and frequency on the percentage of cells exhibiting Ca^{2+} transients in chondrocyte-agarose constructs subjected to 10 cycles of compression between 0% and 10% strain. Statistically significant differences are indicated from the unstrained level based on a Student t test ($p < 0.05$ *). Adapted from (Pingguan-Murphy et al, 2006)

chondrocyte Ca^{2+} signalling by fluid flow does not occur via the same purinergic pathway as that activated by cyclic compression (Yellowley et al, 1999).

Cyclic compression at 1 Hz also produced Ca^{2+} transient shape modulation at strain rates of 25 $\%.s^{-1}$ and 50 $\%.s^{-1}$ but not at a strain rate of 100 $\%.s^{-1}$ (Pingguan-Murphy et al, 2006). However, at a strain rate of 100 $\%.s^{-1}$, shape modulation did occur at the slower, 0.3 Hz frequency. The mechanisms responsible for this complex strain rate and frequency dependent Ca^{2+} signalling modulation are, as yet, unclear.

10.7 Downstream Mechanotransduction Triggered by Purinergic Ca^{2+} Signalling

Mechanically induced intracellular Ca^{2+} signalling may trigger downstream changes in chondrocyte metabolism as part of a chondrocyte mechanotransduction pathway. In a recent study by Chowdhury and Knight, the chondrocyte metabolic response to 24 hours of cyclic compression was examined in the presence of apyrase, gadolinium or suramin, all of which inhibit compression induced purinergic Ca^{2+} signalling (Chowdhury and Knight, 2006). In the absence of inhibitors, 1 Hz cyclic compression at 0–15% sinusoidal strain, stimulated proteoglycan synthesis, reduced nitric oxide production and up-regulated cell proliferation. However, all three inhibiters prevented the compression-induced up-regulation of proteoglycan synthesis (Fig. 10.7). Thus it can be concluded that the purinergic Ca^{2+} signalling pathway summarised in Fig. 10.4 is part of the chondrocyte mechanotransduction pathway controlling proteoglycan synthesis.

Furthermore, inhibition of the pathway also caused a compression induced inflammatory response characterised by a large increase in nitric oxide levels and associated reduction in cell proliferation (Fig. 10.7). Previous studies have reported

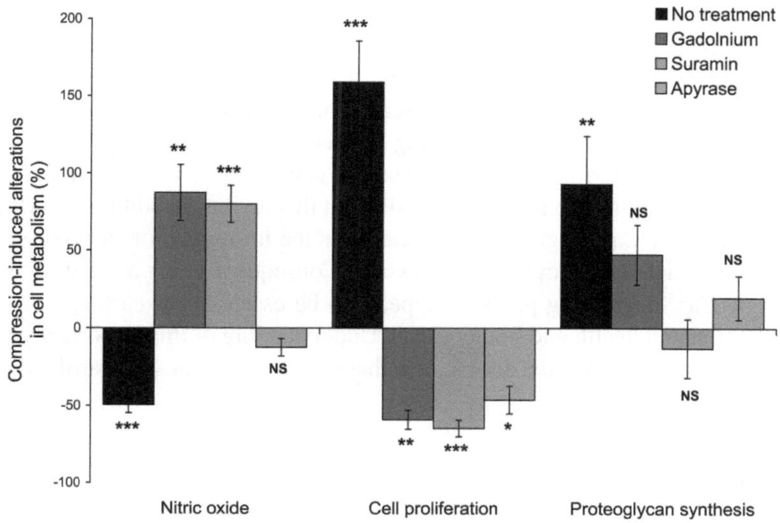

Fig. 10.7 Influence of purinergic Ca^{2+} signalling pathway on mechanically induced alterations in chondrocyte proliferation, protepglycan synthesis and nitric oxide production. Statistical significant differences between loaded and corresponding unloaded cells are indicated based on a Student t test (* $p<0.05$, ** $p<0.01$, *** $p<0.001$). Adapted from (Chowdhury and Knight, 2006)

deformation induced alterations in the chondrocyte actin cytoskeleton, possibly involving a Ca^{2+}-dependent mechanism (Knight et al, 2005; Erickson et al, 2003; Chao et al, 2006). The actin cytoskeleton provides mechanical integrity to the cell and in particular the cell cortex (Ohashi et al, 2006; Trickey et al, 2004). Consequently, it is possible that mechanical stimulation of the purinergic Ca^{2+} pathway causes actin remodelling in order to soften the cell and facilitate cell deformation without cell damage and an associated inflammatory response.

10.8 Conclusions

In conclusion, mechanical deformation of articular chondrocytes activates intracellular Ca^{2+} signalling. This occurs via a purinergic pathway involving the mechanosensitive release of ATP and activation of purine receptors, although the precise details of the pathway have yet to be established. Preliminary studies point to the involvement of the primary cilium (Poole et al, 1985) which mediates mechanically induced intracellular Ca^{2+} signalling in other cell types possible involving a similar purinergic pathway (Praetorius and Spring, 2005). Only a sub population of cells appear to exhibit Ca^{2+} transients in response to mechanical stimuli, possibly reflecting inherent heterogeneity in the expression of purine receptors and/or other components of the signalling pathway. The nature of a mechanical stimuli modulates the Ca^{2+} signalling characteristics through at least three mechanisms: population modulation, i.e the percentage of cells responding, transient frequency modulation, and transient shape modulation. These mechanisms for modulating Ca^{2+} signalling characteristics may represent a means through which chondrocytes differentiate between different loading regimes.

The mechanical activation of purinergic Ca^{2+} signalling is necessary to prevent an inflammatory response with increased nitric oxide production and reduced cell proliferation. It is therefore tempting to speculate that inflammatory joint disease such as osteoarthritis, may be the result of disruption of the mechanosensitive purinergic Ca^{2+} signalling pathway described in this review. In addition, purinergic mechanosensitive Ca^{2+} signalling also mediates the up-regulation of proteoglycan synthesis associated with cyclic compression. Consequently this mechanosensitive purinergic Ca^{2+} signalling pathway appears to be essential for cartilage mechanotransduction and health and homeostasis. Understanding of this important physiological pathway may open the door to new therapeutic treatments for cartilage injury and degradation.

References

Aydelotte M, Keuttner KE (1988). Differences between sub-populations of cultured bovine articular chondrocytes. I. Morphology and cartilage matrix production. Connect Tissue Res 18:205–222.

Behrens F, Kraft EL, Oegema TR (1989). Biochemical changes in articular cartilage after joint immobilisation by casting or external fixation. Journal of Orthopaedic Research 7: 335–343.

Benninghoff A (1925). Form und Bau der Gelenknorpel in ihren Beziehungen zur Funktion. Zeitschrift fur Zellforschung und mikroskopische Anatomie 2:783–789.

Berridge MJ, Bootman MD, Lipp P (1998). Calcium - a life and death signal. Nature 395:645–648.

Berridge MJ, Lipp P, Bootman MD (2000). The versatility and universality of calcium signalling. Nat Rev Mol Cell Biol 1:11–21.

Bodin P, Burnstock G (2001). Purinergic signalling: ATP release. Neurochem Res 26:959–969.

Bootman MD, Lipp P, Berridge MJ (2001). The organisation and functions of local Ca(2+) signals. J Cell Sci 114:2213–2222.

Boudreault F, Grygorczyk R (2002). Cell swelling-induced ATP release and gadolinium-sensitive channels. Am J Physiol Cell Physiol 282:C219–226.

Bourret LA, Rodan GA (1976). The role of calcium in the inhibition of cAMP accunulated in epiphyseal cartilage cells exposed to physiological pressure. Journal of Cellular Physiology 88:353–362.

Boyde A, Jones SJ (1983). Scanning electron microscopy of cartilage. In: Boyde A, editor. Cartilage: Structure, Function and Biochemistry.Acedemic Press. p. 105–148.

Browning JA, Saunders K, Urban JP, Wilkins RJ (2004). The influence and interactions of hydrostatic and osmotic pressures on the intracellular milieu of chondrocytes. Biorheology 41: 299–308.

Buschmann MD, Gluzband YA, Grodzinsky AJ, Hunziker EB (1995). Mechanical compression modulates matrix biosynthesis in chondrocyte/agarose culture. J Cell Sci 108: 1497–1508.

Buschmann MD, Kim YJ, Wong M, Frank E, Hunziker EB, Grodzinsky AJ (1999). Stimulation of aggrecan synthesis in cartilage explants by cyclic loading is localized to regions of high interstitial fluid flow. Arch Biochem Biophys 366:1–7.

Caswell AM, Leong WS, Russell RG (1991). Evidence for the presence of P2-purinoceptors at the surface of human articular chondrocytes in monolayer culture. Biochim Biophys Acta 1074:151–158.

Chao PH, West AC, Hung CT (2006). Chondrocyte Intracellular Calcium, Cytoskeletal Organization and Gene Expression Responses to Dynamic Osmotic Loading. Am J Physiol Cell Physiol. In Press.

Chowdhury TT, Bader DL, Shelton JC, Lee DA (2003). Temporal regulation of chondrocyte metabolism in agarose constructs subjected to dynamic compression. Arch Biochem Biophys 417:105–111.

Chowdhury TT, Knight MM (2006). Purinergic pathway suppresses the release of (.)NO and stimulates proteoglycan synthesis in chondrocyte/agarose constructs subjected to dynamic compression. J Cell Physiol . In Press.

D'Andrea P, Calabrese A, Capozzi I, Grandolfo M, Tonon R, Vittur F (2000). Intercellular Ca^{2+} waves in mechanically stimulated articular chondrocytes. Biorheology 37:75–83.

Edlich M, Yellowley CE, Jacobs CR, Donahue HJ (2001). Oscillating fluid flow regulates cytosolic calcium concentration in bovine articular chondrocytes. J Biomech 34:59–65.

Edlich M, Yellowley CE, Jacobs CR, Donahue HJ (2004). Cycle number and waveform of fluid flow affect bovine articular chondrocytes. Biorheology 41:315–322.

Elfervig MK, Graff RD, Lee GM, Kelley SS, Sood A, Banes AJ (2001). ATP induces Ca(2+) signaling in human chondrons cultured in three-dimensional agarose films. Osteoarthritis Cartilage 9:518–526.

Erickson GR, Alexopoulos LG, Guilak F (2001). Hyper-osmotic stress induces volume change and calcium transients in chondrocytes by transmembrane, phospholipid, and G-protein pathways. J Biomech 34:1527–1535.

Erickson GR, Northrup DL, Guilak F (2003). Hypo-osmotic stress induces calcium-dependent actin reorganization in articular chondrocytes. Osteoarthritis Cartilage 11:187–197.

Fitzgerald JB, Jin M, Dean D, Wood DJ, Zheng MH, Grodzinsky AJ (2004). Mechanical compression of cartilage explants induces multiple time-dependent gene expression patterns and involves intracellular calcium and cyclic AMP. J Biol Chem 279:19502–19511.

Fitzgerald JB, Jin M, Grodzinsky AJ (2006). Shear and compression differentially regulate clusters of functionally related temporal transcription patterns in cartilage tissue. J Biol Chem 281:24095–24103.

Graff RD, Lazarowski ER, Banes AJ, Lee GM (2000). ATP release by mechanically loaded porcine chondrons in pellet culture. Arthritis Rheum 43:1571–1579.

Grodzinsky AJ, Levenston ME, Jin M, Frank EH (2000). Cartilage tissue remodeling in response to mechanical forces. Annu Rev Biomed Eng 2:691–713.

Guilak F, Ratcliffe A, Mow VC (1995). Chondrocyte Deformation and local tissue strain in articular cartilage: A confocal microscopy study. J Orthop Res 13:410–421.

Guilak F, Zell RA, Erickson GR, Grande DA, Rubin CT, McLeod KJ, Donahue HJ (1999). Mechanically induced calcium waves in articular chondrocytes are inhibited by gadolinium and amiloride. J Orthop Res 17:421–429.

Hatori M, Teixeira CC, Debolt K, Pacifici M, Shapiro IM (1995). Adenine nucleotide metabolism by chondrocytes in vitro: role of ATP in chondrocyte maturation and matrix mineralization. J Cell Physiol 165:468–474.

Hauselmann HJ, Fernandes RJ, Mok SS, Schmid TM, Block JA, Aydelotte MB, Kuettner KE, Thonar JMA (1994). Phenotypic stability of bovine articular chondrocytes after long-term culture in alginate beads. Journal of Cell Science 107:17–27.

Hoebertz A, Townsend-Nicholson A, Glass R, Burnstock G, Arnett TR (2000). Expression of P2 receptors in bone and cultured bone cells. Bone 27:503–510.

Hung CT, Mauck RL, Wang CC, Lima EG, Ateshian GA (2004). A paradigm for functional tissue engineering of articular cartilage via applied physiologic deformational loading. Ann Biomed Eng 32:35–49.

Kempson GE, Freeman MAR, Swanson SAV (1971). The determination of creep modulus for articular cartilage from indentation tests on the human femoral head. Journal of Biomechanics 4:239-50.

Kiviranta I, Jurvelin J, Tammi M, Saamanen A, Helminen HJ (1987). Weight bearing controls glycosaminoglycan concentration and articular cartilage thickness in the knee joints of young beagle dogs. Arthritis and Rheumatism 30:801–809.

Knight MM, Bomzon Z, Kimmel E, Sharma A.M., Lee DA, Bader DL (2006). Chondrocyte deformation induces mitochondrial distortion and heterogeneous intracellular strain fields. Biomechanics and Modeling in Mechanobiology 5:180–191.

Knight MM, Lee DA, Bader DL (1998). The influence of elaborated pericellular matrix on the deformation of isolated articular chondrocytes cultured in agarose. Biochim Biophys Acta 1405:67–77.

Knight MM, Lee DA, Bader DL (1996). Distribution of chondrocyte deformation in compressed agarose gel using confocal microscopy. Journal of Cellular Engineering 1:97–102.

Knight MM, Toyoda T, Lee DA, Bader DL (2005). Mechanical compression and hydrostatic pressure induce reversible changes in actin cytoskeletal organisation in chondrocytes in agarose. J Biomech 39:1547–1551.

Kono T, Nishikori T, Kataoka H, Uchio Y, Ochi M, Enomoto K (2006). Spontaneous oscillation and mechanically induced calcium waves in chondrocytes. Cell Biochem Funct 24:103–111.

Koolpe M, Benton HP (1997). Calcium-mobilizing purine receptors on the surface of mammalian articular chondrocytes. J Orthop Res 15:204–212.

Koolpe M, Pearson D, Benton HP (1999). Expression of both P1 and P2 purine receptor genes by human articular chondrocytes and profile of ligand-mediated prostaglandin E2 release. Arthritis Rheum 42:258–267.

Lee DA, Bader DL (1997). Compressive strains at physiological frequencies influence the metabolism of chondrocytes seeded in agarose. J Orthop Res 15:181–188.

Lee DA, Knight MM (2004). Mechanical Loading of Chondrocytes Embedded in 3D Constructs: In Vitro Methods for Assessment of Morphological and Metabolic Response to Compressive Strain. Methods Mol Med 100:307–324.

Lee DA, Knight MM, Bolton JF, Idowu BD, Kayser MV, Bader DL (2000). Chondrocyte deformation within compressed agarose constructs at the cellular and sub-cellular levels. J Biomech 33:81–95.

Leong WS, Russell RG, Caswell AM (1994). Stimulation of cartilage resorption by extracellular ATP acting at P2-purinoceptors. Biochim Biophys Acta 1201:298–304.

Li KW, Klein TJ, Chawla K, Nugent GE, Bae WC, Sah RL (2004). In vitro physical stimulation of tissue-engineered and native cartilage. Methods Mol Med 100:325–352.

Mauck RL, Soltz MA, Wang CC, Wong DD, Chao PH, Valhmu WB, Hung CT, Ateshian GA (2000). Functional tissue engineering of articular cartilage through dynamic loading of chondrocyte-seeded agarose gels. J Biomech Eng 2000:252–260.

McKnight NL, Frangos JA (2003). Strain rate mechanotransduction in aligned human vascular smooth muscle cells. Ann Biomed Eng 31:239–249.

Millward-Sadler SJ, Salter DM (2004). Integrin-dependent signal cascades in chondrocyte mechanotransduction. Ann Biomed Eng 32:435–446.

Millward-Sadler SJ, Wright MO, Flatman PW, Salter DM (2004). ATP in the mechanotransduction pathway of normal human chondrocytes. Biorheology 41:567–575.

Mizuno S (2005). A novel method for assessing effects of hydrostatic fluid pressure on intracellular calcium: a study with bovine articular chondrocytes. Am J Physiol Cell Physiol 288: C329–C337.

Mizuno S, Tateishi T, Ushida T, Glowacki J (2002). Hydrostatic fluid pressure enhances matrix synthesis and accumulation by bovine chondrocytes in three-dimensional culture. J Cell Physiol 193:319–327.

Mobasheri A, Martin-Vasallo P (1999). Epithelial sodium channels in skeletal cells: a role in mechanotransduction ? Cell Biology International 23:237–240.

Ohashi T, Hagiwara M, Bader DL, Knight MM (2006). Intracellular mechanics and mechanotransduction associated with chondrocyte deformation during pipette aspiration. Biorheology 43:201–214.

Palmoski MJ, Brandt KD (1984). Effects of static and cyclic compressive loading on articular cartilage plugs in vitro. Arthritis and Rheumatism 27:675–681.

Pingguan-Murphy B, El-Azzeh M, Bader DL, Knight MM (2006). Cyclic compression of chondrocytes modulates a purinergic calcium signalling pathway in a strain rate- and frequency-dependent manner. J Cell Physiol 209:389–397.

Pingguan-Murphy B, Lee DA, Bader DL, Knight MM (2005). Activation of chondrocytes calcium signalling by dynamic compression is independent of number of cycles. Arch Biochem Biophys 444:45–51.

Poole CA, Flint MH, Beaumont BW (1985). Analysis of the morphology and function of primary cilia in connective tissues: a cellular cybernetic probe? Cell Motil 5:175–193.

Praetorius HA, Spring KR (2005). A physiological view of the primary cilium. Annu Rev Physiol 67:515–529.

Ralevic V, Burnstock G (1998). Receptors for purines and pyrimidines. Pharmacol Rev 50: 413–492.

Roberts SR, Knight MM, Lee DA, Bader DL (2001). Mechanical compression influences intracellular Ca^{2+} signaling in chondrocytes seeded in agarose constructs. J Appl Physiol 90: 1385–1391.

Sah R, Kim Y, Doong J, Grodzinsky AJ, Plaas AHK, Sandy JD (1989). Biosynthetic response of cartilage explants to dynamic compression. J Orthop Res 7:619–636.

Seidel JO, Pei M, Gray ML, Langer R, Freed LE, Vunjak-Novakovic G (2004). Long-term culture of tissue engineered cartilage in a perfused chamber with mechanical stimulation. Biorheology 41:445–458.

Surprenant A, Rassendren F, Kawashima E, North RA, Buell G (1996). The cytolytic P2Z receptor for extracellular ATP identified as a P2X receptor (P2X7). Science 272:735–738.

Tanaka N, Ohno S, Honda K, Tanimoto K, Doi T, Ohno-Nakahara M, Tafolla E, Kapila S, Tanne K (2005). Cyclic mechanical strain regulates the PTHrP expression in cultured chondrocytes via activation of the Ca^{2+} channel. J Dent Res 84:64–68.

Toyoda T, Seedhom BB, Kirkham J, Bonass WA (2003). Upregulation of aggrecan and type II collagen mRNA expression in bovine chondrocytes by the application of hydrostatic pressure. Biorheology 40:79–85.

Trickey WR, Vail TP, Guilak F (2004). The role of the cytoskeleton in the viscoelastic properties of human articular chondrocytes. J Orthop Res 22:131–139.

Urban JPG (1994). The chondrocyte: A cell under pressure. Br J Rheumatol 33:901–908.

Valhmu WB, Stazzone EJ, Bachrach NM, Saed-Nejad F, Fischer SG, Mow VC, Ratcliffe A (1998). Load-controlled compression of articular cartilage induces a transient stimulation of aggrecan gene expression. Arch Biochem Biophys 1998 May 1; 353:29–36.

Wilkins RJ, Fairfax TP, Davies ME, Muzyamba MC, Gibson JS (2003). Homeostasis of intracellular Ca^{2+} in equine chondrocytes: response to hypotonic shock. Equine Vet J 35:439–443.

Wright M, Jobanputra P, Bavington C, Salter DM, Nuki G (1996). Effects of intermittent pressure-induced strain on the electrophysiology of cultured human chondrocytes: evidence for the presence of stretch-activated membrane ion channels. Clinical Science 90:61–71.

Yellowley CE, Hancox JC, Donahue HJ (2002). Effects of cell swelling on intracellular calcium and membrane currents in bovine articular chondrocytes. J Cell Biochem 86:290–301.

Yellowley CE, Jacobs CR, Donahue HJ (1999). Mechanisms contributing to fluid-flow-induced Ca^{2+} mobilization in articular chondrocytes. J Cell Physiol 180:402–408.

Zhang M, Wang JJ, Chen YJ (2005). Effects of mechanical pressure on intracellular calcium release channel and cytoskeletal structure in rabbit mandibular condylar chondrocytes. Life Sci 78:2480–2487.

Chapter 11
Signal Transduction Pathways Involved in Mechanotransduction in Osteoblastic and Mesenchymal Stem Cells

Astrid Liedert, Lutz Claes, and Anita Ignatius

Abstract Bone remodeling, a process in adults that maintains bone mass through the activity of osteoblasts and osteoclasts, is regulated by mechanical forces. Mechanical loading promotes osteoblast function by increasing proliferation and differentiation of these cells. The cellular responses underlying this mechanism are termed mechanotransduction. Mechanotransduction involves various signal transduction pathways, including the activation of ion channels and other mechanoreceptors in the membrane of the bone cell, resulting in gene regulation in the nucleus. Identification and functional characterization of the mechanotransduction components may improve bone tissue engineering.

Key words: Mechanotransduction · Signal transduction pathway · Mechanoreceptor · Ion channel · Bone cell

11.1 Introduction

Mechanical forces play an important role in regulating skeletal homeostasis (Duncan et al, 1995; Harada et al, 2003). In adults this homeostasis is locally maintained by the balance between osteoblastic bone formation and osteoclastic bone resorption. Bone formation and resorption is regulated locally by cytokines and growth factors released from the matrix during resorption and systemically by hormones, such as parathyroid hormone (PTH) and estrogen. Mechanical loading stimulates an anabolic response in osteoblasts by acting together with these factors (Cheng et al, 2002; Ryder et al, 2000; Ryder et al, 2001). The underlying mechanism for this response is termed mechanotransduction (Duncan et al, 1995; Turner et al, 1998). It comprises the detection of the physical stimulus by the cell (mechanoreception, mechanocoupling), the transformation of this stimulus in a biochemical signal (biochemical coupling) and the intracellular signal transduction to the nucleus, wherein gene transcription is modified. In the signal transmission process, osteocytes fulfil an important function by releasing molecular factors, such as nitric oxide (NO) and prostaglandins (PGs), during the early response on mechanical loading (Klein-Nulend et al, 1995; Mikuni-Takagaki, 1999). These paracrine factors

A. Kamkin and I. Kiseleva (eds.), *Mechanosensitive Ion Channels.*

activate osteoblasts on the surface of the bone (bone lining cells), which increase their proliferation and matrix synthesis. The cellular response depends on the type, magnitude, and duration of the mechanical strain (Honda et al, 2003; Rubin et al, 1984; Rubin et al, 1985). Several *in vivo* and *in vitro* studies showed that osteoblasts and osteocytes respond to mechanical loading with biochemical reactions, which may ultimately lead to bone formation (Burger et al, 1999; Hughes-Fulford, 2004; Mikuni-Takagaki, 1999; Weiss et al, 2002). However, the exact molecular mechanisms, which are associated with this anabolic response of bone on mechanical loading are still unknown.

In this review, we focus on mechanoreceptors, particularly with regard to ion channels, and various signaling molecules and pathways, which have been considered as being crucial for the cellular response on mechanical loading of bone. Moreover, we discuss aspects of mechanotransduction in tissue engineering of bone.

11.2 Mechanoreceptors

Various cells in the bone, osteoprogenitor cells, including stromal cells, osteoblasts and osteocytes have been documented to respond to mechanical loading (Rubin et al, 2006). Also other cells, such as endothelial cells and smooth muscle cells in the penetrating vasculature, might contribute to the skeletal adaptive response to mechanical loading. In the cortical bone represented osteocytes, which are enclosed in the calcified tissue, connect to each other and the bone lining cells on the bone surface, through their cellular processes in a network of canaliculi. Mechanical loading generates deformation of the osteoblastic cell membrane by stretching the cell matrix. Shear stress and streaming potentials are also generated by fluid flow in the canaliculi of the osteocytes (Pavalko et al, 2003). The surface proteoglycan layer, the glycocalix, is considered as a primary sensor for the mechanical stimulus. Force can be transmitted to apical structures, such as the plasma membrane or the actin cortical web, where transduction can take place or to remote regions of the cell (Tarbell et al, 2005). In the cholesterol–rich plasma membrane micodomains, lipid rafts and caveolae may serve as signaling complexes in the mechanotransduction process (Ferraro et al, 2004; Shin et al, 2006).

11.2.1 Membrane Ion Channels

Load-induced deformation of the cell membrane and changes to the membrane potential open membrane ion channels. An increase of intracellular calcium ($[Ca^{2+}]i$) is one of the earliest responses elicited by bone cells to mechanical loading. Using patch clamp techniques, at least three classes of mechanosensitive ion channels (MSCC, SA-cat-type) could be shown in human osteoblasts (Davidson et al, 1990). They could be distinguished on the basis of conductance, ion selectivity, and sensitivity to membrane tension. The largest conductance channel (160 pS) was

K^+-selective; one of the two smaller conductances (60 pS and 20 pS) appeared to be non-selective for cations. Focal adhesion kinase pp125FAK associates with the intracellular part of the large conductance calcium-activated hSlo K^+ channel in human osteoblasts (Rezzonico et al, 2003). Numerous studies have indicated the presence of stretch-activated cation channels in osteoblasts (Rawlinson et al, 1996; Ryder et al, 2001). It was shown that radial membrane strains of 800% and tensions of 5×10^{-4} Nm^{-1} were needed to open 50% of the mechanosensitive channels in primary osteoblasts (Charras et al, 2004). Employing an antisense strategy, Duncan et al. demonstrated the involvement of the α_{1C} gene product from the L-type calcium channel family in the swelling activated whole-cell conductance and the activation of the swelling-activated cation channels induced by chronic, intermitted strain (Duncan et al, 1996). Also, the expression of the α-subunit of the epithelial sodium channel (ENaC), which confers touch sensitivity to *Caenorhabditis elegans* was demonstrated in osteoblasts (Kizer et al, 1997). Ma et al. suggested that ATP, acting through a phospholipase C (PLC)-dependent purinergic pathway, masked stretch-induced ENac activation (Ma et al, 2002). In osteoblaststs, the P2X7 nucleotide receptor, which is an ATP-gated ion channel is also involved in skeletal mechanotransduction (Li et al, 2005). GLAST-1, a Na^+-dependent glutamate transporter, which may transport glutamate or operate as a glutamate gated ion channel in bone cells is regulated by mechanical loading in these cells (Mason et al, 2002). Influx of ions, such as Ca^{2+} can alter the membrane potential and activate voltage-sensitive ion channels. The activation of the L-type voltage-sensitive Ca^{2+} channels (VSCC) were shown *in vitro* and *in vivo* (Li et al, 2002; Rawlinson et al, 1996). Ryder and Duncan demonstrated that parathyroid hormone (PTH) enhanced fluid-shear-induced Ca^{2+} signaling in osteoblastic cells through activation of MSCC and VSCC (Ryder et al, 2001). PTH also enhances load-induced bone adaption through involvement of L-type voltage sensitive channels *in vivo* (Li et al, 2003). Chen et al. reported that mechanical loading induced PTHrP expression in osteoblast-like cells and that TREK-2 stretch-activated potassium channels seemed to be involved in this induction (Chen et al, 2005). Using gadolinium III chloride and nifdipine, we demonstrated the involvement of Ca^{2+} channels of the stretch-activated type and the L-type in mechanical regulation of the expression of HB-GAM (Heparin-Binding Growth-Associated Molecule) in osteoblastic cells (Figs. 11.1 and 11.2) (Liedert et al, 2004). HB-GAM, also designated as pleiotrophin, which presents with midkine a family of heparin-binding proteins, is suggested to have a role in osteogenic differentiation (Muramatsu, 2002; Zhou et al, 1992).

11.2.2 Integrins, Focal Contacts, Adhesion and Gap Junctions

In the model of Pavalko et al., multiprotein complexes comprised of focal adhesion (cell matrix contacts)-associated or adherens junction (intercellular junction)-associated proteins and nucleocytoplasmic shuttling DNA-binding proteins, termed mechanosomes, can move between the adhesion complexes and the nucleus (Pavalko et al, 2003). The integrins of the focal adhesion complexes are heterodimeric transmembrane proteins, which interlink the extracellular matrix with the cytoskeleton

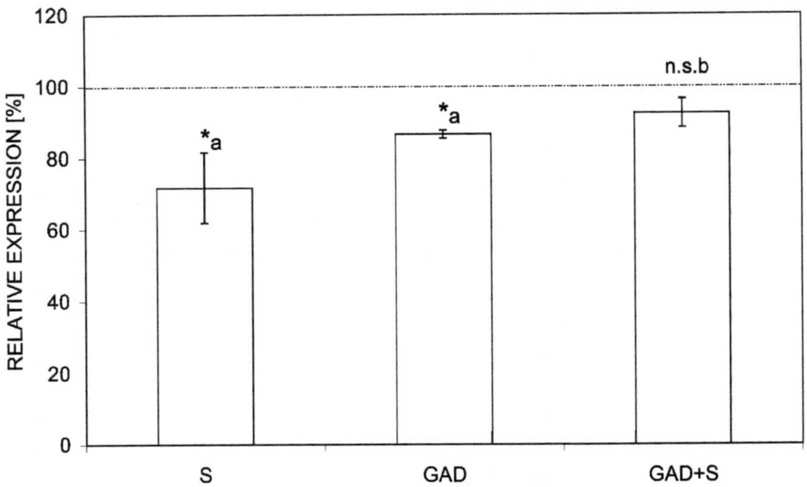

Fig. 11.1 Expression of HB-GAM in the presence of the Ca^{2+} channel (SA-cat-type) blocker gadolinium III chloride (10 μM) without mechanical stimulation and with mechanical stimulation (1000 μstrains, 1 Hz, 1800 cycles) in SaOs-2 cells on day 14 of cultivation. Results are mean ±SEM compared with unstimulated control cells, which are set at 100% (n = 4). a (*$p \leq 0.05$) compared with unstimulated control. b compared with gadolinium III chloride treated control cells (n.s. = not significant). (Reprinted from Mechanical regulation of HB-GAM expression in bone cells, 319 (3), Liedert A, Augat P, Ignatius A, Hausser HJ, Claes L, 951–8, 2004, with permission from Elsevier)

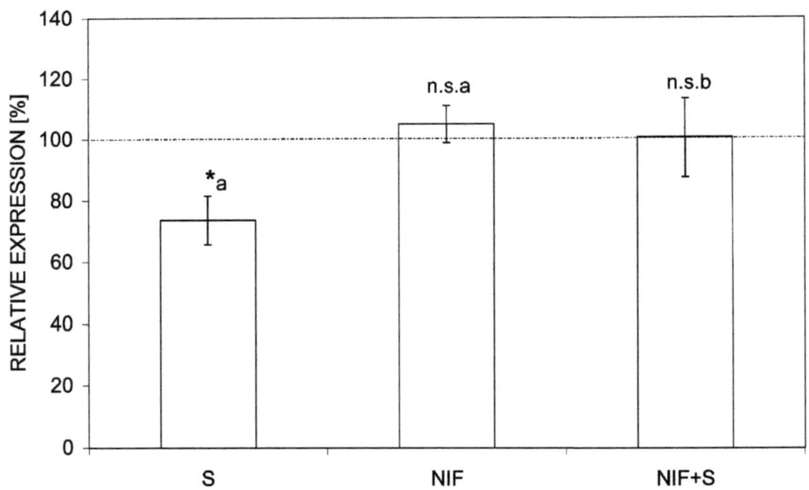

Fig. 11.2 Expression of HB-GAM in the presence of the Ca^{2+} channel (L-type) blocker nifedipine (10 μM) without mechanical stimulation and with mechanical stimulation (1000 μstrains, 1 Hz, 1800 cycles) in SaOs-2 cells on day 14 of cultivation. Stimulated control cells were treated with DMSO vehicle. Results are mean ±SEM compared with unstimulated control cells, which are set at 100% (n = 4). a (*$p \leq 0.05$) compared with unstimulated control. b compared with nifedipine treated control cells (n.s. = not significant). (Reprinted from mechanical regulation of HB-GAM expression in bone cells, 319 (3), Liedert A, Augat P, Ignatius A, Hausser HJ, Claes L, 951–8, 2004, with permission from Elsevier)

through several actin-associated proteins, including α-actin, vinculin, talin, and tensin (Turner et al, 1998). Fluid shear stress (FSS) has been shown to increase β1 subunit expression of integrins and to activate αvβ3 integrins in osteoblasts (Weyts et al, 2002). Mechanical loading prevents apoptosis of osteocytes, which requires integrins and cytoskeletal molecules, together with catalytic molecules, such as Src kinase. Cadherins, N-Cadherins in bone, are single chain transmembrane glycoproteins that interlink the cytoskeletons of neighbouring cells via the calcium-dependent homophilic interaction between cadherin domains (Marie, 2002). Expression of N-cadherin is increased by mechanical strain (Di Palma et al, 2005). There are numerous classical signal transduction molecules that can localize at or in the vicinity of these adhesion complexes (Pavalko et al, 2003). Selected connective membrane skeleton (CMS) proteins are activated by mechanical loading and organized into signaling complexes, which may translocate from the adhesion complexes through the cytoplasma into the nucleus. Thus, Norvell et al. demonstrated that FSS induced translocation of β-catenin from N-cadherin to the nucleus to activate gene transcription together with TCF (T-cell factor) (Norvell et al, 2004).

The expression of connexin (Cx) 43 protein, a major component of gap junctions and the number of functional gap junctions are increased, in part, by the release of PGE2 (Cheng et al, 2001). Cherian et al. suggested that FSS induces the translocation of Cx43 to the membrane surface and that unopposed hemichannels formed by Cx43 play an important role in the release of PGE2 in reponse to mechanical strain (Cherian et al, 2005). Mechanical strain regulates gap junction function through prostaglandin EP2 receptor activation and PKA signaling (Cherian et al, 2003).

11.3 Intracellular Signal Transduction Pathways

Rapid increases of $[Ca^{2+}]$i result from the influx of Ca^{2+} through activated MSCC and/or VSCC in the plasma membrane and from inositol-1,4,5-trisphosphate signaling, which leads to activation of intracellular Ca^{2+}stores (Pavalko et al, 2003). This mobilization of $[Ca^{2+}]$i activates numerous kinase cascades, including Ca^{2+} dependent protein kinase C (PKC), converting the mechanical force that induces conformational changes in the channel, into a biochemical signal (Fig. 11.3). PKC and GTPase Ras may activate the mitogen-activated protein kinase (MAPK) pathway. Activation of this pathway lead to upregulation of c-fos gene expression. The c-fos protein together with c-jun generate the activator protein-1 (AP-1) transcription factor, which binds to the AP-1 transcription factor binding site in the promotor region of mechanosensitive genes (Iqbal et al, 2005). The shear stress response element (SSRE) was identified in the promotor region of genes, including c-fos, inducible nitric oxide synthase (iNOS), and cyclooxygenase-2 (COX-2) (Nomura et al, 2000).

Mechanical force has been shown to upregulate growth factors, which act via paracrine and autocrine mechanisms by binding to their receptors (Ehrlich et al, 2002; Hughes-Fulford, 2004). Various signaling pathways are activated, which may interact with each other, to regulate proliferation and differentiation of the bone cell (Lau et al, 2006). A synergistic effect on the proliferation of osteoblastic cells

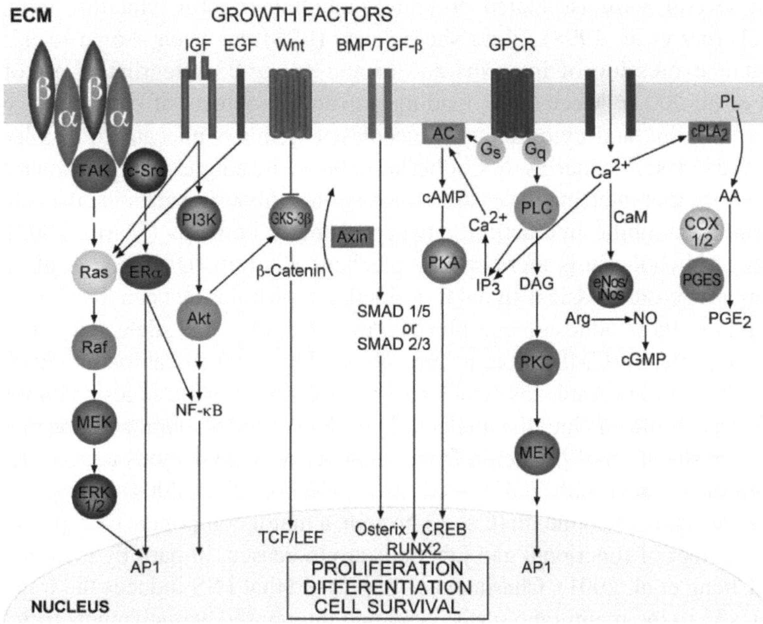

Fig. 11.3 Mechanotransduction pathways shown to be involved in the mechanical response of osteoblastic cells. Abbreviations: AA arachidonic acid, AC adenylate cyclase, Akt akutely transforming (protein kinase B, serin/threonine kinase), AP1 activator protein 1, BMP bone morphogenetic protein, CAM calmodulin, COX1/2 cyclooxygenase 1/2, CREB c-AMP response element binding protein, c-Src tyrosine protein kinase, DAG diacylglycerol, EGF epidermal growth factor, eNOS endothelial nitric oxide synthase, ERα estrogen receptor α, ERK1/2 extracellular signal regulated protein kinase 1/2, FAK focal adhesion kinase, GSK-3β glycogen synthase kinase-3β, G_s stimulatory G protein, GPCR seven-transmembrane-domain G protein-coupled receptor, G_q protein with α_q subunit activates PLC phospholipase C-β, IGF insulin-like growth factor, iNOS inducible nitric oxide synthase, IP3 inositol trisphosphate, LEF lymphoid enhancer-binding factor, MEK mitogen activated protein kinase extracellular signal regulated protein kinase (mitogen activated kinase kinase), NF-κB nuclear factor-κB, PI3K phosphoinositide 3-kinase, PGE_2 prostaglandin E2, PGES prostaglandin synthase, PKA protein kinase A, PKC protein kinase C, PL phospholipid, Raf rat fibrosarcoma serin/threonine protein kinase, SMAD from sma (small) in Caenorhabditis and mad (mother against decapentaplegic) in Drosophila, Ras rat sarcoma monomeric GTP binding-protein, TCF T-cell factor, TGF-β transforming growth factor-β, wnt from wingless in Drosophila, int (integration)-1 in mouse (Reprinted from Signal transduction pathways involved in mechanotransduction in bone cells, 349(1), Liedert A, Kaspar D, Blakytny R, Claes L, Ignatius A, 1–5, 2006, with permission of Elsevier)

has been shown for insulin-like growth factor-1 (IGF-1) and shear stress through integrin-dependent activation of IGF-1 mitogenic pathway (Kapur et al, 2005). Uniaxial cyclic stretch, for example, induces expression of bone morphogenetic proteins (BMPs) and BMP receptors in spinal ligament cells as well as osteogenic differentiation of these cells (Tanno et al, 2003). BMP-2, for example, has been shown to stimulate the expression of the three osteogenic master transcription factors Runx2 (cbfa1) Osterix, and Dlx5 (Lee et al, 2003).

11.4 Nitric Oxide and Prostaglandin Signaling

One of the earliest responses of bone cells is the release of NO and PGs. These molecules are mediators for the mechanical signal within the three dimensional network of bone cells and thus are important for intercellular communication (Bakker et al, 2001; McGarry et al, 2005). Mullender et al. (2006) demonstrated that the release of NO, but not PGE_2 by bone cells depends on the fluid flow frequency (Mullender et al, 2006). PGE_2 mediates opening of the hSlo potassium channel and induces the recruitment of various tyrosine-phosphorylated proteins on the hSlo alpha-subunit of the Ca^{2+}-dependent K^+ channel (BK) (Rezzonico et al, 2002). In osteocytes, the fluid-flow induced PGE_2 production is mediated via the cytoskeleton, activated Ca^{2+} signaling, phospholipase C, PKC, and phospholipase A2 (Ajubi et al, 1999). Although, mechanotransduction does not necessarily require a functional COX-2 gene, upregulation of the expression of COX-2, the inducible isozyme for PGE_2 synthesis, by mechanical loading is promoted by fibronectin-induced formation of focal adhesions in MC3T3-E1 osteoblasts (Ponik et al, 2004). Compressive force upregulates receptor activator of nuclear factor kappaB ligand via PGE synthesis in periodontal ligament (PDL) cells and therefore supports osteoclastogenesis (Kanzaki et al, 2002). It has been shown that strain inhibits receptor activator of nuclear factor kappaB ligand (RANKL) expression and increases endothelial nitric oxide synthase (eNOS) and NO levels through ERK1/2 signaling in immortalized murine calvarial cells (CIMC-4) and in primary bone stromal cells, respectively (Fan et al, 2006). Strain-induced ERK-1 activation in osteoblast-like cells is dependent on calcium mobilization from intracellular stores and production of NO and prostacyclin $(PGI)_2$. ERK-2 activation by fluid-movement involved both PGI_2 and NO production, but its activation by strain does not seem to depend on PGI and NO production (Jessop et al, 2002). The NO synthetase pathway and the COX-2 pathway are involved in the FSS-induced proliferation and differentiation of human osteoblasts (Kapur et al, 2003). The expression of eNOS, the predominant NOS isoform in long bone, is strong in osteocytes and chondrocytes (van't Hof et al, 2001; Zaman et al, 1999). It has been proposed that low concentrations of NO produced by osteocytes and low canalicular flow result in their apoptosis. Whereas, high concentrations of NO produced by enhanced canalicular flow are protective against cell death, and induce osteoclast retraction and detachment from the bone surface (Burger et al, 2003). In an *in vivo* model of the rat hind limb, unloading reloading restored osteocyte apoptosis to control levels and the percentage of inducible NOS (i-NOS)- and eNOS-positive osteocytes increased in reloaded bone, compared to controls (Basso et al, 2006). This was associated with a decrease in osteoclast number. NO signaling can depend on direct actions of the molecule to nitrosylate proteins, as has been shown for NO action to decrease the RANKL/osteoprotegerin (OPG) ratio in stromal cells (Fan et al, 2004). Estrogen receptor (ER) alpha deficient primary osteoblasts show no strain-related increase in NO production (Jessop et al, 2004). Estrogen (E2) and FSS both promote the production of NO and PGE_2 by bone cells from osteoporotic donors in an additive manner (Bakker et al, 2005). Bakker et al. speculated that in primary mouse osteoblastic cells PTH prevents stress-induced NO production via the inhibition of NOS, which will also inhibit the NO-mediated upregulation of PGE_2 by stress, leaving only the NO-independent PGE_2 upregulation by PTH (Bakker et al, 2003).

11.5 Mechanotransduction in Tissue Engineering

Strategies for bone tissue engineering require the collection of a small sample of bone cells from the patient and the cultivation of seeded cells in the appropriate scaffolds *in vitro,* in order for the cells to proliferate, differentiate, and generate extracellular matrix prior to reimplantation in bone defects. Various studies have shown that mechanical loading promotes proliferation and differentiation of bone cells, as a result of activated specific signal transduction pathways and genes in these cells (Kapur et al, 2003; Kaspar et al, 2000; Lau et al, 2006). Therefore, mechanical loading may be an adequate stimulus for affecting cellular function in bone tissue engineering. However, successful mechanotransduction is influenced by a number of factors, including cell-substrate interaction and the biochemical environment. Three-dimensional (3D) scaffolds of various biomaterials and mechanical enviroments have already been proven and discussed for cultivation of mesenchymal and osteoblastic cells as well as to study mechanical effects on these cells (Klein-Nulend et al, 2005; Martin et al, 2006; Rezwan et al, 2006; Tanaka et al, 2005; Wendt et al, 2005; Zhang et al, 2006). Mechanical stimulation led to increased proliferation and differentiation of these cells in 3D constructs (Akhouayri et al, 1999; Cartmell et al, 2003; Datta et al, 2006; Ignatius et al, 2005; Leclerc et al, 2006; Mauney et al, 2004). Recently, it has been demonstrated that the dihydropyridine Bay K8644, a selective agonist of mechanosensitive voltage-operated Ca^{2+} channels (VSCC), enhanced loading effects in human bone-cell seeded constructs (El Haj et al, 2005; Wood et al, 2006). In this study, primary bone cells were seeded in porous 3D mechanoactive, biodegradable poly (L-lactide) acid scaffolds containing Bay K8644. Collagen type I matrix concentration and expression of core binding factor (cbfa)1, a bone cell differentiation marker, was increased in response to mechanical stimulation. Osteopontin (OP) and alpha2delta1 VSCC subunit protein levels were also higher as a result of perfusion-compression conditioning. Released Bay K8644 was stable and maintained its bioactivity following culture of up to 28 days (Wood et al, 2006). These studies demonstrated an example of a model, in which mechanotransduction pathways are considered in the design of scaffolds for bone tissue engineering. However, further research of mechanotransduction pathways and their regulating factors, including the mode of the physical stimulus, nutrient supply and cell-matrix interactions, are required for bone tissue engineering.

11.6 Conclusion

Bone homeostasis, which is maintained by the balance between osteoblastic bone formation and osteoclastic bone resorption, is regulated by mechanical loading. One of the earliest responses to mechanical loading is the opening of membrane ion channels. Ca^{2+} influx and mobilization of $[Ca^{2+}]$i from intracellular stores lead to activation of intracellular Ca^{2+}-dependent protein kinases. Various signaling pathways may interact with each other, to regulate proliferation and differentiation of the bone cell. Mechanotransduction pathways and their regulating factors should be clarified and considered in the stimulation design and development of scaffolds in bone tissue engineering.

References

Ajubi NE, Klein-Nulend J, Alblas MJ, Burger EH, Nijweide PJ (1999) Signal transduction pathways involved in fluid flow-induced PGE2 production by cultured osteocytes. Am J Physiol 276: E171–178

Akhouayri O, Lafage-Proust MH, Rattner A, Laroche N, Caillot-Augusseau A, Alexandre C, Vico L (1999) Effects of static or dynamic mechanical stresses on osteoblast phenotype expression in three-dimensional contractile collagen gels. J Cell Biochem 76: 217–230

Bakker AD, Joldersma M, Klein-Nulend J, Burger EH (2003) Interactive effects of PTH and mechanical stress on nitric oxide and PGE2 production by primary mouse osteoblastic cells. Am J Physiol Endocrinol Metab 285: E608–613

Bakker AD, Klein-Nulend J, Tanck E, Albers GH, Lips P, Burger EH (2005) Additive effects of estrogen and mechanical stress on nitric oxide and prostaglandin E2 production by bone cells from osteoporotic donors. Osteoporos Int 16: 983–989

Bakker AD, Soejima K, Klein-Nulend J, Burger EH (2001) The production of nitric oxide and prostaglandin E(2) by primary bone cells is shear stress dependent. J Biomech 34: 671–677

Basso N, Heersche JN (2006) Effects of hind limb unloading and reloading on nitric oxide synthase expression and apoptosis of osteocytes and chondrocytes. Bone 39: 807–814

Burger EH, Klein-Nulen J (1999) Responses of bone cells to biomechanical forces in vitro. Adv Dent Res 13: 93–98

Burger EH, Klein-Nulend J, Smit TH (2003) Strain-derived canalicular fluid flow regulates osteoclast activity in a remodelling osteon–a proposal. J Biomech 36: 1453–1459

Cartmell SH, Porter BD, Garcia AJ, Guldberg RE (2003) Effects of medium perfusion rate on cell-seeded three-dimensional bone constructs in vitro. Tissue Eng 9: 1197–1203

Charras GT, Williams BA, Sims SM, Horton MA (2004) Estimating the sensitivity of mechanosensitive ion channels to membrane strain and tension. Biophys J 87: 2870–2884

Chen X, Macica CM, Ng KW, Broadus AE (2005) Stretch-induced PTH-related protein gene expression in osteoblasts. J Bone Miner Res 20: 1454–1461

Cheng B, Kato Y, Zhao S, Luo J, Sprague E, Bonewald LF, Jiang JX (2001) PGE(2) is essential for gap junction-mediated intercellular communication between osteocyte-like MLO-Y4 cells in response to mechanical strain. Endocrinology 142: 3464–3473

Cheng MZ, Rawlinson SC, Pitsillides AA, Zaman G, Mohan S, Baylink DJ, Lanyon LE (2002) Human osteoblasts' proliferative responses to strain and 17beta-estradiol are mediated by the estrogen receptor and the receptor for insulin-like growth factor I. J Bone Miner Res 17: 593–602

Cherian PP, Cheng B, Gu S, Sprague E, Bonewald LF, Jiang JX (2003) Effects of mechanical strain on the function of Gap junctions in osteocytes are mediated through the prostaglandin EP2 receptor. J Biol Chem 278: 43146–43156

Cherian PP, Siller-Jackson AJ, Gu S, Wang X, Bonewald LF, Sprague E, Jiang JX (2005) Mechanical strain opens connexin 43 hemichannels in osteocytes: a novel mechanism for the release of prostaglandin. Mol Biol Cell 16: 3100–3106

Datta N, Pham QP, Sharma U, Sikavitsas VI, Jansen JA, Mikos AG (2006) In vitro generated extracellular matrix and fluid shear stress synergistically enhance 3D osteoblastic differentiation. Proc Natl Acad Sci USA 103: 2488–2493

Davidson RM, Tatakis DW, Auerbach AL (1990) Multiple forms of mechanosensitive ion channels in osteoblast-like cells. Pflugers Arch 416: 646–651

Di Palma F, Guignandon A, Chamson A, Lafage-Proust MH, Laroche N, Peyroche S, Vico L, Rattner A (2005) Modulation of the responses of human osteoblast-like cells to physiologic mechanical strains by biomaterial surfaces. Biomaterials 26: 4249–4257

Duncan RL, Kizer N, Barry EL, Friedman PA, Hruska KA (1996) Antisense oligodeoxynucleotide inhibition of a swelling-activated cation channel in osteoblast-like osteosarcoma cells. Proc Natl Acad Sci USA 93: 1864–1869

Duncan RL, Turner CH (1995) Mechanotransduction and the functional response of bone to mechanical strain. Calcif Tissue Int 57: 344–358

Ehrlich PJ, Lanyon LE (2002) Mechanical strain and bone cell function: a review. Osteoporos Int 13: 688–700

El Haj AJ, Wood MA, Thomas P, Yang Y (2005) Controlling cell biomechanics in orthopaedic tissue engineering and repair. Pathol Biol (Paris) 53: 581–589

Fan X, Rahnert JA, Murphy TC, Nanes MS, Greenfield EM, Rubin J (2006) Response to mechanical strain in an immortalized pre-osteoblast cell is dependent on ERK1/2. J Cell Physiol 207: 454–460

Fan X, Roy E, Zhu L, Murphy TC, Ackert-Bicknell C, Hart CM, Rosen C, Nanes MS, Rubin J (2004) Nitric oxide regulates receptor activator of nuclear factor-kappaB ligand and osteoprotegerin expression in bone marrow stromal cells. Endocrinology 145: 751–759

Ferraro JT, Daneshmand M, Bizios R, Rizzo V (2004) Depletion of plasma membrane cholesterol dampens hydrostatic pressure and shear stress-induced mechanotransduction pathways in osteoblast cultures. Am J Physiol Cell Physiol 286: C831–839

Harada S, Rodan GA (2003) Control of osteoblast function and regulation of bone mass. Nature 423: 349–355

Honda A, Sogo N, Nagasawa S, Shimizu T, Umemura Y (2003) High-impact exercise strengthens bone in osteopenic ovariectomized rats with the same outcome as Sham rats. J Appl Physiol 95: 1032–1037

Hughes-Fulford M (2004) Signal transduction and mechanical stress. Sci STKE 2004: RE12

Ignatius A, Blessing H, Liedert A, Schmidt C, Neidlinger-Wilke C, Kaspar D, Friemert B, Claes L (2005) Tissue engineering of bone: effects of mechanical strain on osteoblastic cells in type I collagen matrices. Biomaterials 26: 311–318

Iqbal J, Zaidi M (2005) Molecular regulation of mechanotransduction. Biochem Biophys Res Commun 328: 751–755

Jessop HL, Rawlinson SC, Pitsillides AA, Lanyon LE (2002) Mechanical strain and fluid movement both activate extracellular regulated kinase (ERK) in osteoblast-like cells but via different signaling pathways. Bone 31: 186–194

Jessop HL, Suswillo RF, Rawlinson SC, Zaman G, Lee K, Das-Gupta V, Pitsillides AA, Lanyon LE (2004) Osteoblast-like cells from estrogen receptor alpha knockout mice have deficient responses to mechanical strain. J Bone Miner Res 19: 938–946

Kanzaki H, Chiba M, Shimizu Y, Mitani H (2002) Periodontal ligament cells under mechanical stress induce osteoclastogenesis by receptor activator of nuclear factor kappaB ligand upregulation via prostaglandin E2 synthesis. J Bone Miner Res 17: 210–220

Kapur S, Baylink DJ, Lau KH (2003) Fluid flow shear stress stimulates human osteoblast proliferation and differentiation through multiple interacting and competing signal transduction pathways. Bone 32: 241–251

Kapur S, Mohan S, Baylink DJ, Lau KH (2005) Fluid shear stress synergizes with insulin-like growth factor-I (IGF-I) on osteoblast proliferation through integrin-dependent activation of IGF-I mitogenic signaling pathway. J Biol Chem 280: 20163–20170

Kaspar D, Seidl W, Neidlinger-Wilke C, Claes L (2000) In vitro effects of dynamic strain on the proliferative and metabolic activity of human osteoblasts. J Musculoskelet Neuronal Interact 1: 161–164

Kizer N, Guo XL, Hruska K (1997) Reconstitution of stretch-activated cation channels by expression of the alpha-subunit of the epithelial sodium channel cloned from osteoblasts. Proc Natl Acad Sci USA 94: 1013–1018

Klein-Nulend J, Bacabac RG, Mullender MG (2005) Mechanobiology of bone tissue. Pathol Biol (Paris) 53: 576–580

Klein-Nulend J, Van der Plas A, Semeins C, Nasser E, Frangos J, Nijweide P, Burger E (1995) Sensitivity of osteocytes to biomechanical stress in vitro. FASEB J 9: 441–445

Lau KH, Kapur S, Kesavan C, Baylink DJ (2006) Up-regulation of the Wnt, estrogen receptor, insulin-like growth factor-I, and bone morphogenetic protein pathways in C57BL/6J osteoblasts as opposed to C3H/HeJ osteoblasts in part contributes to the differential anabolic response to fluid shear. J Biol Chem 281: 9576–9588

Leclerc E, David B, Griscom L, Lepioufle B, Fujii T, Layrolle P, Legallaisa C (2006) Study of osteoblastic cells in a microfluidic environment. Biomaterials 27: 586–595

Lee MH, Kwon TG, Park HS, Wozney JM, Ryoo HM (2003) BMP-2-induced Osterix expression is mediated by Dlx5 but is independent of Runx2. Biochem Biophys Res Commun 309: 689–694

Li J, Duncan RL, Burr DB, Gattone VH, Turner CH (2003) Parathyroid hormone enhances mechanically induced bone formation, possibly involving L-type voltage-sensitive calcium channels. Endocrinology 144: 1226–1233

Li J, Duncan RL, Burr DB, Turner CH (2002) L-type calcium channels mediate mechanically induced bone formation in vivo. J Bone Miner Res 17: 1795–1800

Li J, Liu D, Ke HZ, Duncan RL, Turner CH (2005) The P2X7 nucleotide receptor mediates skeletal mechanotransduction. J Biol Chem 280: 42952–42959

Liedert A, Augat P, Ignatius A, Hausser HJ, Claes L (2004) Mechanical regulation of HB-GAM expression in bone cells. Biochem Biophys Res Commun 319: 951–958

Ma HP, Li L, Zhou ZH, Eaton DC, Warnock DG (2002) ATP masks stretch activation of epithelial sodium channels in A6 distal nephron cells. Am J Physiol Renal Physiol 282: F501–505

Marie PJ (2002) Role of N-cadherin in bone formation. J Cell Physiol 190: 297–305

Martin I, Miot S, Barbero A, Jakob M, Wendt D (2006) Osteochondral tissue engineering. J Biomech

Mason DJ, Huggett JF (2002) Glutamate transporters in bone. J Musculoskelet Neuronal Interact 2: 406–414

Mauney JR, Sjostorm S, Blumberg J, Horan R, O'Leary JP, Vunjak-Novakovic G, Volloch V, Kaplan DL (2004) Mechanical stimulation promotes osteogenic differentiation of human bone marrow stromal cells on 3-D partially demineralized bone scaffolds in vitro. Calcif Tissue Int 74: 458–468

McGarry JG, Klein-Nulend J, Prendergast PJ (2005) The effect of cytoskeletal disruption on pulsatile fluid flow-induced nitric oxide and prostaglandin E2 release in osteocytes and osteoblasts. Biochem Biophys Res Commun 330: 341–348

Mikuni-Takagaki Y (1999) Mechanical responses and signal transduction pathways in stretched osteocytes. J Bone Miner Metab 17: 57–60

Mullender MG, Dijcks SJ, Bacabac RG, Semeins CM, Van Loon JJ, Klein-Nulend J (2006) Release of nitric oxide, but not prostaglandin E2, by bone cells depends on fluid flow frequency. J Orthop Res 24: 1170–1177

Muramatsu T (2002) Midkine and pleiotrophin: two related proteins involved in development, survival, inflammation and tumorigenesis. J Biochem (Tokyo) 132: 359–371

Nomura S, Takano-Yamamoto T (2000) Molecular events caused by mechanical stress in bone. Matrix Biol 19: 91–96

Norvell SM, Alvarez M, Bidwell JP, Pavalko FM (2004) Fluid shear stress induces beta-catenin signaling in osteoblasts. Calcif Tissue Int 75: 396–404

Pavalko FM, Norvell SM, Burr DB, Turner CH, Duncan RL, Bidwell JP (2003) A model for mechanotransduction in bone cells: the load-bearing mechanosomes. J Cell Biochem 88: 104–112

Ponik SM, Pavalko FM (2004) Formation of focal adhesions on fibronectin promotes fluid shear stress induction of COX-2 and PGE2 release in MC3T3-E1 osteoblasts. J Appl Physiol 97: 135–142

Rawlinson SC, Pitsillides AA, Lanyon LE (1996) Involvement of different ion channels in osteoblasts' and osteocytes' early responses to mechanical strain. Bone 19: 609–614

Rezwan K, Chen QZ, Blaker JJ, Boccaccini AR (2006) Biodegradable and bioactive porous polymer/inorganic composite scaffolds for bone tissue engineering. Biomaterials 27: 3413–3431

Rezzonico R, Cayatte C, Bourget-Ponzio I, Romey G, Belhacene N, Loubat A, Rocchi S, Van Obberghen E, Girault JA, Rossi B, Schmid-Antomarchi H (2003) Focal adhesion kinase pp125FAK interacts with the large conductance calcium-activated hSlo potassium channel in human osteoblasts: potential role in mechanotransduction. J Bone Miner Res 18: 1863–1871

Rezzonico R, Schmid-Alliana A, Romey G, Bourget-Ponzio I, Breuil V, Breittmayer V, Tartare-Deckert S, Rossi B, Schmid-Antomarchi H (2002) Prostaglandin E2 induces interaction between hSlo potassium channel and Syk tyrosine kinase in osteosarcoma cells. J Bone Miner Res 17: 869–878

Rubin CT, Lanyon LE (1984) Regulation of bone formation by applied dynamic loads. J Bone Joint Surg 66-A: 397–402

Rubin CT, Lanyon LE (1985) Regulation of bone mass by mechanical strain magnitude. Calcif Tissue Int 37: 411–417

Rubin J, Rubin C, Jacobs CR (2006) Molecular pathways mediating mechanical signaling in bone. Gene 367: 1–16

Ryder KD, Duncan RL (2000) Parathyroid hormone modulates the response of osteoblast-like cells to mechanical stimulation. Calcif Tissue Int 67: 241–246

Ryder KD, Duncan RL (2001) Parathyroid hormone enhances fluid shear-induced $[Ca^{2+}]i$ signaling in osteoblastic cells through activation of mechanosensitive and voltage-sensitive Ca^{2+} channels. J Bone Miner Res 16: 240–248

Shin J, Jo H, Park H (2006) Caveolin-1 is transiently dephosphorylated by shear stress-activated protein tyrosine phosphatase mu. Biochem Biophys Res Commun 339: 737–741

Tanaka SM, Sun HB, Roeder RK, Burr DB, Turner CH, Yokota H (2005) Osteoblast responses one hour after load-induced fluid flow in a three-dimensional porous matrix. Calcif Tissue Int 76: 261–271

Tanno M, Furukawa KI, Ueyama K, Harata S, Motomura S (2003) Uniaxial cyclic stretch induces osteogenic differentiation and synthesis of bone morphogenetic proteins of spinal ligament cells derived from patients with ossification of the posterior longitudinal ligaments. Bone 33: 475–484

Tarbell JM, Weinbaum S, Kamm RD (2005) Cellular fluid mechanics and mechanotransduction. Ann Biomed Eng 33: 1719–1723

Turner CH, Pavalko FM (1998) Mechanotransduction and functional response of the skeleton to physical stress: the mechanisms and mechanics of bone adaptation. J Orthop Sci 3: 346–355

Van't Hof RJ, Ralston SH (2001) Nitric oxide and bone. Immunology 103: 255–261

Weiss S, Baumgart R, Jochum M, Strasburger CJ, Bidlingmaier M (2002) Systemic regulation of distraction osteogenesis: a cascade of biochemical factors. J Bone Miner Res 17: 1280–1289

Wendt D, Jakob M, Martin I (2005) Bioreactor-based engineering of osteochondral grafts: from model systems to tissue manufacturing. J Biosci Bioeng 100: 489–494

Weyts FA, Li YS, van Leeuwen J, Weinans H, Chien S (2002) ERK activation and alpha v beta 3 integrin signaling through Shc recruitment in response to mechanical stimulation in human osteoblasts. J Cell Biochem 87: 85–92

Wood MA, Hughes S, Yang Y, El Haj AJ (2006) Characterizing the efficacy of calcium channel agonist-release strategies for bone tissue engineering applications. J Control Release 112: 96–102

Wood MA, Yang Y, Thomas PB, Haj AJ (2006) Using dihydropyridine-release strategies to enhance load effects in engineered human bone constructs. Tissue Eng 12: 2489–2497

Zaman G, Pitsillides AA, Rawlinson SC, Suswillo RF, Mosley JR, Cheng MZ, Platts LA, Hukkanen M, Polak JM, Lanyon LE (1999) Mechanical strain stimulates nitric oxide production by rapid activation of endothelial nitric oxide synthase in osteocytes. J Bone Miner Res 14: 1123–1131

Zhang C, Zhang X, Wu H, Han D, Guan J (2006) Direct compression as an appropriately mechanical environment in bone tissue reconstruction in vitro. Med Hypotheses 67: 1414–1418
Zhou HY, Ohnuma Y, Takita H, Fujisawa R, Mizuno M, Kuboki Y (1992) Effects of a bone lysine-rich 18 kDa protein on osteoblast-like MC3T3-E1 cells. Biochem Biophys Res Commun 186: 1288–1293

Chapter 12
Caveolae

Co-ordinating Centres for Mechanotransduction in the Heart?

Sarah Calaghan

Abstract The heart possesses the intrinsic ability to adjust to short- and long-term haemodynamic demands. These adaptive responses are dependent on the sensation of mechanical stimuli and transduction into cellular events. Recent evidence suggests that caveolae, flask-shaped invaginations of the cell membrane, may be an important part of the mechanotransductive pathway in the cardiac cell. Caveolae are 'signalosomes', microdomains enriched in components of signal transduction cascades, ion channels and exchangers, which are known to control some elements of cell signalling. The marker protein for caveolae, caveolin, acts as a scaffold for macromolecular signalling complexes, and can also regulate the activity of proteins with which it interacts. In this review, the morphological, biochemical and functional evidence to support a role for caveolae in mechanosensation and mechanotransduction will be presented. Although there is a paucity of direct evidence in the cardiac myocyte, the available data support the idea that caveolae are an integral part of downstream stretch-activated signalling, and that they are essential for the proper integration and co-ordination of mechanosensitive signalling pathways.

Key words: Lipid rafts · Caveolae · Cardiac · Stretch · Swelling

12.1 Introduction

12.1.1 Mechanotransduction in the Heart

Mechanotransduction, the conversion of a mechanical stimulus into a cellular response, plays a fundamental role in a variety of processes from fertilization to the senses of hearing, touch and balance (Knoll *et al.*, 2003). Mechanotransduction is vital for the normal function of the heart. The myocardium is stretched under physiological conditions (by diastolic filling) and is also mechanically loaded in the

A. Kamkin and I. Kiseleva (eds.), *Mechanosensitive Ion Channels.*

disease state (during pressure- or volume-overload of the ventricle). Another form of mechanical stimulation experienced by cardiac myocytes is osmotic swelling (a consequence of episodes of ischaemia-reperfusion). Although we are beginning to understand the cellular consequences of mechanical stimulation, the identity of the mechanosensors and the process of mechanotransduction are still not known. This review will explore the role of caveolae in mechanotransduction in the heart, focusing on the slow inotropic response to stretch, the volume-regulated anionic current $I_{Cl,swell}$, and pathways implicated in cardiac hypertrophy. Caveolae are flask-shaped membrane microdomains which organise signalling by their ability to sequester and regulate signal components. There is considerable support for the regulation of $[Ca^{2+}]_i$ handling and control of G protein-coupled receptor signalling by caveolae, but the idea that they are involved in mechanosensation or in the co-ordination of downstream stretch-activated signalling is still under investigation. At this stage most of the evidence for a role of caveolae in mechanotransduction comes from non-cardiac cells (particularly endothelial and vascular smooth muscle cells), or neonatal cardiac myocytes.

12.1.1.1 Chronotropic and Inotropic Response to Stretch

Stretch of the myocardium is a potent stimulus for the chronotropic and inotropic response of the heart, as first recognized by Bainbridge (1915), and Frank and Starling (Frank, 1895; Patterson & Starling, 1914). During exercise when venous return to the heart increases, increased right atrial filling will accelerate heart rate. This is due, in part at least, to opening of stretch-activated channels in the sinoatrial node myocytes (Cooper *et al.*, 2000). Likewise, an increase in end-diastolic volume (or stretch) of the ventricle will enhance contractility. This occurs in a biphasic manner; there is an immediate increase in force followed by a slow response which takes place over several minutes (see Fig. 12.1 and Calaghan *et al.*, 2003 for a review). The immediate response is due to an increase in the probability of strong-binding crossbridge formation, possibly mediated through titin strain (Cazorla *et al.*, 2001; Moss & Fitzsimons, 2002). By contrast, the slow force response arises as a result of a corresponding slow increase in the amplitude of the $[Ca^{2+}]_i$ transient (Allen & Kurihara, 1982; Kentish & Wrzosek, 1998), which may, in part, be secondary to increased $[Na^+]_i$ (Alvarez *et al.*, 1999; Perez *et al.*, 2001; Isenberg *et al.*, 2003). Mechanisms which have been linked with the slow response include opening of cationic stretch-activated channels which conduct Na^+ and Ca^{2+} (Calaghan & White, 2004), autocrine/paracrine release of endothelin 1 (ET 1) (Alvarez *et al.*, 1999; Calaghan & White, 2001), activation of the NaH exchanger (NHE) which may be dependent on, or independent of, ET 1 release (Alvarez *et al.*, 1999; von Lewinski *et al.*, 2003; Calaghan & White, 2004), increased cyclic AMP (Calaghan *et al.*, 1999; Bardswell & Kentish, 2005), and activation of eNOS through phosphoinositide 3-kinase (PI3 kinase) (Vila Petroff *et al.*, 2001; Bardswell & Kentish, 2004). The role of cAMP and eNOS in the slow response remain controversial (see Calaghan *et al.*, 2003).

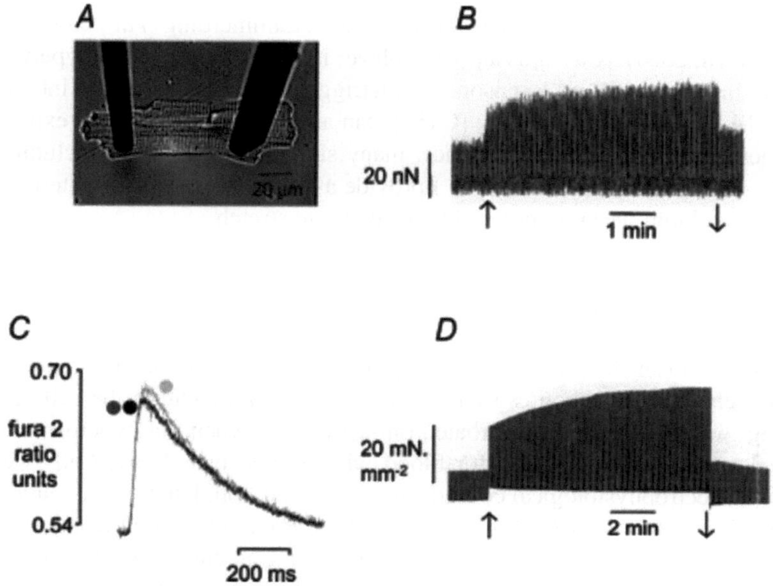

Fig. 12.1 The biphasic inotropic response to stretch seen in single ventricular myocytes and intact cardiac muscle provides an example of mechanotransduction which allows the heart to match cardiac output to demand. A. Rat ventricular myocyte attached to a pair of carbon fibres shows an immediate and slow increase in developed force (B.) when stretched from a sarcomere length (SL) of 1.85 to 2.00 μm. C. In a representative cell loaded with fura-2 and held at a SL of 1.81 μm (dark grey circle), no increase in $[Ca^{2+}]_i$ transient is seen at 15 s after stretch to 1.90 μm (mid grey; traces are superimposed), but at 5 min after stretch (light grey), $[Ca^{2+}]_i$ transient amplitude is increased. D. A similar biphasic response to stretch is seen in rat papillary muscle stretched from 88% to 98% of the length at which maximum force is developed. Taken from Calaghan & White (2004) with permission

12.1.1.2 Cardiac Hypertrophy

Clearly the chronotropic and inotropic response of the heart to stretch is vital to match cardiac output to the circulatory demands of the body. However, stretch can also have pathological consequences and has been linked with the development of hypertrophy, the basic adaptive response of the heart to haemodynamic loading (Ruwhof & van der Laarse, 2000). The process by which mechanical stress is transduced into biochemical signals that regulate protein synthesis (growth), and ultimately disturb myocyte function, is not fully understood. The myocyte itself contains the machinery which allows it to respond to increased load (Sadoshima & Izumo, 1997), but evidence points to an additional role for paracrine signalling involving fibroblasts or endothelial cells (Harada *et al.*, 1997). Signalling components which have been shown to be involved in the hypertrophic process include ET 1, heteromeric G-Proteins ($G_{\alpha q}$), protein kinase C (PKC), the NHE, low molecular weight GTPases (Ras, RhoA and Rac-1), protein kinase A (PKA), mitogen activated protein (MAP) kinases, PI3 kinase, and focal adhesion kinase (FAK) (see (Yamazaki *et al.*, 1997; Yamazaki *et al.*, 1995; Lammerding *et al.*, 2004; Fliegel & Karmazyn, 2004). $[Ca^{2+}]_i$ is a pivotal player in the development of hypertrophy: Calmodulin acts as a $[Ca^{2+}]_i$ sensor and can trigger hypertrophy through interaction with CaM kinase and calcineurin; $[Ca^{2+}]_i$ can also directly affect gene expression (Lammerding *et al.*, 2004). In essence, many signalling molecules (including ET 1, PKC, NHE, PKA, PI3K and Ca^{2+}) provide a link between the short-term (slow response) and long-term (hypertrophic) response to stretch.

12.1.1.3 Cell Swelling

In the heart, cell swelling can occur during acute episodes of ischaemia-reperfusion. During ischaemia, metabolites such as lactate accumulate within the cell causing swelling, which is further exacerbated on reperfusion when the hyperosmotic extracellular milieu is exchanged for blood with normal osmolarity. Cell swelling has many electrophysiological consequences and can modulate Cl^-, K^+ and Ca^{2+} currents (Vandenberg *et al.*, 1996; Okada, 1999). The main current responsible for the regulatory volume decrease after swelling is $I_{Cl,swell}$, but the molecular identity and mechanisms of gating of this channel are still not understood. $I_{Cl,swell}$ is also activated by agents that alter membrane tension and by direct stretch (Baumgarten & Clemo, 2003). Modulation of $I_{Cl,swell}$ by swelling is not instantaneous, and several kinases have been shown to be involved in its activation (tyrosine kinases, PKA, PKC and MAP kinase; see Baumgarten & Clemo, 2003 for a review). $I_{Cl,swell}$ can be activated in ventricular myocytes through pulling on surface integrins, and FAK and Src are involved in this process (Browe & Baumgarten, 2003). $I_{Cl,swell}$ is clearly relevant to cardiac disease; it is persistently activated in failure (e.g. Clemo *et al.*, 1999) and, because its activation shortens action potential duration, it has been implicated in arrythmogenesis (Baumgarten & Clemo, 2003).

Work to date has identified a number of signalling processes which are involved in the response of the cardiac cell to mechanical stimulation, yet the way in which these are activated by stretch is not understood. Over the years, suggestions for components which play a role in the early detection or transduction of mechanical stimuli have included the extracellular matrix-integrin axis, cytoskeleton, and stretch-activated channels (Lammerding *et al.*, 2004). Some 3 decades ago, (Prescott & Brightman, 1976) proposed that caveolae, flask-shaped invaginations of the cell membrane, could act as mechanotransducers. However, it is only recently that the first functional evidence to support a role for caveolae in the mechanotransductive process in the heart has been presented (Kawamura *et al.*, 2003).

12.1.2 Lipid Rafts and Caveolae

Cholesterol and sphingolipids are not uniformly distributed through the cell membrane, and areas of membrane where these are concentrated are more ordered and less fluid (See Simons & Toomre, 2000). These membrane domains are known as

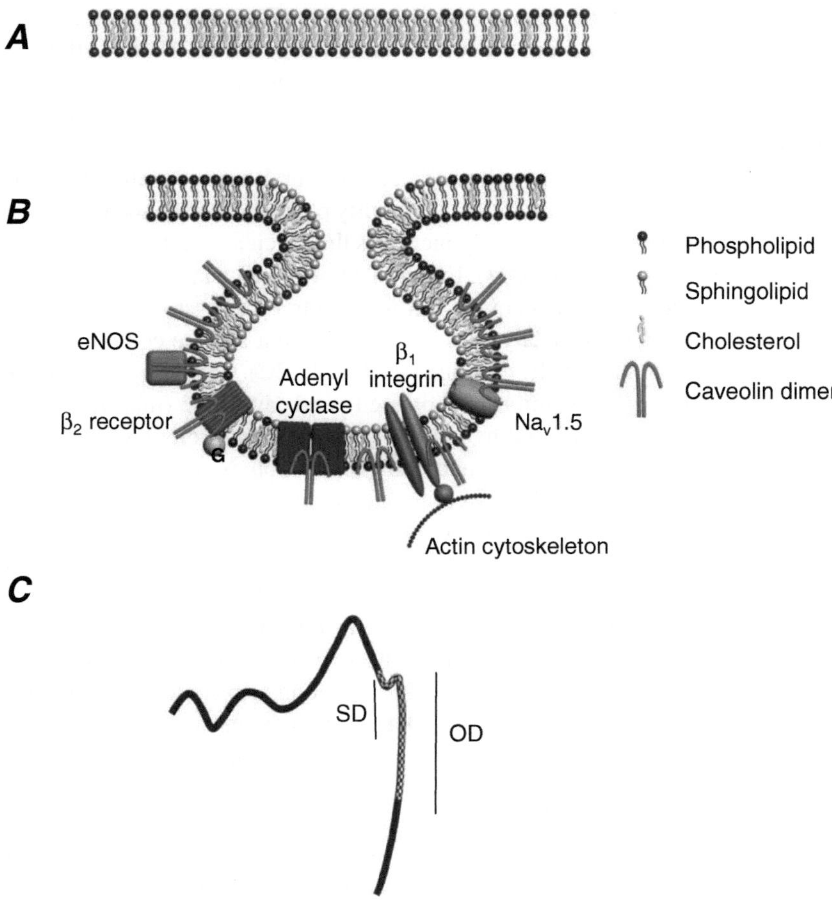

Fig. 12.2 Lipid rafts, caveolae and caveolin. A. Lipids such as cholesterol and sphingolipids are not uniformly distributed throughout the cell membrane but are concentrated in lipid rafts. B. Caveolae are invaginated lipid rafts, distinguished by the presence of caveolae which lines the internal surface of the raft and may be responsible for their typical flask-like shape. A variety of receptors, signalling molecules, ion channels, adapter and structural proteins are enriched in caveolae, suggesting their role as signalosomes. Examples of proteins known to be enriched in caveolae in the adult cardiac myocyte are $Na_V1.5$, β_2 adrenoceptors, $G_{\alpha s}$ proteins, adenyl cyclase and eNOS. In non-cardiac cells, β_1 integrins, actin binding proteins and cytoskeletal actin are also found in association with caveolae. C. Caveolin is a cholesterol-binding protein which adopts a hairpin-like configuration in the membrane. Caveolin forms oligomers within the caveolae via its oligomerisation domain (OD) and interacts with many different signalling molecules through its scaffolding domain (SD). With thanks to Mr Tim Lee, Faculty of Biological Sciences, University of Leeds

lipid rafts (Fig. 12.2A). Some lipid rafts are enriched in the protein caveolin which promotes the typical flask-shaped invagination of the membrane; this subset of lipid rafts are known as caveolae ('little caves'; see Fig. 12.2B and reviews by (Razani *et al.*, 2002; Cohen *et al.*, 2004; Maguy *et al.*, 2006).

Caveolae were first identified by electron microscopists in the 1950s, but it was not until the identification of caveolins that their full significance began to be realised. Caveolae have been assigned many functions including endocytosis, cholesterol homeostasis and tumorigenesis (see Cohen *et al.*, 2004), but perhaps one of their most interesting roles is as 'signalosomes' – orchestrators of signal transduction events. During the 1990s it became clear that a variety of pivotal signalling molecules, including tyrosine kinases, components of G protein-coupled receptor cascades, ion channels, adapter proteins and structural proteins, are enriched in rafts/caveolae (see Razani *et al.*, 2002). The ability of caveolae to concentrate (or exclude) signalling molecules ultimately increases the efficiency and fidelity of signal transduction (Ostrom & Insel, 2004). Enrichment of proteins in caveolae can occur because of the lipid environment of the raft which attracts myristoylated, palmitoylated or prenylated proteins (Zacharias *et al.*, 2002), but a more complex level of regulatory control is provided by caveolin (see Fig. 12.2C). Caveolin exists as 3 isoforms; caveolin 1 and 2 (ubiquitously expressed), and caveolin 3 (muscle-specific). All caveolins have a short central hydrophobic domain that promotes the adoption of a hairpin configuration within the membrane so that both N and C termini face the cytoplasm (Dupree *et al.*, 1993). Caveolins interact with many different signalling molecules via the 20 amino acid scaffolding domain. Direct interactions between caveolin and G protein α subunits, Src family tyrosine kinases, PKA, PKC, and eNOS have been shown (Couet *et al.*, 1997; Li *et al.*, 1996; Oka *et al.*, 1997; Smart *et al.*, 1999). This interaction with caveolin can regulate activity, generally (but not universally) stabilising molecules in an inactive conformation (Okamoto *et al.*, 1998; Razani *et al.*, 2002; Cohen *et al.*, 2004). Caveolin also contains a domain that results in the oligomerisation of 14–16 caveolin monomers which allows for the organisation of a number of elements of a cascade in a macromolecular signalling complex (Sargiacomo *et al.*, 1995).

Most cells contain both lipid rafts and caveolae, yet we do not know the relative contribution of non-caveolar lipid rafts and caveolae to cellular function. Because it is possible to identify caveolae morphologically, and because they can be distinguished by the presence of caveolin, much research has focused on this subset of lipid rafts. However, methods used to disrupt caveolae in order to pinpoint their functional relevance often rely on cholesterol depletion, a technique which will target both caveolar and non-caveolar rafts. Although we have no direct evidence in the adult cardiac cell, it has been reported in neonatal cardiac myocytes that the majority of proteins are concentrated in caveolae rather than non-caveolar rafts (Morris *et al.*, 2006), which suggests a major role for caveolae in these cells.

12.2 Role of Caveolae in Ca^{2+} Regulation

There is a body of evidence linking caveolae with the regulation of $[Ca^{2+}]_i$, and this has implications for the basal function of the myocyte, as well as its response to stimuli (such as stretch) which increase $[Ca^{2+}]_i$. The idea that caveolae regulate $[Ca^{2+}]_i$ was first addressed by Popescu *et al.* (1974) who showed that Ca^{2+} was enriched in smooth muscle cell caveolae, and proposed that caveolae are involved in Ca^{2+} entry during contraction. Likewise, in endothelial cells, Isshiki *et al.* (1998)

demonstrated that Ca^{2+} waves originate at membrane sites rich in caveolin. It has been suggested that the invaginated structure of the caveola could promote efficient Ca^{2+}-induced Ca^{2+} release by decreasing the distance between L-type Ca^{2+} channels and the adjacent ryanodine receptors (RyR) of the sarcoplasmic reticulum (Lohn *et al.*, 2000). Although this may be the case in cells without t-tubules (including neonatal ventricular and adult atrial myocytes), in adult ventricular myocytes it has been shown that the majority of L-type Ca^{2+} channels do not co-localise with caveolin, and are therefore outside caveolae (Scriven *et al.*, 2005). However, this does not preclude a role for caveolae in Ca^{2+} regulation in the adult myocyte. Caveolin 3 is adjacent to a group of extra-dyadic RyR (Scriven *et al.*, 2005), and disruption of caveolae by cholesterol depletion has been shown to reduce the efficiency of the Ca^{2+}-induced Ca^{2+} release process in the adult ventricular myocyte, perhaps through modulation of these extra-dyadic RyRs (Calaghan & White, 2006).

Other proteins which control Ca^{2+}, including the Ca^{2+}-ATPase and the NaCa exchanger (NCX) have been shown to be enriched in caveolae in a variety of cell types (Fujimoto, 1993; Bossuyt *et al.*, 2002). Specifically in the adult heart, most NCX is found in the lipid raft fraction of the cell, and can be co-immunoprecipitated by antibodies to caveolin. This association is disturbed following depletion of membrane cholesterol, resulting in a reduction in NCX activity (Bossuyt *et al.*, 2002). The authors suggest that interaction of caveolin with the XIP region of the NCX prevents endogenous inhibition of activity produced by internal interaction of the XIP region with another part of the exchanger. These data which show that caveolae enhance NCX activity are of potential relevance to the study of cardiac mechanotransductive pathways because the NCX has been implicated in the slow increase in $[Ca^{2+}]_i$ secondary to raised $[Na^+]_i$ seen following stretch (Perez *et al.*, 2001).

The idea of caveolae as a Ca^{2+} microdomain is attractive because of the number of proteins including adenyl cyclase, protein kinase C, and eNOS which are found in caveolae and whose activity is dependent on $[Ca^{2+}]_i$ (Ishikawa & Homcy, 1997; Steinberg *et al.*, 1995; Michel *et al.*, 1997). For example, eNOS is highly enriched in caveolae where interaction with caveolin clamps it in an inactive conformation (Feron *et al.*, 1996). In endothelial cells, Ca^{2+}-calmodulin disrupts the inhibitory interaction between caveolin and eNOS (Michel *et al.*, 1997). If caveolae are sites of Ca^{2+} entry or control, this has implications for the regulation of Ca^{2+}-dependent proteins enriched therein.

12.3 Role of Caveolae in G Protein-Coupled Receptor Signal Cascades

Some of the most persuasive evidence for the role of caveolae as signalosomes comes from the study of G protein-coupled receptor (GPCR) cascades. The distribution of components of these cascades between different membrane microdomains in the adult cardiac cell gives clear support to the idea that many of these pathways are controlled by caveolae. Although the methodology used in this field is prone to artefacts and misinterpretation (see Calaghan & Taggart, 2006), it is generally agreed that β_2 and α_1 adrenoceptor, ET 1 and muscarinic M2 pathways are regulated by caveolae (Rybin *et al.*, 2003; Head *et al.*, 2005; Calaghan & White, 2006; Calaghan, 2006; Balijepalli *et al.*, 2006; Fujita *et al.*, 2001; Boivin *et al.*, 2005; Foell *et al.*, 2006; Feron *et al.*, 1997). Many components of the β adrenoceptor pathway (including β_2 receptors, $G_{\alpha s}$, adenyl cyclase 5/6, PKA RII, and phosphatase 2A) are enriched in the buoyant caveolae-containing fraction of the adult myocyte (Head *et al.*, 2005; Yarbrough *et al.*, 2002; Rybin *et al.*, 2003; Balijepalli *et al.*, 2006). Evidence also places α_1 receptors, $G_{\alpha q}$ protein and phospholipase C exclusively in caveolae (Fujita *et al.*, 2001; Boivin *et al.*, 2005), and Foell *et al.* (2006) report that the ET_A receptor is enriched in caveolae, although this is not found universally (Boivin *et al.*, 2005). The dynamic nature of protein localisation in caveolae is often not considered in such studies, although it is clearly relevant. For example, the M2 receptor moves to caveolae following M2 stimulation (Feron *et al.*, 1997).

Few groups have used a functional approach to study caveolae and GPCR signalling, and of these, most use neonatal cardiac myocytes (Rybin *et al.*, 2000; Xiang *et al.*, 2002; Balijepalli *et al.*, 2006). One recent study in the adult ventricular myocyte, however, has shown that stimulation of $I_{Ca,L}$ by β_2 adrenoceptor stimulation is enhanced by disruption of caveolae through cholesterol depletion, and by disruption of caveolin binding by intracellular dialysis with an antibody to caveolin 3 (Calaghan & White, 2006). These data are consistent with the idea that $G_{\alpha i}$ activation following β_2 stimulation requires caveolae. The $G_{\alpha i}$ pathway is responsible for restriction of the cAMP-dependent signal to the sarcolemma following β_2 adrenoceptor stimulation (Chen-Izu *et al.*, 2000), and indeed, disruption of caveolae in the adult myocyte has been shown to promote a more diffuse β_2 signal which targets the sarcoplasmic reticulum (Calaghan, 2006).

With reference to mechanotransduction, these data are of interest because many stretch-activated signalling pathways converge on G protein coupled receptor cascades. For example the slow response to stretch has been shown to be dependent on a compartmentalised increase in cAMP (Calaghan *et al.*, 1999), and on ET 1 acting through the ET_A receptor (Calaghan & White, 2001; Alvarez *et al.*, 1999; Perez *et al.*, 2001).

12.4 Role of Caveolae in Mechanotransduction

12.4.1 Effect of Stretch on Caveolar Structure

The idea that caveolae are involved in mechanotransduction is an attractive one because caveolae are essentially reservoirs of extra membrane which can be recruited when cells are stretched. Prescott & Brightman (1976) suggested that caveolae may serve as stretch receptors because stretch makes signalling molecules located in the caveolae more accessible. This idea is still under consideration (see Kawamura *et al.*, 2003). An alternative scheme is that deformation of caveolae compromises the organised caveolar structure, removing the constitutive inhibitory influence placed by caveolin on a plethora of signalling molecules (Kawabe *et al.*, 2004; Bellott *et al.*, 2005). Both models assume that the degree of membrane stretch experienced by cells in vivo has the potential to deform caveolae, but early work in this field was not promising. In skeletal muscle, opening of caveolae was shown only to occur following stretch to non-physiological sarcomere lengths (Dulhunty & Franzini-Armstrong, 1975), and alteration in the size or number of caveolae was not observed in stretched smooth muscle preparations (Gabella & Blundell, 1978). However, a recent study by Kohl *et al.* (2003) in adult rabbit myocardium showed evidence of caveolae incorporating into surface membrane during stretch within the physiological range and in response to hypo-osmotic swelling (see Fig. 12.3). Conversely, an increase in the size and number of caveolae is seen in atrial muscle exposed to hypertonic extracellular solution (Kordylewski *et al.*, 1993).

It seems likely that tissue variations in caveolar density underlie the differing morphological sensitivity of individual caveolae to stretch. For example, in rat ventricular muscle caveolar density is lower than in many other tissues (6 per μm^2 of membrane compared with 70–90/μm^2 in endothelial cells, 30/μm^2 in smooth muscle, and 20/μm^2 in skeletal muscle) (Dulhunty & Franzini-Armstrong, 1975; Gabella, 1976; Gabella & Blundell, 1978). For the ventricular myocyte, it has been estimated that including caveolar membrane in the estimate of membrane area will increase this value by 27% (Gabella, 1978). From this we can estimate that 40% of caveolae would open upon stretch from slack length to the length at which maximum tension is generated; this would increase membrane surface area by $\approx$10% assuming that the myocyte is an elliptical cylinder (Boyett *et al.*, 1991). Comparable data in skeletal muscle suggest that only 10–20% of caveolae open upon stretch (Bellott *et al.*, 2005).

Another consideration which will alter the effect of stretch on caveolar conformation is the distribution of caveolae between surface and t-tubular membrane.

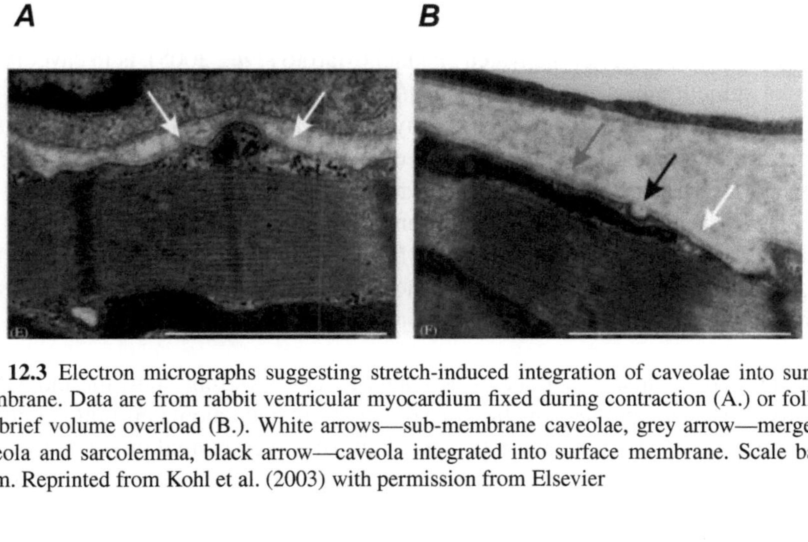

Fig. 12.3 Electron micrographs suggesting stretch-induced integration of caveolae into surface membrane. Data are from rabbit ventricular myocardium fixed during contraction (A.) or following brief volume overload (B.). White arrows—sub-membrane caveolae, grey arrow—merger of caveola and sarcolemma, black arrow—caveola integrated into surface membrane. Scale bar is 1 μm. Reprinted from Kohl et al. (2003) with permission from Elsevier

It seems likely that surface caveolae will be more directly responsive to membrane stretch. In the adult myocardium, caveolae have been visualised in both surface and t-tubular sarcolemma by electron microscopy, but with lower density in t-tubular membranes (Page, 1978). This is supported by confocal microscopy of myocytes immunolabelled for caveolin 3 which show brighter surface than intracellular staining (Fig. 12.4; Calaghan & White, 2006; Scriven *et al.*, 2005).

At present, there is no direct evidence that morphological changes in caveolae are central to mechanotransduction in the cardiac cell, but what we know of the density and distribution of caveolae in these cells means that we cannot exclude physical deformation of the caveolae from a role in the sensing and transduction of mechanical stimuli.

12.4.2 Location of Stretch/Swelling Related Components

A variety of evidence from a range of cells supports a role for caveolae in stretch-dependent signalling. Much of this relates to the raft- or caveolae-based location of components implicated in the response to stretch. Where available, data obtained in the adult cardiac myocyte will be reported.

If we consider the signal cascades which have been shown to contribute to the slow inotropic response to stretch, many elements of these are found in lipid rafts/caveolae or interact with caveolin. In adult cardiac myocytes various components of the cAMP signalling pathway, the ET_A receptor and eNOS have been reported to be enriched in caveolae (Foell *et al.*, 2006; Shaul *et al.*, 1996; Feron *et al.*, 1996). Localisation of eNOS in caveolae is essential for its activation (although under basal conditions interaction with caveolin clamps it in an inactive configuration (Sbaa *et al.*, 2005). PI3 kinase and Akt implicated in the activation of eNOS by axial stretch in the adult cardiac cell (Vila Petroff *et al.*, 2001) have also been shown to be enriched in caveolae (albeit in non-cardiac cells (Liu *et al.*, 1997; Baron *et al.*, 2003). Preferential location of the NHE in lipid rafts is reported in cultured cell lines (Willoughby *et al.*, 2005), and TRPC1, proposed as a component of a vertebrate stretch-activated channel (Maroto *et al.*, 2005), is in caveolae in vascular smooth muscle cells (Bergdahl *et al.*, 2003).

During the development of hypertrophy, signals arising from growth factors and the extracellular matrix-integrin axis converge on tyrosine kinase and small GTPase pathways (see review by Ross, 2004). Tyrosine kinases are greatly enriched in the caveolar fraction of a plethora of different types of cell (Anderson, 1998). In the (neonatal) cardiac myocyte, components linked with the development of hypertrophy, including the MAP kinase ERK, RhoA and Rac1, are found in caveolae, and for RhoA and Rac1 this location is vital for their stretch activation (Kawamura *et al.*, 2003).

In terms of cellular components activated by cell swelling, until the molecular identity of the channel responsible for volume-regulated anionic current is known, its membrane distribution cannot be described. However, several proteins involved in the regulation of $I_{Cl,swell}$, including the Src family tyrosine kinases, Rho and annexin II, are known to be enriched in caveolae (see Martin, 2006), and there is persuasive functional evidence linking caveolae/caveolin and activation of $I_{Cl,swell}$ (see Section 4.4.4).

12.4.3 Caveolae, Caveolin, Integrins and the Cytoskeleton

The extracellular matrix-integrin-cytoskeleton axis provides a bidirectional pathway for signalling between the extra- and intra-cellular environment. Integrins are well established as mechanotransducers in a variety of cells, including the cardiac myocyte (see Ross, 2004). Integrins do not possess enzymatic activity but instead rely on activation of downstream signalling molecules at focal adhesions, including nonreceptor tyrosine kinases such as Src and FAK, or small GTPases such as Rho or Rac (Davis *et al.*, 2001; Ross & Borg, 2001). Recent evidence suggests that integrins and caveolin may interact to facilitate mechanotransductive signalling. β_1 integrins are enriched in caveolae in vascular smooth muscle cells and it has been shown that the integrin-caveolin complex moves from caveolae to non-caveolar membrane upon stretch (Kawabe *et al.*, 2004). Indeed, given the shape of caveolae, for integrins linked to caveolin to form a viable mechanotransductive pathway with the extracellular matrix, it seems likely that this takes place outside caveolae.

Associations between caveolae and the actin cytoskeleton, another cellular structure implicated in mechanotransduction, have also been reported. Lipid rafts and caveolae interact with actin, and this interaction is disrupted by depletion of cellular cholesterol (Harder *et al.*, 1997). In non-cardiac cells, the actin cytoskeleton has been implicated in the organisation and restraint of caveolae within the membrane (Stahlhut & van Deurs, 2000; Thomsen *et al.*, 2002), and in the translocation of caveolin out of caveolae (Kawabe *et al.*, 2004). Likewise, Okada (1999) has suggested that rearrangement of actin microfilaments in response to swelling may release or redistribute the $I_{Cl,swell}$ protein, or its regulators, from the caveolar microdomain. These data suggest that the actin cytoskeleton may play a role in the structure of, and the transport from, caveolae. However, caution should be applied in extrapolating this to the adult cardiac cell in which a highly developed actin cytoskeleton is not evident (see Calaghan *et al.*, 2000).

12.4.4 Functional Evidence for a Role of Caveolae and Caveolin in Mechanotransduction

To date there has been no direct functional assessment of the part that caveolae or caveolin play in mechanotransduction in the adult cardiac myocyte. However, these data are available from other studies many of which have used the vasculature as a model system of mechanotransduction. Blood vessels are continually being subjected to mechanical forces, in the form of shear stress (sensed by the endothelial cells which line the vasculature) and cyclic mechanical strain (as a result of pulsatile blood flow), and these stimuli are important modulators of cell function and morphology under physiological and pathological conditions (see Davies, 1995; Shaw & Xu, 2003).

12.4.4.1 Endothelial and Vascular Smooth Muscle Cells

A close association between shear stress and caveolae is well established. Chronic exposure of cultured endothelial cells to laminar shear stress (flow pre-conditioning) alters caveolin expression and distribution, and leads to an increase in the density of caveolae at the luminal surface. This has the potential to alter, in turn, endothelial cell mechanosensitivity. Shear stress has been shown to increase phosphorylation of a range of proteins including caveolin 1, eNOS, Akt and MAP kinases such as ERK (Boyd *et al.*, 2003; Rizzo *et al.*, 2003). In flow pre-conditioned cells, the shear stress-induced phosphorylation of eNOS and Akt is enhanced, suggesting that the caveolar microdomain facilitates activation of these signalling molecules (Boyd *et al.*, 2003; Rizzo *et al.*, 2003). By contrast, flow pre-conditioning blunts ERK phosphorylation and activation in response to shear stress (Boyd *et al.*, 2003; Rizzo *et al.*, 2003). Although these data are consistent with the negative regulation of ERK by its interaction with caveolin within caveolae, disruption of caveolae by cholesterol depletion and inhibition of caveolin binding by a neutralising antibody to caveolin 1 also attenuate shear activation of ERK (Park *et al.*, 1998; Park *et al.*, 2000). It is possible that there is an optimum range of caveolin levels that allow the proper activation of ERK in response to shear stress, and that deviation from this compromises the mechanotransductive pathway (Boyd *et al.*, 2003).

Functional evidence also supports a role for caveolae in stretch-activated signalling in vascular smooth muscle. Activation of PI3 kinase and Akt in response to cyclic stretch of vascular smooth muscle cells is attenuated by disruption of caveolae by cholesterol depletion or down-regulation of caveolin with antisense oligonucleotides, and this mechanotransductive pathway is almost abolished in vasculature from caveolin knockout animals (Sedding *et al.*, 2005). By contrast to the situation in endothelial cells, in vascular smooth muscle cells in which caveolae are disrupted or caveolin expression is down-regulated, stretch activation of ERK is markedly enhanced (Kawabe *et al.*, 2004). Surprisingly this is not due to a reduction in caveolar caveolin releasing the constitutive brake placed by caveolin on ERK, but rather to an increase in non-caveolar caveolin (see below).

In studies from endothelial cells, data suggest that the phosphorylation of caveolin 1 at Tyr^{14} plays a pivotal role in mechanotransductive signalling. Phosphorylation of caveolin has several potential consequences. One is a disruption of the binding of caveolin through its scaffolding domain (even though this is in a different part of the molecule to the phosphorylation site). It has been proposed that increased Tyr phosphorylation of caveolin 1 seen with shear stress activates eNOS by decreasing the interaction between caveolin and eNOS (Rizzo *et al.*, 1998; Rizzo *et al.*, 2003). Alternatively, the $pTyr^{14}$ site can function as an SH2 binding domain for the docking of signalling molecules, allowing caveolin to act as a scaffold for a macromolecular stretch-activated signalling complex (Cao *et al.*, 2002). A third option is that phosphorylation of caveolin provides a stimulus for the translocation of caveolin out of caveolae (Venema *et al.*, 1997; del Pozo *et al.*, 2005). In endothelial cells, shear stress results in Tyr^{14} phosphorylation of caveolin 1 which promotes its recruitment to β_1 integrin sites (Radel & Rizzo, 2005). At present it is not clear how these data can be applied to the situation in the adult cardiac myocyte; although these cells have been shown to express caveolin 1 and 2 (Head *et al.*, 2005), the predominant isoform is the muscle specific caveolin 3 which lacks the N-terminal Tyr residue of caveolin 1.

Evidence does support the idea that translocation of proteins (including caveolin) from caveolae, as a result of phosphorylation or mechanical disruption of the caveolar structure, is an important step in the mechanotransductive process. In vascular smooth muscle cells, cyclic stretch causes translocation of caveolin, β_1 integrins and Fyn from caveolae to non-caveolar membrane (independent of caveolin phosphorylation), which ultimately results in the activation of ERK (Kawabe *et al.*, 2004). Pivotally, these authors showed that it is the increase in non-caveolar caveolin that is responsible for stretch activation of ERK, highlighting a role for caveolin which is independent of caveolae.

12.4.4.2 Cardiac Myocytes

The only study which provides functional evidence for caveolae/caveolin-dependent mechanotransduction in the cardiac myocyte focused on the activation of the small G proteins RhoA and Rac1 in response to maintained equibiaxial stretch in the neonatal cell. As illustrated in Fig. 12.5, this study showed that stretch promoted activation of RhoA and Rac1 and translocation of these G proteins, but not caveolin 3, from the caveolae compartment to other membrane sites (Kawamura *et al.*, 2003). The intact caveolar compartment was essential for stretch-activation of RhoA and

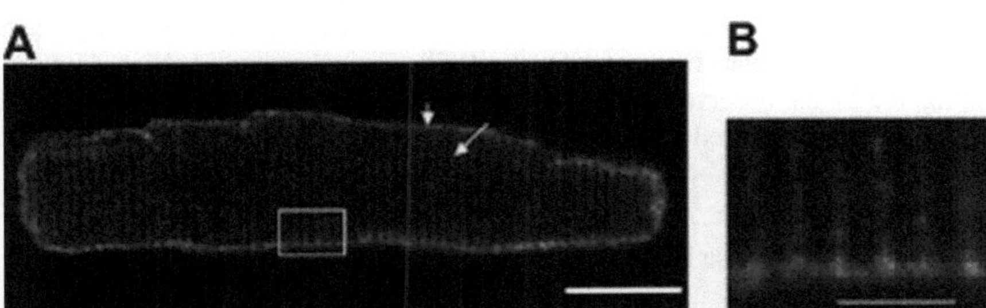

Fig. 12.4 Caveolin 3 is more densely distributed at the surface, rather than t-tubular, membrane in the adult ventricular myocyte. A. Representative confocal image of the mid-plane of a rat ventricular myocyte immunolabelled with an antibody to caveolin 3 showing both surface and t-tubular staining. Scale bar is 20 μm. B. Comparison of surface and t-tubular membranes from the area defined in A. at a higher magnification suggests that caveolae density is higher at the surface membrane. This is supported by data from (Page, 1978; Scriven et al., 2005). Scale bar is 5 μm. Reprinted from Calaghan & White (2006) with permission from European Society of Cardiology

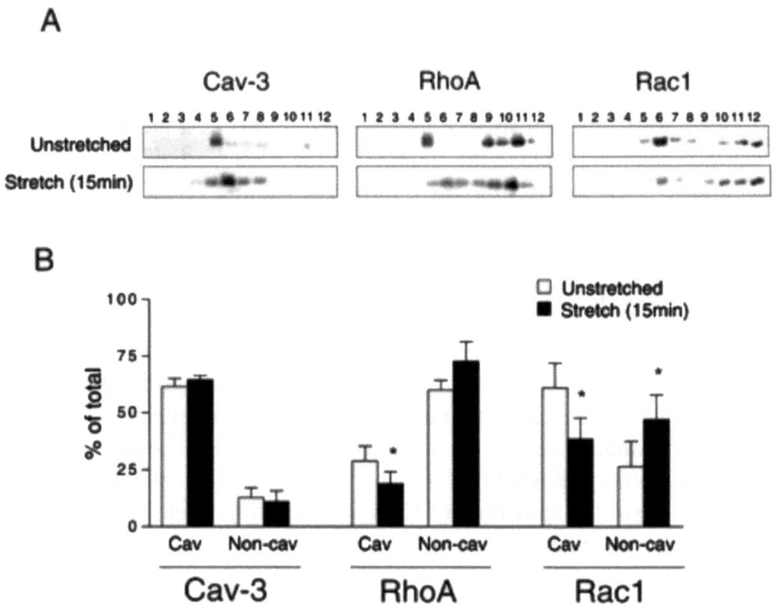

Fig. 12.5 Stretch causes translocation of RhoA and Rac1, but not caveolin 3, from caveolae in the rat neonatal cardiac myocyte. A. Cells were plated on silicon membranes and subject to 20% equibiaxial stretch for 15 min. Proteins in 12 fractions obtained by detergent-free sucrose density gradient fractionation were subject to SDS-PAGE and Western blotting. B. Densitometric analysis of the distribution of caveolin 3, RhoA and Rac1 between caveolar fractions (5,6) and non-caveolar fractions (9–12). Data are mean + S.E.M ($n \geq 3$). * $P < 0.05$ vs. stretch. Reprinted with permission from Kawamura et al. (2003)

Rac1; in cells in which caveolae were disrupted by cholesterol depletion, stretch did not result in G protein activation (Fig. 12.6; Kawamura *et al.*, 2003). These authors suggest that localization of RhoA and Rac1 in caveolae is essential for sensing externally applied force and that stretch, by altering caveolar morphology, renders RhoA and Rac1 in caveolae more accessible to guanine nucleotide exchange facts that catalyse the replacement of GDP by GTP. The relevance of these data was highlighted by the observation that disruption of caveolae by cholesterol depletion prevented stretch-induced hypertrophy in these cells.

12.4.4.3 Slow Response Pathways

There is some functional evidence to support a contribution from caveolae or caveolin to several mediators associated with the slow inotropic response to stretch. Studies which have manipulated the number of caveolae or the expression of caveolin provide clear evidence for a role in PI3 kinase, Akt and eNOS activation in response to shear stress and cyclic stretch (Rizzo *et al.*, 1998; Rizzo *et al.*, 2003; Boyd *et al.*, 2003; Sedding *et al.*, 2005). The response to ET 1 has also been shown to be dependent on caveolae. In vascular smooth muscle, caveolae are vital for the

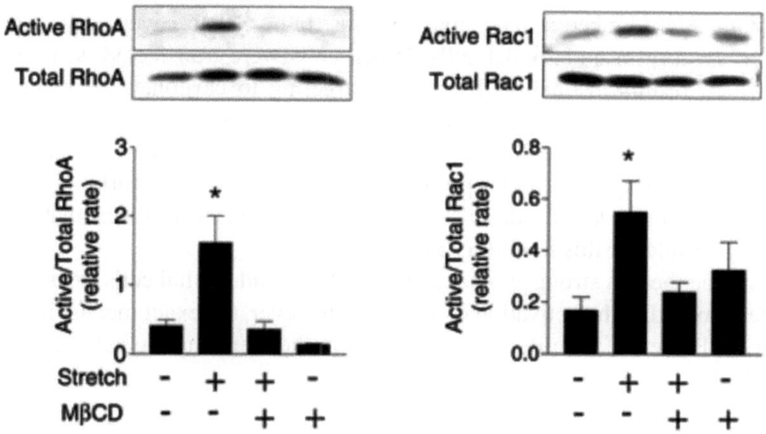

Fig. 12.6 Disruption of caveolae with methyl β cyclodextrin (MβCD) prevents stretch activation of RhoA and Rac1 in rat neonatal cardiac myocytes. Myocytes were plated on silicon membranes and subject to 20% equibiaxial stretch for 4 min. Pulldown assays were performed and blots were probed for RhoA or Rac1. Caveolae were disrupted using a 1h exposure to 5 mM MβCD. Bars are mean + S.E.M (n≥4). * P< 0.05 vs. stretched. Reprinted with permission from Kawamura et al. (2003)

contractile response to ET 1 (Bergdahl *et al.*, 2003), and for the stretch-induced activation of ERK1/2 and increased protein synthesis which is dependent on autocrine ET 1 production (Zeidan *et al.*, 2003). Direct functional evidence in the adult cardiac myocyte is limited, although a recent study presented data consistent with a caveolar-based signalling complex of ET_A receptors, tyrosine kinases, and the L-type Ca^{2+} channel in these cells (Foell *et al.*, 2006).

12.4.4.4 $I_{Cl,swell}$

There is persuasive evidence for a role for caveolae/caveolin in activation of $I_{Cl,swell}$. Cholesterol depletion in endothelial cells enhances swelling-activated current activity in response to modest osmotic challenge, suggesting a negative regulation by caveolae/rafts (Levitan *et al.*, 2000). This is potentially at odds with data showing that caveolin is required for $I_{Cl,swell}$ activation. Expression of caveolin 1 in a caveolin-deficient cell line upregulates $I_{Cl,swell}$ (Trouet *et al.*, 1999), whereas expression of a dominant negative caveolin 1 mutant in an endothelial cell line (with high endogenous caveolin 1 expression) inhibits its activation (Trouet *et al.*, 2001b). It is possible that the effects of cholesterol depletion on channel activation are due to effects on bilayer thickness, mechanical properties and monolayer curvature equilibrium of the membrane rather than on caveolae (see Klausen *et al.*, 2006).

Evidence from adult cardiac myocytes and endothelial cells has shown that $I_{Cl,swell}$ activation in response to a hyposmotic challenge depends on tyrosine phosphorylation (Sorota, 1995; Voets *et al.*, 1998). In a human T-lymphocyte cell line, the tyrosine kinase responsible belongs to the Src family (Lepple-Wienhues *et al.*,

1998). One study looked at the impact of targeting c-Src to caveolae for the activation of $I_{Cl,swell}$ in endothelial cells. When cells were transfected with Ser^3Cys c-Src which results in dual acylation and targeting to caveolae, the response of $I_{Cl,swell}$ to hyposmotic challenge was abrogated (Trouet *et al.*, 2001a). The authors suggest that the location of c-Src in caveolae clamps the kinase in an inactive conformation through a caveolar protein-protein interaction network, although the effect of caveolin on $I_{Cl,swell}$ described by Trouet *et al.* (1999) suggest that caveolin is not responsible for this inhibitory influence.

In essence there is strong evidence (at least from endothelial cells) that the caveolae are involved in the activation of $I_{Cl,swell}$. However, the exact mechanism is still unclear, and whether a similar scenario exists in the adult cardiac cell has yet to be established.

12.5 Mechanotransduction, Caveolae and Disease

Further evidence for a role for caveolae in mechanotransduction comes from the study of human cardiac disease. Disruption in the caveolin 3 gene underlies a family of disorders known as 'caveolinopathies' which present a range of phenotypes, varying in severity, onset and outcome. These include limb girdle muscular dystrophies (Galbiati *et al.*, 2001; Hagiwara *et al.*, 2000), and rippling muscle disease (Betz *et al.*, 2001). Such diseases are thought to reflect, in part, the failure of the mechanotransductive processes vital for skeletal muscle growth and development (see Razani *et al.*, 2002; Cohen *et al.*, 2004).

In the heart, the importance of caveolae/caveolin for the development of cardiac hypertrophy is well illustrated by findings that both caveolin 1 and caveolin 3 null mice develop cardiac hypertrophy (Cohen *et al.*, 2003; Woodman *et al.*, 2002). This has been ascribed to a hyperactivation of the p42/44 MAP kinase cascade either in the cardiac fibroblast (for caveolin 1 knockout) or in the cardiac myocyte itself (for caveolin 3 knockout). The implication of these data is that caveolin normally plays a pivotal role in placing a constitutive brake on this signal cascade, such that loss of caveolin and its ability to negatively regulate the MAP kinase pathway induces the molecular program leading to hypertrophy. Although this evidence suggests that caveolae have the potential to control the response to a hypertrophic stimulus such as hypertension, this has yet to be demonstrated directly in the human heart.

12.6 Conclusions

Mechanotransduction is vital for the normal function of the healthy heart, and has also been linked with the development of cardiac disease such as hypertrophy. We are still trying to understand the process by which mechanical stimuli are sensed and transduced into cellular responses. A body of evidence in a range of tissues supports a role for caveolae in mechanotransduction. Caveolae are well established as microdomains which increase the efficiency or fidelity of signalling by sequestering or excluding signal components; the ability of caveolin oligomers to act as macromolecular scaffolds which directly regulate protein activity adds another level of control to this. The fact that many stretch-activated components have been shown to be enriched in the caveolar fraction of cells, and to interact with caveolin, is good evidence that caveolae are involved in mechanotransduction. A range of functional data lends support to this, although the exact contribution of caveolae to mechanotransductive processes in the cardiac cell is not yet known.

It is possible that mechanical stimuli are physically sensed by caveolae. In the adult cardiac myocyte, data concerning the density and location of caveolae are consistent with the idea that caveolae are effectively a reservoir of membrane that can be incorporated into surface membrane upon stretch. This underpins two models that rely on physical sensation of mechanical stimuli by caveolae. The first of these is that stretch activates signalling pathways by increasing the accessibility of signalling molecules enriched in caveolae; the second that disruption of the organised caveolar structure by stretch is the first step in mechanotransductive signalling. Whilst these are attractive models, there is as yet no conclusive evidence to demonstrate their involvement in mechanotransduction in the cardiac cell.

For several stretch-related signalling molecules, we do know that their location in caveolae is essential for their activation, and that they are regulated by direct activation with caveolin. This arrangement can be disturbed by phosphorylation of caveolin which could represent an early step in the mechanotransductive pathway, leading to disruption of caveolin binding, a change in the scaffolding function of caveolin, or translocation of proteins from caveolae. Functional evidence supports this in non cardiac cells. An alternative scenario places caveolae further downstream in the stretch-activated signalling, where their ability to concentrate and regulate cascade component promotes efficient signal transduction.

Ultimately, we do not know whether caveolae are involved in mechanosensation or further downstream in stretch-activated signalling pathways. However, current evidence lends more weight to the hypothesis that they are essential for the proper integration and co-ordination of mechanosensitive signalling pathways, than acting as the primary cellular mechanosensor.

Acknowledgment Sarah Calaghan is supported by the British Heart Foundation.

References

Allen DG & Kurihara S (1982). The effects of muscle length on intracellular calcium transients in mammalian cardiac muscle. *J Physiol* **327:** 79–94.

Alvarez BV, Perez NG, Ennis IL, Camilion de Hurtado MC, & Cingolani HE (1999). Mechanisms underlying the increase in force and Ca(2+) transient that follow stretch of cardiac muscle: a possible explanation of the Anrep effect. *Circ Res* **85**, 716–722.

Anderson RG (1998). The caveolae membrane system. *Annu Rev Biochem* **67:** 199–225.

Bainbridge FA (1915). The influence of venous filling upon the rate of the heart. *J Physiol* **50**, 65–84.

Balijepalli RC, Foell JD, Hall DD, Hell JW, & Kamp TJ (2006). Localization of cardiac L-type Ca(2+) channels to a caveolar macromolecular signaling complex is required for beta(2)-adrenergic regulation. *Proc Natl Acad Sci U S A* **103**, 7500–7505.

Bardswell S & Kentish J (2004). A role for NO in the slow force response to a length change in cardiac muscle. *J Physiol* **557P**, PC6.

Bardswell S & Kentish J (2005). Effect of beta-adrenoceptor stimulation on the slow force response to a length change in cardiac muscle. *J Mol Cell Cardiol* **39**, 179–180.

Baron W, Decker L, Colognato H, & Ffrench-Constant C (2003). Regulation of integrin growth factor interactions in oligodendrocytes by lipid raft microdomains. *Curr Biol* **13**, 151–155.

Baumgarten CM & Clemo HF (2003). Swelling-activated chloride channels in cardiac physiology and pathophysiology. *Prog Biophys Mol Biol* **82**, 25–42.

Bellott AC, Patel KC, & Burkholder TJ (2005). Reduction of caveolin-3 expression does not inhibit stretch-induced phosphorylation of ERK2 in skeletal muscle myotubes. *J Appl Physiol* **98**, 1554–1561.

Bergdahl A, Gomez MF, Dreja K, Xu SZ, Adner M, Beech DJ, Broman J, Hellstrand P, & Sward K (2003). Cholesterol depletion impairs vascular reactivity to endothelin-1 by reducing store-operated Ca2+ entry dependent on TRPC1. *Circ Res* **93**, 839–847.

Betz RC, Schoser BG, Kasper D, Ricker K, Ramirez A, Stein V, Torbergsen T, Lee YA, Nothen MM, Wienker TF, Malin JP, Propping P, Reis A, Mortier W, Jentsch TJ, Vorgerd M, & Kubisch C (2001). Mutations in CAV3 cause mechanical hyperirritability of skeletal muscle in rippling muscle disease. *Nat Genet* **28**, 218–219.

Boivin B, Villeneuve LR, Farhat N, Chevalier D, & Allen BG (2005). Sub-cellular distribution of endothelin signaling pathway components in ventricular myocytes and heart: lack of preformed caveolar signalosomes. *J Mol Cell Cardiol* **38**, 665–676.

Bossuyt J, Taylor BE, James-Kracke M, & Hale CC (2002). The cardiac sodium-calcium exchanger associates with caveolin-3. *Ann N Y Acad Sci* **976:** 197–204.

Boyd NL, Park H, Yi H, Boo YC, Sorescu GP, Sykes M, & Jo H (2003). Chronic shear induces caveolae formation and alters ERK and Akt responses in endothelial cells. *Am J Physiol Heart Circ Physiol* **285**, H1113-H1122.

Boyett MR, Frampton JE, & Kirby MS (1991). The length, width and volume of isolated rat and ferret ventricular myocytes during twitch contractions and changes in osmotic strength. *Exp Physiol* **76**, 259–270.

Browe DM & Baumgarten CM (2003). Stretch of beta 1 integrin activates an outwardly rectifying chloride current via FAK and Src in rabbit ventricular myocytes. *J Gen Physiol* **122**, 689–702.

Calaghan S (2006). Disruption of caveolae converts local beta2 adrenoceptor signalling to a more diffuse global signal in the rat ventricular myocyte. Proc Physiol Soc **3**, C54

Calaghan S & White E (2004). Activation of Na+-H+ exchange and stretch-activated channels underlies the slow inotropic response to stretch in myocytes and muscle from the rat heart. *J Physiol* **559**, 205–214.

Calaghan S & White E (2006). Caveolae modulate excitation-contraction coupling and beta2-adrenergic signalling in adult rat ventricular myocytes. *Cardiovasc Res* **69**, 816–824.

Calaghan SC, Belus A, & White E (2003). Do stretch-induced changes in intracellular calcium modify the electrical activity of cardiac muscle? *Prog Biophys Mol Biol* **82**, 81–95.

Calaghan SC, Colyer J, & White E (1999). Cyclic AMP but not phosphorylation of phospholamban contributes to the slow inotropic response to stretch in ferret papillary muscle. *Pflugers Arch* **437**, 780–782.

Calaghan SC & Taggart MJ (2006). Compartmentalized signaling in cardiomyocyte lipid domains-Do structure and function match up? *J Mol Cell Cardiol* **41**, 1–3.

Calaghan SC & White E (2001). Contribution of angiotensin II, endothelin 1 and the endothelium to the slow inotropic response to stretch in ferret papillary muscle. *Pflugers Arch* **441**, 514–520.

Calaghan SC, White E, Bedut S, & Le Guennec JY (2000). Cytochalasin D reduces Ca2+ sensitivity and maximum tension via interactions with myofilaments in skinned rat cardiac myocytes. *J Physiol* **529 Pt 2:** 405–411.

Cao H, Courchesne WE, & Mastick CC (2002). A phosphotyrosine-dependent protein interaction screen reveals a role for phosphorylation of caveolin-1 on tyrosine 14: recruitment of C-terminal Src kinase. *J Biol Chem* **277**, 8771–8774.

Cazorla O, Wu Y, Irving TC, & Granzier H (2001). Titin-based modulation of calcium sensitivity of active tension in mouse skinned cardiac myocytes. *Circ Res* **88**, 1028–1035.

Chen-Izu Y, Xiao RP, Izu LT, Cheng H, Kuschel M, Spurgeon H, & Lakatta EG (2000). G(i)-dependent localization of beta(2)-adrenergic receptor signaling to L-type Ca(2+) channels. *Biophys J* **79**, 2547–2556.

Clemo HF, Stambler BS, & Baumgarten CM (1999). Swelling-activated chloride current is persistently activated in ventricular myocytes from dogs with tachycardia-induced congestive heart failure. *Circ Res* **84**, 157–165.

Cohen AW, Hnasko R, Schubert W, & Lisanti MP (2004). Role of caveolae and caveolins in health and disease. *Physiol Rev* **84**, 1341–1379.

Cohen AW, Park DS, Woodman SE, Williams TM, Chandra M, Shirani J, Pereira de SA, Kitsis RN, Russell RG, Weiss LM, Tang B, Jelicks LA, Factor SM, Shtutin V, Tanowitz HB, & Lisanti MP (2003). Caveolin-1 null mice develop cardiac hypertrophy with hyperactivation of p42/44 MAP kinase in cardiac fibroblasts. *Am J Physiol Cell Physiol* **284**, C457-C474.

Cooper PJ, Lei M, Cheng LX, & Kohl P (2000). Selected contribution: axial stretch increases spontaneous pacemaker activity in rabbit isolated sinoatrial node cells. *J Appl Physiol* **89**, 2099–2104.

Couet J, Li S, Okamoto T, Ikezu T, & Lisanti MP (1997). Identification of peptide and protein ligands for the caveolin-scaffolding domain. Implications for the interaction of caveolin with caveolae-associated proteins. *J Biol Chem* **272**, 6525–6533.

Davies PF (1995). Flow-mediated endothelial mechanotransduction. *Physiol Rev* **75**, 519–560.

Davis MJ, Wu X, Nurkiewicz TR, Kawasaki J, Gui P, Hill MA, & Wilson E (2001). Regulation of ion channels by protein tyrosine phosphorylation. *Am J Physiol Heart Circ Physiol* **281**, H1835-H1862.

del Pozo MA, Balasubramanian N, Alderson NB, Kiosses WB, Grande-Garcia A, Anderson RG, & Schwartz MA (2005). Phospho-caveolin-1 mediates integrin-regulated membrane domain internalization. *Nat Cell Biol* **7**, 901–908.

Dulhunty AF & Franzini-Armstrong C (1975). The relative contributions of the folds and caveolae to the surface membrane of frog skeletal muscle fibres at different sarcomere lengths. *J Physiol* **250**, 513–539.

Dupree P, Parton RG, Raposo G, Kurzchalia TV, & Simons K (1993). Caveolae and sorting in the trans-Golgi network of epithelial cells. *EMBO J* **12**, 1597–1605.

Feron O, Belhassen L, Kobzik L, Smith TW, Kelly RA, & Michel T (1996). Endothelial nitric oxide synthase targeting to caveolae. Specific interactions with caveolin isoforms in cardiac myocytes and endothelial cells. *J Biol Chem* **271**, 22810–22814.

Feron O, Smith TW, Michel T, & Kelly RA (1997). Dynamic targeting of the agonist-stimulated m2 muscarinic acetylcholine receptor to caveolae in cardiac myocytes. *J Biol Chem* **272**, 17744–17748.

Fliegel L & Karmazyn M (2004). The cardiac Na-H exchanger: a key downstream mediator for the cellular hypertrophic effects of paracrine, autocrine and hormonal factors. *Biochem Cell Biol* **82**, 626–635.

Foell JD, Balijepalli RC, Lomax J, Shi M, Wang J, Walker JW, Hell JW, & Kamp TJ (2006). Regulation of cardiac L-type calcium channel by non-receptor tyrosine kinases localised to caveolae. *Biophys J (abs)* 779-Plat.

Frank O (1895). Zur dynamik des herzmusckels. *K Biol* **32**, 370–447.

Fujimoto T (1993). Calcium pump of the plasma membrane is localized in caveolae. *J Cell Biol* **120**, 1147–1157.

Fujita T, Toya Y, Iwatsubo K, Onda T, Kimura K, Umemura S, & Ishikawa Y (2001). Accumulation of molecules involved in alpha1-adrenergic signal within caveolae: caveolin expression and the development of cardiac hypertrophy. *Cardiovasc Res* **51**, 709–716.

Gabella G (1976). Quantitative morphological study of smooth muscle cells of the guinea-pig taenia coli. *Cell Tissue Res* **170**, 161–186.

Gabella G (1978). Inpocketings of the cell membrane (caveolae) in the rat myocardium. *J Ultrastruct Res* **65**, 135–147.

Gabella G & Blundell D (1978). Effect of stretch and contraction on caveolae of smooth muscle cells. *Cell Tissue Res* **190**, 255–271.

Galbiati F, Razani B, & Lisanti MP (2001). Caveolae and caveolin-3 in muscular dystrophy. *Trends Mol Med* **7**, 435–441.

Hagiwara Y, Sasaoka T, Araishi K, Imamura M, Yorifuji H, Nonaka I, Ozawa E, & Kikuchi T (2000). Caveolin-3 deficiency causes muscle degeneration in mice. *Hum Mol Genet* **9**, 3047–3054.

Harada M, Itoh H, Nakagawa O, Ogawa Y, Miyamoto Y, Kuwahara K, Ogawa E, Igaki T, Yamashita J, Masuda I, Yoshimasa T, Tanaka I, Saito Y, & Nakao K (1997). Significance of ventricular myocytes and nonmyocytes interaction during cardiocyte hypertrophy: evidence for endothelin-1 as a paracrine hypertrophic factor from cardiac nonmyocytes. *Circulation* **96**, 3737–3744.

Harder T, Kellner R, Parton RG, & Gruenberg J (1997). Specific release of membrane-bound annexin II and cortical cytoskeletal elements by sequestration of membrane cholesterol. *Mol Biol Cell* **8**, 533–545.

Head BP, Patel HH, Roth DM, Lai NC, Niesman IR, Farquhar MG, & Insel PA (2005). G-protein-coupled receptor signaling components localize in both sarcolemmal and intracellular caveolin-3-associated microdomains in adult cardiac myocytes. *J Biol Chem* **280**, 31036–31044.

Isenberg G, Kazanski V, Kondratev D, Gallitelli MF, Kiseleva I, & Kamkin A (2003). Differential effects of stretch and compression on membrane currents and [Na+]c in ventricular myocytes. *Prog Biophys Mol Biol* **82**, 43–56.

Ishikawa Y & Homcy CJ (1997). The adenylyl cyclases as integrators of transmembrane signal transduction. *Circ Res* **80**, 297–304.

Isshiki M, Ando J, Korenaga R, Kogo H, Fujimoto T, Fujita T, & Kamiya A (1998). Endothelial Ca2+ waves preferentially originate at specific loci in caveolin-rich cell edges. *Proc Natl Acad Sci U S A* **95**, 5009–5014.

Kawabe J, Okumura S, Lee MC, Sadoshima J, & Ishikawa Y (2004). Translocation of caveolin regulates stretch-induced ERK activity in vascular smooth muscle cells. *Am J Physiol Heart Circ Physiol* **286**, H1845-H1852.

Kawamura S, Miyamoto S, & Brown JH (2003). Initiation and transduction of stretch-induced RhoA and Rac1 activation through caveolae: cytoskeletal regulation of ERK translocation. *J Biol Chem* **278**, 31111–31117.

Kentish JC & Wrzosek A (1998). Changes in force and cytosolic Ca2+ concentration after length changes in isolated rat ventricular trabeculae. *J Physiol* **506**, 431–444.

Klausen TK, Hougaard C, Hoffmann EK, & Pedersen SF (2006). Cholesterol modulates the volume-regulated anion current in Ehrlich-Lettre ascites cells via effects on Rho and F-actin. *Am J Physiol Cell Physiol* **291**, C757-C771.

Knoll R, Hoshijima M, & Chien K (2003). Cardiac mechanotransduction and implications for heart disease. *J Mol Med* **81**, 750–756.

Kohl P, Cooper PJ, & Holloway H (2003). Effects of acute ventricular volume manipulation on in situ cardiomyocyte cell membrane configuration. *Prog Biophys Mol Biol* **82**, 221–227.

Kordylewski L, Goings GE, & Page E (1993). Rat atrial myocyte plasmalemmal caveolae in situ. Reversible experimental increases in caveolar size and in surface density of caveolar necks. *Circ Res* **73**, 135–146.

Lammerding J, Kamm RD, & Lee RT (2004). Mechanotransduction in cardiac myocytes. *Ann N Y Acad Sci* **1015:** 53–70.

Lepple-Wienhues A, Szabo I, Laun T, Kaba NK, Gulbins E, & Lang F (1998). The tyrosine kinase p56lck mediates activation of swelling-induced chloride channels in lymphocytes. *J Cell Biol* **141**, 281–286.

Levitan I, Christian AE, Tulenko TN, & Rothblat GH (2000). Membrane cholesterol content modulates activation of volume-regulated anion current in bovine endothelial cells. *J Gen Physiol* **115**, 405–416.

Li S, Couet J, & Lisanti MP (1996). Src tyrosine kinases, Galpha subunits, and H-Ras share a common membrane-anchored scaffolding protein, caveolin. Caveolin binding negatively regulates the auto-activation of Src tyrosine kinases. *J Biol Chem* **271**, 29182–29190.

Liu J, Oh P, Horner T, Rogers RA, & Schnitzer JE (1997). Organized endothelial cell surface signal transduction in caveolae distinct from glycosylphosphatidylinositol-anchored protein microdomains. *J Biol Chem* **272**, 7211–7222.

Lohn M, Furstenau M, Sagach V, Elger M, Schulze W, Luft FC, Haller H, & Gollasch M (2000). Ignition of calcium sparks in arterial and cardiac muscle through caveolae. *Circ Res* **87**, 1034–1039.

Maguy A, Hebert TE, & Nattel S (2006). Involvement of lipid rafts and caveolae in cardiac ion channel function. *Cardiovasc Res* **69**, 798–807.

Maroto R, Raso A, Wood TG, Kurosky A, Martinac B, & Hamill OP (2005). TRPC1 forms the stretch-activated cation channel in vertebrate cells. *Nat Cell Biol* **7**, 179–185.

Martin S (2006). Caveolae and cell swelling. Focus on "Stimulation by caveolin-1 of the hypotonicity-induced release of taurine and ATP at basolateral, but not apical, membrane of Caco-2 cells". *Am J Physiol Cell Physiol* **290**, C1273-C1274.

Michel JB, Feron O, Sacks D, & Michel T (1997). Reciprocal regulation of endothelial nitric-oxide synthase by $Ca2^{+}$-calmodulin and caveolin. *J Biol Chem* **272**, 15583–15586.

Morris JB, Huynh H, Vasilevski O, & Woodcock EA (2006). Alpha1-adrenergic receptor signaling is localized to caveolae in neonatal rat cardiomyocytes. *J Mol Cell Cardiol* **41**, 17–25.

Moss RL & Fitzsimons DP (2002). Frank-Starling relationship: long on importance, short on mechanism. *Circ Res* **90**, 11–13.

Oka N, Yamamoto M, Schwencke C, Kawabe J, Ebina T, Ohno S, Couet J, Lisanti MP, & Ishikawa Y (1997). Caveolin interaction with protein kinase C. Isoenzyme-dependent regulation of kinase activity by the caveolin scaffolding domain peptide. *J Biol Chem* **272**, 33416–33421.

Okada Y (1999). A scaffolding for regulation of volume-sensitive Cl- channels. *J Physiol* **520 Pt 1**, 2.

Okamoto T, Schlegel A, Scherer PE, & Lisanti MP (1998). Caveolins, a family of scaffolding proteins for organizing "preassembled signaling complexes" at the plasma membrane. *J Biol Chem* **273**, 5419–5422.

Ostrom RS & Insel PA (2004). The evolving role of lipid rafts and caveolae in G protein-coupled receptor signaling: implications for molecular pharmacology. *Br J Pharmacol* **143**, 235–245.

Page E (1978). Quantitative ultrastructural analysis in cardiac membrane physiology. *Am J Physiol* **235**, C147-C158.

Park H, Go YM, Darji R, Choi JW, Lisanti MP, Maland MC, & Jo H (2000). Caveolin-1 regulates shear stress-dependent activation of extracellular signal-regulated kinase. *Am J Physiol Heart Circ Physiol* **278**, H1285-H1293.

Park H, Go YM, St John PL, Maland MC, Lisanti MP, Abrahamson DR, & Jo H (1998). Plasma membrane cholesterol is a key molecule in shear stress-dependent activation of extracellular signal-regulated kinase. *J Biol Chem* **273**, 32304–32311.

Patterson S & Starling EH (1914). On the mechanical factors which determine the output of the ventricles. *J Physiol* **48**, 337–357.

Perez NG, de Hurtado MC, & Cingolani HE (2001). Reverse mode of the Na+-Ca2+ exchange after myocardial stretch: underlying mechanism of the slow force response. *Circ Res* **88**, 376–382.

Popescu LM, Diculescu I, Zelck U, & Ionescu N (1974). Ultrastructural distribution of calcium in smooth muscle cells of guinea-pig taenia coli. A correlated electron microscopic and quantitative study. *Cell Tissue Res* **154**, 357–378.

Prescott L & Brightman MW (1976). The sarcolemma of Aplysia smooth muscle in freeze-fracture preparations. *Tissue Cell* **8**, 248–258.

Radel C & Rizzo V (2005). Integrin mechanotransduction stimulates caveolin-1 phosphorylation and recruitment of Csk to mediate actin reorganization. *Am J Physiol Heart Circ Physiol* **288**, H936-H945.

Razani B, Woodman SE, & Lisanti MP (2002). Caveolae: from cell biology to animal physiology. *Pharmacol Rev* **54**, 431–467.

Rizzo V, Morton C, DePaola N, Schnitzer JE, & Davies PF (2003). Recruitment of endothelial caveolae into mechanotransduction pathways by flow conditioning in vitro. *Am J Physiol Heart Circ Physiol* **285**, H1720-H1729.

Rizzo V, Sung A, Oh P, & Schnitzer JE (1998). Rapid mechanotransduction in situ at the luminal cell surface of vascular endothelium and its caveolae. *J Biol Chem* **273**, 26323–26329.

Ross RS (2004). Molecular and mechanical synergy: cross-talk between integrins and growth factor receptors. *Cardiovasc Res* **63**, 381–390.

Ross RS & Borg TK (2001). Integrins and the myocardium. *Circ Res* **88**, 1112–1119.

Ruwhof C & van der Laarse A (2000). Mechanical stress-induced cardiac hypertrophy: mechanisms and signal transduction pathways. *Cardiovasc Res* **47**, 23–37.

Rybin VO, Pak E, Alcott S, & Steinberg SF (2003). Developmental changes in beta2-adrenergic receptor signaling in ventricular myocytes: the role of Gi proteins and caveolae microdomains. *Mol Pharmacol* **63**, 1338–1348.

Rybin VO, Xu X, Lisanti MP, & Steinberg SF (2000). Differential targeting of beta -adrenergic receptor subtypes and adenylyl cyclase to cardiomyocyte caveolae. A mechanism to functionally regulate the cAMP signaling pathway. *J Biol Chem* **275**, 41447–41457.

Sadoshima J & Izumo S (1997). The cellular and molecular response of cardiac myocytes to mechanical stress. *Annu Rev Physiol* **59:** 551–571.

Sargiacomo M, Scherer PE, Tang Z, Kubler E, Song KS, Sanders MC, & Lisanti MP (1995). Oligomeric structure of caveolin: implications for caveolae membrane organization. *Proc Natl Acad Sci U S A* **92**, 9407–9411.

Sbaa E, Frerart F, & Feron O (2005). The double regulation of endothelial nitric oxide synthase by caveolae and caveolin: a paradox solved through the study of angiogenesis. *Trends Cardiovasc Med* **15**, 157–162.

Scriven DR, Klimek A, Asghari P, Bellve K, & Moore ED (2005). Caveolin-3 is adjacent to a group of extradyadic ryanodine receptors. *Biophys J* **89**, 1893–1901.

Sedding DG, Hermsen J, Seay U, Eickelberg O, Kummer W, Schwencke C, Strasser RH, Tillmanns H, & Braun-Dullaeus RC (2005). Caveolin-1 facilitates mechanosensitive protein kinase B (Akt) signaling in vitro and in vivo. *Circ Res* **96**, 635–642.

Shaul PW, Smart EJ, Robinson LJ, German Z, Yuhanna IS, Ying Y, Anderson RG, & Michel T (1996). Acylation targets emdothelial nitric-oxide synthase to plasmalemmal caveolae. *J Biol Chem* **271**, 6518–6522.

Shaw A & Xu Q (2003). Biomechanical stress-induced signaling in smooth muscle cells: an update. *Curr Vasc Pharmacol* **1**, 41–58.

Simons K & Toomre D (2000). Lipid rafts and signal transduction. *Nat Rev Mol Cell Biol* **1**, 31–39.

Smart EJ, Graf GA, McNiven MA, Sessa WC, Engelman JA, Scherer PE, Okamoto T, & Lisanti MP (1999). Caveolins, liquid-ordered domains, and signal transduction. *Mol Cell Biol* **19**, 7289–7304.

Sorota S (1995). Tyrosine protein kinase inhibitors prevent activation of cardiac swelling-induced chloride current. *Pflugers Arch* **431**, 178–185.

Stahlhut M & van Deurs B (2000). Identification of filamin as a novel ligand for caveolin-1: evidence for the organization of caveolin-1-associated membrane domains by the actin cytoskeleton. *Mol Biol Cell* **11**, 325–337.

Steinberg SF, Goldberg M, & Rybin VO (1995). Protein kinase C isoform diversity in the heart. *J Mol Cell Cardiol* **27**, 141–153.

Thomsen P, Roepstorff K, Stahlhut M, & van DB (2002). Caveolae are highly immobile plasma membrane microdomains, which are not involved in constitutive endocytic trafficking. *Mol Biol Cell* **13**, 238–250.

Trouet D, Carton I, Hermans D, Droogmans G, Nilius B, & Eggermont J (2001a). Inhibition of VRAC by c-Src tyrosine kinase targeted to caveolae is mediated by the Src homology domains. *Am J Physiol Cell Physiol* **281**, C248-C256.

Trouet D, Hermans D, Droogmans G, Nilius B, & Eggermont J (2001b). Inhibition of volume-regulated anion channels by dominant-negative caveolin-1. *Biochem Biophys Res Commun* **284**, 461–465.

Trouet D, Nilius B, Jacobs A, Remacle C, Droogmans G, & Eggermont J (1999). Caveolin-1 modulates the activity of the volume-regulated chloride channel. *J Physiol* **520 Pt 1,** 113–119.

Vandenberg JI, Rees SA, Wright AR, & Powell T (1996). Cell swelling and ion transport pathways in cardiac myocytes. *Cardiovasc Res* **32,** 85–97.

Venema VJ, Ju H, Zou R, & Venema RC (1997). Interaction of neuronal nitric-oxide synthase with caveolin-3 in skeletal muscle. Identification of a novel caveolin scaffolding/inhibitory domain. *J Biol Chem* **272,** 28187–28190.

Vila Petroff MG, Kim SH, Pepe S, Dessy C, Marban E, Balligand JL, & Sollott SJ (2001). Endogenous nitric oxide mechanisms mediate the stretch dependence of Ca2+ release in cardiomyocytes. *Nat Cell Biol* **3,** 867–873.

Voets T, Manolopoulos V, Eggermont J, Ellory C, Droogmans G, & Nilius B (1998). Regulation of a swelling-activated chloride current in bovine endothelium by protein tyrosine phosphorylation and G proteins. *J Physiol* **506,** 341–352.

von Lewinski D, Stumme B, Maier LS, Luers C, Bers DM, & Pieske B (2003). Stretch-dependent slow force response in isolated rabbit myocardium is Na+ dependent. *Cardiovasc Res* **57,** 1052–1061.

Willoughby D, Masada N, Crossthwaite AJ, Ciruela A, & Cooper DM (2005). Localized Na+/H+ exchanger 1 expression protects Ca2+-regulated adenylyl cyclases from changes in intracellular pH. *J Biol Chem* **280,** 30864–30872.

Woodman SE, Park DS, Cohen AW, Cheung MW, Chandra M, Shirani J, Tang B, Jelicks LA, Kitsis RN, Christ GJ, Factor SM, Tanowitz HB, & Lisanti MP (2002). Caveolin-3 knock-out mice develop a progressive cardiomyopathy and show hyperactivation of the p42/44 MAPK cascade. *J Biol Chem* **277,** 38988–38997.

Xiang Y, Rybin VO, Steinberg SF, & Kobilka B (2002). Caveolar localization dictates physiologic signaling of beta 2-adrenoceptors in neonatal cardiac myocytes. *J Biol Chem* **277,** 34280–34286.

Yamazaki T, Komuro I, Kudoh S, Zou Y, Shiojima I, Mizuno T, Takano H, Hiroi Y, Ueki K, Tobe K, & . (1995). Mechanical stress activates protein kinase cascade of phosphorylation in neonatal rat cardiac myocytes. *J Clin Invest* **96,** 438–446.

Yamazaki T, Komuro I, Zou Y, Kudoh S, Mizuno T, Hiroi Y, Shiojima I, Takano H, Kinugawa K, Kohmoto O, Takahashi T, & Yazaki Y (1997). Protein kinase A and protein kinase C synergistically activate the Raf-1 kinase/mitogen-activated protein kinase cascade in neonatal rat cardiomyocytes. *J Mol Cell Cardiol* **29,** 2491–2501.

Yarbrough TL, Lu T, Lee HC, & Shibata EF (2002). Localization of cardiac sodium channels in caveolin-rich membrane domains: regulation of sodium current amplitude. *Circ Res* **90,** 443–449.

Zacharias DA, Violin JD, Newton AC, & Tsien RY (2002). Partitioning of lipid-modified monomeric GFPs into membrane microdomains of live cells. *Science* **296,** 913–916.

Zeidan A, Broman J, Hellstrand P, & Sward K (2003). Cholesterol dependence of vascular ERK1/2 activation and growth in response to stretch: role of endothelin-1. *Arterioscler Thromb Vasc Biol* **23,** 1528–1534.

Chapter 13
Multimodal Activation and Regulation of Neuronal Mechanosensitive Cation Channels

Mario Pellegrino, Cristina Barsanti, and Monica Pellegrini

Abstract Recent findings show that mechanosensitive cation channels are expressed in the central nervous system. These molecules can be found not only, as expected, in mechanosensory cells but also in neurons not involved in sensory mechanotransduction. The expression of these channels in nonspecialized neurons is related to the need for cells to deal with general functions such as volume and electrolyte homeostasis as well as cell movement regulation. Since adhesion and advance of nerve growth cones are associated with changes in membrane tension and with oscillations of intracellular calcium concentration, mechanosensitive cation channels may play critical roles in neurite growth. In keeping with this, elementary mechanosensitive cation currents can be recorded from membranes of neuronal growth cones.

Large conductance mechanosensitive cation channels have been investigated in central neurons of the leech. A multimodal activation, by membrane potential, intracellular calcium and pH, as well as a powerful modulation by adenosine nucleotides have been recently established. These features are discussed as possible mechanisms enabling these channels to contribute to neurite remodeling.

Key words: Mechanosensitive ion channels · Single channel recording · TRP channel · Adenosine nucleotides

13.1 Introduction

Mechanosensitive ion channels (MS) have been found in nerve and muscle cells, as well as in most other cell types (Sackin, 1995; Sachs and Morris, 1998; Martinac, 2004). The presence of these channels also in neurons not involved in sensory mechanoreception has been related to general cellular functions such as volume and electrolyte homeostasis. In particular, mechanical interaction between cell and substrate is quite relevant for extending neurites of developing or regenerating neurons since the nature of the substrate on which growth cones navigate has a powerful effect on the remodeling of the nerve cell processes (Grumbacher-Reinert, 1989). The involvement of stretch-activated K^+ channels in the neurite growth,

A. Kamkin and I. Kiseleva (eds.), *Mechanosensitive Ion Channels.*

as tension-dependent modulators of membrane potential, has been put forward (Sigurdson and Morris, 1989). However, the failure to evoke macroscopic currents in response to mechanical stimuli applied to the growth cones was found not consistent with this hypothesis (Morris and Horn, 1991). On the other hand, adhesion and dynamics of nerve growth cones are associated both with detectable changes in membrane tension (Lamoureux et al, 1989) and with transient oscillations of intracellular free calcium (Gomez and Spitzer, 1999). Therefore, more recently, a role of stretch-activated cation channels (SACCs) in neurite growth has been newly supported by the observation that gentamicin, which blocks SACCs, affects neurite outgrowth (Calabrese et al, 1999). A role of SACCs as regulators of cell movement has been also found in non-neuronal cells (Lee et al, 1999). The picture emerging in the last few years not only confirms a key role of neuronal MS channels in neurite growth but also shows that their susceptibility is under local and complex regulation (Jacques-Fricke et al, 2006). Furthermore, both the molecular identification (Li et al, 2005; Shim et al, 2005; Wang and Poo, 2005) and the functional expression (Pellegrini et al, 2001) of some of these channels are revealing their quality of being multimodally activated (Barsanti et al, 2006a-b). SACCs of leech neurons are similar to typical transduction channels of hair cells in vertebrates and have biophysical features which are consistent with a TRP (Transient Receptor Potential) channel molecular structure. This review will describe the multimodal sensitivity of these channels to various cellular signals and their powerful regulation by adenosine nucleotides. These properties will be discussed to support the idea that neuronal SACCs can act as molecular integrators of chemical and physical stimuli and might participate in the shaping of the calcium signals and in the intracellular Ca^{2+}/cAMP cross-talk.

13.2 Stretch-Activated Channels of Leech Neurons

This section is addressed to summarize the properties of SACCs of leech neurons. Large conductance stretch-activated channels were identified in the soma membrane of leech central neurons using single-channel recording. Both cell-attached and inside-out patches displaying unitary currents, sensitive to negative pressure applied to the pipette, were first studied in symmetrical high K^+ (120 mM) solutions (Pellegrino et al, 1990). The current-voltage relation exhibited an outward rectification and the single channel slope conductance at depolarizing membrane potentials was about 200 pS. Experiments of ion substitution indicated a cation selectivity and the single-channel currents in symmetrical Na^+ were about 70% of those measured in symmetrical K^+ solutions.

Ca^{2+} can carry current through these channels. Single channel activity was recorded from inside-out patches by filling the pipette tip with 120 mM KCl and backfilling with 110 mM $CaCl_2$, according to the technique described by Auerbach (1991). This approach allowed an unambiguously identification of the unitary currents carried by K^+ at the beginning of the experiment and the measurement of their changes while Ca^{2+} was diffusing to the electrode tip. In the chosen experimental conditions, the time taken by $Ca^{2+}$to nearly completely substitute K^+ was about 20 minutes and the current amplitude decreased to reach a steady state value corresponding to about 65 pS. Similar values were measured with pipettes filled with high calcium solution (110 mM) (Calabrese et al, 1999). The ability of these channels to contribute to the intracellular calcium homeostasis was measured by fura-2 imaging. Cell bodies of identified leech neurons were mechanically isolated, plated onto concanavalin A and maintained in culture for a few days. After loading, ratiometric measurements of intracellular free calcium in single cell bodies were carried out, while transiently perfusing with hypotonic solution, at a physiological external calcium concentration (1.8 mM). Macroscopic reversible $[Ca^{2+}]_i$ elevations were induced by hypotonic swelling associated with an approximately 10% increase of cell surface. Two components contribute to the calcium response: one is independent of external calcium and is presumably due to the ion release from internal stores, the other component is sensitive to $Gd^{+3}$and to gentamicin (Barsanti et al, 2006b). Leech mechanosensitive channels are sensitive to the extracellular application of the three main non specific blockers (Hamill and McBride, 1996). Gd^{+3} induces a fast flickery voltage-independent block of the single channel activity. Amiloride causes a partial and voltage-dependent block. Gentamicin does not significantly affect the single channel current at positive membrane potentials, whereas at negative potentials it completely blocks the pore (Calabrese et al, 1999). This aminoglycoside, unlike Gd^{+3}, is unable to block voltage-dependent Ca^{2+} channels of leech neurons at a dose of 200 μM, which completely blocks mechanosensitive currents.

In cell-attached configuration, slow hypotonic perfusion of the whole preparation induces cell swelling and parallel channel activation, while slow negative pressure applied to the patch pipette performs a rough stimulation of leech channels in inside-out patches (Pellegrino et al, 1990). To overcome the limitation due to the slow rate of change of applied pressure, SACC activity was analysed by applying to the patch pipette fast pressure changes, according to Hurwitz and Segal (2001): fast channel activation and deactivation were observed in keeping with a direct activation mechanism (Hamill and McBride, 1996).

The neurite outgrowth of identified leech neurons can be studied in culture for few weeks. The neuronal growth cones express mechanosensitive channels in high density. Usually, their open probability is high, even in the absence of applied pressure to the patch electrode, moreover, they are sensitive to stretch both in cell-attached and in inside-out configuration. Using time-lapse video recording, gentamicin was found to induce a transient increase in the number of neurite segments of the same mean length, between 10 and 20 hours after cell plating (Calabrese et al, 1999). Although this was the first evidence that SACCs down-regulated axon outgrowth, an accurate confirmation of this role has recently come with the use of either gentamicin or a highly specific peptide, isolated from Grammostola spatulata spider venom, on Xenopus spinal neurons (Jacques-Fricke et al, 2006).

The functional properties of leech SACCs were comparatively studied in membranes of freshly desheathed ganglia and of their counterparts in culture; clear-cut differences in the activity pattern of channels from quiescent and growing neurons were found. Two distinctive activity modes named spike-like (SL) and multiconductance (MC) could be characterized (Pellegrini et al, 2001). Channel activity in patches excised from freshly desheathed cell bodies typically consisted of bursts of activity with transition frequencies of up to tens of Hz and brief dwell times. The all points histogram showed a single open level corresponding to 115 pS in symmetrical Na^+ solutions. The mean channel open time never exceeded 10 ms and the distributions of the open times could be fitted by the sum of two exponentials, with time constants shorter than 2 and 10 ms. Conversely, in the same experimental conditions, current traces from patches of growing neurons displayed a typical pattern consisting of long openings occurring at frequencies lower than 5 Hz. Open current fluctuated between a number of distinct levels. The all points histogram presented well defined peaks, indicating the presence of a main level corresponding to 115 pS, conductance sublevels corresponding to about 40, and 80 pS, along with a superlevel of 150 pS. The mean channel open time exceeded 200 ms and the exponentials fitting the distribution of open dwell times had time constants of about 15 and 150 ms.

Since the two activity modes share main conductance, outward rectification, activation by stretch and depolarization as well as specific type of blocking by Gd^{+3} and gentamicin, it is likely that these modes belong to the same ion channel. The attractive hypothesis that SACCs might exhibit different modes as a consequence of facing different microenvironmental conditions has been considered. However, with the available evidence we cannot establish the physiological meaning of the two activity modes. They might reveal a different response to excision, while it is unlikely that soluble modulating factors are relevant, because of the absence of rundown phenomena. One of the simplest hypotheses to account for the two modes is that the assembly of membrane-cytoskeleton complex of cultured neurons, affected by cell adhesion and growth, might be favourable to MC mode. The possibility of molecular complexes with different subunits underlying the two modes will be discussed below.

13.3 Polymodal Activation

SACCs excised from identified cell bodies of desheathed ganglia or from the same neurons in culture can be slowly and reversibly activated by depolarizing membrane potentials, in the absence of applied pressure (Menconi et al, 2001). This is in agreement with the observations in Xenopus oocytes reported by Silberberg and Magleby (1997). Moreover, we established that modifying basal membrane tension to inside-out patches the concurrent voltage-induced channel activation was affected, negative pressure enhancing whereas positive pressure reducing channel activity. The delay of voltage-induced activation was widely variable from patch to patch but inversely related to the mean level of channel activity, as expected, assuming that basal channel activity is a mirror of the resting tension in the membrane patch. Voltage-induced channel activation was observed both in inside-out and outside-out configuration, at negative and positive reference potentials respectively, confirming that membrane depolarization was equivalent to convex membrane curving. An interesting outcome of this study was that SACCs in membrane patches from growing neurons showed significantly shorter voltage-induced activation delay than their counterparts from quiescent neurons, suggesting an enhanced channel mechano-susceptibility in regenerating neurons (Menconi et al, 2001). The nature of the mechanism underlying the slow voltage-dependent activation of SACCs is still unclear. On one hand, a direct quantitative evaluation of voltage-induced membrane movements was made and the converse flexoelectric effect was studied by combining atomic force microscopy with whole-cell patch clamp. Depolarization was found to move up the cantilever and the peak displacement was found to be linear with voltage (Petrov et al, 1993; Mosbacher et al, 1998; Zhang et al, 2001). On the other hand, a direct effect of membrane potential on the channel or on structures associated with it should be expected in view of the fact that the superfamily of TRP channels, to which leech SACCs presumably belong, are structurally equipped to exhibit a peculiar, slow voltage dependence (Nilius et al, 2005).

SACCs were found to be affected by intracellular pH. The effects of intracellular acidification was studied in inside-out membrane patches under perfusion, in the absence of pressure stimulation. Mean channel open time and opening frequency were reversibly increased by a pH_i reduction from 7.2 to 6.2 both in SL and in MC activity mode. Interestingly, channels in SL mode displayed a stronger activation whereas those in MC mode were inhibited by a further reduction of pH_i to 5.5. Moreover, intracellular acidosis depressed the single-channel conductance only in SL mode (Barsanti et al, 2006b).

While the two modes were both affected by pH_i , although with some differences, they showed a clear-cut difference in the sensitivity to changes of calcium concentration in the solution bathing the internal side. Unlike MC mode, SL mode was strongly activated by intracellular calcium. The dose-response curve had a threshold of 1 μM and saturation at 10 μM. The selective activation of intracellular calcium of SACCs was confirmed in membrane patches containing channels exhibiting the two modes (Barsanti et al, 2006b). In section 2 it has been reported that leech SACCs exhibit a high calcium conductance and contribute to the swelling-induced calcium response. The calcium concentrations required to activate SL channels are far from the resting level and can be probably reached when calcium is released by the internal stores. Although this does not appear as a true store-operated mechanism (Parekh and Putney, 2005), it can attain the same final goal. Thus, SACCs are expected to be recruited under calcium release and can work as amplifiers of calcium signals both directly, through calcium-activated calcium currents, and indirectly, by depolarization and consequent activation of voltage-gated calcium channels. The scheme in Fig. 13.1 shows two positive feedback mechanisms which might yield the signal amplification.

Mechanosensitive channels belong to unrelated molecular families. Well-characterized mechanosensitive channels in eukaryotes have been identified among the two pore-domain K^+ channels TREK and TRAAK (Patel et al, 1998; Maingret et al, 1999), DEG/ENaC channels (Tavernarakis and Driscoll, 1997) and TRP cation channels (Lin and Corey, 2005). The multimodal activation is emerging as a notable common property of some mechanosensitive channels. For example, TREK members of 2P-domain channels, which have been extensively studied, respond to various intra- and extracellular signals, such as stretch or swelling, low intracellular

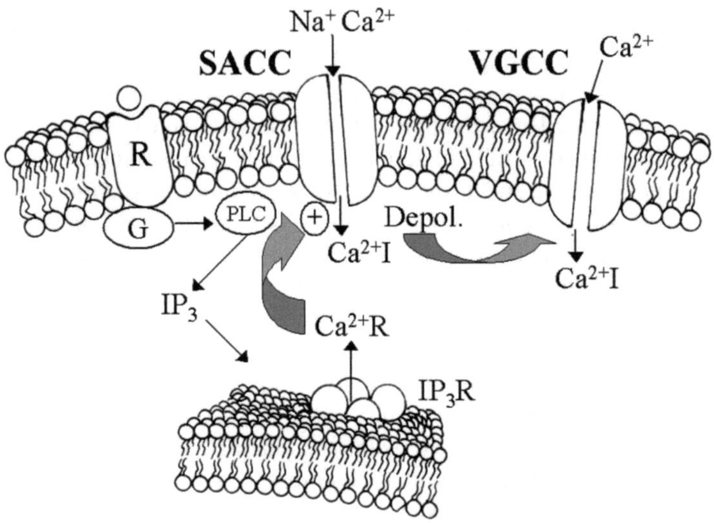

Fig. 13.1 Schematic drawing that shows the possible role of leech SACCs in the amplification of calcium signals. Receptor activation of calcium release (Ca^{2+}R) from the intracellular stores causes opening of SACCs. This results in both calcium inflow (Ca^{2+}I) and membrane depolarization which, in turn, can recruit voltage-gated calcium channels (VGCC) to increase Ca^{2+}I. Abbreviations also include: R, G, PLC and IP_3R for receptor, G-protein, phospholipase C and inositol 1,4,5-trisphosphate receptors

pH, intracellular nucleotides, heat, volatile anesthetics and polyunsaturated fatty acids (Patel and Honoré, 2001; Xu and Enyeart, 2001; Tan et al, 2002). These properties make them remarkable molecular integrators. TRP channels fulfill the integrative function through a modular architecture (Clapham, 2003; Pedersen et al, 2005). They can be expressed with a rich variety of cytoplasmic domains and exhibit hetero-multimerization (Schaefer, 2005), so providing cells with transducers of several intra- and extracellular signals, of both physical and chemical nature.

The molecular structure of leech neuronal SACCs is not yet available. The high single-channel conductance in K^+ solutions and the low sensitivity to amiloride make it unlikely that leech SACCs belong to DEG/ENaC family (Ishikawa et al, 1998; O'Hagan et al, 2005). Rather, the biophysical and pharmacological properties are similar to those described for stretch-activated channels in Xenopus, recently identified as TRPC1 (Maroto et al, 2005), and to the typical transduction channels of vertebrate hair cells (Strassmaier and Gillespie, 2002) structurally characterized as TRPA1 (Corey et al, 2004). Cation selectivity, conductance of about 100 pS, calcium permeability, blockage by Gd^{+3}, amiloride and gentamicin, activation by mechanical stimuli and weak voltage-dependence are shared features of SACCs in leech neurons and hair cells, consistent with those of TRP channels.

The question arises why some pH_i and $[Ca^{2+}]_i$ differently affect the two activity modes. We do not have a ready answer but some hypotheses can be put forward. The characteristics of the two modes might be due to different subunit composition, since the expression of wide range of channel subtypes is typical of TRP channels (Hofmann et al, 2002; Strubing et al, 2003). TRPC1 and TRPC3 channels co-immunoprecipitate and co-localize with caveolins and all members of TRPCs have binding domains for calmodulin, PLC and scaffolding proteins (Kiselyov et al, 2005), suggesting their localization within Ca^{2+} signaling microdomains (Ambudkar, 2006). Moreover, activation of some TRPC members consists in rapid (within few minutes) and reversible translocation of vesicles containing constitutively active channels into the plasma membrane (Bezzerides et al, 2004). MC SACCs of leech growth cones have spontaneous activity in the absence of applied pressure and very often are found in clusters (Calabrese et al, 1999). Thus a different association with cytoskeletal, scaffolding or modulation molecules might account for the different sensitivities of the two activity modes to pH_i and $[Ca^{2+}]_i$.

One of the main arguments raised against the role of stretch-sensitive channels as mechano-transducers in non-specialized cells is the low susceptibility to external mechanical stimuli found in some preparations (Wan et al, 1999). Although SACCs exhibit intrinsic sensitivity to mechanical stimuli, as shown by their prompt activation in membrane blebs lacking a cortical cytoskeleton (Zhang et al, 2000), extrinsic factors such as membrane infolding (Zhang and Hamill, 2000a-b) and submembrane shock absorber cytoskeletal structures can reduce the fraction of mechanical energy forcing the channels (Ko and McCulloch, 2000). In addition, the multiplicity of activation mechanisms reported above makes TRPs suitable to integrate different signals, meeting the requirements for context-dependent sensors. This is specially relevant in cell regions, as growth cones, where a variety of simultaneous environmental guidance cues is likely to confront the plasma membrane.

13.4 Modulation by Adenosine Nucleotides

Other powerful activators and modulators of SACCs activity are the adenosine nucleotides, this section will deal with their effects. Since MC channels are expressed in high density in growth cones of leech neurons, and their block with gentamicin, which does not affect voltage-gated calcium currents in this preparation, increases the neurite extension in culture, the modulation of this activity mode was further investigated. In particular, the effects of adenosine nucleotides were studied in inside-out membrane patches while various compounds were applied by a rapid solution changer. The application of MgATP to the internal side of the membrane patches, robustly and reversibly increased the mean current with an activation delay of few seconds. This effect was absent in SL mode. MC channels up-regulated by ATP maintained their mechano-sensitivity as well as their voltage-dependence. The effects of ATP were dose-dependent, with a micromolar threshold concentration and a saturation attained at millimolar doses. The ATP-induced up-modulation was observed regardless of the membrane potential, the intracellular pH or the concentration of calcium bathing the internal side of the membrane patch. The short activation delay suggested a mechanism that does not involve ATP hydrolysis. After some negative findings to demonstrate a direct action of ATP as ligand, the use of the non-hydrolyzable ATP analog AMP-PNP allowed us to demonstrate that the effect of this compound on the channel is indistinguishable from that of MgATP.

Moreover, ATP without Mg^{2+}, ADP or adenosine were found effective at enhancing the MC activity (Barsanti et al, 2006a). These results confirm that the channel is activated by non-hydrolytic binding of adenosine or its nucleotides.

These molecules are reported to affect many different ion channels, including mechanosensitive channels. Non-hydrolytic activation has been found for TREK-like channels of rat ventricular myocytes (Tan et al, 2002), TREK of adrenal cortical cells (Xu and Enyeart, 2001), $K_{Ca,ATP}$ of chick ventricular myocytes (Kawakubo et al, 1999), cation channels of frog renal tubules (Robson and Hunter, 2000), ENaCs expressed in NIH-3T3 (Ishikawa et al, 2003) and rat neuronal TRPV1 channels (Kwak et al, 2000). Non-hydrolytic inhibition has been reported for ATP-sensitive K^+ channels (Noma, 1983), cation mechanogated channels of the amphibian kidney (Hurwitz et al, 2002), and TRPM4 channels expressed in HEK-293 cells (Nilius et al, 2004).

Thus, leech cation channels are quickly upmodulated by micromolar doses of ATP. Adenosine nucleotides, which are normally present in neurons at least at millimolar concentration (Traut, 1994; Erecinska et al, 2004), are capable of maintaining a powerful channel activation. The physiological meaning of this was really difficult to imagine until the effects of cAMP on this channel were investigated. Perfusion of the cytoplasmic side of excised membrane patches with cAMP alone induced just a weak and rapid increase of activity, but when this cyclic nucleotide was applied in the presence of MgATP it produced a complete inhibition. This effect took minutes to develop as well as to be removed, suggesting the involvement of an endogenous protein kinase (Barsanti et al, 2006a). It is remarkable that in the presence of ATP at a saturating concentration (1 mM), cAMP was capable to slowly overcome this activation to completely inhibit the channel. Since adenosine and related nucleotides provide a rapid activating bias, whereas cAMP promotes a slow

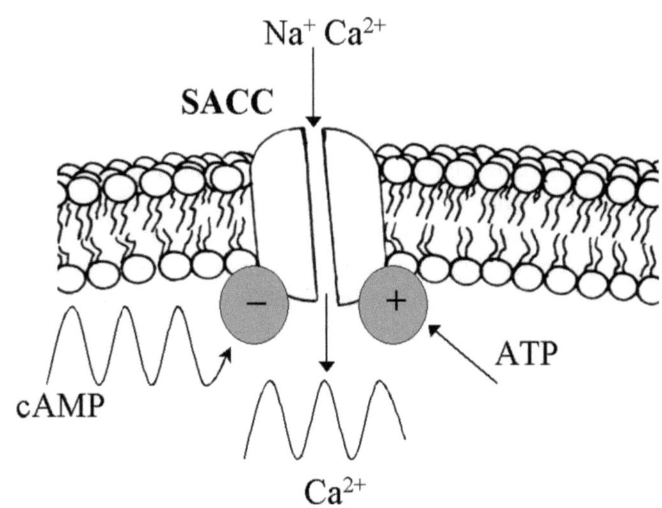

Fig. 13.2 Schematic showing that SACCs can contribute to crosstalk between cAMP and Ca^{2+}. The steady activation due to ATP can be cyclically interrupted by slow oscillations of $[cAMP]_i$ inducing a periodical calcium inflow through SACCs

but strong inhibition of the channels, in principle it is conceivable that oscillations of cAMP concentration on a timescale of minutes might induce the channel activity and the associated calcium influx to oscillate (Fig. 13.2).

A general physiological relevance of this mechanism is emerging from recent work addressed to understand the spatial and temporal patterning of intracellular messengers. As a matter of fact, spontaneous slow oscillations of intracellular cAMP concentration and their dynamic interactions with intracellular calcium transients have been revealed and modelled in embryonic spinal neurons (Gorbunova and Spitzer, 2002). The general hypothesis that the crosstalk between Ca^{2+} and cAMP oscillations represents a new paradigm for signal transduction has been put forward (Zaccolo et al, 2002). The properties of modulation of leech SACCs here reported are consistent with their role in the shaping of the calcium signals in neurons.

13.5 Conclusion

Summing up, leech SACCs exhibit typical features of TRP channels, they are multimodally activated and powerfully modulated by adenosine nucleotides. They are expressed in neuronal growth cones and their block enhances neurite outgrowth.

Increasing evidence indicates that some members of TRP channel family are essential in axon pathfinding, with the general role of Ca^{2+} signal amplifiers (Li et al, 2005; Shim et al, 2005; Wang and Poo, 2005). Some of these channels are molecular integrators of various stimuli and can be translocated in membrane signal complexes to generate intense calcium responses in restricted membrane domains (Bezzerides et al, 2004). Furthermore, intracellular calcium oscillations have been reported to occur in association with growth cone dynamics, both in vitro and in vivo. These calcium transients are inversely related to the growth rate and probably represent a general signal regulating cell motility (Gomez et al, 1995; Gomez and Spitzer, 1999; Henley and Poo, 2004). Intracellular calcium elevations are produced both by intracellular release and by influx, through different types of ion channels. Experimentally induced changes in the cAMP transients modify the frequency of calcium oscillations; on the other hand, cAMP transients are generated only in the presence of specific patterns of calcium oscillations (Gorbunova and Spitzer, 2002). The notion that the levels of intracellular cyclic nucleotides have a pivotal role in determining growth cone behaviour has been also established (Nishiyama et al, 2003). The complexity of the neurite outgrowth and guidance is such that a differential control of calcium influx and release is needed, as recently highlighted. In particular, calcium influx through mechanosensitive channels has been confirmed to act as inhibitor of neurite outgrowth, in opposition to other calcium sources (Jacques-Fricke et al, 2006). It is conceivable, therefore, that the role of TRP channels in the temporary structures of neuronal growth cones is to realize a shaping of the calcium signals, presumably in concert with cAMP dynamics. Accordingly, the multimodal activation of TRP channels might enable them to act as context-dependent integrators which are capable to amplify the calcium signals only in membrane microdomains where meaningful association of appropriate environmental cues occur.

Despite the recent impressive progress in this field, next steps are to identify the code of Ca^{2+} and cAMP oscillation patterns, as well as to understand how cytoskeletal components translate these signals into neurite remodeling.

Acknowledgment The authors gratefully acknowledge all the colleagues that contributed to the experiments on the SACCs of the leech neurons in our laboratory and P. Orsini, F. Montanari for their expert technical assistance.

References

Ambudkar IS (2006) Ca^{2+} signaling microdomains: platforms for the assembly and regulation of TRPC channels. Trends Pharmacol Sci 27:25–32

Auerbach A (1991) Single channel dose-response studies in single cell-attached patches. Biophys J 60:660–670

Barsanti C, Pellegrini M, Pellegrino M (2006a) Regulation of the mechanosensitive cation channels by ATP and cAMP in leech neurons. BBA Biomembranes 1758:666–672

Barsanti C, Pellegrini M, Ricci D, Pellegrino M (2006b) Effects of intracellular pH and Ca^{2+} on the activity of stretch-sensitive cation channels in leech neurons. Eur J Physiol 452:435–443

Bezzerides VJ, Ramsey IS, Kotecha S, Greka A, Clapham DE (2004) Rapid vesicular translocation and insertion of TRP channels. Nat Cell Biol 6:709–720

Calabrese B, Manzi S, Pellegrini M, Pellegrino M (1999) Stretch-activated cation channels of leech neurons: characterization and role in neurite outgrowth. Eur J Neurosci 11:2275–2284

Clapham DE (2003) TRP channels as cellular sensors. Nature 426:517–524

Corey DP, Garcia-Anoveros J, Holt JR, Kwan KY, Lin SY, Vollrath MA, Amalfitano A, Cheung EL, Derfler BH, Duggan A, Geleoc GS, Gray PA, Hoffman MP, Rehm HL, Tamasauskas D, Zhang DS (2004) TRPA1 is a candidate for the mechanosensitive transduction channel of vertebrate hair cells. Nature 432:723–730

Erecinska M, Cherian S, Silver IA (2004) Energy metabolism in mammalian brain during development. Prog Neurobiol 73:397–445

Gomez TM, Snow DM, Letourneau PC (1995) Characterization of spontaneous calcium transients in nerve growth cones and their effect on growth cone migration. Neuron 14:1233–1246

Gomez TM, Spitzer NC (1999) In vivo regulation of axon extension and pathfinding by growth-cone calcium transients. Nature 397:350–355

Gorbunova YV, Spitzer NC (2002) Dynamic interactions of cyclic AMP transients and spontaneous Ca^{2+} spikes. Nature 418:93–96

Grumbacher-Reinert S (1989) Local influence of substrate molecules in determining distinctive growth patterns of identified neurons in culture. Proc Natl Acad Sci USA 86:7270–7274

Hamill OP, McBride DW Jr (1996) The pharmacology of mechanogated membrane ion channels. Pharmacol Rev 48:231–252

Henley J, Poo MM (2004) Guiding neuronal growth cones using Ca^{2+} signals. Trends Cell Biol 14:320–330

Hofmann T, Schaefer M, Schultz G, Gudermann T (2002) Subunit composition of mammalian transient receptor potential channels in living cells. Proc Natl Acad Sci USA 99:7461–7466

Hurwitz CG, Segal AS (2001) Application of pressure steps to mechanosensitive channels in membrane patches: a simple, economical, and fast system. Eur J Physiol 442:150–156

Hurwitz CG, Hu VY, Segal AS (2002) A mechanogated nonselective cation channel in proximal tubule that is ATP sensitive. Am J Physiol Renal Physiol 283:F93–F104

Ishikawa T, Marunaka Y, Rotin D (1998) Electrophysiological characterization of the rat epithelial Na^+ channel (rENaC) expressed in MDCK cells. Effects of Na^+ and Ca^{2+}. J Gen Physiol 111:825–846

Ishikawa T, Jiang C, Stutts MJ, Marunaka Y, Rotin D (2003) Regulation of the epithelial Na^+channel by cytosolic ATP. J Biol Chem 278:38276–38286

Jacques-Fricke BT, Seow Y, Gottlieb PA, Sachs F, Gomez TM (2006) Ca^{2+} influx through mechanosensitive channels inhibits neurite outgrowth in opposition to other influx pathways and release from intracellular stores. J Neurosci 26:5656 –5664

Kawakubo T, Naruse K, Matsubara T, Hotta N, Sokabe M (1999) Characterization of a newly found stretch-activated $K_{Ca,ATP}$ channel in cultured chick ventricular myocytes. Am J Physiol 276: H1827–H1838

Kiselyov K, Kim JY, Zeng W, Muallem S (2005) Protein-protein interaction and functionTRPC channels. Eur J Physiol 451:116–124

Ko KS, McCulloch CA (2000) Partners in protection: interdipendence of cytoskeleton and plasma membrane in adaptations to applied forces. J Membr Biol 174:85–95

Kwak J, Wang MH, Hwang SW, Kim T, Lee S, Oh U (2000) Intracellular ATP increases capsaicin-activated channel activity by interacting with nucleotide-binding domains. J Neurosci 20:8298–8304

Lamoureux P, Buxbaum RE, Heidemann SR (1989) Direct evidence that growth cones pull. Nature 340:159–162

Lee J, Ishihara A, Oxford G, Johnson B, Jacobson K (1999) Regulation of cell movement is mediated by stretch-activated calcium channels. Nature 400:382–386

Li Y, Jia YC, Cui K, Li N, Zheng ZY, Wang YZ, Yuan XR (2005) Essential role of TRPC channels in the guidance of nerve growth cones by brain-derived neurotrophic factor. Nature 434:894–898

Lin SY, Corey DP (2005) TRP channels in mechanosensation. Curr Opin Neurobiol 15:350–357

Maingret F, Fosset M, Lesage F, Lazdunski M (1999) TRAAK is a mammalian neuronal mechanogated K^+ channel. J Biol Chem 274:1381–1387

Maroto R, Raso A, Wood TG, Kurosky A, Martinac B, Hamill OP (2005) TRPC1 forms the stretch-activated cation channel in vertebrate cells. Nat Cell Biol 7:179–185

Martinac B (2004) Mechanosensitive ion channels: molecules of mechanotransduction. J Cell Sci 117:2449–2460

Menconi MC, Pellegrini M, Pellegrino M (2001) Voltage-induced activation of mechanosensitive cation channels of leech neurons. J Membr Biol 180:65–72

Morris CE, Horn R (1991) Failure to elicit neuronal macroscopic mechanosensitive currents anticipated by single-channel studies. Science 251:1246–1249

Mosbacher J, Langer M, Horber JK, Sachs F (1998) Voltage-dependent membrane displacements measured by atomic force microscopy. J Gen Physiol 111:65–74

Nilius B, Prenen J, Voets T, Droogmans G (2004) Intracellular nucleotides and polyamines inhibit Ca^{2+}-activated cation channel TRPM4b. Eur J Physiol 448:70–75

Nilius B, Talavera K, Owsianik G, Prenen J, Droogmans G, Voets T (2005) Gating of TRP channels: a voltage connection? J Physiol 567:35–44.

Nishiyama M, Hoshino A, Tsai L, Henley JR, Goshima Y, Tessier-Lavigne M, Poo MM, Hong K (2003) Cyclic AMP/GMP-dependent modulation of Ca^{2+} channels sets the polarity of nerve growth-cone turning. Nature 423:990–995

Noma A (1983) ATP-regulated K^+ channels in cardiac muscle. Nature 305:147–148

O'Hagan R, Chalfie M, Goodman MB (2005) The MEC-4 DEG/ENaC channel of Caenorhabditis elegans touch receptor neurons transduces mechanical signals. Nat Neurosci 8:43–50

Parekh AB, Putney JW Jr (2005) Store-operated calcium channels. Physiol Rev 85: 757–810

Patel AJ, Honoré E, Maingret F, Lesage F, Fink M, Duprat F, Lazdunski M (1998) A mammalian two pore domain mechano-gated S-like K^+ channel. EMBO J 17:4283–4290

Patel A, Honoré E (2001) Properties and modulation of mammalian 2P domain K^+ channels. Trends Neurosci 24:339–346

Pedersen SF, Owsianik KG, Nilius B (2005) TRP channels: an overview. Cell Calcium 38:233–252
Pellegrini M, Menconi MC, Pellegrino M (2001) Stretch-activated cation channels of leech neurons exhibit two activity modes. Eur J Neurosci 13:503–511
Pellegrino M, Pellegrini M, Simoni A, Gargini C (1990) Stretch-activated cation channels with large unitary conductance in leech central neurons. Brain Res 525:322–326
Petrov AG, Miller BA, Hristova K, Usherwood PN (1993) Flexoelectric effects in model and native membranes containing ion channels. Eur Biophys J 22:289–300
Robson L, Hunter M (2000) An intracellular ATP-activated, calcium-permeable conductance on the basolateral membrane of single renal proximal tubule cells isolated from Rana temporaria. J Physiol 523:301–311
Sachs F, Morris CE (1998) Mechanosensitive ion channels in nonspecialized cells. Rev Physiol Biochem Pharmacol 132:1–77
Sackin H (1995) Mechanosensitive channels. Annu Rev Physiol 57:333–353
Schaefer M (2005) Homo- and heteromeric assembly of TRP channel subunits. Eur J Physiol 451:35–42
Shim S, Goh EL, Ge S, Sailor K, Yuan JP, Roderick HL, Bootman MD, Worley PF, Song H, Ming G (2005) XTRPC1-dependent chemotropic guidance of neuronal growth cones. Nature Neurosci 8:730–735
Sigurdson WJ, Morris CE (1989) Stretch-activated ion channels in growth cones of snail neurons. J Neurosci 9:2801–2808
Silberberg SD, Magleby KL (1997) Voltage-induced slow activation and deactivation of mechanosensitive channels in Xenopus oocytes. J Physiol (London) 505: 551–569
Strassmaier M, Gillespie PG (2002) The hair cell's transduction channel. Curr Opin Neurobiol 12:380–386
Strubing C, Krapivinsky G, Krapivinsky L, Clapham DE (2003) Formation of novel TRPC channels by complex subunit interactions in embryonic brain. J Biol Chem 278:39014–39019
Tan JH, Liu W, Saint DA (2002) Trek-like potassium channels in rat cardiac ventricular myocytes are activated by intracellular ATP. J Membr Biol 185:201–207
Tavernarakis N, Driscoll M (1997) Molecular modelling of mechanotransduction in the nematode Caenorhabditis elegans. Annu Rev Physiol 59:659–689
Traut TW (1994) Physiological concentrations of purines and pyrimidines. Mol Cell Biochem 140: 1–22
Wan X, Juranka P, Morris CE (1999) Activation of mechanosensitive currents in traumatized membrane. Am J Physiol 276:C318–C327
Wang XP, Poo MM (2005) Requirement of TRPC channels in netrin-1-induced chemotropic turning of nerve growth cones. Nature 434:898–904
Xu L, Enyeart JJ (2001) Properties of ATP-dependent K^+ channels in adrenocortical cells. Am J Physiol Cell Physiol 280:C199–C215
Zaccolo M, Magalhaes P, Pozzan T (2002) Compartmentalisation of cAMP and Ca^{2+} signals. Curr Opin Cell Biol 14:160–166
Zhang PC, Keleshian AM, Sachs F (2001) Voltage-induced membrane movement. Nature 413:428–432
Zhang Y, Gao F, Popov VL, Wen JW, Hamill OP (2000) Mechanically gated channel activity in cytoskeleton-deficient plasma membrane blebs and vescicles from Xenopus oocytes. J Physiol (London) 523: 117–130
Zhang Y, Hamill OP (2000a) Calcium-, voltage- and osmotic stress-sensitive currents in Xenopus oocytes and their relationship to single mechanically gated channels. J Physiol (London) 523: 83–99
Zhang Y, Hamill OP (2000b) On the discrepancy between whole-cell and membrane patch mechanosensitivity in Xenopus oocytes. J Physiol (London) 523: 101–115.

Chapter 14
Regulation of Intracellular Signal Transduction Pathways by Mechanosensitive Ion Channels

Aladin M. Boriek and Ashok Kumar

Abstract Only in the past five decades have we begun to understand, on the chemical level, what and how mechanosensitive signaling molecules are involved in the physiological regulation of downstream events in response to mechanical stimulation in health and disease. Currently, the forefront of mechanotransduction is focused on mapping an enormous number of signalling and regulatory signalling pathways that are responsible for sensing and transducing mechanical forces. Mechanosensitive ion channels (MSC) are one of the major classes of molecules involved in mechanosensitive signal transduction. MSC have been described in a wide variety of cell types in many different organisms, ranging from bacteria to mammals. MSC participate in several physiological processes such as touch and pain sensation, salt and fluid balance, blood pressure control, cell volume regulation, and turgor control. Abnormal regulation of the structure and function of MSC may contribute to the pathogenesis of quite a few diseases including neuronal degeneration, muscular dystrophy, cardiac arrhythmias, hypertension, kidney disease, and glioma. Accumulating evidence from our and other groups suggests that MSC may play an important role in the activation of several intracellular mechanosensitive signaling pathways. This review summarizes the recent developments and the state of the current thinking regarding the role of MSC in the regulation of different mechanosensitive signaling proteins and signaling pathways

Key words: Mechanosensitive ion channels · Mechanical stretch · Signal transduction · Mitogen-activated protein kinase · NF-kappa B · akt

Abbreviations: AP-1, activator protein-1; ERK1/2, extracellular signal-related kinase1/2; IκB, I kappa B; IKK, IκB kinase; MAPKs, Mitogen-activated protein kinases; MSC, mechanosensitive ion channels; PI3K, phosphatidylinositol 3-kinase; SAC, stretch-activated channels; SMC, smooth muscle cells; TRP, transient receptor potential

A. Kamkin and I. Kiseleva (eds.), *Mechanosensitive Ion Channels.*

14.1 Introduction

Whether it is to sense a touch, arterial pressure, or an osmotic gradient across a cell membrane, virtually all living organisms require the ability to detect mechanical force. With the exception of slow processes such as tissue remodeling that rely on cell adhesion molecules such as integrins, the majority of the reactions to force are attributed to mechanosensitive ion channels (MSC), also known as stretch-activated ion channels (SAC). MSC are membrane proteins whose gating can be altered by mechanical forces leading to the generation of an ionic current and subsequent transformation of the mechanical stimulus into an electrical or biochemical response (Hamill and Martinac, 2001). Alternatively, the conformational change encountered by MSC upon mechanical stimulation might, in its own right, serve as a catalytic event for signaling cascades involved in mechanoreception and mechanotransduction (Kung, 2005). Changes in membrane tension or alteration in mechanical stretch can be induced under experimental conditions by suction applied to the rear of a patch pipette. The activation of the channels shows a gradual increase of channel activity with respect to the degree of membrane stretch dictated by membrane stiffness. This highlights the importance of the different cellular components in resisting suction into the pipette.

Over the last decade, patch clamp studies and fluorescence imaging data have revealed that MSC are present ubiquitously in various types of cells. MSC have been identified in most organisms including archaea, several prokaryote species, fungi, plants, and mammals (Duggan et al., 2000; Gustin et al., 1988; Pivetti et al., 2003; Tavernarakis and Driscoll, 2001). Among the MS channels studied to date, the best characterized are bacterial MscL and MscS channels, the MSC of large (L) and small (S) conductance (Martinac, 2004). Their three-dimensional (3D) structure was determined by X-ray crystallography, allowing for in-depth studies of the structure and function, of the gating mechanism in these channels. In particular, the structure, function and structural dynamics of MscL channel has been well characterized by a number of techniques, including the patch-clamp technique, electron paramagnetic resonance spectroscopy, molecular dynamics simulations, and most recently, fluorescence resonance energy transfer spectroscopy (Anishkin et al., 2003; Corry et al., 2005; Perozo et al., 2002a; Tsai et al., 2005). On the other hand, the study of MSC in plants and animals lags behind, partly because of their anatomical complexities. Nonetheless, recent findings indicate the involvement of lipids in MSC gating of worms, flies, frog oocytes and mammals. The same type of channel proteins, transient receptor potential (TRP) proteins, for example, have been found to sense vibration, touch, and osmotic membrane stretch (Sukharev and Anishkin, 2004).

These studies challenged two different views with regard to the functional significance of MSC in living cells. On the one hand, MSC appeared to mediate a variety of functions in excitable and nonexcitable tissues of organisms. For example, in animals and humans they play a role in hearing, touch, proprioception or regulation of blood pressure; in plants they may allow gravitropism; and in bacteria they prevent excessive water inflow and turgor pressure build-up by acting as mechano-electrical switches which open in response to cell membrane deformations caused by osmotic forces under hypotonic conditions (Martinac, 2004). On the other hand, the patch clamping technique prompted some researcher to interpret MSC activities as artifacts because of the way mechanical force was usually applied to cell membranes in such experiments (Hamill and Martinac, 2001; Morris and Horn, 1991). Nevertheless, following the cloning and characterization of several prokaryotic MSC proteins, interest in the MSC research was revived (Hamill and Martinac, 2001; Sukharev et al., 1993). Further support came from the discovery that the efflux of solutes from *E. coli* cells in response to a lowering of the external osmolarity could be prevented by gadolinium ions (Gd^{3+}), which are classical inhibitors of MSC in higher organisms (Berrier et al., 1992).

Broad generalizations of MSC have been extremely difficult to make due to the diversities of these channels and modes of action under specific stimuli. Attempts to build unifying concepts have resulted in some physical models that are supportive but unrelated to any specific molecular structure (Sachs, 1992). MSC have been broadly divided into channels that normally require an increase in cell volume for activation (volume-activated channels; VAC) and those that respond to cell deformation in the absence of cell volume changes (stretch-activated channels; SAC). SAC and VAC can be further subdivided into subgroups defined by their ion selectivity, such as cation-non-selective, potassium-selective, or chloride-selective channels. The ion selectivity also determines characteristic electrophysiological properties of MSC, such as their reversal potential, which is negative to the resting membrane potential for potassium selective channels and lies somewhere between resting and action potential plateau levels for cation-non-selective MSC. Thus, activation of potassium-selective channels will generally cause membrane repolarization/hyperpolarization, while activation of cation-non-selective channels will tend to depolarize resting cells and repolarize cells at more positive membrane potentials (Kohl et al., 2006).

At the cellular level, stimulation of MSC influences wide array of functions, e.g. proliferation, apoptosis, migration, permeability, remodeling, and gene expression, in lung cells, endothelial cells, vascular smooth cells, chondrocytes, keratinocytes, and fibroblasts (Ivanchenko and Markwardt, 2005; Liu et al., 1995a; Liu et al., 1995b; Liu et al., 1994; Shaw and Xu, 2003; Stula et al., 2000; Sun and Cho, 2004; Wu and Chen, 2000). We are not yet sure how the cell interprets the multitude of signals in the environment – and respond appropriately. MscL and other prokaryotic MSC are gated by bilayer deformation forces indicating that the mechanism of mechanotransduction in these channels is defined by both the local and/or the global asymmetries in the transbilayer pressure profile and/or bilayer curvature at the lipid protein interface (Perozo et al., 2002a; Perozo et al., 2002b). Moreover, eukaryotic MSC found in non-specialized mechanotransducer cells, such as TREK-1 (Patel et al., 2001) and TRPC1 (Maroto et al., 2005), are gated by membrane tension developed solely in the lipid bilayer. These findings suggest the possibility that the lipid bilayer may actively modulate the specificity and fidelity of signaling by membrane proteins.

14.2 Gating of MSC

Mechanical forces can affect the permeability of the cellular membrane to various ions through the gating of MSC. Most of the studies on the MSC have focused on stretch-activated channels (SAC), which are the most frequently encountered type of MSC in electrophysiological recording from membrane patches. MSC that are inhibited by the application of pressure stimulus (stretch-inactivated channels) have been detected less frequently, although they are found in a variety of cell types (Franco and Lansman, 1990; Morris and Sigurdson, 1989; Oliet and Bourque, 1993). It is generally believed that pressure stimuli open MSC by deforming the sarcolemmal membrane patch, which increases the tension in the plane of the membrane (Gustin et al., 1988; Martinac et al., 1990; Sokabe et al., 1991). Though the precise mechanisms that lead to the gating of MSC in response to membrane tension remains enigmatic, two models have been proposed that can explain MSC gating in response to mechanical stimulation (Martinac, 2004). The bilayer model proposed for gating of MS channels in *E. coli* suggests that lipid bilayer tension alone is adequate to gate the MSC directly (Fig. 14.1A). In this model, the energy input is equal to (tension modulus) $\times$ (area of extension). This model is supported by the observations that purified MscL, MscS and other prokaryotic MSC are still mechanosensitive when reconstituted into liposomes (Hase et al., 1995; Kloda and Martinac, 2001a; b; Martinac et al., 1990; Perozo and Rees, 2003; Sukharev et al., 1994). The tethered model proposes direct interactions between MSC with the extracellular matrix and/or or the cytoskeletal proteins (Fig. 14.1B). Channel gating requires relative disarticulation of the channel gate with respect to the cytoskeleton or extracellular matrix protein (Gillespie and Walker, 2001; Hamill and Martinac, 2001; Hamill and McBride, 1997). Although most investigators favor the tethered model (Garcia-Anoveros and Corey, 1997; Hamill and Martinac, 2001; Markin and Hudspeth, 1995), mechanosensitive properties can also be attributed to MscL, MscS, and even gramicidin when these channels are studied without any

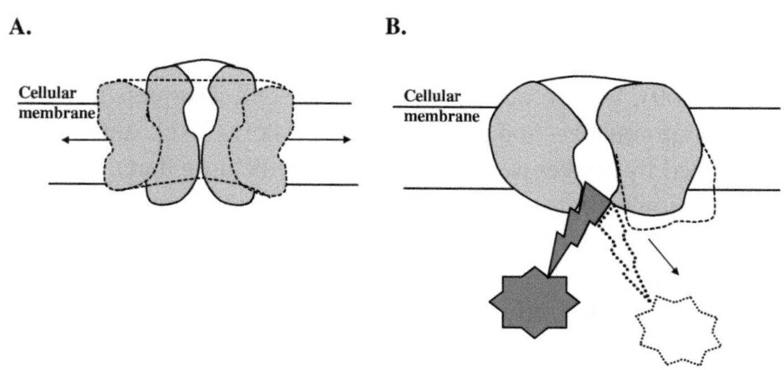

Fig. 14.1 Schematic representation of MSC gating models. **A).** The tension model in which the simultaneous displacement of MSC subunits upon membrane stretch is responsible for the opening of the channel pore. **B).** The tethered model proposes that a swing-like movement of the gating proteins is the major component responsible for the opening of the channel

apparent attachments (Blount, 2003; Hamill and Martinac, 2001; Suchyna et al., 2004; Sukharev and Anishkin, 2004).

14.3 Cytoskeletal and Extracellular Matrix Proteins in MSC Activation

In living tissues, mechanical stresses are normally distributed to cells through the extracellular matrix (ECM) scaffolds that provide mechanical support to the cells. Mechanical signals that propagate from the ECM are expected to converge on either integrins or MSC. Some investigators have hypothesized that membrane deformation is the main mediator of cellular mechanotransduction (Gudi et al., 1998). In this case, stretching of the cell membranes modulates the cation-transporting activity of the MSC by producing conformational changes. This would result in distortion of the associated lipid bilayer, altering the MSC opening or closing rates. It is important to note that the function of MSC channels involve the intracellular cytoskeleton and molecular motors (Huang et al., 2004; Janmey, 1998). However, it remains unclear whether MSC are activated by forces transmitted directly via extracellular contacts or tension in the cell membrane. A third possibility is membrane forces transmit into the cell through integrin receptors to the cytoskeleton and then to the channel proteins. Furthermore, the mechanisms involved in the control of open/closed ion channel conformations by mechanical forces still remain obscure (Lehoux and Tedgui, 2003). There are reports that the cytoskeleton has no effect on channel activity (Opsahl and Webb, 1994); on the other hand, there are interesting hypotheses that actin filaments and microtubules could directly transmit ion fluxes because of their polyelectrolyte nature (Lin and Cantiello, 1993). It remains a challenge to demonstrate possible direct effects of the cytoskeleton on the regulation of mechanosensitive signaling pathways.

14.4 Activation of MSC in Mammalian Cells

MSC activities have been found in variety of cell types. Increased activity of MSC in response to mechanical strain has been proposed as one of the initial events that lead to the activation of several downstream signaling pathways. Ca^{2+} is one of the most common molecules that mediates a large number of cellular responses by binding to specific intracellular proteins, which are considered as Ca^{2+} receptors (Soderling and Stull, 2001; Stull, 2001). The mobilization of Ca^{2+} through MSC may account for the majority of cellular responses including, the activation of different signaling molecules in response to mechanical strain. However, the physiological roles of MSC have been mostly studied using their nonselective blockers, such as gadolinium (Caldwell et al., 1998; Yang and Sachs, 1989) and streptomycin (Gannier et al., 1994; Salmon et al., 1997), which block many types of MSC, irrespective of their origin, conductance, or selectivity. More recently, a peptide GsTMx-4 from the venom of *Grammostola spatulata* has been isolated which specifically inhibits the conductance of MSC in cultured mammalian cells, opening the new way for studying MSC in physiological and biochemical processes (Oswald et al., 2002; Suchyna et al., 2000; Suchyna et al., 2004).

Although application physical forces can lead to the activation of MSC in almost all cell types, most of the studies have been performed in the cells and tissues that are routinely subjected to mechanical strain *in vivo*. For example, MSC activity has been noticed in cells that make lung tissue; linear stretch of rat pulmonary arterial smooth muscle cells increased both Ca^{2+} influx and efflux (Bialecki et al., 1992). Stretch-stimulated Ca^{2+} influx does not require Na^{+} influx and is mediated in part by a pathway sensitive to both gadolinium (MSC inhibitor) and verapamil (an inhibitor of L-type calcium channel). Stretch-stimulated Ca^{2+} efflux, however requires both Ca^{2+} influx via a gadolinium-sensitive pathway and the mobilization of intracellular calcium stores (Bialecki et al., 1992). Biaxial strain of pulmonary arterial endothelial cells also resulted in an increase in cell Ca^{2+} through increased influx and release from intracellular stores (Winston et al., 1993). Cyclic intermittent stretch induced a rapid Ca^{2+} influx and DNA synthesis via gadolinium-sensitive stretch-activated ion channels in fetal rat lung cells (Liu et al., 1994). In lung venular capillaries, pressure elevations increased mean endothelial Ca^{2+} concentration by augmenting Ca^{2+} influx through gadolinium-inhabitable channels (Kuebler et al., 2002).

Shear stress did not change Ca^{2+} concentration of calf pulmonary arterial endothelial cells (Schilling et al., 1992) but enhanced a transient increase in membrane K^{+} permeability (Alevriadou et al., 1993; Schilling et al., 1992). Ca^{2+}-conducting channels in airway epithelial cells were opened when mechanical stimulation was applied to the cellular membrane via a micropipette (Boitano et al., 1994). However, it was later demonstrated that mechanical stimulation caused a rapid depolarization of the stimulated cell which, in turn, activated voltage-sensitive channels (Boitano et al., 1995). Therefore, although mechanical stimulation can induce Ca^{2+} influx and/or efflux, the types of ion channels affected may vary.

Stretch-activated ion channels (SAC), including stretch-activated Ca^{2+} channels and Na^{+} channels, exist in arterial smooth muscle cells (SMC) (Kirber et al., 1988). Stretch induces a rise in intracellular calcium that depends primarily on Ca^{2+} extracellular sources. Calcium influx across the plasma membrane could occur through a stretch-activated channel or voltage-gated Ca^{2+} channel activation as a result of SAC-induced depolarization. The increased Ca^{2+} influx can be blocked by gadolinium but not by L-type calcium channel blockers (Davis et al., 1992a; Davis et al., 1992b; McCarron et al., 1997). In isolated vascular SMC, membrane stretching by applying suction through the patch electrode enhanced the opening frequency of channels permeable to Ca^{2+}(Ohya et al., 1998). In addition, the activity of stretch-activated channels is enhanced in arterial smooth muscle cell from spontaneously hypertensive rats compared to controls (Ohya et al., 1998). The altered SAC in arterial SMC may contribute to the development of hypertrophy and remodeling of arterial tissue in hypertension.

MSC activities have also been studied in bone cell cultures (Rubin et al., 2006). Gadolinium ions were found to block load-related increases in prostaglandin E2 and nitric oxide (Rawlinson et al., 1996). Patch-clamp techniques have demonstrated the existence of at least three classes of MSC in human osteoblasts (Davidson et al., 1990). Through these channels, mechanical stimuli could induce membrane hyper and depolarization or an intricate multiphasic response. In fact, two MSC have been identified in bone cells (Duncan et al., 1996; Kizer et al., 1997). Chronically strained osteoblasts had significantly larger increases in whole-cell conductance when subjected to additional mechanical strain than unstrained controls (Duncan and Hruska, 1994). It was recently reported that a radial membrane strains of 800% is necessary to open half of the MSC in bone cells (Charras et al., 2004). MSC have also been implicated in the response of bone cells to fluid shear stress (Ryder and Duncan, 2001). In addition to direct activation of intracellular signaling cascades, influx of a charged species such as Ca^{2+} can alter membrane potential and activate voltage sensitive channels that are not directly mechanosensitive. For example, the L-type voltage sensitive calcium channel has been implicated in mechanosensitivity *in vivo* (Li et al., 2002).

MSC have been shown to play an important role in the function of cardiac muscle, skeletal muscle, and SMC response to mechanical strain (Kent et al., 1989; Komuro et al., 1996; Sadoshima and Izumo, 1993) . An example of one of the earliest studies that demonstrated the presence of MSC in cardiac muscle was a study by Hu and Sachs (Hu and Sachs, 1996; 1997). The authors showed that acutely isolated embryonic chick heart cells can exhibit whole-cell mechanosensitive currents. They provided experimental evidence that the activity of MSC was correlated with the presence of whole-cell currents, which was blocked by Gd^{3+}. In addition, mechanical stimulation of spontaneously active cells increased the beating rate and Gd^{3+} blocked this effect. The experiments clearly demonstrated that physiologically active mechanosensitive currents arose from the MSC (Hu and Sachs, 1996; 1997).

Ca^{2+} influx is known to be prerequisite for membrane fusion of myoblasts. Accumulating evidence from our group (Wedhas et al., 2005) and others (Lansman, 1990; Park et al., 2002; Shin et al., 1996) suggest that Ca^{2+} influx in myoblasts prior to their differentiation into myotubes occurs through the activation of MSC. Shin et al (Shin et al., 1996) showed that Gd^{3+} inhibited phloretin-induced precocious fusion of chick embryonic myoblasts in normal growth medium. Similarly, we showed that formation of multinucleated myotubes in C2C12 or primary myoblast cultures, was significantly reduced in the presence of gadolinium upon induction of differentiation. In addition, we found that the inhibition of MSC in turn inhibited the expression of specific skeletal muscle proteins such myosin heavy chain and creatine kinase in myoblast cultures upon induction of differentiation. This further suggests that MSC regulates early events during myogenic differentiation (Wedhas et al., 2005).

McBride et al. (McBride et al., 2000) also demonstrated a physiologic role of the MSC in skeletal muscle. In the predominately fast-twitch rat tibialis anterior skeletal muscles, this group tested the hypothesis that eccentric contractions activate the MSC in skeletal muscles (McBride et al., 2000). Their data demonstrate clearly that eccentric contractions induced a significant, prolonged depolarization of the muscle. They also proposed that the activation of MSC may serve as an initial step in muscle adaptation and possibly in the development of muscle hypertrophy during training involving eccentric contractions (McBride et al., 2000). Another study that demonstrates a crucial physiologic role of the MSC in smooth muscle cells was one by Strege et al. (Strege et al., 2003). The study was conducted in the interstitial smooth muscle cells of Cajal, which are known to generate the electrical slow wave that is required for normal gastrointestinal motility. The authors demonstrated the presence of a mechanosensitive Na^+ current in these cells and their data support the contention that the MSC channel plays an important role in the normal physiological control of human intestinal motor function. This could occur by contributing to the setting of the membrane potential, the rate of rise of the slow wave, and mechanosensitive regulation of slow wave frequency (Strege et al., 2003). Taken together, these studies show that there exists a unique physiologic role of MSC in wide variety of cell type.

14.5 MSC Role in Intracellular Signal Transduction Cascades

Different types of physical forces applied to the same cell may activate different signal transduction pathways to mediate diverse biological responses. These mechanisms involve the activation and interplay of various intracellular events. These include the generation of second-messenger molecules, activation of specific protein kinases, phosphorylation and activation of participating signaling molecules, amplification through enzymatic cascades, and modulation of gene expression. In essence, mechanical forces have been shown to activate every type of signal transduction cascade, including increases in intracellular cAMP (Lavandero et al., 1993), inositol 1,4,5-trisphosphate and intracellular calcium (Dassouli et al., 1993; Li et al., 2004), guanine (G) regulatory proteins (Gudi et al., 2003), and mitogen-activated protein kinases (MAPK) (Rubin et al., 2002). In the nucleus, physical forces can exert their effects by influencing expression of immediate early response genes, c-fos, c-jun, c-myc, JE, ETS-like protein-1, activation protein-1 (AP-1), specificity protein-1, nuclear factor-kappa B (NF-κB), and early growth response-1, which encode proteins related to transcriptional factors and signal transduction. The extent that there is similarity in the cellular response to diverse physical signals suggests that a common molecular event such as a conformational change in a transducer protein is induced by all of these signals. Activation of MSC is considered to be of the earliest events in the process of mechanotransduction (Fig. 14.2). Recent evidence further suggests that activation of MSC is prerequisite for the activation of various signal transduction pathways in many cell types. Although the number of studies implicating MSC in activation of various signal transduction pathways in mammalian cells are limited, we provide here a succinct review of the literature regarding the potential role of MSC in activation of various signal transduction pathways and nuclear gene expression.

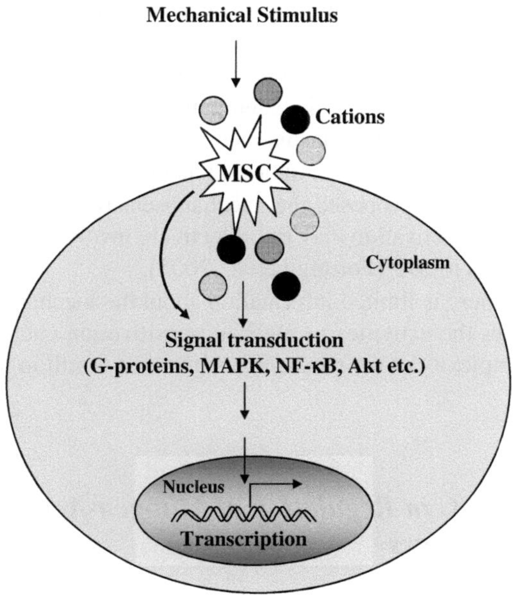

Fig. 14.2 Putative mechanisms of action of MSC in regulation of intracellular signal transduction pathways. Mechanical stretch applied to the cell membrane leads to the activation of MSC which allows transport of various ions including Ca^{2+} across the cell membrane. An alteration in cytoplasmic levels of these ions may lead to the activation of different downstream signal transduction pathways either directly or indirectly. Mechanical stretch can also produce conformational changes in the MSC protein resulting in the initiation of downstream signaling events. Increased activation of various mechanosensitive signaling molecules and mechanoresponsive transcription factors result in the increased expression of different genes culminating in specific biological response. MSC, mechanosensitive ion channels; MAPK, mitogen-activated protein kinase; NF-κB, nuclear factor-kappa B

14.5.1 Regulation of MSC by G Proteins

MSC are one class of membrane proteins regulated at least in part by the G-protein-coupled receptors. G-proteins serve as signal transducers, linking extracellularly oriented receptors to membrane-bound effectors. The precise role of the G-protein in the modulation of mechanosensitive ion channels and their physiological importance in mammals is not well understood. One of the early events implicated in regulatory chondrocytes volume changes and mechanotransduction is an increase of intracellular Ca^{2+} through MSC. Recently, Erickson et al (2001) established that G-proteins are involved in the osmotic stress-induced initiation of intracellular calcium signaling in chondrocytes. These investigators discovered that hyper-osmotic stress induced cell volume change and calcium transients by transmembrane, phospholipid, and G-protein pathways (Erickson et al., 2001).

An exciting study by Formigli et al. (Formigli et al., 2005) investigated whether the Rho signaling pathway is involved in Sphingosine 1-phosphate (S1P)-induced cytoskeletal reorganization in C2C12 myoblasts. The study also established that stress fibers modulated ion current through stretch-activated channels. Interestingly, it was revealed by whole-cell patch-clamp that there was an S1P-induced stress fiber formation associated with increased ion currents and conductance via SAC through Rho- and phospholipase D (PLD)-mediated pathways (Formigli et al., 2005). Experiments aimed at stretching the plasma membrane of C2C12 cells, using the cantilever of an atomic force microscope indicated that there was a Ca^{2+} influx through putative SAC. The authors proposed that mechanosensitive Rho signaling pathway cytoskeleton, and SAC activation may be collectively involved in the differentiative response of myoblasts to S1P (Formigli et al., 2005).

Taken together, there is limited information about the mechanisms by which the G-protein-modulates the activities of mechanosensitive ion channels and research uncovering the complexity of the mechanisms involved is still in the infancy stage.

14.5.2 Role of MSC in Regulation of Mitogen-Activated Protein Kinase Pathways

The mitogen-activated protein kinases (MAPKs) pathways constitute a family of serine/threonine kinases that mediate the transduction of external stimuli into intracellular signals. These signals regulate cellular responses such as proliferation, differentiation, and survival. In mammalian cells, three parallel MAPK pathways have been described: extracellular signal-related kinase (ERK1/2), c-Jun-N-terminal kinases (JNKs), and p38 MAPK (Chang and Karin, 2001). Activation of MAPKs requires phosphorylation on both a threonine and tyrosine residue and, thus, need the activity of dual specificity kinases known as MAP/ERK kinases (MEKs). MEKs, in turn, are substrates for phosphorylation by MEK kinases (MEKKs), which are serine/threonine kinases. In addition, small G proteins (i.e. Ras, Rac, and Cdc42) and specific kinases that may act as MAPK kinase kinase kinases (MEKKKs) regulate the activity of MEKKs (i.e., Raf, MEKK 1–4, transforming growth factor-beta activated kinase 1, and p21 activated kinase) and control the activation of specific MAPK pathway (Garrington and Johnson, 1999; Gutkind, 1998; Li et al., 1999; Widmann et al., 1998; Widmann et al., 1999). MAPKs act via regulation of the activity of many downstream transcription factors, including AP-1, c-myc, Elk1, and CCAAT/enhancer binding protein-β (Chang and Karin, 2001; Peyssonnaux and Eychene, 2001; Whitmarsh and Davis, 1996).

The MAPK signaling cascades play a crucial role in the response of many cell types to mechanical stress. Sequential activation of protein kinases within the MAPK cascades is a common mechanism of mechanosensitive signal transduction in many cellular processes. Recently we investigated whether MAPKs are involved in mechanotransduction in normal skeletal muscle, skeletal muscle lacking dystrophin, and normal airway SMC (Kumar et al., 2002). We have shown that mechanical stretch activates the ERK1/2 pathway in the diaphragm muscle, and demonstrated for the first time that mechanical signal transduction is dependent of the direction of applied stretch (Kumar et al., 2002). Interestingly, we observed that the activation of MSC is not required for the mechanical stretch-induced activation of ERK1/2 in normal diaphragm muscles. In contrast, the activation of MSC and influx of Ca^{2+} from extracellular source was responsible for the spurious activation of ERK1/2 and its downstream transcription factor AP-1 in response to mechanical stretch of muscle fibers lacking the protein dystrophin (Kumar et al., 2004). We have demonstrated that mechanical stretch activates MAPKs in human airway smooth muscle cells (Kumar et al., 2003a). In particular, we have demonstrated that mechanical stretch activated p38 kinase and the stretch-induced activation of p38 MAPK required the extracellular Ca^{2+} influx, small GTPase proteins as well as the activation of SAC. MAPK signaling pathways have been implicated in the mechanical regulation of many physiologic processes (e.g. cardiac hypertrophy and smooth muscle growth) *via* mechanogated ion membrane channels. For example, the mechanisms involved in the mechanical stretch-induced ERK1/2 activation and protein synthesis in cultured rat vascular SMC have been investigated (Iwasaki et al., 2000). Interestingly, those authors produced experimental evidence that suggest that ERK1/2 activation by mechanical stretch requires calcium-sensitive epidermal growth factor receptor activation mainly via SAC, thereby leading to vascular SMC growth (Iwasaki et al., 2000). Other investigators have demonstrated a role of stretch-induced activation of Na^+/H^+ exchanger in modulating the hypertrophic responses of cardiomyocytes (Yamazaki et al., 1999). In an earlier report, the same group of investigators reported that mechanical stretch of cardiomyocytes sequentially activated Raf-1 and MAPK, followed by an increase in protein synthesis (Yamazaki et al., 1995). It is important to note that the mechanosensitive signaling pathways leading to MAPK activation could be different depending on cell types. In cardiac fibroblasts, AngII activated MAPK via the $G\beta^{\gamma}$ subunit of Gi, Src, Shc, Grb2, and Ras, whereas Gq and protein kinase C were vital mechanosensitive signaling molecules in cardiomyocytes (Yamazaki et al., 1999). Other experimental evidences showed that stretch-induced ERK1/2 activation may be dependent or independent of MSC in a cell-specific manner. For example, stretch-induced ERK1/2 activation appears to be independent of SAC, nifedipine-sensitive Ca^{2+}, or amiloride-sensitive Na^+ influxes (Correa-Meyer et al., 2002). It was observed that in primary lung alveolar epithelial cells, cyclic stretch activated ERK1/2 *via* G proteins and epidermal growth factor receptor in Na^+ and Ca^{2+} influxes and Grb2-SOS-, Ras-, and Raf-1-independent pathways (Correa-Meyer et al., 2002).

A group of investigators have assessed whether MAPK are involved in sensing hypotonic cell swelling in cardiac myocytes (Sadoshima et al., 1996). The authors discovered that tyrosine kinase activation is the earliest detectable cell response and plays an essential role in hypotonic swelling-induced ERK1/2 activation in cardiac myocytes. It is important to note that hypotonic swelling causes stretching of the cell membrane and possibly affecting the activities of the MSC. Though the activities of MSC were not measured, this mechanistic study highlights a novel role of tyrosine kinase activation in modulating the effect of cell membrane stretch leading to the activation of the ERK1/2 signaling pathway (Sadoshima et al., 1996).

Recently, Chess et al. (Chess et al., 2005) designed an *in vitro* study that addressed mechanosensitive proliferative signaling pathway in pulmonary epithelial cells. The investigators established that mechanosensitive signaling pathways clearly link MSC activation to ERK1/2 activation. More precisely, their data revealed that mechanical strain in pulmonary epithelial cells initiated induction of reactive oxygen species (ROS) production, which required mechanosensitive ion channel activation, and that this ROS induction regulated strain-induced ERK1/2 activity, both of which were necessary for strain-induced proliferation in lung epithelial cells.

JNK was identified recently as a mechanosensing kinase in smooth muscle cells. In particular, Kushida et al. examined early signaling events provoked by sustained mechanical stretch using primary bladder SMC cultured on deformable silicon dishes (Kushida et al., 2001). Their data demonstrated signaling pathway for stretch-induced activation of JNK in bladder SMC: mechanical stretch evokes Ca^{2+} influx via Gd^{3+}-sensitive MSC, resulting in JNK activation under regulation in part by calmodulin and calcineurin. Interestingly, ERK1/2 was not activated by mechanical stretch in this study (Kushida et al., 2001). Taken together, these studies support the contention that the MAPK signaling cascade appears to be central in mediating the effects of cell stretching. Furthermore, mechanically induced MAPK activation may occur through an early signal in the mechanotransduction pathway *via* influx of ions through the MSC (Fig. 14.2).

14.5.3 Role of MSC in Regulation of Nuclear Factor-Kappa B (NF-κB) Pathways

The nuclear factor-kappa B (NF-κB) transcription factor regulates the expression of a plethora of genes, especially those involved in the inflammatory and acute stress responses. The NF-κB family consists of NF-κB1 (p105/p50), NF-κB2 (p100/p52), RelA (p65), c-Rel, and RelB, which form various homo- and heterodimers. In resting cells, the NF-κB dimers reside in the cytoplasm in an inactive form bound to inhibitory proteins known as IkappaB (IκB). Many stimuli activate NF-κB, mostly through IκB kinase (IKK)-dependent phosphorylation and subsequent degradation of IκB proteins. The IKK complex consists of two highly homologous kinase subunits, IKKα and IKKβ, and a nonenzymatic regulatory component, IKKγ/NEMO. The liberated NF-κB dimers enter the nucleus where they regulate transcription of diverse genes encoding antiapoptotic proteins, cytokines, growth factors, and cell adhesion molecules (Ghosh and Karin, 2002; Karin and Delhase, 2000; Karin and Lin, 2002; Luo et al., 2005).

Several investigators were interested in uncovering the mechanosensitive intracellular signaling cascade that links the MSC activation to NF-κB translocation. For example, Inoh et al (Inoh et al., 2002) established a novel mechanism of the effect of mechanical stretch on the activation of NF-κB in a human fibroblast cells. They showed that direct Ca^{2+} influx through the MSC is required for mechanical stretch-activated NF-κB in cultured human lung fibroblasts. Gong et al. 2001) reported that elevation in the intracellular Ca^{2+} possibly through MSC lead to an increased reactive oxygen species (ROS) production in mitochondria that in turn leads to the translocation of NF-κB from cytoplasm to nucleus. More recently, Amma et al.

(2005) tested the hypothesis that ROS production by intracellular Ca^{2+} increases caused by MSC activation in response to mechanical stretch of lung fibroblasts would induce IκB phosphorylation, leading to the translocation of NF-κB to the nucleus. It was found that the NF-κB activation by mechanical stretch of the fibroblast cells is mediated by the following mechanosensitive signal cascade: MSC activation → intracellular Ca^{2+} increase → production of ROS → activation of IKK → phosphorylation of IκB → NF-κB translocation to the nucleus (Amma et al., 2005). This was the first study to show that Ca^{2+}-dependent increase in oxidative stress is involved in the MSC-dependent NF-κB activation. Although mechanosensitive signaling cascade that linking MSC activation to NF-κB in lung fibroblasts have been identified, little information is available on such signaling cascade in skeletal muscles. For example, we investigated the role of Ca^{2+} influx in mechanical stretch-induced activation of NF-κB in the diaphragm muscle and found only a marginal inhibition in stretch-induced activation of NF-κB in the presence of MSC inhibitor or removal of Ca^{2+} from the medium (Kumar et al., 2003b). Taken together, these studies clearly link the MSC activation to NF-κB activation through mechanosensitive intracellular signaling pathways.

14.5.4 Role of MSC in Regulating Phosphatidylinositol 3-kinase (PI3K)/Akt Pathway

The PI3K/Akt signaling pathway is now recognized as one of the most critical pathways involved in regulation of cell viability and protein synthesis (Brunet et al., 1999; Datta et al., 1997; Kandarian and Jackman, 2005; Song et al., 2005). Akt is a serine/threonine kinase that belongs to the AMP-dependant protein kinase A/ protein kinase G/ protein kinase C (AGC) super family of protein kinases (Lawlor and Alessi, 2001; Song et al., 2005). Akt is activated in response to a wide variety of stimuli including mechanical stretch via mechanisms involving PI3K and phosphoinositide-dependent kinase (PDK-1) (Alessi et al., 1997; Franke et al., 1997; Kulik et al., 1997). PI3K catalyzes the production of lipid molecules, including phosphatidylinositol-3,4,5-triphosphate (PIP3). PIP3 lipids trigger the attachment of Akt to the plasma membrane where it subsequently becomes phosphorylated at two key sites, threonine 308 and serine 473, resulting in its full activation. Once Akt is fully activated, it dissociates from the plasma membrane and phosphorylates both cytoplasmic and nuclear target proteins, notably glycogen synthase kinase-3β, p27Kip, mammalian target of rapamycin, p70S6kinase (p70S6K), and forkhead transcription factors (Craig and Pardo, 1983; Datta et al., 1999; Lawlor and Alessi, 2001; Scott et al., 1998). Published reports also suggested that Akt can be activated in some cell systems by a mechanism independent of PI3K activation (Moule et al., 1997; Sakaue et al., 1997; Yano et al., 1998). Accumulating evidence further suggests that activation of PI3K/Akt pathway is essential for anti-apoptotic action of NF-κB transcription factor (Choi and Jeong, 2005; Madrid et al., 2001; Madrid et al., 2000; Ozes et al., 1999; Romashkova and Makarov, 1999).

It has been reported that cyclic stretch causes biphasic activation of endothelial nitric oxide synthase (eNOS) and Akt kinase activation in bovine arterial endothelial cells. Interestingly, Gd^{3+} and depletion of external Ca^{2+} exclusively inhibited the first peaks of eNOS and Akt activity but exerted little effect on the second peak (Takeda et al., 2006). Kim et al. have demonstrated that mechanical stress-induced expression of vascular endothelial growth factor and hypoxia-inducible factor-1 (HIF-1) are involved the activation of SAC and PI3K/Akt pathway (Kim et al., 2002). Interestingly, mechanical stretch was shown to be an important stimulus for the activation of different mechanosensitive signaling pathways in skin. Yano et al. (Yano et al., 2006) showed that fl 20% stretching of human keratinocytes led to increased phosphorylation and activation of Akt. The stretch-induced activation of Akt in response to mechanical stretch was blunted by inhibition of MSC with Gd^{3+} (Yano et al., 2006). In addition to keratinocytes, cyclic mechanical stretch was reported to activate PI3K/Akt pathway through the recruitment of MSC in osteoblasts (Danciu et al., 2003).

MSC have also been implicated in activation of PI3K/Akt signaling pathway in skeletal muscle in response to eccentric contraction leading to hypertrophy. *In vivo* administration of MSC inhibitors such as streptomycin or Gd^{3+} in rats was found to block the EC-induced muscle hypertrophy (McBride, 2003) and the activation of Akt and its downstream phosphorylation target p70S6K in tibial anterior muscle (Spangenburg and McBride, 2006). In contrast, Hornberger et al. (2005) have reported that in cultured myotubes stretch-induced activation of p70S6K was independent of MSC. These evidences thus suggest that MSC may play an important role in cell survival and maintenance through the activation of PI3K/Akt signaling pathway.

14.5.5 MSC role in Activation of Other Mechanosensitive Signaling Molecules

Besides the above signaling cascades, activation of MSC can also activate other signaling molecules that can influence cell behavior via direct or indirect mechanisms. For example, mechanical stretch was found to augment nitric oxide (NO) signaling and cardiomyocyte apoptosis, at least in part, through the recruitment of MSC (Liao et al., 2006). Gysembergh et al. (1998) have produced experimental evidence that supports the idea that activation of mechanosensitive ion channels protect rabbit hearts from ischemic injury through a mechanism that involves downstream activation of protein kinase C, adenosine receptors, and/or K^+ATP channels (Gysembergh et al., 1998). On the other hand, potassium channels distinct from the non-selective gadolinium-sensitive ion channels may also participate in transduction of mechanical stress (Olesen et al., 1988). In contrast, it appears that in neonatal rat cardiomyocytes, stretch induced-MAPK, and Raf-1 kinase activation are not affected by stretch-sensitive cation channels, ATP-sensitive K^+ channels, or hyperpolarization-activated inward channels. The Na^+-H^+ exchanger, however, markedly mediated stretch-induced activation of Raf-1 kinase and MAPK, implying that membrane proteins, such as ion channels and exchangers, can first receive extracellular stimuli and evoke intracellular signals (Yamazaki et al., 1998a). Although mechanical stretch was found to rapidly activate janus kinase (JAK)/signal transducers and activators of transcription (STAT) pathway in cardiomyocytes, stretch-induced activation of this pathway does not appear to involve MSC (Pan et al., 1999).

Uniaxial stretching of human endothelial cells led to a time-dependent activation of c-src, a kinase associated with focal adhesion complex (Naruse et al., 1998a). The activation of c-src in response to stretch was completely blocked by either inhibition of MSC using Gd^{3+} or through the depletion of extracellular Ca^{2+} from the culture medium. This suggests that Ca^{2+} influx through MSC is important for the activation of c-Src in response to stretch (Naruse et al., 1998a). Stretch was also shown to induce phospholipase C activation, leading to increased inositol 1,4,5-trisphosphate [Ins(1,4,5)P3] and inositol 1,4-bisphosphate [Ins(1,4)P2] contents of rabbit aortic muscles, which involves influx of Ca^{2+} via gadolinium-sensitive ion channels (Matsumoto et al., 1995). In endothelial cells, the MSC appeared to regulate stretch-specific tyrosine phosphorylation of paxillin, focal adhesion kinase (pp125FAK), and pp130CAS, resulting in cell shape change and cytoskeletal remodeling (Malek and Izumo, 1996; Naruse et al., 1998b).

14.6 MSC as Drug Targets

The pharmacology of vertebrate mechanoreceptors was reviewed more than four decades ago by Paintal (Paintal, 1964). His major conclusion was that drugs known to inhibit or stimulate mechanosensation acted not on mechanotransduction itself but processes such as the action potential or muscular/vascular tone. Almost a decade ago, the pharmacology of the MSC was reviewed in depth by Hamill and McBride (Hamill and McBride, 1996). The authors discussed the molecular mechanisms of various ions and drugs that interact with MSC. They concluded that MSC can be chemically blocked or activated by a range of compounds requiring alteration of Paintal's statement that mechanotransduction is a process with a low susceptibility to chemical influence.

The role of SAC in modulating contractile dysfunction in dilated cardiomyopathy was investigated recently (Nicolosi et al., 2004). The investigators tested the hypothesis that SAC antagonists would enhance contractile function in a hamster model of dilated cardiomyopathy. They presented experimental evidence that SAC antagonists enhanced contractile function in dilated cardiomyopathy to equal that of normal controls, and SAC antagonists had no effect on contractile function in normal muscle. They concluded that SAC plays a major role in contractile dysfunction of dilated cardiomyopathy and further suggested that SAC antagonists may represent a possibility of a novel therapy in heart failure.

Besides cardiomyopathy, a dysfunction of MSC causes stretch-induced arrhythmias (Franz et al., 1992; Hansen et al., 1990), neuronal degeneration (Driscoll and Chalfie, 1991; Hong and Driscoll, 1994), and muscular degeneration (Gottlieb et al., 2004). These channels were blocked by the stretch-activated ion channel blocker gadolinium (Gd^{3+}) at 10–20 μm and by aminoglycoside antibiotics such as streptomycin at 100–200 μM (Nicolosi et al., 2004). Neither of these agents are specific for SAC, but the spider venom toxin GsMTx4 is more potent (effective at 5–10 μM) and appears to be more specific (Suchyna et al., 2000). In a study of stretch-induced arrhythmias, GsMTx4 blocked atrial fibrillation in Langendorff-perfused rabbit hearts (Gottlieb et al., 2004). Gd^{3+} has also been shown to attenuate ischemic ST-segment elevation in canine myocardium (Shimada et al., 1999). In addition, Gd^{3+} was shown to attenuate the upward shift of the left ventricular diastolic pressure-volume relation during pacing-induced ischemia (Takano and Glantz, 1995). Similarly, Gd^{3+} was also shown to reduce short-term stretch-induced muscle damage in isolated *mdx* mice (a mouse model of Duchenne muscular dystrophy) muscle fibers (Yeung et al., 2003). Recently, Yeung et al. (2005) demonstrated that treatment of mdx myofibers with the blockers of SAC such as streptomycin and GsMTx4 peptide ameliorates the force reduction following stretched contractions. Furthermore, they found that the incidence of central nuclei (a measure of muscle pathology) in myofibers was significantly reduced by treatment of mdx mice with streptomycin (Yeung et al., 2005). Inhibitors of SAC have also been reported to slow down glioma development (Gottlieb et al., 2004).

Collectively, above evidences suggest that blockers of SAC have potential to be used as drugs for treatment of various SAC-related disorders. However, more investigations are required to understand the mechanisms through which activation of SAC leads to pathological states in various diseases. It is a possibility that activation of MSC leads to the aberrant activation of a common signal transduction pathway that might be responsible for the pathogenesis. Blocking such pathways that are activated on activation of SAC provide alternative drug targets for the treatment of SAC-related disorders.

14.7 Conclusion

Despite complexities of intracellular mechanosensitive signaling pathways, there is a possible common mechanism for mechanotransduction through mechanogated ion channels. Mechanical stretch causes Ca^{2+} and other ions influx through the MSC and this leads to the activation of signaling cascades leading to activation of several kinases and phosphatases. The activation of these intermediate signaling molecules leads to the activation of transcription factors such as AP-1, C/EBP, and NF-κB that cause increased expression of mechanosensitive genes. Furthermore, an increased Ca^{2+} influx in response to mechanical stretch could lead to the activation of these transcription factors via Ca^{2+}-responsive proteins, such as protein kinase C. Future work could resolve the question of whether mechanical stretch directly alters the stiffness of the cell membrane leading to altered receptor/G-protein conformation, whether altered state of the gating of the MSC in response to mechanical stretch would result in altered stiffness of the membrane leading to the activation of

G-protein and/or its receptors, or mechanical stretch could lead in altered state of membrane stiffness resulting in simultaneous activation of the MSC and G-protein and its receptors. Whether or not MSC channels can directly interact with the cytoskeleton and, thereby, intrinsically sense alteration in mechanical stiffness by detecting mechanical stretch of the cell membrane.

Acknowledgment This work was supported by grants from the Muscular Dystrophy Association, USA (to AMB and AK) and the National Institute of Health (HL63134 to AMB and AG29623 to AK).

References

Alessi, D. R., James, S. R., Downes, C. P., Holmes, A. B., Gaffney, P. R., Reese, C. B. and Cohen, P. (1997) Characterization of a 3-phosphoinositide-dependent protein kinase which phosphorylates and activates protein kinase Balpha. Curr Biol 7: 261–269.

Alevriadou, B. R., Eskin, S. G., McIntire, L. V. and Schilling, W. P. (1993) Effect of shear stress on 86Rb+ efflux from calf pulmonary artery endothelial cells. Ann Biomed Eng 21: 1–7.

Amma, H., Naruse, K., Ishiguro, N. and Sokabe, M. (2005) Involvement of reactive oxygen species in cyclic stretch-induced NF-kappaB activation in human fibroblast cells. Br J Pharmacol 145: 364–373.

Anishkin, A., Gendel, V., Sharifi, N. A., Chiang, C. S., Shirinian, L., Guy, H. R. and Sukharev, S. (2003) On the conformation of the COOH-terminal domain of the large mechanosensitive channel MscL. J Gen Physiol 121: 227–244.

Berrier, C., Coulombe, A., Szabo, I., Zoratti, M. and Ghazi, A. (1992) Gadolinium ion inhibits loss of metabolites induced by osmotic shock and large stretch-activated channels in bacteria. Eur J Biochem 206: 559–565.

Bialecki, R. A., Kulik, T. J. and Colucci, W. S. (1992) Stretching increases calcium influx and efflux in cultured pulmonary arterial smooth muscle cells. Am J Physiol 263: L602-L606.

Blount, P. (2003) Molecular mechanisms of mechanosensation: big lessons from small cells. Neuron 37: 731–734.

Boitano, S., Sanderson, M. J. and Dirksen, E. R. (1994) A role for Ca(2+)-conducting ion channels in mechanically-induced signal transduction of airway epithelial cells. J Cell Sci 107: 3037–3044.

Boitano, S., Woodruff, M. L. and Dirksen, E. R. (1995) Evidence for voltage-sensitive, calcium-conducting channels in airway epithelial cells. Am J Physiol 269: C1547-C1556.

Brunet, A., Bonni, A., Zigmond, M. J., Lin, M. Z., Juo, P., Hu, L. S., Anderson, M. J., Arden, K. C., Blenis, J. and Greenberg, M. E. (1999) Akt promotes cell survival by phosphorylating and inhibiting a Forkhead transcription factor. Cell 96: 857–868.

Caldwell, R. A., Clemo, H. F. and Baumgarten, C. M. (1998) Using gadolinium to identify stretch-activated channels: technical considerations. Am J Physiol 275: C619-C621.

Chang, L. and Karin, M. (2001) Mammalian MAP kinase signalling cascades. Nature 410: 37–40.

Charras, G. T., Williams, B. A., Sims, S. M. and Horton, M. A. (2004) Estimating the sensitivity of mechanosensitive ion channels to membrane strain and tension. Biophys J 87: 2870–2884.

Chess, P. R., O'Reilly, M. A., Sachs, F. and Finkelstein, J. N. (2005) Reactive oxidant and p42/44 MAP kinase signaling is necessary for mechanical strain-induced proliferation in pulmonary epithelial cells. J Appl Physiol 99: 1226–1232.

Choi, Y. S. and Jeong, S. (2005) PI3-kinase and PDK-1 regulate HDAC1-mediated transcriptional repression of transcription factor NF-kappaB. Mol Cells 20: 241–246.

Correa-Meyer, E., Pesce, L., Guerrero, C. and Sznajder, J. I. (2002) Cyclic stretch activates ERK1/2 via G proteins and EGFR in alveolar epithelial cells. Am J Physiol Lung Cell Mol Physiol 282: L883-L891.
Corry, B., Rigby, P., Liu, Z. W. and Martinac, B. (2005) Conformational changes involved in MscL channel gating measured using FRET spectroscopy. Biophys J 89: L49-L51.
Craig, S. W. and Pardo, J. V. (1983) Gamma actin spectrin and intermediate. Cell Motil 3: 449–462.
Danciu, T. E., Adam, R. M., Naruse, K., Freeman, M. R. and Hauschka, P. V. (2003) Calcium regulates the PI3K-Akt pathway in stretched osteoblasts. FEBS Lett 536: 193–197.
Dassouli, A., Sulpice, J. C., Roux, S. and Crozatier, B. (1993) Stretch-induced inositol trisphosphate and tetrakisphosphate production in rat cardiomyocytes. J Mol Cell Cardiol 25: 973–982.
Datta, S. R., Brunet, A. and Greenberg, M. E. (1999) Cellular survival: a play in three Akts. Genes Dev 13: 2905–2927.
Datta, S. R., Dudek, H., Tao, X., Masters, S., Fu, H., Gotoh, Y. and Greenberg, M. E. (1997) Akt phosphorylation of BAD couples survival signals to the cell-intrinsic death machinery. Cell 91: 231–241.
Davidson, R. M., Tatakis, D. W. and Auerbach, A. L. (1990) Multiple forms of mechanosensitive ion channels in osteoblast-like cells. Pflugers Arch 416: 646–651.
Davis, M. J., Donovitz, J. A. and Hood, J. D. (1992a) Stretch-activated single-channel and whole cell currents in vascular smooth muscle cells. Am J Physiol 262: C1083-C1088.
Davis, M. J., Meininger, G. A. and Zawieja, D. C. (1992b) Stretch-induced increases in intracellular calcium of isolated vascular smooth muscle cells. Am J Physiol 263: H1292-H299.
Driscoll, M. and Chalfie, M. (1991) The mec-4 gene is a member of a family of Caenorhabditis elegans genes that can mutate to induce neuronal degeneration. Nature 349: 588–593.
Duggan, A., Garcia-Anoveros, J. and Corey, D. P. (2000) Insect mechanoreception: what a long, strange TRP it's been. Curr Biol 10: R384-R387.
Duncan, R. L. and Hruska, K. A. (1994) Chronic, intermittent loading alters mechanosensitive channel characteristics in osteoblast-like cells. Am J Physiol 267: F909-F916.
Duncan, R. L., Kizer, N., Barry, E. L., Friedman, P. A. and Hruska, K. A. (1996) Antisense oligodeoxynucleotide inhibition of a swelling-activated cation channel in osteoblast-like osteosarcoma cells. Proc Natl Acad Sci U S A 93: 1864–1869.
Erickson, G. R., Alexopoulos, L. G. and Guilak, F. (2001) Hyper-osmotic stress induces volume change and calcium transients in chondrocytes by transmembrane, phospholipid, and G-protein pathways. J Biomech 34: 1527–1535.
Formigli, L., Meacci, E., Sassoli, C., Chellini, F., Giannini, R., Quercioli, F., Tiribilli, B., Squecco, R., Bruni, P., Francini, F. and Zecchi-Orlandini, S. (2005) Sphingosine 1-phosphate induces cytoskeletal reorganization in C2C12 myoblasts: physiological relevance for stress fibres in the modulation of ion current through stretch-activated channels. J Cell Sci 118: 1161–1171.
Franco, A., Jr. and Lansman, J. B. (1990) Calcium entry through stretch-inactivated ion channels in mdx myotubes. Nature 344: 670–673.
Franke, T. F., Kaplan, D. R. and Cantley, L. C. (1997) PI3K: downstream AKTion blocks apoptosis. Cell 88: 435–437.
Franz, M. R., Cima, R., Wang, D., Profitt, D. and Kurz, R. (1992) Electrophysiological effects of myocardial stretch and mechanical determinants of stretch-activated arrhythmias. Circulation 86: 968–978.
Gannier, F., White, E., Lacampagne, A., Garnier, D. and Le Guennec, J. Y. (1994) Streptomycin reverses a large stretch induced increases in [Ca2+]i in isolated guinea pig ventricular myocytes. Cardiovasc Res 28: 1193–1198.
Garcia-Anoveros, J. and Corey, D. P. (1997) The molecules of mechanosensation. Annu Rev Neurosci 20: 567–594.
Garrington, T. P. and Johnson, G. L. (1999) Organization and regulation of mitogen-activated protein kinase signaling pathways. Curr Opin Cell Biol 11: 211–218.
Ghosh, S. and Karin, M. (2002) Missing pieces in the NF-kappaB puzzle. Cell 109 Suppl: S81–96.

Gillespie, P. G. and Walker, R. G. (2001) Molecular basis of mechanosensory transduction. Nature 413: 194–202.

Gong, G., Waris, G., Tanveer, R. and Siddiqui, A. (2001) Human hepatitis C virus NS5A protein alters intracellular calcium levels, induces oxidative stress, and activates STAT-3 and NF-kappa B. Proc Natl Acad Sci U S A 98: 9599–9604.

Gottlieb, P. A., Suchyna, T. M., Ostrow, L. W. and Sachs, F. (2004) Mechanosensitive ion channels as drug targets. Curr Drug Targets CNS Neurol Disord 3: 287–295.

Gudi, S., Huvar, I., White, C. R., McKnight, N. L., Dusserre, N., Boss, G. R. and Frangos, J. A. (2003) Rapid activation of Ras by fluid flow is mediated by Galpha(q) and Gbetagamma subunits of heterotrimeric G proteins in human endothelial cells. Arterioscler Thromb Vasc Biol 23: 994–1000.

Gudi, S., Nolan, J. P. and Frangos, J. A. (1998) Modulation of GTPase activity of G proteins by fluid shear stress and phospholipid composition. Proc Natl Acad Sci U S A 95: 2515–2519.

Gustin, M. C., Zhou, X. L., Martinac, B. and Kung, C. (1988) A mechanosensitive ion channel in the yeast plasma membrane. Science 242: 762–765.

Gutkind, J. S. (1998) The pathways connecting G protein-coupled receptors to the nucleus through divergent mitogen-activated protein kinase cascades. J Biol Chem 273: 1839–1842.

Gysembergh, A., Margonari, H., Loufoua, J., Ovize, A., Andre-Fouet, X., Minaire, Y. and Ovize, M. (1998) Stretch-induced protection shares a common mechanism with ischemic preconditioning in rabbit heart. Am J Physiol 274: H955-H964.

Hamill, O. P. and Martinac, B. (2001) Molecular basis of mechanotransduction in living cells. Physiol Rev 81: 685–740.

Hamill, O. P. and McBride, D. W., Jr. (1996) The pharmacology of mechanogated membrane ion channels. Pharmacol Rev 48: 231–252.

Hamill, O. P. and McBride, D. W., Jr. (1997) Induced membrane hypo/hyper-mechanosensitivity: a limitation of patch-clamp recording. Annu Rev Physiol 59: 621–631.

Hansen, D. E., Craig, C. S. and Hondeghem, L. M. (1990) Stretch-induced arrhythmias in the isolated canine ventricle. Evidence for the importance of mechanoelectrical feedback. Circulation 81: 1094–1105.

Hase, C. C., Le Dain, A. C. and Martinac, B. (1995) Purification and functional reconstitution of the recombinant large mechanosensitive ion channel (MscL) of Escherichia coli. J Biol Chem 270: 18329–18334.

Hong, K. and Driscoll, M. (1994) A transmembrane domain of the putative channel subunit MEC-4 influences mechanotransduction and neurodegeneration in C. elegans. Nature 367: 470–473.

Hornberger, T. A., Armstrong, D. D., Koh, T. J., Burkholder, T. J. and Esser, K. A. (2005) Intracellular signaling specificity in response to uniaxial vs. multiaxial stretch: implications for mechanotransduction. Am J Physiol Cell Physiol 288: C185-C194.

Hu, H. and Sachs, F. (1996) Mechanically activated currents in chick heart cells. J Membr Biol 154: 205–216.

Hu, H. and Sachs, F. (1997) Stretch-activated ion channels in the heart. J Mol Cell Cardiol 29: 1511–1523.

Huang, H., Kamm, R. D. and Lee, R. T. (2004) Cell mechanics and mechanotransduction: pathways, probes, and physiology. Am J Physiol Cell Physiol 287: C1-C11.

Inoh, H., Ishiguro, N., Sawazaki, S., Amma, H., Miyazu, M., Iwata, H., Sokabe, M. and Naruse, K. (2002) Uni-axial cyclic stretch induces the activation of transcription factor nuclear factor kappaB in human fibroblast cells. FASEB J 16: 405–407.

Ivanchenko, E. and Markwardt, F. (2005) Characterization of large-conductance Ca2+-dependent and -independent K^+ channels in HaCaT keratinocytes. Skin Pharmacol Physiol 18: 115–122.

Iwasaki, H., Eguchi, S., Ueno, H., Marumo, F. and Hirata, Y. (2000) Mechanical stretch stimulates growth of vascular smooth muscle cells via epidermal growth factor receptor. Am J Physiol Heart Circ Physiol 278: H521-H529.

Janmey, P. A. (1998) The cytoskeleton and cell signaling: component localization and mechanical coupling. Physiol Rev 78: 763–781.

Kandarian, S. C. and Jackman, R. W. (2006) Intracellular signaling during skeletal muscle atrophy. Muscle Nerve 33: 155–165.

Karin, M. and Delhase, M. (2000) The I kappa B kinase (IKK) and NF-kappa B: key elements of proinflammatory signaling. Semin Immunol 12: 85–98.
Karin, M. and Lin, A. (2002) NF-kappaB at the crossroads of life and death. Nat Immunol 3: 221–227.
Kent, R. L., Hoober, J. K. and Cooper, G. t. (1989) Load responsiveness of protein synthesis in adult mammalian myocardium: role of cardiac deformation linked to sodium influx. Circ Res 64: 74–85.
Kim, C. H., Cho, Y. S., Chun, Y. S., Park, J. W. and Kim, M. S. (2002) Early expression of myocardial HIF-1alpha in response to mechanical stresses: regulation by stretch-activated channels and the phosphatidylinositol 3-kinase signaling pathway. Circ Res 90: E25-E33.
Kirber, M. T., Walsh, J. V., Jr. and Singer, J. J. (1988) Stretch-activated ion channels in smooth muscle: a mechanism for the initiation of stretch-induced contraction. Pflugers Arch 412: 339–345.
Kizer, N., Guo, X. L. and Hruska, K. (1997) Reconstitution of stretch-activated cation channels by expression of the alpha-subunit of the epithelial sodium channel cloned from osteoblasts. Proc Natl Acad Sci U S A 94: 1013–1018.
Kloda, A. and Martinac, B. (2001a) Mechanosensitive channel of Thermoplasma, the cell wall-less archaea: cloning and molecular characterization. Cell Biochem Biophys 34: 321–347.
Kloda, A. and Martinac, B. (2001b) Structural and functional differences between two homologous mechanosensitive channels of Methanococcus jannaschii. EMBO J 20: 1888–1896.
Kohl, P., Bollensdorff, C. and Garny, A. (2006) Effects of mechanosensitive ion channels on ventricular electrophysiology: experimental and theoretical models. Exp Physiol 91: 307–321.
Komuro, I., Kudo, S., Yamazaki, T., Zou, Y., Shiojima, I. and Yazaki, Y. (1996) Mechanical stretch activates the stress-activated protein kinases in cardiac myocytes. FASEB J 10: 631–636.
Kuebler, W. M., Ying, X. and Bhattacharya, J. (2002) Pressure-induced endothelial Ca(2+) oscillations in lung capillaries. Am J Physiol Lung Cell Mol Physiol 282: L917-L923.
Kulik, G., Klippel, A. and Weber, M. J. (1997) Antiapoptotic signalling by the insulin-like growth factor I receptor, phosphatidylinositol 3-kinase, and Akt. Mol Cell Biol 17: 1595–1606.
Kumar, A., Chaudhry, I., Reid, M. B. and Boriek, A. M. (2002) Distinct signaling pathways are activated in response to mechanical stress applied axially and transversely to skeletal muscle fibers. J Biol Chem 277: 46493–46503.
Kumar, A., Khandelwal, N., Malya, R., Reid, M. B. and Boriek, A. M. (2004) Loss of dystrophin causes aberrant mechanotransduction in skeletal muscle fibers. FASEB J 18: 102–13.
Kumar, A., Knox, A. J. and Boriek, A. M. (2003a) CCAAT/enhancer-binding protein and activator protein-1 transcription factors regulate the expression of interleukin-8 through the mitogen-activated protein kinase pathways in response to mechanical stretch of human airway smooth muscle cells. J Biol Chem 278: 18868–18876.
Kumar, A., Lnu, S., Malya, R., Barron, D., Moore, J., Corry, D. B. and Boriek, A. M. (2003b) Mechanical stretch activates nuclear factor-kappaB, activator protein-1, and mitogen-activated protein kinases in lung parenchyma: implications in asthma. FASEB J 17: 1800–1811.
Kung, C. (2005) A possible unifying principle for mechanosensation. Nature 436: 647–654.
Kushida, N., Kabuyama, Y., Yamaguchi, O. and Homma, Y. (2001) Essential role for extracellular Ca^{2+} in JNK activation by mechanical stretch in bladder smooth muscle cells. Am J Physiol Cell Physiol 281: C1165-C1172.
Lansman, J. B. (1990) Blockade of current through single calcium channels by trivalent lanthanide cations. Effect of ionic radius on the rates of ion entry and exit. J Gen Physiol 95: 679–696.
Lavandero, S., Cartagena, G., Guarda, E., Corbalan, R., Godoy, I., Sapag-Hagar, M. and Jalil, J. E. (1993) Changes in cyclic AMP dependent protein kinase and active stiffness in the rat volume overload model of heart hypertrophy. Cardiovasc Res 27: 1634–1638.
Lawlor, M. and Alessi, D. (2001) PKB/Akt: a key mediator of cell proliferation, survival and insulin responses? J Cell Sci 114: 2903–2910.
Lehoux, S. and Tedgui, A. (2003) Cellular mechanics and gene expression in blood vessels. J Biomech 36: 631–643.

Li, C., Hu, Y., Mayr, M. and Xu, Q. (1999) Cyclic strain stress-induced mitogen-activated protein kinase (MAPK) phosphatase 1 expression in vascular smooth muscle cells is regulated by Ras/Rac-MAPK pathways. J Biol Chem 274: 25273–25280.

Li, J., Duncan, R. L., Burr, D. B. and Turner, C. H. (2002) L-type calcium channels mediate mechanically induced bone formation in vivo. J Bone Miner Res 17: 1795–1800.

Li, Y. J., Batra, N. N., You, L., Meier, S. C., Coe, I. A., Yellowley, C. E. and Jacobs, C. R. (2004) Oscillatory fluid flow affects human marrow stromal cell proliferation and differentiation. J Orthop Res 22: 1283–1289.

Liao, X., Liu, J. M., Du, L., Tang, A., Shang, Y., Wang, S. Q., Chen, L. Y. and Chen, Q. (2006) Nitric oxide signaling in stretch-induced apoptosis of neonatal rat cardiomyocytes. FASEB J 20: 1883–1885.

Lin, E. C. and Cantiello, H. F. (1993) A novel method to study the electrodynamic behavior of actin filaments. Evidence for cable-like properties of actin. Biophys J 65, 1371–1378.

Liu, M., Xu, J., Liu, J., Kraw, M. E., Tanswell, A. K. and Post, M. (1995a) Mechanical strain-enhanced fetal lung cell proliferation is mediated by phospholipase C and D and protein kinase C. Am J Physiol 268: L729-L738.

Liu, M., Xu, J., Souza, P., Tanswell, B., Tanswell, A. K. and Post, M. (1995b) The effect of mechanical strain on fetal rat lung cell proliferation: comparison of two- and three-dimensional culture systems. In Vitro Cell Dev Biol Anim 31: 858–866.

Liu, M., Xu, J., Tanswell, A. K. and Post, M. (1994) Inhibition of mechanical strain-induced fetal rat lung cell proliferation by gadolinium, a stretch-activated channel blocker. J Cell Physiol 161: 501–507.

Luo, J. L., Kamata, H. and Karin, M. (2005) IKK/NF-kappaB signaling: balancing life and death–a new approach to cancer therapy. J Clin Invest 115: 2625–2632.

Madrid, L. V., Mayo, M. W., Reuther, J. Y. and Baldwin, A. S., Jr. (2001) Akt stimulates the transactivation potential of the RelA/p65 Subunit of NF-kappa B through utilization of the Ikappa B kinase and activation of the mitogen-activated protein kinase p38. J Biol Chem 276: 18934–18940.

Madrid, L. V., Wang, C. Y., Guttridge, D. C., Schottelius, A. J., Baldwin, A. S., Jr. and Mayo, M. W. (2000) Akt suppresses apoptosis by stimulating the transactivation potential of the RelA/p65 subunit of NF-kappaB. Mol Cell Biol 20: 1626–1638.

Malek, A. M. and Izumo, S. (1996) Mechanism of endothelial cell shape change and cytoskeletal remodeling in response to fluid shear stress. J Cell Sci 109: 713–726.

Markin, V. S. and Hudspeth, A. J. (1995) Gating-spring models of mechanoelectrical transduction by hair cells of the internal ear. Annu Rev Biophys Biomol Struct 24: 59–83.

Maroto, R., Raso, A., Wood, T. G., Kurosky, A., Martinac, B. and Hamill, O. P. (2005) TRPC1 forms the stretch-activated cation channel in vertebrate cells. Nat Cell Biol 7: 179–185.

Martinac, B. (2004) Mechanosensitive ion channels: molecules of mechanotransduction. J Cell Sci 117: 2449–2460.

Martinac, B., Adler, J. and Kung, C. (1990) Mechanosensitive ion channels of E. coli activated by amphipaths. Nature 348: 261–263.

Matsumoto, H., Baron, C. B. and Coburn, R. F. (1995) Smooth muscle stretch-activated phospholipase C activity. Am J Physiol 268: C458-C465.

McBride, T. A. (2003) Stretch-activated ion channels and c-fos expression remain active after repeated eccentric bouts. J Appl Physiol 94: 2296–2302.

McBride, T. A., Stockert, B. W., Gorin, F. A. and Carlsen, R. C. (2000) Stretch-activated ion channels contribute to membrane depolarization after eccentric contractions. J Appl Physiol 88: 91–101.

McCarron, J. G., Crichton, C. A., Langton, P. D., MacKenzie, A. and Smith, G. L. (1997) Myogenic contraction by modulation of voltage-dependent calcium currents in isolated rat cerebral arteries. J Physiol 498: 371–379.

Morris, C. E. and Horn, R. (1991) Failure to elicit neuronal macroscopic mechanosensitive currents anticipated by single-channel studies. Science 251: 1246–1249.

Morris, C. E. and Sigurdson, W. J. (1989) Stretch-inactivated ion channels coexist with stretch-activated ion channels. Science 243: 807–809.

Moule, S. K., Welsh, G. I., Edgell, N. J., Foulstone, E. J., Proud, C. G. and Denton, R. M. (1997) Regulation of protein kinase B and glycogen synthase kinase-3 by insulin and beta-adrenergic agonists in rat epididymal fat cells. Activation of protein kinase B by wortmannin-sensitive and -insensitive mechanisms. J Biol Chem 272: 7713–7719.

Naruse, K., Sai, X., Yokoyama, N. and Sokabe, M. (1998a) Uni-axial cyclic stretch induces c-src activation and translocation in human endothelial cells via SA channel activation. FEBS Lett 441: 111–115.

Naruse, K., Yamada, T., Sai, X. R., Hamaguchi, M. and Sokabe, M. (1998b) Pp125FAK is required for stretch dependent morphological response of endothelial cells. Oncogene 17: 455–463.

Nicolosi, A. C., Kwok, C. S. and Bosnjak, Z. J. (2004) Antagonists of stretch-activated ion channels restore contractile function in hamster dilated cardiomyopathy. J Heart Lung Transplant 23: 1003–1007.

Ohya, Y., Adachi, N., Nakamura, Y., Setoguchi, M., Abe, I. and Fujishima, M. (1998) Stretch-activated channels in arterial smooth muscle of genetic hypertensive rats. Hypertension 31: 254–258.

Olesen, S. P., Clapham, D. E. and Davies, P. F. (1988) Haemodynamic shear stress activates a K+ current in vascular endothelial cells. Nature 331: 168–170.

Oliet, S. H. and Bourque, C. W. (1993) Mechanosensitive channels transduce osmosensitivity in supraoptic neurons. Nature 364: 341–343.

Opsahl, L. R. and Webb, W. W. (1994) Transduction of membrane tension by the ion channel alamethicin. Biophys J 66: 71–74.

Oswald, R. E., Suchyna, T. M., McFeeters, R., Gottlieb, P. and Sachs, F. (2002) Solution structure of peptide toxins that block mechanosensitive ion channels. J Biol Chem 277: 34443–34450.

Ozes, O. N., Mayo, L. D., Gustin, J. A., Pfeffer, S. R., Pfeffer, L. M. and Donner, D. B. (1999) NF-kappaB activation by tumour necrosis factor requires the Akt serine-threonine kinase. Nature 401: 82–85.

Paintal, A. S. (1964) Effects of Drugs on Vertebrate Mechanoreceptors. Pharmacol Rev 16: 341–380.

Pan, J., Fukuda, K., Saito, M., Matsuzaki, J., Kodama, H., Sano, M., Takahashi, T., Kato, T. and Ogawa, S. (1999) Mechanical stretch activates the JAK/STAT pathway in rat cardiomyocytes. Circ Res 84: 1127–1136.

Park, J. Y., Lee, D., Maeng, J. U., Koh, D. S. and Kim, K. (2002) Hyperpolarization, but not depolarization, increases intracellular Ca(2+) level in cultured chick myoblasts. Biochem Biophys Res Commun 290: 1176–1182.

Patel, A. J., Lazdunski, M. and Honore, E. (2001) Lipid and mechano-gated 2P domain K(+) channels. Curr Opin Cell Biol 13: 422–428.

Perozo, E., Cortes, D. M., Sompornpisut, P., Kloda, A. and Martinac, B. (2002a) Open channel structure of MscL and the gating mechanism of mechanosensitive channels. Nature 418: 942–948.

Perozo, E., Kloda, A., Cortes, D. M. and Martinac, B. (2002b) Physical principles underlying the transduction of bilayer deformation forces during mechanosensitive channel gating. Nat Struct Biol 9: 696–703.

Perozo, E. and Rees, D. C. (2003) Structure and mechanism in prokaryotic mechanosensitive channels. Curr Opin Struct Biol 13, 432–42.

Peyssonnaux, C. and Eychene, A. (2001) The Raf/MEK/ERK pathway: new concepts of activation. Biol Cell 93: 53–62.

Pivetti, C. D., Yen, M. R., Miller, S., Busch, W., Tseng, Y. H., Booth, I. R. and Saier, M. H., Jr. (2003) Two families of mechanosensitive channel proteins. Microbiol Mol Biol Rev 67: 66–85.

Rawlinson, S. C., Pitsillides, A. A. and Lanyon, L. E. (1996) Involvement of different ion channels in osteoblasts' and osteocytes' early responses to mechanical strain. Bone 19: 609–614.

Romashkova, J. A. and Makarov, S. S. (1999) NF-kappaB is a target of AKT in anti-apoptotic PDGF signalling. Nature 401: 86–90.

Rubin, J., Murphy, T. C., Fan, X., Goldschmidt, M. and Taylor, W. R. (2002) Activation of extracellular signal-regulated kinase is involved in mechanical strain inhibition of RANKL expression in bone stromal cells. J Bone Miner Res 17: 1452–1460.

Rubin, J., Rubin, C. and Jacobs, C. R. (2006) Molecular pathways mediating mechanical signaling in bone. Gene 367: 1–16.

Ryder, K. D. and Duncan, R. L. (2001) Parathyroid hormone enhances fluid shear-induced [Ca2+]i signaling in osteoblastic cells through activation of mechanosensitive and voltage-sensitive Ca2+ channels. J Bone Miner Res 16: 240–248.

Sachs, F. (1992) Stretch-sensitive ion channels: an update. Soc Gen Physiol Ser 47: 241–260.

Sadoshima, J. and Izumo, S. (1993) Mechanical stretch rapidly activates multiple signal transduction pathways in cardiac myocytes: potential involvement of an autocrine/paracrine mechanism. EMBO J 12: 1681–1692.

Sadoshima, J., Qiu, Z., Morgan, J. P. and Izumo, S. (1996) Tyrosine kinase activation is an immediate and essential step in hypotonic cell swelling-induced ERK activation and c-fos gene expression in cardiac myocytes. EMBO J 15: 5535–5546.

Sakaue, H., Ogawa, W., Takata, M., Kuroda, S., Kotani, K., Matsumoto, M., Sakaue, M., Nishio, S., Ueno, H. and Kasuga, M. (1997) Phosphoinositide 3-kinase is required for insulin-induced but not for growth hormone- or hyperosmolarity-induced glucose uptake in 3T3-L1 adipocytes. Mol Endocrinol 11: 1552–1562.

Salmon, A. H., Mays, J. L., Dalton, G. R., Jones, J. V. and Levi, A. J. (1997) Effect of streptomycin on wall-stress-induced arrhythmias in the working rat heart. Cardiovasc Res 34: 493–503.

Schilling, W. P., Mo, M. and Eskin, S. G. (1992) Effect of shear stress on cytosolic Ca2+ of calf pulmonary artery endothelial cells. Exp Cell Res 198: 31–35.

Scott, P. H., Brunn, G. J., Kohn, A. D., Roth, R. A. and Lawrence, J. C., Jr. (1998) Evidence of insulin-stimulated phosphorylation and activation of the mammalian target of rapamycin mediated by a protein kinase B signaling pathway. Proc Natl Acad Sci U S A 95: 7772–7777.

Shaw, A. and Xu, Q. (2003) Biomechanical stress-induced signaling in smooth muscle cells: an update. Curr Vasc Pharmacol 1: 41–58.

Shimada, M., Nakamura, Y., Iwanaga, S., Asakura, K., Hattori, S., Takahashi, M. and Ogawa, S. (1999) Stretch-activated ion channel blocker gadolinium attenuates ischemic ST-segment elevation in canine myocardium. Jpn Circ J 63: 624–628.

Shin, K. S., Park, J. Y., Ha, D. B., Chung, C. H. and Kang, M. S. (1996) Involvement of K(Ca) channels and stretch-activated channels in calcium influx, triggering membrane fusion of chick embryonic myoblasts. Dev Biol 175: 14–23.

Soderling, T. R. and Stull, J. T. (2001) Structure and regulation of calcium/calmodulin-dependent protein kinases. Chem Rev 101: 2341–2352.

Sokabe, M., Sachs, F. and Jing, Z. Q. (1991) Quantitative video microscopy of patch clamped membranes stress, strain, capacitance, and stretch channel activation. Biophys J 59: 722–728.

Song, G., Ouyang, G. and Bao, S. (2005) The activation of Akt/PKB signaling pathway and cell survival. J Cell Mol Med 9: 59–71.

Spangenburg, E. E. and McBride, T. A. (2006) Inhibition of stretch-activated channels during eccentric muscle contraction attenuates p70S6K activation. J Appl Physiol 100: 129–135.

Strege, P. R., Ou, Y., Sha, L., Rich, A., Gibbons, S. J., Szurszewski, J. H., Sarr, M. G. and Farrugia, G. (2003) Sodium current in human intestinal interstitial cells of Cajal. Am J Physiol Gastrointest Liver Physiol 285: G1111-G1121.

Stula, M., Orzechowski, H. D., Gschwend, S., Vetter, R., von Harsdorf, R., Dietz, R. and Paul, M. (2000) Influence of sustained mechanical stress on Egr-1 mRNA expression in cultured human endothelial cells. Mol Cell Biochem 210: 101–108.

Stull, J. T. (2001) Ca2+-dependent cell signaling through calmodulin-activated protein phosphatase and protein kinases minireview series. J Biol Chem 276: 2311–2312.

Suchyna, T. M., Johnson, J. H., Hamer, K., Leykam, J. F., Gage, D. A., Clemo, H. F., Baumgarten, C. M. and Sachs, F. (2000) Identification of a peptide toxin from Grammostola spatulata spider venom that blocks cation-selective stretch-activated channels. J Gen Physiol 115: 583–598.

Suchyna, T. M., Tape, S. E., Koeppe, R. E., 2nd, Andersen, O. S., Sachs, F. and Gottlieb, P. A. (2004) Bilayer-dependent inhibition of mechanosensitive channels by neuroactive peptide enantiomers. Nature 430: 235–240.

Sukharev, S. and Anishkin, A. (2004) Mechanosensitive channels: what can we learn from 'simple' model systems? Trends Neurosci 27: 345–351.

Sukharev, S. I., Blount, P., Martinac, B., Blattner, F. R. and Kung, C. (1994) A large-conductance mechanosensitive channel in E. coli encoded by mscL alone. Nature 368: 265–268.

Sukharev, S. I., Martinac, B., Arshavsky, V. Y. and Kung, C. (1993) Two types of mechanosensitive channels in the Escherichia coli cell envelope: solubilization and functional reconstitution. Biophys J 65: 177–183.

Sun, S. and Cho, M. (2004) Human fibroblast migration in three-dimensional collagen gel in response to noninvasive electrical stimulus. II. Identification of electrocoupling molecular mechanisms. Tissue Eng 10: 1558–1565.

Takano, H. and Glantz, S. A. (1995) Gadolinium attenuates the upward shift of the left ventricular diastolic pressure-volume relation during pacing-induced ischemia in dogs. Circulation 91: 1575–1587.

Takeda, H., Komori, K., Nishikimi, N., Nimura, Y., Sokabe, M. and Naruse, K. (2006) Bi-phasic activation of eNOS in response to uni-axial cyclic stretch is mediated by differential mechanisms in BAECs. Life Sci 79: 233–239.

Tavernarakis, N. and Driscoll, M. (2001) Mechanotransduction in Caenorhabditis elegans: the role of DEG/ENaC ion channels. Cell Biochem Biophys 35: 1–18.

Tsai, I. J., Liu, Z. W., Rayment, J., Norman, C., McKinley, A. and Martinac, B. (2005) The role of the periplasmic loop residue glutamine 65 for MscL mechanosensitivity. Eur Biophys J 34: 403–412.

Wedhas, N., Klamut, H. J., Dogra, C., Srivastava, A. K., Mohan, S. and Kumar, A. (2005) Inhibition of mechanosensitive cation channels inhibits myogenic differentiation by suppressing the expression of myogenic regulatory factors and caspase-3 activity. FASEB J 19: 1986–1997.

Whitmarsh, A. J. and Davis, R. J. (1996) Transcription factor AP-1 regulation by mitogen-activated protein kinase signal transduction pathways. J Mol Med 74: 589–607.

Widmann, C., Gerwins, P., Johnson, N. L., Jarpe, M. B. and Johnson, G. L. (1998) MEK kinase 1, a substrate for DEVD-directed caspases, is involved in genotoxin-induced apoptosis. Mol Cell Biol 18: 2416–2429.

Widmann, C., Gibson, S., Jarpe, M. B. and Johnson, G. L. (1999) Mitogen-activated protein kinase: conservation of a three-kinase module from yeast to human. Physiol Rev 79: 143–180.

Winston, F. K., Thibault, L. E. and Macarak, E. J. (1993) An analysis of the time-dependent changes in intracellular calcium concentration in endothelial cells in culture induced by mechanical stimulation. J Biomech Eng 115: 160–168.

Wu, Q. Q. and Chen, Q. (2000) Mechanoregulation of chondrocyte proliferation, maturation, and hypertrophy: ion-channel dependent transduction of matrix deformation signals. Exp Cell Res 256: 383–391.

Yamazaki, T., Komuro, I., Kudoh, S., Zou, Y., Nagai, R., Aikawa, R., Uozumi, H. and Yazaki, Y. (1998a) Role of ion channels and exchangers in mechanical stretch-induced cardiomyocyte hypertrophy. Circ Res 82: 430–437.

Yamazaki, T., Komuro, I., Kudoh, S., Zou, Y., Shiojima, I., Mizuno, T., Takano, H., Hiroi, Y., Ueki, K., Tobe, K. and et al. (1995) Mechanical stress activates protein kinase cascade of phosphorylation in neonatal rat cardiac myocytes. J Clin Invest 96: 438–446.

Yamazaki, T., Komuro, I., Shiojima, I. and Yazaki, Y. (1999) The molecular mechanism of cardiac hypertrophy and failure. Ann N Y Acad Sci 874: 38–48.

Yang, X. C. and Sachs, F. (1989) Block of stretch-activated ion channels in Xenopus oocytes by gadolinium and calcium ions. Science 243: 1068–1071.

Yano, S., Komine, M., Fujimoto, M., Okochi, H. and Tamaki, K. (2006) Activation of Akt by mechanical stretching in human epidermal keratinocytes. Exp Dermatol 15: 356–361.

Yano, S., Tokumitsu, H. and Soderling, T. R. (1998) Calcium promotes cell survival through CaM-K kinase activation of the protein-kinase-B pathway. Nature 396: 584–587.

Yeung, E. W., Head, S. I. and Allen, D. G. (2003) Gadolinium reduces short-term stretch-induced muscle damage in isolated mdx mouse muscle fibres. J Physiol 552: 449–458.

Yeung, E. W., Whitehead, N. P., Suchyna, T. M., Gottlieb, P. A., Sachs, F. and Allen, D. G. (2005) Effects of stretch-activated channel blockers on [Ca^{2+}]i and muscle damage in the mdx mouse. J Physiol 562: 367–380.

Part III
Cell Mechanobiology

Chapter 15
The Effects of Mechanical Stimulation on Vertebrate Hearts

A Question of Class

Holly A. Shiels and Ed White

Abstract All vertebrate cardiac muscle responds intrinsically to mechanical stimulation which can lead to changes in both the inotropic and chronotropic state of the heart. However the magnitude and physiological relevance of these mechanically-induced responses differ between vertebrate classes. This review will discuss the differences and similarities in the response of vertebrate hearts to stretch. It will focus on responses to mechanical stimulation that have been well characterised in mammals, and discuss what is known about them in non-mammalian vertebrates. Specifically we focus on the Frank- Starling response or length-tension relationship, stretch acceleration of heart rate (the Bainbridge effect) and mechanically-induced effects on cardiac rhythm. Although they have not been categorically studied, these three basic mechanical and electrical responses to stretch are likely present in all vertebrate classes. For example, in a manner similar to mammals, one of the earliest vertebrates, the hagfish (*Myxine glutinosa*), shows a remarkable increase (up to 150%) in heart rate in response to cardiac stretch and in amphibian hearts, modification of action potential profiles and mechanically triggered action potentials have been observed. These commonalities are interesting given the differences in whole heart, cellular and sub-cellular morphology and working environments between early and later vertebrates. These differences may have led to fundamental differences in cardiovascular design between classes. For instance, the exquisite sensitivity of the Starling response in the heart of the rainbow trout (*Oncorhynchus mykiss*) may explain why fish increase predominantly stroke volume rather than heart rate when modulating cardiac output. We speculate that despite the variety of vertebrate hearts, mechanosensitivity is fundamentally similar, if subtly different, across classes. Shared mechanisms, such as mechanosensitive channels (possibly TRP), make early vertebrate hearts useful models for the study of cardiac mechanosensitivity. Finally, given that early vertebrates are known to rely on intrinsic regulation to a greater extent than later vertebrates they may provide useful systems in which to study the role of mechanosensitivity in the evolution of cardiac function and cardiac regulation.

Key words: Mechanosensitivity · Mammals · Birds · Reptiles · Amphians · Fish

A. Kamkin and I. Kiseleva (eds.), *Mechanosensitive Ion Channels.*

15.1 Introduction

Mechanosensitivity is found in all types of life and most tissues (Hamill & Martinac, 2001; Martinac & Kloda, 2003). Interest in certain forms of organisms has been intense, for example, in bacteria with regard to the structure and function of mechanosensitive channels (MSCs) (Perozo *et al.*, 2002; Betanzos *et al.*, 2002); in nematodes, in order to understand how a 'relatively simple' mechanically modulated system can affect whole organism behaviour (Volgis & Tavernarakis, 2005) and in mammalian hearts in order to better understand physiological and pathological regulation of the human heart (Kamkin & Kiseleva, 2005; Kohl *et al.*, 2005). In contrast, mechanical regulation of non-mammalian vertebrate hearts has (with a few exceptions) received relatively little attention. This is unfortunate because in addition to the inherent comparative and evolutionary interest, the differences in heart and myocyte structure could reveal novel insights into the mechanisms of cardiac mechanosensitivity and transduction. Indeed, one would predict intrinsic mechanical mechanisms to be relatively more important in early vertebrates where neural and hormonal regulation of the heart is less well developed than mammals (Burggren *et al.*, 1997).

Of the non-mammalian vertebrate systems that have been studied, amphibian cardiac muscle has played a central role in understanding the effect of mechanical stimulation on both electrical and mechanical activity of cardiac muscle, in a manner similar to the key findings of the length-dependence of skeletal muscle contraction (Gordon *et al.*, 1966). This situation has probably arisen for reasons other than a specific interest in amphibian physiology. Amphibian hearts, typically from frogs and toads have several advantages that have encouraged their use. They are more robust than mammalian hearts, ambient temperature is 'physiological' and there is no coronary circulation, so superfusion of tissue or a whole heart is sufficient to maintain oxygenation. Additionally it is thought that the excitation-contraction coupling (ECC) process is simplified by the minimal involvement of Ca^{2+} release from the sarcoplasmic reticulum (SR) (Klitzner & Morad, 1983) though see (Ju & Allen, 2000). When work on the mechanics of single cardiac myocytes began, the long (up to 400 μm) thin (approx. 5–10 μm) nature of amphibian ventricular and atrial myocytes made them a useful model for attaching to force transducers, either by suction into micropipettes or by wrapping around glass beams. To assist those interested in amphibian cardiac physiology, computer models for the electrophysiology of amphibian pacemaker myocytes (Rasmusson *et al.*, 1990) and ventricular myocytes (Riemer *et al.*, 1998) have been developed.

Two avian models have also received attention; a dilated cardiomyopathy model induced by furazolidone in turkeys (Czarnecki, 1984) which is interesting with regard to mechanical stimuli as it models the effect of chronic ventricular dilation; and the embryonic chick cardiac myocyte model which, as will be described, has greatly advanced our understanding of cardiac MSCs.

The response of the reptilian heart to mechanical stimuli has received little attention which is regrettable as this vertebrate group represent the transition from ectothermy to endothermy and the advent of pressure separation in the ventricle and in the systemic and pulmonary circulations. In contrast, much is known about the effect of mechanical stimuli on the heart of fishes (Satchell, 1991; Farrell, 1991), although the majority of this work is at the level of the whole heart and as such, many of the underlying cellular mechanisms are still poorly understood.

15.2 Differences in Vertebrate Heart and Myocyte Structure

The role of the heart is the same in each vertebrate class; to pump blood around the body facilitating gas, nutrient and waste exchange. However morphological and functional differences between vertebrate hearts are important to consider here as they may underlie some of the varying responses to stretch. Birds and mammals have independently evolved a 4-chambered heart divided into 2 sides so that oxygenated blood from the lungs is anatomically separated from deoxygenated blood from the body. The 3-chambered heart of amphibians and non-crocodilian reptiles represents a more flexible system where pulmonary blood from the left atria and systemic blood from the right atria are able to mix in the common ventricle. This mixing is a controlled process and allows shunting of blood between circulations which is advantageous during activities such as feeding and diving (Wang *et al.*, 2003). In crocodiles, the ventricles are separate but a short-circuit unique to crocodiles called the foramen of Panizza, allows shunting of blood from the pulmonary to the systemic circuit. The fish heart is composed of 4-chambers in series contained within a rigid pericardium: the sinus venosus which receives venous blood from the body, the contractile atrium, the powerful ventricle, and the bulbous arteriosus which acts as a dampener, smoothing out the pulsatile flow exiting the ventricle to protect the delicate microcirculation of the gills. For a more comprehensive account of cardiovascular differences between vertebrates see the review by Burggren *et al.* (1997).

In general terms, as one moves from ectothermic to endothermic vertebrates there is an increase in the density of the myocardium with the proportion of trabeculate myocardium decreasing. Detailed 3D structure of any type of vertebrate hearts is limited, even for human hearts (Anderson *et al.*, 2005) but the spongy ventricles of most lower vertebrates lack the complex sheet structure of mammalian hearts. Active fish species are a notable exception, where a significant proportion of the heart is made of compact myocardium with fibre bundles arranges in layered sheets (Farrell & Jones, 1992; Sanchez-Quintana *et al.*, 1996). The big eye tuna (*Thunnus obesus*) may represent the extreme example where the compact myocardium comprises up to 73% of the heart and is supplied with blood by a well developed coronary circulation (Santer & Greer Walker, 1980). Within the air-breathing vertebrates the extent of the coronary circulation increases with endothermy, as does heart rate. Additionally, there is a change in the principle regulator of cardiac output from stroke volume to heart rate (Burggren *et al.*, 1997; Lillywhite *et al.*, 1999). As both stroke volume (the Frank-Starling law of the heart) and heart rate (the Bainbridge effect) are mechanically modulated it might be predicted that the way stretch influences cardiac function is also altered as one moves from ectothermic to endothermic vertebrates.

Table 15.1 Comparative morphometric data for vertebrate ventricular myocytes

	Trout	Frog	Turtle[d]	Lizard[d]	Turkey[e]	Chicken[i]	Rat[a]
t-tubulated	No	No	No	No	No	No	Yes
Cell length (μm)	159.8 ± 11.5[c]	300[f]	189.1 ± 10.3	151.2 ± 23.8	136 ± 5	201.9 ± 8.6	141.9
Cell width (μm)	9.9 ± 1.7[c]	5[f]	7.2 ± 0.4	5.9 ± 0.9	8.7 ± 0.3	9.0 ± 0.7	32.0
Cell depth (μm)	5.7 ± 0.3[c]	n/a	5.4 ± 0.3	5.6 ± 0.4	n/a	n/a	13.3
Cell capacitance (pF)	46.0[b]	75[g]	42.4 ± 1.9	41.2 ± 2.3	25.9 ± 0.6	n/a	170.2
Cell volume (pL)	6.5 ± 0.6[c]	2.9[h]	2.3 ± 2.3	2.3 ± 0.1	4.0[j]	13.8 ± 2.6	34.4
SA/V ratio (pF/pL)	18.2[b]	25.8[h]	18.3	18.2	6.5[j]	n/a	~5

Data are means. (**a**) (Satoh et al., 1996; Brette & Orchard, 2003); (**b**) (Vornanen, 1998); (**c**) (Shiels & White, 2005); (**d**) for turtle (Galli et al., 2006); for varanid lizard Varanus exanthematicus, n = 4; GLJ Galli, unpublished observations; (**e**) (Kim et al., 2000); (**f**) (Goaillard et al., 2001); (**g**) (Bean et al., 1984); (**i**) from (Li et al., 1997), (**h**) derived from (**f**) and (**g**) derived from (**e**) assuming a 2:1 elliptical cross section.

Morphological differences across vertebrate classes are also apparent at the level of the single cardiac myocyte. In all non-mammalian vertebrates studied to date, myocytes from both the atria and the ventricle have an extended length:width when compared with mammals, and lack t-tubules: fish (Vornanen, 1997); (Shiels & White, 2005); frogs (Goaillard *et al.*, 2001; Tarr *et al.*, 1981); reptiles (Lustig *et al.*, 1996; Galli *et al.*, 2006); birds (Sommer *et al.*, 1991; Li *et al.*, 1997) (Table 15.1). Thus, only the mammalian ventricular myocyte has an extensive and regular t-tubular network. This t-tubular network allows the myocytes to have a greater cross-sectional area and still produce a uniform Ca^{2+} transient. The thicker cells undoubtedly provide greater active force, but may also have implications for passive properties with respect to stretch.

15.3 The Effect of Stretch on Cardiac Force – The Frank-Starling Response

15.3.1 Introduction

The Frank-Starling response is an intrinsic property of vertebrate hearts which ensures that an increase in venous return, which stretches the myocardium, results in a more forceful contraction. Axial stretch causes a biphasic increase in contractility in mammalian myocardium. The largest and immediate effect is associated with an increase in myofilament Ca^{2+} sensitivity, functionally this means that force is increased without an increase in the intracellular Ca^{2+} transient, because of an increase in the number of strong binding cross-bridges. The way this happens is not fully understood but seems to involve the amount of thick and thin filament overlap and their lattice spacing, that is, myofibrillar longitudinal and cross-sectional geometry. Following this rapid increase in force a slower, smaller increase occurs associated with an increase in the size of the intracellular Ca^{2+} transients, via mechanisms discussed later (see section 3.6). For reviews see (Allen & Kentish, 1985; Calaghan & White, 2005).

15.3.2 Fish

Fish are distinguished from other vertebrates in that they rely more heavily on changes in stroke volume than heart rate to modulate cardiac output. Not surprisingly then, fish hearts are remarkably sensitive to stretch. This has been elegantly demonstrated through the use of an *in situ* heart preparation (Farrell *et al.*, 1986; Farrell *et al.*, 1989; Graham & Farrell, 1989; Farrell *et al.*, 1988; Farrell, 1991) where the sinus venosus and ventral aorta are cannulated, without disrupting the rigid pericardial cavity, and the response of the heart to different filling pressures (preload) and output pressures (afterload) are investigated. *In situ* heart preparations are capable of producing cardiac outputs and power outputs comparable to *in vivo* values, and, over the *in vivo* range of filling pressures, only very small increments (0.2–0.3 kPa) result in very large (3-fold) increases in stroke volume (Farrell & Olson, 2000). Much larger increases in filling pressure are required for smaller increases in stroke volume in mammals. Thus, the fish heart has a large capacity for volume-regulation (Farrell & Jones, 1992).

If one considers the geometry of a pyramid-shaped heart such as that of the trout, 3-fold increases in stroke volume require substantial increases in the major axes of the heart (Franklin & Davie, 1992). In the absence of some kind of myocardial sheet slippage, significant elongation of the individual myocytes is predicted.

We have recently investigated the cellular mechanism that might allow the fish heart to accommodate such large changes in blood volume whilst retaining robust pumping ability (Shiels *et al.*, 2006). By attaching single trout ventricular myocytes to carbon fibres of known compliance (Cazorla *et al.*, 2000b; Calaghan & White, 2004) we were able to stretch these cells along their longitudinal axis and simultaneously record passive and active tension and diastolic sarcomere length (SL). We observed that active tension continued to increase at longer SLs than those previously demonstrated for intact mammalian myocytes (Cazorla *et al.*, 2000b) (Fig. 15.1) or small multicellular preparations (Kentish *et al.*, 1986). This results in a 2-fold extension of the functional ascending limb of the length-tension relationship (Fig. 15.1). Importantly, actin filament length is not different in fish and mammals (~ 0.95 μm), which means that active tension increases in the fish heart past the length for optimum overlap of myofilaments. The passive tension curve of trout myocytes remains relatively shallow and essentially linear compared with active tension with SL extension up to 40%, while in both intact mammalian tissue (e.g. Kentish *et al.*, 1986) and skinned single myocytes (e.g. Cazorla *et al.*, 2000a;

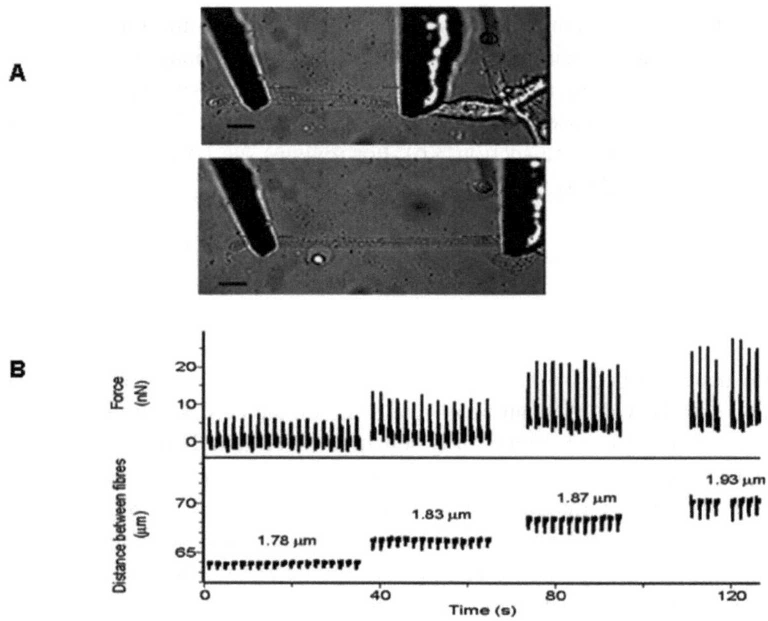

Fig. 15.1 Axial stretch of single trout ventricular myocytes. **(A)** Single trout ventricular myocyte held between two carbon fibres. Images show the myocyte at slack length (top; sarcomere length (SL) 1.80 µm), and following stretches to SL 2.50 µm (bottom). Scale bar is 10 µm in each image. **(B)** Data from a different cell showing how progressive stretching increases passive and active tension (upper trace), when the distance between the fibres is increased (lower trace; numbers show diastolic SL at each stretch). Reproduced from (Shiels et al., 2006) with copyright permission of the Rockfeller University Press, Journal of General Physiology

Wu *et al.*, 2000), large, exponential increases of passive tension are typical for stretches of 20% but see (Fabiato & Fabiato, 1978).

The capacity for greater sarcomere extension in fish myocardium may be linked to low resting tension developed during stretch (Fig 15.2A). Fish, like all lower vertebrates have long thin cardiac myocytes in comparison with mammals (Table 15.1) and hence a much smaller cross-sectional area which may facilitate lower levels of passive tension per unit area compared with mammalian myocytes. Additionally, the properties of titin, which is the main protein determining passive tension on the ascending limb of the length-tension relationship may favour low passive tension in fish, at present titin isomers are unknown for fish (see section 15.3.7).

15.3.3 Amphibians

In a manner similar to that discussed for fish myocardium, early work in the frog myocardium demonstrated an extended SL-active tension curve, when compared to mammalian myocardium. Frog cardiac muscle strips were reported to exhibit SLs on the ascending limb of the Frank-Starling curve of up to 3 µm, although

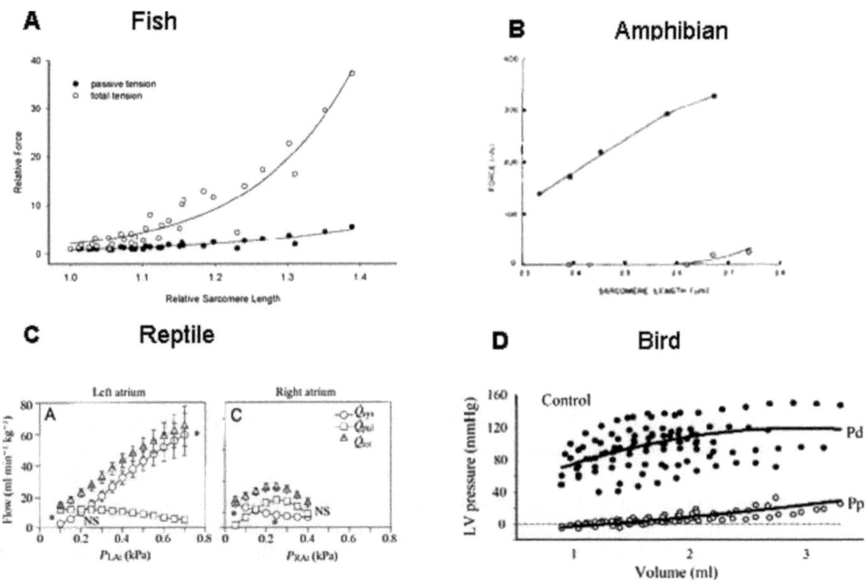

Fig. 15.2 Frank-Starling relationships from vertebrates. **(A)** and **(B)** Sarcomere length (SL)-tension relationships for passive and total (active + passive) tension. **(A)** Single trout ventricular myocytes (n = 9 myocytes from 5 fish) normalized to relative scales where 1.0 = 1.84 ± 0.04 μm for x-axis and 1.0 = 0.20 ± 0.04 nN·μm^{-2} for y-axis. Passive tension at slack length (i.e. where 1.0 = 1.84 ± 0.04 μm) is set to zero. **(B)** A single frog atrial myocyte. Note that active tension (the difference between the 2 curves) continues to increase above SLs of 2.2–2.3 μm and that resting tension is low. **(C)** The systemic and the pulmonary sides of the python (*python molurus*) heart respond very differently to stretch. Figure shows effects of increasing input pressure to left atria (PLAt) and right atria (PRAt) on stroke volume in each circulation. In each case, when input pressure was raised in one atria it was held constant in the other. All values are means ± S.E.M. (n = 7); the asterisk indicates significant differences ($P < 0.05$) between systemic (Q_{sys}) or pulmonary (Q_{pul}) flow rates at a given input pressure. NS, not significantly different. **(D)** Diastolic (Pp) and developed (Pd) pressure in isolated turkey left ventricles as diastolic volume is increased. Peak Pd (equivalent to active tension in parts **A** and **B**) was achieved at a SL of 2.2 μm, this relationship is more similar to Frank-Starling responses in mammals than to the others shown in this figure. **A** Reproduced from (Shiels et al., 2006) with copyright permission of the Rockfeller University Press, Journal of General Physiology, **B** Reproduced from (Tarr et al., 1981) with copyright permission of Circulation Research, **C** Reproduced from (Wang et al., 2002) with copyright permission of the Company of Biologists Ltd and J. Exp. Biol., **D** Reproduced from (Wu et al., 2004) with copyright permission from Elsevier and Journal of Molecular and Cellular Cardiolog

the SL homogeneity of these preparations was uncertain (Nassar *et al.*, 1974; Allen & Blinks, 1978). However similar observations were seen in skinned (Fabiato & Fabiato, 1978) and intact (Tarr *et al.*, 1981) frog myocytes (Fig. 15.2B). It was shown that the length-dependent increase in force was due to an increase in myofilament Ca^{2+} sensitivity. This was demonstrated by a leftward shift in the force-pCa curve in skinned ventricular myocytes as SL was increased (Fabiato & Fabiato, 1978) (Fig 15.3A), and in the first recordings of vertebrate intracellular calcium transients which showed stretch-increased force without an increase in the intracellular Ca^{2+} transient (Allen & Blinks, 1978) (Fig. 15.3B).

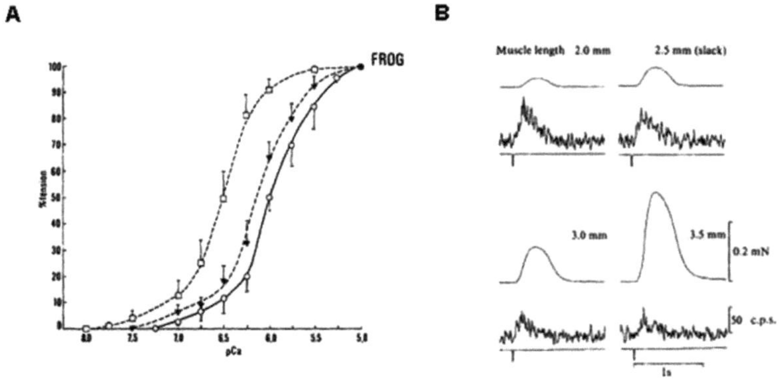

Fig. 15.3 Axial stretch increases force via an increase in myofilament Ca^{2+} sensitivity. **(A)** Force PCa curves in skinned frog ventricular myocytes. As sarcomere length (SL) was increased there is a leftward shift in the curve (SL = ○, 2.2–2.3 μm; ▼, 2.6–2.7 μm; □, 3.0–3.1 μm). **(B)** Simultaneous measurement of isometric force and intracellular Ca^{2+} transients in an intact frog atrial trabecula. When muscle length was increased there was an increase in force but a decrease in the Ca^{2+} transient amplitude. Both observations indicate stretch causes an increase in the myofilament Ca^{2+}sensitivity. **A** Reproduced from (Fabiato & Fabiato, 1978) with copyright permission of the Rockfeller University Press, Journal of General Physiology, **B** Reproduced from (Allen & Blinks, 1978) with copyright permission of the Nature Publishing Group

One observation from amphibians that differs from both fish and mammals is that the resting SL of amphibian myocytes is consistently reported in the region of 2–2.3 μm (Fabiato & Fabiato, 1978; Tarr *et al.*, 1979; Tarr *et al.*, 1981; Tung & Morad, 1988; Tung & Zou, 1995). This is not accompanied by longer myofilaments, which were described as similar in length for cat papillary and frog atrial muscle (Page, 1974). This means resting SL in amphibians is close to the optimal SL for mammals. Although resting SL may be greater in amphibians, a SL of 3 μm at L_{max} (the length at which maximum active tension is developed) still represents an extended ascending limb, compared to mammals.

15.3.4 Reptiles

Although single cell experiments have not been preformed in reptiles, whole heart work shows that turtles (*Chrysemys scripta*, Farrell *et al.*, 1994; *Emydura signata,* Franklin, 1994), snakes (*Python molurus*, Wang *et al.*, 2002) and crocodilian hearts (*Crocodylus porosus*; Franklin & Axelsson, 1994) also possess a robust Frank-Starling response (Fig. 15.2C). Similar to mammals, the right side of the crocodile heart shows a greater Frank-Starling response than the left side, although maximum stroke volume is the same in both chambers (Franklin & Axelsson, 1994; Burggren *et al.*, 1997). The opposite was found in the python where the left side of the heart was more responsive to filling pressure than the right (Wang *et al.*, 2002). Further experiments are necessary to determine whether these left and right differences in cardiac stretch-sensitivity are related to shunt patterns (Burggren *et al.*, 1997). In any case, from the limited number of reptiles that have been investigated it is clear that stretch can dramatically increase cardiac force. Despite this, unlike fish, reptiles tend to increase heart rate to a greater extent than stroke volume to elevate cardiac output during exercise (e.g. Butler *et al.*, 2002). This is particularly interesting as reptiles do increase stroke volume dramatically under conditions such as apnea (breath-hold). For example, during a dive, the heart rate of the turtle (*C. Scripta*) may fall to 5 bpm (from 30 bpm at rest at the surface). During this diving bradycardia, stroke volume can increase up to 5-fold resulting in an overall increase in cardiac output (Burggren *et al.*, 1997; Farrell *et al.*, 1994; Franklin, 1994). The cellular mechanisms which enable turtles to accommodate such large volume changes await investigation. Moreover, the reason for utilizing frequency to a greater extent than stroke volume for modulating cardiac output, when they clearly have the capacity for large volume changes are unknown.

15.3.5 Birds

In turkey, the left ventricular developed pressure and passive pressure increased with volume in a manner similar to mammalian cardiac tissue (Wu *et al.*, 2004) (Fig. 15.2D). In tissue fixed at different ventricular volumes, SL at peak developed pressure was about 2.2 μm and about 1.95 μm at 80% of maximum volume. While these values are not directly comparable with those taken from single myocytes, because of factors such as connective tissue and fibre orientation affecting both SL and pressure development, they are more akin to findings in mammalian hearts than to fish or amphibians. With respect to the balance of heart rate and stroke volume in regulating cardiac output, although stroke volume can increase with exercise in birds it was reported that the exercise-induced increase in cardiac output in ducks was due to increased heart rate with no increase in stroke volume (Grubb, 1982). Similarly, pigeon heart rate can increase more than 6-fold between rest and flight with no change in stroke volume (Peters *et al.*, 2005). A very interesting recent finding has shown that the Frank-Starling mechanism is intrinsic to the embryonic chick myocyte, this was demonstrated using engineered heart tissue, i.e. myocytes grown in a collagen matrix (Asnes *et al.*, 2006).

15.3.6 The Slow Increase in Force

In mammalian myocardium the positive inotropic effect of stretch is biphasic. Following the initial or rapid increase there is secondary slower increase that is associated with an increase in the intracellular Ca^{2+} transient. We have shown that this response is (at least in part) intrinsic to the individual myocyte and is regulated by MSCs, probably non-specific cationic channels, and by the Na^+/H^+ exchanger

(Calaghan & White, 2004; Calaghan & White, 2005). We are unaware of any studies that have specifically investigated whether this mechanism exists in other vertebrate classes, although some information may be available. For example, stroke volume was increased immediately upon elevating preload in the eel (*Anguilla anguilla*) heart but did not increase further over the next 30 mins (Imbrogno *et al.*, 2001). In contrast there is an intriguing possibility that the slow response is present in chick embryonic myocytes growth in collagen matrices (see Asnes *et al.*, 2006, their Fig. 6).

15.3.7 Titin

Species differences in the length of collagen fibres (Hanley *et al.*, 2006) may be important in explaining differences at the level of the tissue or the whole heart. It seems that both fish and amphibian cardiac muscle, in contrast to mammalian and bird cardiac muscle, have relatively compliant myocytes and an ascending limb of the length-tension curve extending beyond optimal overlap of the myofilaments. This enhanced extensibility is likely to be related to relatively low levels of resting tension which increases steadily with SL (Tarr *et al.*, 1979; Shiels *et al.*, 2006, see Fig. 15.2A) rather than abruptly as in mammalian tissue (Kentish *et al.*, 1986). Low levels of resting tension could be due to the lower number of parallel sarcomeres in these thin myocytes, although rabbit SAN cells are of comparable cross-sectional area but are very difficult to extend with the same carbon fibres used in our fish experiments (P.Cooper and P. Kohl, personal communication).

Both resting tension and the compression of the myofilament lattice spacing (an important mechanism in the 'length-dependant' increase in force) are largely determined by the giant elastic protein, titin. Six titin molecules span each half sarcomere (Granzier & Irving, 1995; Fukuda *et al.*, 2003; Cazorla *et al.*, 2000a). It is therefore very interesting to note that turkey ventricular myocardium (Wu *et al.*, 2004) only expresses the shorter, stiffer isoform of titin. In mammals the expression of the longer and more compliant isoform increases with heart size; < rabbit < bovine, with the rat expressing only the short form (Cazorla *et al.*, 2000a) (Fig. 15.4). There are also regional variations in titin expression in large mammalian hearts between atrial and ventricular tissue and within the ventricle itself, with sub-epicardial < sub-endocardial myocardium (Cazorla *et al.*, 2000a). Long and short titin isoforms are expressed in Zebrafish (Seeley *et al.*, 2007). At present the titin isoforms expressed in amphibians and reptiles are unknown but the finding from turkey suggests that cardiac myocyte size and working environment may be more important than phylogeny. Additionally, modulators of titin-dependent passive tension like Ca^{2+} (Labeit *et al.*, 2003) or phosphorylation (Fukuda et al., 2005) may play a large role in passive properties in eary vertebrate hearts. Indeed, Labeit *et al.*, (2003) found that titin's Ca^{2+}-sensitive tension may explain Ca^{2+}-induced tension in sarcomeres stretched beyond optimal thick and thin filament overlap. Whether such a mechanism plays a role in the large active force and low passive force at long SLs in fish (Shiels et al., 2006) and frogs (Tarr *et al.*, 1979; Tung & Zou, 1995) remains to be discovered.

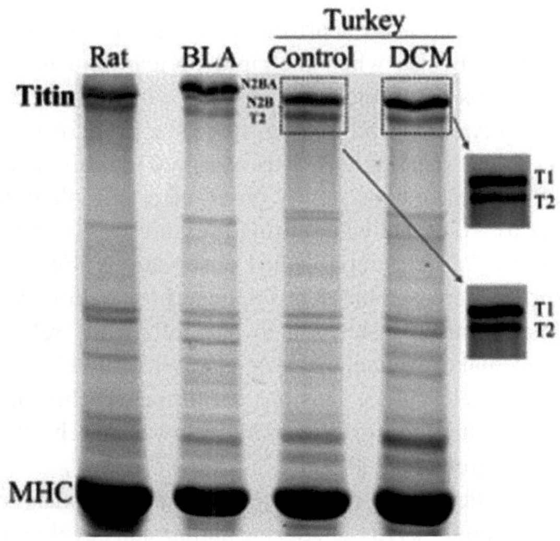

Fig. 15.4 Isoforms of titin in mammalian and avian myocardium. SDS-PAGE (Coomassie-blue stained) of rat, bovine left atria (BLA) and turkey normal and failing (DCM) myocardium. Titin is found at the top of the figure as a N2BA band which represents the longer more compliant isoform and the N2B band, the shorter stiffer form. The rat expresses only N2B titin while BLA expresses both, but predominantly N2BA. Both normal and failing avian myocardium expressed only N2B, thus titin expression is not rigidly linked to phylogeny. MHC: myosin heavy chain. Reproduced from (Wu et al., 2004) with copyright permission from Elsevier and Journal of Molecular and Cellular Cardiology

15.4 The Effect of Mechanical Stimulation on Electrical Activity

15.4.1 Heart Rate

Stretch-acceleration of heart rate has been documented in every vertebrate class. The extent of stretch-acceleration depends on the initial stretch, the applied stretch and the optimal stretch (Pathak, 1973).

Similar to the Frank-Starling response, stretch-acceleration (Bainbridge effect), mechanically links cardiac output with venous return. Pathak (1966) argued strongly that increased rate was an intrinsic response of the pacemaker to stretch, rather than an 'autonomic reflex', and indeed it has subsequently been shown (in rabbit) that the response is intrinsic to the single adult pacemaker myocyte upon axial stretch (Cooper *et al.*, 2000). It has been suggested that this intrinsic stretch-regulated mechanism for elevating heart rate is more important in lower vertebrates (and invertebrates) than in higher vertebrates where it has been superseded by neuro-hormonal control (Jensen, 1961; Pathak, 1973; Burggren *et al.*, 1997). Indeed, in the aneural heart of the Atlantic hagfish *(M. glutinosa)* stretch of pacemaker tissue, through increased venous return, leads to a 50–150% increase in heart rate (Satchell, 1991). However, this is not a universal response amongst cyclostomes as the Pacific hagfish (*Eptatretus stouti*) showed much smaller increases (0–10%) in heart rate (Forster, 1989).

Stretch-induced acceleration of heart rate has been observed in isolated hearts from some teleost fish species when the rigid pericardium is disrupted and the atrium and sinus venosus are stretched with large volumes. In eel (*A. Anguilla*), heart rate increased between 7 and 15% in response to increased filling pressure (Cerra *et al.*, 2004) but no such increase was observed in seabream (*Sparus auratus* (Icardo et al., 2005). We have consistently observed a more than doubling of heart rate from ~0.25 Hz to ~0.45 Hz when filling pressure was doubled in the isolated rainbow trout heart (S. Patrick, unpublished observations; Fig. 15.5A). However, when filling pressure is increased in the *in situ* heart preparation, which better maintains sinus venosus and atrial distension to physiologically relevant levels, there is no increase in pacemaker frequency associated with stretch in rainbow trout or other teleost hearts (Farrell, 1991). This has led to the conclusion that mechanical regulation of heart rate, though intrinsic, even in lower vertebrates may have little *in vivo* relevance (Burggren *et al.*, 1997).

Evidence for stretch-dependent modulation of heart rate exists for adult amphibians (as reviewed by (Pathak, 1973) and during amphibian development (Burggren & Doyle, 1986). In amphibian hearts pacemaking is centred in the sinus venous, indeed pacemaking cells have been isolated (e.g. Ju & Allen, 2000), though none to date have been stretched. In a manner similar to mammalian pacemakers, increase in transmural pressure increases heart rate (Pathak, 1958; Pathak, 1976).

Some reptiles like the tortoise (sp unknown) clearly show stretch-acceleration of heart rate with increases between 5–20% (Keatinge, 1959). However, in turtles

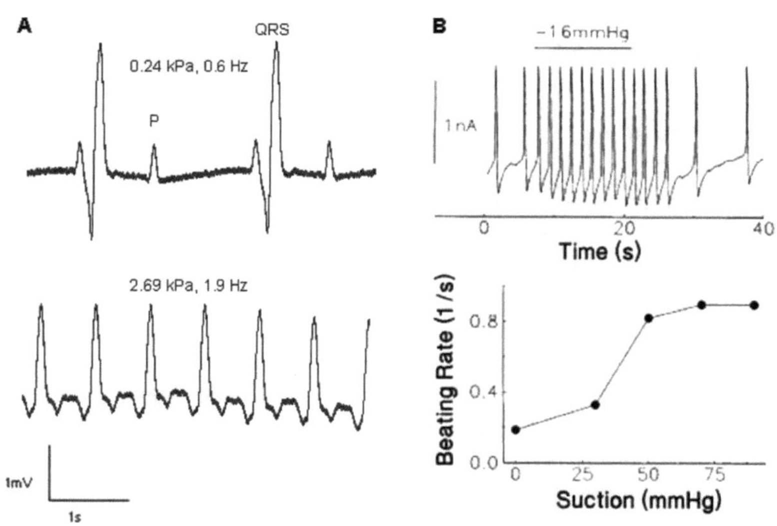

Fig. 15.5 Mechanical stimulation increases beating frequency. **(A)**. Effect of increasing filling pressure (as indicated) on the surface ECG and heart rate (as indicated) of the isolated trout heart. Similar observations were seen in 5 out of 5 hearts. **(B)** Upper panel shows the effect of suction from a loosed-sealed patch pipette on the spontaneous beating frequency of a single cultured embryonic chick myocyte. The lower panel shows the response is dependant upon the level of suction. Suction increased beating frequency. **A** unpublished data, S. Patrick, **B** Reproduced from (Hu & Sachs, 1996) with copyright permission of Springer Science and Business Media

(*C. scripta* Farrell *et al.*, 1994; *E. signata* Franklin, 1994) and crocodiles (*C. porosus*) increased filling pressure leads to only small (5%) or insignificant increases in heart rate (Franklin & Axelsson, 1994). Stretch-acceleration has not been observed in snakes (*P. molurus*; Wang *et al.*, 2002). Together these data suggest that this mechanism plays a minor role in modulating heart rate *in vivo* in reptiles.

In developing chick embryos (5-days old), Faber (1968) observed a 27% increase in heart rate in response to stretch. This response too is intrinsic to the myocyte, as suction applied by a loose-sealed patch pipette increases beating frequency of single embryonic chick myocytes (Hu & Sachs, 1996) (Fig. 15.5B). Both this study and the single rabbit SAN myocyte work of Cooper *et al.* (2000) recorded mechanically-activated non-specific cationic currents that might explain the increased rate of pacemaker depolarisation and hence the increase in rate.

15.4.2 Heart Rhythm

We use the term heart rhythm to cover mechanically induced changes in electrical activity other than those discussed in the previous section.

Early work characterizing the effect of stretch and release of stretch on ventricular muscle was performed by Max Lab on frog hearts using microelectrodes or contact electrodes to measure intracellular membrane potential or monophasic action potentials (MAPs) (Lab, 1978; Lab, 1980). Work on ventricular strips found that early stretches shortened the action potential duration (APD) or depressed the plateau and shortened the Q-T interval of the ECG. Stretches later in the plateau had smaller effects. Releases from long lengths prolonged the action potential. These effects were timing dependant, with stretches earlier in the AP plateau having a greater effect than when imposed later. However, the effects of release were less prominent in frog than mammalian tissue. Observations in intact ventricles and isolated muscle strips reported stretch-induced cross-over effects on the APD (Fig. 15.6A), and stretch-induced action potentials (Fig. 15.6B)

The cross-over effect describes a stretch-induced shortening of the early AP plateau and late AP lengthening. This is often explained by the inward and outward flow of cations through MSCs with the current flow dependant upon a cell's membrane potential and the reversal potential of the MSCs. Action potentials triggered purely by stretch are a major source of evidence for the existence of stretch-activated arrhythmias in human hearts (Taggart et al., 1999). The lengthening effect of release on the APD in frogs is broadly similar to those seen in mammalian tissue and reported in the seminal mechano-electric feedback (MEF) work of (Lab *et al.*, 1984). In this study release-induced prolongation of the Ca^{2+} transient was mimicked by late APD prolongation. This suggested a modulating role for the Ca^{2+} released from de-activating myofilaments, on the electrical activity of cardiac muscle. This was thought to occur via Ca^{2+}-dependant ion channels and exchangers such as the Na^+/Ca^{2+} exchange. Therefore within these 2 papers on frog (Lab, 1978; Lab, 1980) there are observations of many of the basic phenomena that have been used to define MEF in mammalian cardiac tissue.

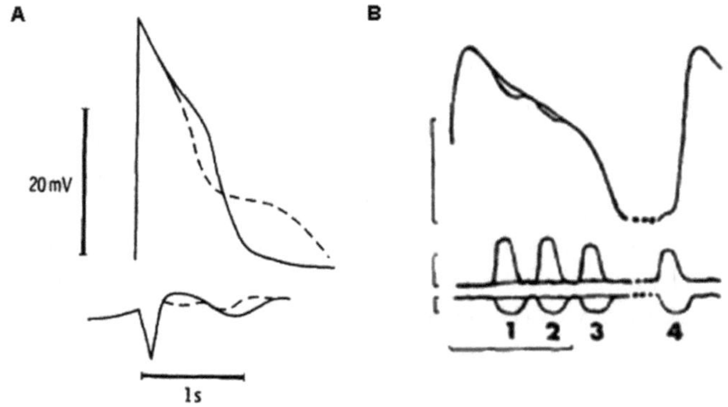

Fig. 15.6 Frog hearts demonstrate basic mechano-electric feedback phenomema. (A) Monophasic action potential (MAP) recordings (upper trace) and T wave from an ECG (lower trace) from the surface of an isolated frog ventricle. Ventricular dilation caused a cross-over of the MAP. Cross-over effects are explained in terms of activation of mechanosensitive ion channels with reversal potentials mid-way between resting and action potential plateau membrane potentials. **(B)** In frog ventricular strips, brief stretches at various points during the action potential caused various perturbation of the MAP (1–3). The stretch at 4, during the diastolic period, results in a mechanically triggered action potential (upper trace, MAP calibration 20 mV; middle trace, tension 10 mn/mm^2, lower trace indicates stretch). Reproduced from (Lab, 1978) with copyright permission of Circulation Research

More recently work by Leslie Tung's group has shown stretch lowers the threshold of excitability of amphibian muscle sheets and single myocytes (Fasciano & Tung, 1999; Tung & Zou, 1995). Stretch was able to consistently activate tissue (i.e. trigger contractions) (Fasciano & Tung, 1999). However this was not the case in single myocytes, where the action potential profile was rather insensitive to stretch (Riemer & Tung, 2003). This finding was in contrast to the effects predicted by modelling (Riemer *et al.*, 1998). The stretches applied to cells were quite large, 10–25% and comparable to those given to tissue sheets. Given the difficulty in stretching mammalian myocytes, including thin SAN cells, by more than 10%, the extensibility of fish and amphibian myocytes might suggest them as useful models for research, particularly given the similar responses of amphibian tissue and mammalian tissue to stretch and the similarity in APD profile of both fish and amphibian myocytes to most mammals (i.e. a high APD plateau potential, rather than a rapidly repolarising plateau as found in rats and mice). Unfortunately the findings of (Riemer & Tung, 2003), which are the only data available, are not encouraging. Possibly the very extensibility and low levels of resting tension in these single cells may counter the development of bilayer tension, which is thought to be important in the activation of MSCs (Hamill & Martinac, 2001).

An isolated myocyte is not in contact with non-myocyte cells that might influence the response to stretch. There is evidence that cardiac fibroblasts although not electrically excitable, are subject to mechanical modulation of their membrane potential (Kamkin *et al.*, 2005). Fibroblasts are electrically coupled to each other and there is evidence for coupling to atrial myocytes. Measurements in single fibroblasts have revealed mechanically activated whole cell currents sensitive to Gd^{3+}. Thus the hypothesis that mechanosensitive fibroblasts might influence the electrical activity of the atria was formed. The initial work characterising electrical properties of cardiac fibroblasts were carried out in frog hearts (reviewed in English by Kamkin *et al.*, 2005) and see (Kohl *et al.*, 1992).

Another important aspect of mechanical stimulation of the heart is the modulation of intracellular second messengers. The signalling pathway of the second messenger cAMP is implicated in the response of cardiac muscle to stretch (Watson, 1991) this was first demonstrated in frog ventricle (Singh, 1982).

The embryonic chick model has been used by the lab of Fred Sachs to accumulate an important body of knowledge on cardiac MSCs. Activation of whole cell and single channel kinetics of non-specific cationic MSCs by pipette suction (Bustamante *et al.*, 1991), steady (Hu & Sachs, 1996) or sinusoidal compression (Bett & Sachs, 2000) have been reported. These channels are sensitive to Gd^{3+}, aminoglycosidic antibiotics and the venom of the Chile Rose Tarantula, from which a more specific toxin (GsMTx4) has been isolated (Hu & Sachs, 1996; Sokabe *et al.*, 1993; Suchyna *et al.*, 2000). Thus, these channels exhibit the same pharmacology as MSCs and mechanically-induced events in mammalian tissue (White, 2006). This cell has also demonstrated local increases in sub-membrane Ca^{2+} induced by cell prodding (Sigurdson *et al.*, 1992) and the presence of both K^+ selective MSCs and non-selective cationic MSCs (Ruknudin *et al.*, 1993).

Embryonic chick myocytes have also been used by Lieberman to investigate the response to hypo-osmotic swelling, a complex stimulus that contains some mechanical components. They have reported the activation of a chloride conductance (Zhang & Lieberman, 1996) similar to that described in adult mammalian myocytes (Baumgarten & Clemo, 2003). However, some aspects of volume regulation in the embryonic model, such as the role of the actin cytoskeleton and microtubules (Zhang *et al.*, 1997; Larsen *et al.*, 2000), seem different in adult mammalian myocytes (Calaghan *et al.*, 2004).

It is possible to question the extrapolation of physiological importance from the chick embryonic myocyte to the avian or mammalian adult animal. However, this model possess mechanically induced increases in both force and rate (see above). Furthermore, to date, the information on the function of MSCs gained from the chick embryonic model has not been contradicted by findings in adult models and has greatly enhanced our understanding of MSCs. This is not surprising if one considers that the most recent candidate for non-specific cationic MSCs are Transient Receptor Potential (TRP) channels (Maroto *et al.*, 2005). These channels seem to be as common as mechanosensitivity itself (Saito & Shingai, 2006; Pedersen *et al.*, 2005) with one of their many roles being to act as MSCs (Barritt & Rychkov, 2005).

15.5 Conclusion

Although there are many gaps in our knowledge it seems that variations in cardiac mechanosensitivity exist within a framework of basic physiological parameters, i.e. mechanical stimulation increases contractility, pacemaker frequency and action potential profile. In ectothermic vertebrates, changes in heart rate in response to changes in demand are relatively small compared to birds and mammals. In fish, in particular, the effect of cardiac distension is greater upon contractility (stroke volume) than that upon rate (pacemaking) and the passive mechanical properties of fish myocytes may explain this. However, amphibian myocytes seem to share these properties but *in vivo* cardiac output is regulated more by rate rather than rhythm. However, as many amphibians are vulnerable to desiccation, volume regulation may not be as practical as in fully aquatic organisms. If the differences in the emphasis of mechanical modulation are the result of the cardiac environment, species that span the normal distinctions between vertebrate classes - lizards with high body temperatures, tuna with high heart rates and mammals that undergo torpor may prove particularly interesting.

Acknowledgment Supported by The University of Leeds and The University of Manchester. The authors thank Simon Patrick and Gina Galli for use of unpublished data.

References

Allen DG & Blinks JR (1978). Calcium transients in aequorin-injected frog cardiac muscle. *Nature* 273, 509–513.

Allen DG & Kentish JC (1985). The cellular basis of the length-tension relation in cardiac muscle. *J Mol Cell Cardiol* 17, 821–840.

Anderson RH, Ho SY, Redmann K, Sanchez-Quintana D, & Lunkenheimer PP (2005). The anatomical arrangement of the myocardial cells making up the ventricular mass. *Eur J Cardiothorac Surg* 28, 517–525.

Asnes CF, Marquez JP, Elson EL, & Wakatsuki T (2006). Reconstitution of the Frank-Starling mechanism in engineered heart tissues. *Biophys J* 91, 1800–1810.

Barritt G & Rychkov G (2005). TRPs as mechanosensitive channels. *Nat Cell Biol* 7, 105–107.

Baumgarten CM & Clemo HF (2003). Swelling-activated chloride channels in cardiac physiology and pathophysiology. *Prog Biophys Mol Biol* 82, 25–42.

Bean BP, Nowycky MC, & Tsien RW (1984). [beta]-Adrenergic modulation of calcium channels in frog ventricular heart cells. *Nature* 307, 371–375.

Betanzos M, Chiang CS, Guy HR, & Sukharev S (2002). A large iris-like expansion of a mechanosensitive channel protein induced by membrane tension 145. *Nat Struct Biol* 9, 704–710.

Bett GC & Sachs F (2000). Activation and inactivation of mechanosensitive currents in the chick heart. *J Membr Biol* 173, 237–254.

Brette F & Orchard C (2003). T-Tubule Function in Mammalian Cardiac Myocytes. *Circ Res* 92, 1182–1192.

Burggren W, Farrell A, & Lillywhite H (1997). Vertebrate cardiovascular systems. In *The Handbook of Physiology*, ed. W.H. Dantzler, pp. 215–308. Oxford University Press, New York, Oxford.

Burggren W & Doyle M (1986). Ontogeny of heart rate regulation in the bullfrog, *Rana catesbeiana. AJP - Regulatory, Integrative and Comparative Physiology* 251, R231–R239.

Bustamante JO, Ruknudin A, & Sachs F (1991). Stretch-activated channels in heart cells: relevance to cardiac hypertrophy. *J Cardiovasc Pharmacol* 17 Suppl 2, S110–S113.

Butler PJ, Frappell PB, Wang T, & Wikelski M (2002). The relationship between heart rate and rate of oxygen consumption in Galapagos marine iguanas (*Amblyrhynchus cristatus*) at two different temperatures. *J Exp Biol* 205, 1917–1924.

Calaghan S & White E (2004). Activation of Na+-H+ exchange and stretch-activated channels underlies the slow inotropic response to stretch in myocytes and muscle from the rat heart. *J Physiol* 559, 205–214.

Calaghan SC, Le Guennec JY, & White E (2004). Cytoskeletal modulation of electrical and mechanical activity in cardiac myocytes. *Prog Biophys Mol Biol* 84, 29–59.
Calaghan SC & White E (2005). Mechanical modulation of intracellular ion concentrations: mechanisms and electrical consequences. In *Mechanosensitivity in cells and tissues*, eds. Kamkin A & Kiseleva I, pp. 230–254. Academia Publishing House Ltd., Moscow.
Cazorla O, Freiburg A, Helmes M, Centner T, McNabb M, Wu Y, Trombitas K, Labeit S, & Granzier H (2000a). Differential expression of cardiac titin isoforms and modulation of cellular stiffness. *Circ Res* 86, 59–67.
Cazorla O, Le Guennec JY, & White E (2000b). Length-tension relationships of sub-epicardial and sub-endocardial single ventricular myocytes from rat and ferret hearts. *Journal of Molecular and Cellular Cardiology* 32, 735–744.
Cerra MC, Imbrogno S, Amelio D, Garofalo F, Colvee E, Tota B, & Icardo JM (2004). Cardiac morphodynamic remodelling in the growing eel (*Anguilla anguilla L.*). *J Exp Biol* 207, 2867–2875.
Cooper PJ, Lei M, Cheng LX, & Kohl P (2000). Selected contribution: axial stretch increases spontaneous pacemaker activity in rabbit isolated sinoatrial node cells. *J Appl Physiol* 89, 2099–2104.
Czarnecki CM (1984). Animal models of drug-induced cardiomyopathy. *Comp Biochem Physiol C* 79, 9–14.
Faber JJ (1968). Mechanical function of the septating embryonic heart. *Am J Physiol* 214, 475–481.
Fabiato A & Fabiato F (1978). Myofilament-generated tension oscillations during partial calcium activation and activation dependence of the sarcomere length-tension relation of skinned cardiac cells. *J Gen Physiol* 72, 667–699.
Farrell A, Franklin C, Arthur P, Thorarensen H, & Cousins K (1994). Mechanical performance of an in situ perfused heart from the turtle *chrysemys scripta* during normoxia and anoxia at 5 C and 15 C. *J Exp Biol* 191, 207–229.
Farrell AP (1991). From hagfish to tuna: a perspective on cardiac-function in fish. *Physiological Zoology ;* 64, 1137–1164.
Farrell AP, Johansen JA, & Graham MS (1988). The role of the pericardium in cardiac-performance of the trout *(Salmo Gairdneri*). *Physiological Zoology;* 61, 213–221.
Farrell AP & Jones DR (1992). The Heart. In *The Cardiovascular System*, eds. Hoar WS, Randall DJ, & Farrell AP, pp. 1–88. Academic Press, San Diego, CA.
Farrell AP, MacLeod K, & Chancey B (1986). Intrinsic mechanical properties of the perfused rainbow trout heart and the effects of catecholamines and extracellular calcium under control and acidotic conditions. *J Exp Biol* 125, 319–345.
Farrell AP & Olson KR (2000). Cardiac natriuretic peptides: a physiological lineage of cardioprotective hormones? *Physiol Biochem Zool* 73, 1–11.
Farrell AP, Small S, & Graham MS (1989). Effect of Heart-Rate and Hypoxia On the Performance of a Perfused Trout Heart. *Canadian Journal of Zoology;* 67, 274–280.
Fasciano RW & Tung L (1999). Factors governing mechanical stimulation in frog hearts. *Am J Physiol* 277, H2311–H2320.
Forster EM. Performance of the heart of the hagfish, *Eptatretus cirrhatus*. Fish Physiology and Biochemistry 6, 327–331. 1989.
Franklin C & Axelsson M (1994). The intrinsic properties of an in situ perfused crocodile heart. *J Exp Biol* 186, 269–288.
Franklin CE. Intrinsic properties of an in situ turtle heart (*Emydura signata*) preparation perfused via both atria. *Comp Biochem Physiol* 107, 501–507. 1994.
Franklin CE & Davie PS (1992). Dimensional analysis of the ventricle of an in situ perfused trout heart using echocardiography. *J Exp Biol* 166, 47–60.
Fukuda N, Wu Y, Farman G, Irving TC, & Granzier H (2003). Titin isoform variance and length dependence of activation in skinned bovine cardiac muscle. *J Physiol (Lond)* 553, 147–154.
Fukuda, N.; Wu, Y.; Nair, P.; Granzier, H.L. (2005). Phosphorylation of titin modulates passive stiffness of cardiac muscle in a titin isoform-dependent manner. *J Gen Physiol* 125; 257–271.

Galli GLJ, Gesser H, Taylor EW, Shiels HA, & Wang T (2006). The role of the sarcoplasmic reticulum in the generation of high heart rates and blood pressures in reptiles. *J Exp Biol* 209, 1956–1963.

Goaillard JM, Vincent PV, & Fischmeister R (2001). Simultaneous measurements of intracellular cAMP and L-type Ca^{2+} current in single frog ventricular myocytes. *J Physiol* 530, 79–91.

Gordon AM, Huxley AF, & Julian FJ (1966). The variation in isometric tension with sarcomere length in vertebrate muscle fibres. *J Physiol* 184, 170–192.

Graham MS & Farrell AP (1989). The effect of temperature-acclimation and adrenaline on the performance of a perfused trout heart. *Physiol Zool* 62, 38–61.

Granzier HL & Irving TC (1995). Passive tension in cardiac muscle - contribution of collagen, titin, microtubules, and intermediate filaments. *Biophys J* 68, 1027–1044.

Grubb BR (1982). Cardiac output and stroke volume in exercising ducks and pigeons. *J Appl Physiol* 53, 207–211.

Hamill OP & Martinac B (2001). Molecular basis of mechanotransduction in living cells. *Physiol Rev* 81, 685–740.

Hanley PJ, Young AA, LeGrice IJ, Edgar SG, & Loiselle DS (2006). 3-Dimensional configuration of perimysial collagen fibres in rat cardiac muscle at resting and extended sarcomere lengths. *J Physiol* 517, 831–837.

Hu H & Sachs F (1996). Mechanically activated currents in chick heart cells. *J Membr Biol* 154, 205–216.

Icardo J, Imbrogno S, Gattuso A, Colvee, & Tota B (2005). The heart of *Sparus auratus*: a reappraisal of cardiac functional morphology in teleosts. *Journal of Experimental Zoology* 303A, 665–675.

Imbrogno S, De IL, Mazza R, & Tota B (2001). Nitric oxide modulates cardiac performance in the heart of *Anguilla anguilla*. *J Exp Biol* 204, 1719–1727.

Jensen D (1961). Cardioregulation in an aneural heart. *Comp Biochem Physiol* 2, 181–201.

Ju YK & Allen DG (2000). The mechanisms of sarcoplasmic reticulum Ca^{2+} release in toad pacemaker cells. *J Physiol* 525 Pt 3, 695–705.

Kamkin A & Kiseleva I (2005). *Mechanosensitivity in cells and tissues*, first ed., pp. 1–465. Academia Publishing House Ltd, Moscow.

Kamkin A, Kiseleva I, Lozinsky I, Wagner KD, Isenberg G, & Scholz H (2005). The role of mechanosensitive fibroblasts in the heart. In *Mechanosensitivity in cells and tissues*, eds. Kamkin A & Kiseleva I, pp. 203–229. Academia Publishing House Ltd, Moscow.

Keatinge WR (1959). The effect of increased filling pressure on rhythmicity and atrioventricular conduction in isolated hearts. *J Physiol* 149, 193–208.

Kentish JC, ter Keurs HE, Ricciardi L, Bucx JJ, & Noble MI (1986). Comparison between the sarcomere length-force relations of intact and skinned trabeculae from rat right ventricle. Influence of calcium concentrations on these relations. *Circ Res* 58, 755–768.

Kim CS, Davidoff AJ, Maki TM, Doye AA, & Gwathmey JK (2000). Intracellular calcium and the relationship to contractility in an avian model of heart failure. *J Comp Physiol [B]* 170, 295–306.

Klitzner T & Morad M (1983). Excitation-contraction coupling in frog ventricle. Possible Ca2+ transport mechanisms. *Pflugers Arch* 398, 274–283.

Kohl P, Kamkin AG, Kiseleva IS, & Streubel T (1992). Mechanosensitive cells in the atrium of frog heart. *Exp Physiol* 77, 213–216.

Kohl P, Sachs F, & Franz MR (2005). *Cardiac mechano-electric feedback & arrhythmias, from patient to pipette*, pp. 1–423. Saunders Elsevier, Philadelphia.

Lab MJ (1978). Mechanically dependent changes in action potentials recorded from the intact frog ventricle. *Circ Res* 42, 519–528.

Lab MJ (1980). Transient depolarisation and action potential alterations following mechanical changes in isolated myocardium. *Cardiovasc Res* 14, 624–637.

Lab MJ, Allen DG, & Orchard CH (1984). The effects of shortening on myoplasmic calcium concentration and on the action potential in mammalian ventricular muscle. *Circ Res* 55, 825–829.

Labeit D, Watanabe K, Witt C, Fujita H, Wu Y, Lahmers S, Funck T, Labeit S, Granzier H (2003). Calcium-dependent molecular spring elements in the giant protein titin. *Proc Natl Acad Sci* U.S.A 100;13716–13721.

Larsen TH, Dalen H, Boyle R, Souza MM, & Lieberman M (2000). Cytoskeletal involvement during hypo-osmotic swelling and volume regulation in cultured chick cardiac myocytes *Histochem Cell Biol* 113, 479–488.

Li F, McNelis MR, Lustig K, & Gerdes AM (1997). Hyperplasia and hypertrophy of chicken cardiac myocytes during posthatching development. *Am J Physiol* 273, R518–R526.

Lillywhite HB, Zippel KC, & Farrell AP (1999). Resting and maximal heart rates in ectothermic vertebrates. *Comp Biochem Physiol A Mol Integr Physiol* 124, 369–82.

Lustig KH, Gerdes AM, & Capasso JM (1996). Characterization of enzymically isolated myocytes from the turtle, *Chrysemys picta*. *Comp Biochem Physiol B Biochem Mol Biol* 115B, 457–464.

Maroto R, Raso A, Wood TG, Kurosky A, Martinac B, & Hamill OP (2005). TRPC1 forms the stretch-activated cation channel in vertebrate cells. *Nat Cell Biol* 7, 179–185.

Martinac B & Kloda A (2003). Evolutionary origins of mechanosensitive ion channels 110. *Prog Biophys Mol Biol* 82, 11–24.

Nassar R, Manring A, & Johnson EA (1974). Light diffraction of cardiac muscle:sarcomere motion during contraction. In *The Physiological Basis of Starling's Law of the Heart*, eds. Porter R & Fitzsimons DW, pp. 57–91. Elsevier, Amsterdam.

Page SG (1974). Measurement of structural parameters in cardiac muscle. In *The Physiological basis of Starling's law of the heart*, eds. Porter R & Fitzsimons DW, pp. 13–30. Elsevier, Amsterdam.

Pathak CL (1958). Effect of stretch on formation and conduction of electrical impulses in the isolated sinoauricular chamber of frog's heart. *Am J Physiol* 192, 111–113.

Pathak CL (1966). The fallacy of the Bainbridge reflex. *Am Heart J* 72, 577–581.

Pathak CL (1973). Autoregulation of chronotropic response of the heart through pacemaker stretch. *Cardiology* 58, 45–64.

Pathak CL (1976). Transmural pressure as a determinant of basic intrinsic heart rate. *Experientia* 32, 1295–1297.

Pedersen SF, Owsianik G, & Nilius B (2005). TRP channels: an overview. *Cell Calcium* 38, 233–252.

Perozo E, Cortes DM, Sompornpisut P, Kloda A, & Martinac B (2002). Open channel structure of MscL and the gating mechanism of mechanosensitive channels 143. *Nature* 418, 942–948.

Peters GW, Steiner DA, Rigoni JA, Mascilli AD, Schnepp RW, & Thomas SP (2005). Cardiorespiratory adjustments of homing pigeons to steady wind tunnel flight. *J Exp Biol* 208, 3109–3120.

Rasmusson RL, Clark JW, Giles WR, Robinson K, Clark RB, Shibata EF, & Campbell DL (1990). A mathematical model of electrophysiological activity in a bullfrog atrial cell. *Am J Physiol* 259, H370–H389.

Riemer TL, Sobie EA, & Tung L (1998). Stretch-induced changes in arrhythmogenesis and excitability in experimentally based heart cell models. *Am J Physiol* 275, H431–H442.

Riemer TL & Tung L (2003). Stretch-induced excitation and action potential changes of single cardiac cells. *Prog Biophys Mol Biol* 82, 97–110.

Ruknudin A, Sachs F, & Bustamante JO (1993). Stretch-activated ion channels in tissue-cultured chick heart. *Am J Physiol* 264, H960–H972.

Saito S & Shingai R (2006). Evolution of thermoTRP ion channel homologs in vertebrates. *Physiol Genomics*. 27, 219–230.

Sanchez-Quintana, D., García-Martínez, V., Climent, V., Hurlé, J.M. (1996). Myocardial fiber and connective tissue architecture in the fish heart ventricle. *J Exp Zool* **275**; 112–224.

Santer RM & Greer Walker M. Morphological studies on the ventricle of teleost and elasmobranch hearts. *J Zool* 190, 259–272. 1980.

Satchell GH (1991). *Physiology and form of fish circulation*, pp. 235. Cambridge University Press, Cambridge.

Satoh H, Delbridge LM, Blatter LA, & Bers DM (1996). Surface:volume relationship in cardiac myocytes studied with confocal microscopy and membrane capacitance measurements: species-dependence and developmental effects. *Biophys J* 70, 1494–1504.

Seeley M, Huang W, Chen Z, Wolff WO, Lin X & Xu X (2007). Depletion of Zebrafish titin reduces cardiac contractility by disrupting the assembly of Z-discs and A-bands. *Circ Res* 100, 238–45.

Shiels HA, Calaghan SC, & White E (2006). The cellular basis for enhanced volume-modulated cardiac output in fish hearts. *J Gen Physiol* 128, 37–44.

Shiels HA & White E (2005). Temporal and spatial properties of cellular Ca2+ flux in trout ventricular myocytes. *Am J Physiol* 288, R1756–R1766.

Sigurdson W, Ruknudin A, & Sachs F (1992). Calcium imaging of mechanically induced fluxes in tissue-cultured chick heart: role of stretch-activated ion channels. *Am J Physiol* 262, H1110–H1115.

Singh J (1982). Stretch stimulates cyclic nucleotide metabolism in the isolated frog ventricle. *Pflugers Arch* 395, 162–164.

Sokabe M, Hasegawa N, & Yamamori K (1993). Blockers and activators for stretch-activated ion channels of chick skeletal muscle 824. *Ann N Y Acad Sci* 707, 417–420.

Sommer JR, Bossen E, Dalen H, Dolber P, High T, Jewett P, Johnson EA, Junker J, Leonard S, Nassar R, & . (1991). To excite a heart: a bird's view. *Acta Physiol Scand Suppl* 599, 5–21.

Suchyna TM, Johnson JH, Hamer K, Leykam JF, Gage DA, Clemo HF, Baumgarten CM, & Sachs F (2000). Identification of a peptide toxin from Grammostola spatulata spider venom that blocks cation-selective stretch-activated channels. *J Gen Physiol* 115, 583–598.

Taggart, P.; Sutton, P.M. (1999) Cardiac mechano-electric feedback in man: clinical relevance. *Prog Biophys Mol Biol* 139–154

Tarr M, Trank JW, Goertz KK, & Leiffer P (1981). Effect of initial sarcomere length on sarcomere kinetics and force development in single frog atrial cardiac cells. *Circ Res* 49, 767–774.

Tarr M, Trank JW, Leiffer P, & Shepherd N (1979). Sarcomere length-resting tension relation in single frog atrial cardiac cells. *Circ Res* 45, 554–559.

Tung L & Morad M (1988). Contractile force of single heart cells compared with muscle strips of frog ventricle. *Am J Physiol* 255, H111–H120.

Tung L & Zou S (1995). Influence of stretch on excitation threshold of single frog ventricular cells. *Exp Physiol* 80, 221–235.

Volgis G & Tavernarakis N (2005). Mechanotransduction in the nematode *Caenorhabditis elegans*. In *Mechanosensitivity in cells and tissues*, eds. Kamkin A & Kiseleva I, pp. 22–56. Academia Publishing House Ltd, Moscow.

Vornanen M (1997). Sarcolemmal Ca influx through L-type Ca channels in ventricular myocytes of a teleost fish. *Am J Physiol* 41, R1432–R1440.

Vornanen M (1998). L-type Ca^{2+} current in fish cardiac myocytes: Effects of thermal acclimation and beta-adrenergic stimulation. *J Exp Biol* 201, 533–547.

Wang T, Altimiras J, & Axelsson M (2002). Intracardiac flow separation in an in situ perfused heart from Burmese python Python molurus. *J Exp Biol* 205, 2715–2723.

Watson PA (1991). Function follows form: generation of intracellular signals by cell deformation. *FASEB J* 5, 2013–2019.

White E (2006). Mechanosensitive channels:Therapeutic targets in the Myocardium? *Curr Pharm Des* 12, 3645–3663.

Wu Y, Cazorla O, Labeit D, & Granzier H (2000). Changes in titin and collagen underlie diastolic stiffness diversity of cardiac muscle. *J Mol Cell Cardiol* 32, 2151–2162.

Wu Y, Tobias AH, Bell K, Barry W, Helmes M, Trombitas K, Tucker R, Campbell KB, & Granzier HL (2004). Cellular and molecular mechanisms of systolic and diastolic dysfunction in an avian model of dilated cardiomyopathy. *J Mol Cell Cardiol* 37, 111–119.

Zhang J, Larsen TH, & Lieberman M (1997). F-actin modulates swelling-activated chloride current in cultured chick cardiac myocytes. *Am J Physiol* 273, C1215–C1224.

Zhang J & Lieberman M (1996). Chloride conductance is activated by membrane distension of cultured chick heart cells. *Cardiovasc Res* 32, 168–179.

Chapter 16
Mechanobiology of Fibroblasts

Bhavani P. Thampatty and James H-C. Wang

Abstract Extensive research has shown that connective tissues such as tendons and ligaments are able to respond to mechanical forces by changing their structure, composition, and function. This mechanical adaptation is made possible largely by fibroblasts, the major cell types responsible for maintaining, repairing, and remodeling extracellular matrix (ECM) in connective tissues. This review focuses on mechanobiological responses of tendon, ligament, and skin fibroblasts in terms of their ECM gene expression and protein synthesis and secretion. The mechanobiological responses of fibroblasts in tissue engineering constructs and in wound healing are also discussed, followed by a review of the roles of major cellular components including integrins, the cytoskeleton, and stretch-activated ion channels in cellular mechanotransduction mechanisms. Finally, the review concludes with a brief discussion of future research directions in fibroblast mechanobiology.

Key words: Connective tissue · Fibroblasts · ECM · Mechanical loading · Uniaxial stretching · Biaxial stretching · Growth factors · Collagen · MMPs · Tissue engineering · Mechanotransduction · Integrins · Cytoskeleton · SACs

16.1 Technical Terms/Abbreviations

2-D	Two-dimensional
3-D	Three-dimensional
ACL	Anterior cruciate ligament
α-SMA	α-smooth muscle actin
BATs	Bioartificial tissues
bFGF	Basic fibroblast growth factor
CAM kinase	Calcium/calmodulin-dependent kinase
COX-2	Cyclooxygenase-2
CREB	cAMP-response element binding protein
CTGF	Connective tissue growth factor
ECM	Extracellular matrix
FACs	Focal adhesion complexes
FAKs	Focal adhesion kinases

A. Kamkin and I. Kiseleva (eds.), *Mechanosensitive Ion Channels.*

FPCGs	Fibroblast-populated collagen gels
FSPC	Fibroblast-seeded polyurethane constructs
GTPases	Guanosine triphosphatases
IGF-I	Insulin-like growth factor-I
IGF-II	Insulin-like growth factor-II
IL-1	Interleukin-1
MAPK	Mitogen activated protein kinase
MCL	Medial collateral ligament
MMPs	Matrix metalloproteinases
MSCs	Mesenchymal stem cells
NF-κB	Nuclear factor-κB
PCL	Poly e-carprolactone
PDGF	Platelet-derived growth factor
PDLF	Periodontal ligament fibroblast
PGE_2	Prostaglandin E_2
PKC	Protein kinase C
ROCK	Rho-associated kinase
SACs	Stretch-activated ion channels
TGF-β	Transforming growth factor-β
TIMPs	Tissue inhibitors of metalloproteinases
VEGF	Vascular endothelial growth factor

16.2 Introduction

The human body is continually exposed to mechanical forces originating from gravity, blood flow, and bodily movement. The connective tissue systems are well known examples of organs designed to withstand and transmit mechanical forces such as tension, compression, and shear stress. For example, tendons and ligaments are constantly subjected to mechanical tension, while skin tissue stretches and bends. The cells in these systems translate mechanical forces into biochemical signals by numerous mechanisms that are not well understood. Fibroblasts of connective tissues are such mechanoresponsive cells. They sense mechanical cues and alter their extracellular matrix (ECM) gene and protein expression to maintain tissue structure and function. Wound healing represents a situation in which dermal fibroblasts develop tensile forces that act on the granulation tissue matrix. Mechanical forces provide normal functioning and tissue homeostasis to connective tissues since they are known to experience large mechanical stresses. The repair and maintenance of connective tissues are mainly performed by mesenchymal cells or fibroblasts. Mechanical loads can regulate various cellular functions including proliferation, gene expression, and protein secretion.

Connective tissues such as tendons and ligaments depend on a constant input of mechanical forces to maintain homeostasis. Both tissues function primarily in tension and have similar structural characteristics. However, abnormal mechanical loading on these tissues can result in pathological conditions. For example, tendon overuse injuries, or tendinopathy, occur frequently in athletes as well as in the general population due to excessive mechanical loading conditions placed on tendons. It is also well known that appropriate mechanical stress is essential for tendons and ligaments, as stress deprivation such as that caused by immobilization can result in inferior biochemical and biomechanical properties and affect tendon healing (Akeson et al., 1973; Amiel et al., 1982 ; Woo et al., 1987 ; Yasuda et al., 2000).

Both external and internal forces affect mechanobiology of the skin. External forces include a normal force when skin is compressed or stretched in tension and shear forces that result from friction. Internal forces in the skin exist as passive tension in the collagen fibrils. Mechanical stress is considered important in the secretion of matrix proteins, and subsequent matrix deposition contributes to normal healing and scar formation. Excess deposition can lead to fibrotic diseases such as hypertrophic scars of skin and fibrosis.

Numerous *in vitro* models have been developed to study the role of mechanical loading effects on tissues and cells. Most of these models make use of flexible materials or cell-secreted matrices to which cells are adhered. In addition, three dimensional (3-D) collagen matrices, which better simulate the *in vivo* tissue environment, have also been developed. The effects of externally applied forces and those generated by cells have been investigated using such models. Likewise, cell-populated biomaterial scaffolds are used in conjunction with mechanical loading toward developing functional tissue engineering constructs. Cells respond to such mechanical forces and are able to translate them into biochemical events including gene expression and protein secretion. This phenomenon is termed mechanotransduction.

Various mechanisms have been proposed to explain mechanotransduction at the cellular and molecular levels. Cells are equipped with numerous receptors that allow them to detect and respond to mechanical forces. In addition, the cytoskeleton and other structural components are able to transmit and modulate cellular tension *via* integrins and focal adhesion sites that link with ECM. Mechanosensitive ion channels, receptor tyrosine kinases, and G proteins are all implicated in cellular mechanotransduction. However, the precise mechanotransduction mechanisms have yet to be understood completely.

This review focuses on *in vitro* mechanical loading effects on tendons, ligaments, and skin. It also describes various proposed mechanotransduction mechanisms in the literature. Finally, future directions in cell mechanobiology are discussed.

16.3 Fibroblasts and ECM

Fibroblasts are highly heterogeneous mesenchymal cells that lack a specific marker (Camelliti et al., 2005; Chang et al., 2002). They are the major cell types that produce collagen in connective tissues such as tendons and ligaments. Fibroblasts are embedded within the fibrillar ECM of connective tissues which consist largely of type I collagen and fibronectin. They interact with the surrounding microenvironment through integrins. Various growth factors (e.g., TGF-β , PDGF, bFGF), hormones (e.g., IGF-1, IGF-II), and cytokines (e.g., IL-1, TNF-α) are secreted by fibroblasts (MacKenna et al., 2000). These cells maintain a balance between synthesis and degradation of growth factors, cytokines, matrix metalloproteinases (MMPs), and tissue inhibitors of metalloproteinases (TIMPs). Fibroblasts can acquire an activated phenotype termed "myofibroblasts" which express α-smooth muscle actin (α-SMA) (Tomasek et al., 2002). The activated phenotype is associated with increased proliferative activity and enhanced secretion of ECM components such as type I collagen, fibronectin, and tenascin-C. Thus, fibroblasts play a crucial role in wound healing and tissue remodeling.

ECM of connective tissues is composed of a highly organized protein network consisting of a mixture of collagens, glycoproteins, and proteoglycans providing structural support and mechanical strength to the tissues (Aumailley and Gayraud, 1998; Bosman and Stamenkovic, 2003; Labat-Robert et al., 1990). In addition, ECM serves as an attachment site for cell surface receptors. Several signaling molecules that modulate various cell functions such as migration and differentiation also reside in ECM. Collagen, the major component of ECM, is arranged in fibrils that form superhelices and cross-links in tissues such as tendons and ligaments to withstand tensile, shear, and compressive forces (Wang, 2006). At present, at least 20 types of collagens have been characterized, of which type I collagen is the most predominant in tendon ECM (Canty and Kadler, 2002). Fibril forming collagens such as type I, III, and V interact with fibril associated collagens such as collagen XII. Proteoglycans of ECM provide receptor sites for various hormones and growth factors (Ruoslahti, 1988). Many adhesive glycoproteins such as fibronectin, laminin, elastin, and tenascin-C are also constituents of connective tissue ECM (Labat-Robert et al., 1990). Fibronectin provides cell attachment, while laminin is involved in mediating cell-ECM interactions (Mostafavi-Pour et al., 2003). Turnover of ECM is dependent on the balance between MMPs and their inhibitors, TIMPs (Visse and Nagase, 2003).

16.4 Mechanobiological Responses of Fibroblasts

Various two-dimensional (2-D) devices that make use of fibroblasts on flexible substrates such as silicone elastomers have been in use to apply mechanical stretching to fibroblasts *in vitro.* The silicone surfaces are usually coated with ECM proteins such as type I collagen or fibronectin to aid cell attachment. The two major types of mechanical stretching used are uniaxial stretching, where strain is applied in only one direction, and biaxial stretching, where strain is applied in two orthogonal directions (Banes et al., 1995a; Lee et al., 1996; Wang and Thampatty, 2006; Yost et al., 2000). Equibiaxial stretches can also be applied, in which cells are equally stretched in all directions (Fig. 16.1). In some cases, the silicone surfaces are modified to form microgrooves so that cell alignment and orientation *in vivo* can be mimicked. Using such a system, various responses of tendon fibroblasts to cyclic mechanical stretching, including the production of inflammatory agents (e.g., PGE_2) and collagen secretion, have been investigated (Wang et al., 2003; Wang et al., 2005; Yang et al., 2004; Yang et al., 2005).

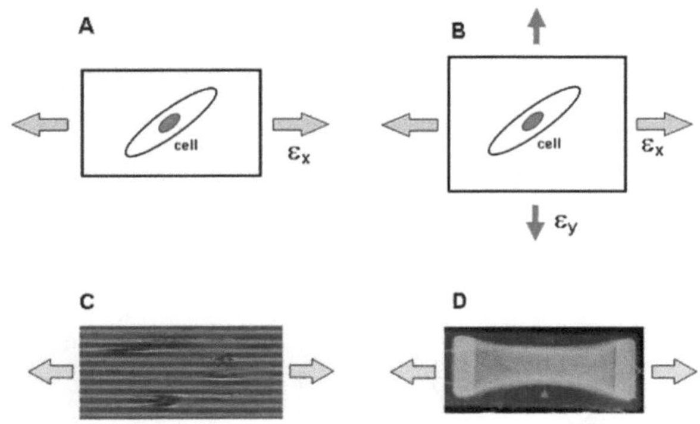

Fig. 16.1 Several manners of applying mechanical stretching to cells *in vitro*. **A.** Application of uniaxial stretching to cells on a 2-D substrate with a smooth culture surface. The cells are stretched in the horizontal direction (arrows) but compressed in the perpendicular direction; **B**. Application of biaxial stretching to cells on a 2-D substrate with a smooth surface. Cells are stretched in two perpendicular directions. In the case that the two stretching magnitudes (ε_x and ε_y) are equal, the cells are said to be subjected to equi-biaxial stretching; **C**. Application of uniaxial stretching to cells on a microgrooved substrate. All cells are aligned with the microgroove direction and stretched in the same direction; and **D**. Mechanical loading of cells in a 3-D matrix. Cells in the collagen gel produce tension, and as a result, the gel changes from rectangular (arrow) to parabolic shape (triangle). Also, a uniaxial mechanical stretching can be applied to the gel (arrows)

Although 2-D systems have provided much information on mechanical loading effects, such systems can not replicate the *in vivo* tissue environment where cells are surrounded by matrix that can be remodeled. As an alternative, 3-D matrices using type I collagen have been developed where fibroblasts reside in a more *in vivo*-like environment, which better preserves cell phenotype. Such matrices populated with fibroblasts or fibroblast-populated collagen gels (FPCGs) are attached to the dish to generate internal tension, and tension dissipates when they are released from the dish (Grinnell, 2000; Grinnell, 2003). External mechanical loading also can be applied to FPCGs, similar to that used in the development of bioartificial tissues (BATs) for the fabrication of tissue engineering constructs (Garvin et al., 2003).

Mechanobiological responses of fibroblasts from tendons, ligaments, and skin mainly fall into two categories, anabolic and catabolic (Fig. 16.2), and they are summarized in Table 16.1.

16.4.1 Mechanobiological Responses of Tendon Fibroblasts

In tendons, resident fibroblasts are responsible for synthesizing collagen and other macromolecules (Koob, 2002). They assemble these molecules into a well-organized unit and organize the fibrous phase in parallel with the direction of tensile load (Wang, 2006). Both cultured tendon explants and isolated fibroblasts retain the capacity to respond to mechanical loads (Koob et al., 1992; Ralphs et al., 2002; Yamamoto et al., 2002). ECM turnover in tendons is affected by fibroblast response

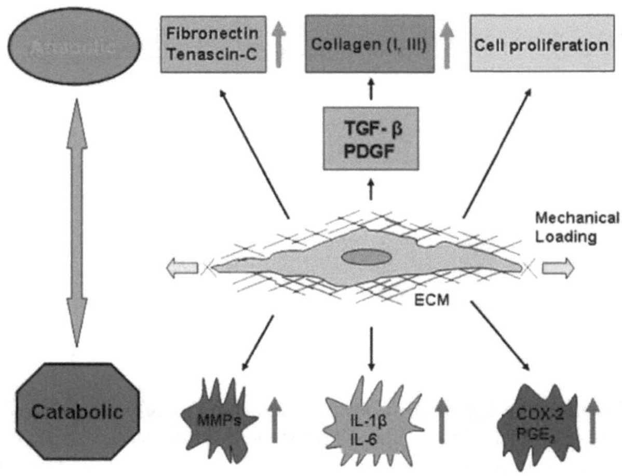

Fig. 16.2 Mechanobiological responses of fibroblasts. Both anabolic and catabolic responses are observed under cyclic uniaxial and biaxial stretching and tension in 3-D collagen gels. The responses largely depend on cell source and also on the type, magnitude, frequency, and duration of mechanical loading. In general, anabolic effects such as increased collagen I and tenascin-C expression are mainly TGF-β1-mediated, and catabolic effects such as increased expression of MMPs are mainly IL-1β-mediated

to mechanical loads. ECM turnover is also closely tied with clinical conditions of tendinopathy (Riley, 2004; Riley, 2005). *In vitro* studies have provided evidence that mechanical loading can induce either anabolic or catabolic effects depending on the type and magnitude of loading applied to tendon fibroblasts. In addition, the presence of growth factors and cytokines during loading has profound effects on ECM turnover by tendon fibroblasts. Since tendons are largely tensile load-bearing tissues, *in vitro* experiments are focused on application of mechanical stretching, such as cyclic biaxial and uniaxial stretching, which are the two major types of mechanical stretching used in studies of tendon fibroblasts' response to mechanical loads.

Tendon fibroblasts under mechanical loading secrete soluble molecules as autocrine or paracrine factors that affect tendon remodeling. Application of cyclic biaxial mechanical stretching (5%, 1Hz) for 60 min has been shown to induce expression of various growth factors such as TGF-β, PDGF, and bFGF in human tendon fibroblasts (Skutek et al., 2001b). Increased secretion of growth factors can have a positive influence on tendon matrix through stimulation of cell proliferation and collagen secretion. It was shown previously that the presence of growth factors is necessary for tendon fibroblasts to generate a load-induced mitogenic response (Banes et al., 1995a). For example, chicken tendon fibroblasts subjected to cyclic biaxial stretching (5%, 1Hz) for 8 hr, in presence of PDGF and IGF-1 showed increased DNA synthesis. The combined mechanical loading and PDGF-treatment increased thymidine incorporation by 10-fold compared to unloaded or PDGF alone-treated cultures. Similarly, mechanical loading in the presence of IGF-1 caused 5-fold increase in thymidine incorporation compared to the values of non-loaded or non-IGF-1-treated cultures (Banes et al., 1995a). In addition, load and

Table 16.1 Mechanobiological response of fibroblasts

Synthetic Events

Type of cells	Type of loading	Loading conditions	Response	References
Human tendon fibroblasts	Cyclic biaxial	5 % strain, 1Hz, 1hr	Induction of growth factors, TGF-β1, PDGF, bFGF	Skutek et al., 2001a
Chicken tendon fibroblasts	Cyclic biaxial	5% strain, 1Hz, 8hr	Increase in DNA synthesis in presence of PDGF, IGF-1	Banes et al., 1995
Human tendon fibroblasts	Cyclic uniaxial	8% strain, 0.5Hz, 4hr	Increase in cell proliferation and TGF-β1 induction	Yang et al., 2004
Human tendon fibroblasts	Cyclic biaxial	5% strain, 1Hz, 6, 24hr	Increase in cell proliferation	Zeichen et al., 2000
Human tendon fibroblasts	Cyclic biaxial	5% strain, 1Hz, 24 hr	Increase in cell proliferation	Barkhausen et al., 2003
Chicken tendon fibroblasts	Cyclic biaxial	5% strain, 1Hz, 6hr	Increase in transcription levels of type I collagen	Banes et al., 1999
Human tendon fibroblasts	Cyclic uniaxial	4% and 8% strains, 0.5Hz, 4hr	Stretching magnitude-dependent increase in collagen type I mRNA and protein and TGF-β1 mRNA	Yang et al., 2004
Rat tail tendon fibroblasts	3-D collagen gels	N/A	Upregulation of type I collagen mRNA, release of tension inhibition of type I collagen expression	Lavagnino, 2005
Avian tendon fibroblasts	3-D collagen gels and cyclic uniaxial stretching	1% strain, 1Hz, 1hr/day	Expression of type I collagen and fibronectin genes	Garvin et al., 2003
Human ACL fibroblasts	Cyclic uniaxial	10% strain, 0.1Hz, 24hr	Increased gene expression of type I and III collagen with increased release of TGF-β1	Kim et al., 2002
Human ACL fibroblasts	Cyclic biaxial	5% strain, 1Hz, 24 hr	Increase in type I collagen mRNA expression with slight increase in type III collagen	Hsieh et al., 2000

(continued)

Table 16.1 (continued)

Synthetic Events

Type of cells	Type of loading	Loading conditions	Response	References
Human MCL fibroblasts	Cyclic biaxial	5% strain, 1Hz, 24 hr	Increase in type III collagen expression with decrease in type I collagen mRNA expression	Hsieh et al., 2000
Human PDL fibroblasts	Cyclic equibiaxial	5% strain, 0.05 Hz, 24hr	Increase in DNA synthesis and TGF-β1 secretion	Kimoto et al., 1999
Human PDL fibroblasts	Cyclic biaxial	5% strain, 0.5Hz, 24hr	Increase in type I collagen and fibronectin synthesis with decrease in tropoelastin production	Howard et al., 1998
		10% strain, 0.5Hz, 24hr	Similar response in fibronectin and tropoelastin production with no change in collagen synthesis	
Human PDL fibroblasts	Cyclic equibiaxial	10% strain, 0.5Hz, 24hr	Increase in type I collagen, total protein, col1a1 mRNA, MMP-2 and TIMP-2 mRNA	He et al., 2004
Human PDL fibroblasts	Cyclic equibiaxial	20% strain, 0.1Hz, 48 hr	Increase in TIMP-1 and -2mRNA without change in MMP-1 and -2 mRNA	Tsuji et al., 2004
Human dermal fibroblasts	Cyclic equibiaxial	20% strain, 1Hz, 48hr	Increase in procollagen synthesis and procollagen mRNA levels in presence of TGF-β1	Parsons et al., 1999
Chicken skin fibroblasts	Cyclic equibiaxial	10% strain, 0.3Hz, 6hr	Increase in tenascin-C mRNA and protein, additional induction in presence of serum, TGF-β1, and PDGF	Chiquet et al., 2004; Sarasa-renedo et al., 2006
Chicken skin fibroblasts	3-D collagen gels	N/A	Increase in tenascin-C expression in stressed gels, decrease in free-floating gel	Chiquet-Ehrismann et al., 1994

Table 16.1 (continued)

Synthetic Events

Type of cells	Type of loading	Loading conditions	Response	References
Chicken skin fibroblasts	3-D collagen gels	N/A	Increase in collagen XII and fibronectin production in tensed gels	Fluck et al., 2003
Chicken skin fibroblasts	3-D collagen gels	N/A	Increase in collagen XII and tenascin-C production in tensed gels without change in MMP-2 mRNA	Trachslin et al., 1999
Human dermal fibroblasts	3-D collagen gels	N/A	Induction of several ECM genes such as collagens, fibronectin, tenascin-C and growth factor genes, CTGF, TGF-β, and VEGF	Kesseler et al., 2001
Human dermal fibroblasts	Cyclic and static equibiaxial	20% strain, 0.1Hz, 24hr	Increase in α 1(I) collagen expression	Kesseler et al., 2001
Degradative Events				
Rabbit tendon fibroblasts	Cyclic biaxial	5% strain, 0.33hz, 6hr	Induction of MMP-3 gene and protein expression in presence of IL-1β	Archambault et al., 2002
Human tendon fibroblasts	Cyclic biaxial	3.5%, 1Hz, 2hr	Gene inductions of IL-1β, COX-2, and MMP-3	Tsuzaki et al., 2003
Human tendon fibroblasts	Cyclic uniaxial	4% and 8% strains, 0.5Hz, 4hr	Decreased COX-2 and MMP-1 gene expression and PGE_2 production at 4% stretch, increased COX-2 and MMP-1 gene expression and PGE_2 production at 8% in presence of IL-1β	Yang et al., 2005

(continued)

Table 16.1 (continued)

Catabolic Events

Type of cells	Type of loading	Loading conditions	Response	References
Human tendon fibroblasts	Cyclic uniaxial	4, 8, 12 % strains, 0.5Hz, 24hr	Induction of COX-2 gene expression and PGE_2 production at 8% and 12%, no change at 4%	Wang et al., 2003; Wang et al., 2004; Li et al., 2004
Human tendon fibroblasts	Cyclic biaxial	5% strain, 1Hz, 15min	Secretion of IL-6	Skutek et al., 2001b
Human ACL/MCL fibroblasts	Cyclic equibiaxial	6–14% strain, 4–24 hr	Increase in Pro-MMP-2 level, higher levels of pro-MMP-2 in ACL than MCL	Zhou et al., 2005
Human PDL fibroblasts	Cyclic compressional	10%, 0.5Hz, 24 hr	Decrease in type I collagen, fibronectin, col1a1 mRNA, increase in MMP-2 mRNA and protein with no change in TIMP-2 mRNA	He et al., 2004
Human PDL fibroblasts	3-D collagen gel free-floating	N/A	Presence and activity of MMP-2 and -9, collagen content decrease, evidence of matrix degradation by histology	Von den Hoff, 2003
Human PDL fibroblasts	Cyclic equibiaxial	9–24% strain, 5 days	6-fold increase in PGE_2 at 9% strain and 25-fold increase at 24% strain	Yamaguchi et al., 1994

growth factors together induced protein kinase activity, while mechanical load alone could not. Cell division did not occur unless a sufficient amount of growth factor was present, suggesting a mechanism to downregulate the responsiveness to load to avoid cell division. However, cell proliferation in absence of growth factors has been reported in human tendon fibroblasts subjected to uniaxial and biaxial stretching conditions (Barkhausen et al., 2003; Yang et al., 2004; Zeichen et al., 2000). In one of these studies, however, TGF-β1 increased in parallel with the increase in cell proliferation (Yang et al., 2004). The tissue types and loading conditions in these studies may have affected the differential cellular response to mechanical stretching.

In addition to affecting cell proliferation, mechanical stretching has been shown to positively modulate ECM synthesis in tendon fibroblasts. For example, in mechanically-loaded chicken tendons and in cyclically-stretched (5%, 1Hz, 6hr) tendon fibroblasts, transcription levels of type I collagen were increased regardless of the presence or absence of growth factors (Banes et al., 1999). Similar induction of type I collagen mRNA and protein was observed in human tendon fibroblasts in stretching magnitude-dependent manner under uniaxial stretching regimens (4% and 8%, 0.5Hz, 4 hr) under serum-free conditions (Yang et al., 2004). Similarly, TGF-β1 mRNA was upregulated by stretching, and anti-TGF-β1 antibody suppressed the collagen induction, indicating a TGF-β1-mediated collagen synthesis in response to mechanical load. It should be noted that the fibroblasts were grown on microgrooved flexible surfaces to closely mimic the *in vivo* orientation and alignment of tendon fibroblasts in this study, which is in contrast to other studies where smooth culture surfaces were used in cell stretching experiments. The advantage of using microgrooved culture surfaces is that they can control cell responses to mechanical stretching. For example, tendon fibroblasts grown on microgrooves were elongated in shape, aligned with microgroove direction, and remained so under cyclic mechanical stretching conditions (Wang et al., 2005). In addition, application of cyclic uniaxial stretching parallel to the cells' orientation increased α-SMA expression levels compared to those of non-stretched cells, but this was not true when stretching was applied perpendicular to the cells' orientation. The results suggest that mechanobiological response of tendon fibroblasts is cell orientation-dependent.

Similar to the upregulation of ECM genes in fibroblasts on 2-D substrates under mechanical stretching, upregulation of type I collagen mRNA was observed in rat tendon cells in 3-D collagen matrices (Lavagnino and Arnoczky, 2005). Release of gel tension resulted in the inhibition of type I collagen mRNA expression. Application of uniaxial stretching (1%, 1 Hz, 1hr/day) to 3-D collagen matrices populated with avian tendon fibroblasts showed expression of several ECM genes such as type I collagen and fibronectin similar to that of native tendons (Garvin et al., 2003). The results indicate that tendon cells fabricated in a mechanically loaded collagen gel construct can assume a similar phenotype to that of native tendons in terms of their histologic features, cell alignment, and expression of ECM genes. Moreover, the constructs were mechanically stronger than the non-loaded counterparts (Garvin et al., 2003).

Several *in vivo* studies support the previous *in vitro* findings that mechanical loading can upregulate ECM gene expression that is mediated *via* growth factors.

For example, mechanical loading of human tendons during exercise such as uphill treadmill running elevates type I collagen synthesis that is mediated *via* IGF-1 or TGF-β1 (Kjaer et al., 2005; Langberg et al., 1999; Olesen et al., 2006). Although physical training promotes both synthesis and degradation of collagen, anabolic effects dominate, resulting in net increase in type I collagen in human tendons. Application of mechanical loading to tendons with chronic tendinopathy can relieve pathological symptoms, although the exact mechanisms for such loading effects are unclear (Alfredson et al., 1998; Davidson et al., 1997).

Mechanical loading not only causes anabolic effects but catabolic effects as well. Several *in vitro* studies reveal elevated levels of MMPs and release of inflammatory mediators (e.g., PGE_2) and cytokines (e.g., IL-1β) from stretched tendon fibroblasts. For instance, cyclic biaxial stretching (5%, 0.33 Hz, 6hr) in presence of inflammatory cytokine, IL-1β has been shown to induce significantly high MMP-3 gene and protein levels in rabbit tendon fibroblasts compared to levels in cells treated with IL-1β without stretching (Archambault et al., 2002). Similar stretching with a slightly different loading regimen (3.5%, 1Hz, 2hr) in human tendon fibroblasts induced gene expression of IL-1β , COX-2, and MMP-3 without increasing TIMP-1 and -2 mRNA expression (Tsuzaki et al., 2003a). Imbalance between MMP and TIMP could favor MMP activity and matrix weakening (Tsuzaki et al., 2003b). In addition, IL-1β may establish a positive feedback loop in triggering fibroblast-mediated cytokine-MMP matrix destruction (Tsuzaki et al., 2003b). Recently, we have shown that exogenous addition of IL-1β to human tendon fibroblasts initiated catabolic effects such as elevated protein expression of $cPLA_2$ and COX-2 and PGE_2 production. Moreover, IL-1β induced increased gene expression of MMP-1 and MMP-3 and decreased mRNA expression of type I collagen (Thampatty et al., 2007). Also, differential gene inductions of COX-2, MMP-1, and PGE_2 production were observed in human tendon fibroblasts subjected to uniaxial stretching at different magnitudes in presence of IL-1β (Yang et al., 2005). Interestingly, a 4% stretching (0.5Hz, 4hr) decreased IL-1β-induced COX-2, MMP-1 gene induction, and PGE_2 production; while 8% stretching further increased them. The results suggest that a small magnitude stretching (close to physiologic level) can be anti-inflammatory and less tissue destructive, while 8% stretching (excessive load) can be pro-inflammatory and catabolic to tendon matrix. In addition, 8% and 12% uniaxial stretching for 24 hr caused significant induction of COX-2 gene expression and PGE_2 production compared to unstretched tendon fibroblasts, while 4% stretching did not have any significant effect on COX-2 gene expression and PGE_2 production (Wang et al., 2003). Similar studies (8% strain, 0.5Hz, 4hr) in tendon fibroblasts showed upregulation of cytosolic and secretory phospholipase A_2 (upstream regulators of PGE_2 production), COX-2, and PGE_2 (Li et al., 2004; Wang et al., 2004). Enhanced secretion of IL-6 due to cyclic biaxial stretching (5% strain, 1Hz, 15 min) of tendon fibroblasts also has been reported (Skutek et al., 2001a). Taken together, these studies suggest that certain mechanical loading conditions induce inflammatory mediators such as PGE_2, various matrix degrading enzymes, and inflammatory cytokines, which can accelerate tendon degeneration.

16.4.2 Mechanical Loading Effects in Ligament Fibroblasts

Like tendons, ligaments are dense connective tissues. They contain rows of fibroblasts within parallel bundles of ECM composed mainly of type I collagen and a small proportion of type III collagen (Amiel et al., 1984). Mechanical stretching increases collagen type I and III gene expression in anterior cruciate ligament (ACL) cells, and the loading-induced gene expression is mediated by TGF-β1. For example, uniaxial stretching (10%, 0.1Hz, 24 hr) of ACL increased the gene expression of both type I and III collagen, which corresponded with an increased release of TGF-β1 into the medium (Kim et al., 2002). Moreover, anti-TGF-β1 antibody inhibited the stretch-induced elevated gene expression of collagens. The findings suggest that mechanical stretching is a positive regulator of ligament healing and remodeling, which are mediated *via* an autocrine mechanism of TGF-β1. It is noted that cyclic biaxial stretching (5%, 1Hz, 24 hr) of ACL fibroblasts increased only type I collagen mRNA expression with a slight increase in type III collagen (Hsieh et al., 2000). However, medial collateral ligament (MCL) fibroblasts under similar stretching conditions responded differently with an increase in type III collagen mRNA expression and a decrease in type I collagen mRNA levels. The different healing potentials of ACL and MCL are well known. An injured ACL does not heal satisfactorily, while an injured MCL can heal relatively well. An initial expression of type III collagen is important for healing and remodeling (Woo et al., 1999). Therefore, the differential gene expression of type III collagen in ACL and MCL in response to stretching may explain the differential healing potential between the two types of ligaments. In addition, as a response to mechanical stretch, differential MMP-2 activity was observed in ACL and MCL fibroblasts, which may further explain their different healing potentials. For example, equibiaxial stretching at various stretching magnitudes (6–14%) and time periods (4-24 hr) has shown that pro-MMP-2 levels increased in a stretch-magnitude dependent manner in ACL (Zhou et al., 2005). The levels of pro-MMP-2 were much higher than that of MCL. Furthermore, more ACL pro-MMP-2 was converted into an active form.

Similar to ACL and MCL fibroblasts, periodontal ligament fibroblasts (PDLFs) respond to mechanical loading by changing their ECM expression to maintain, repair, and remodel the tissue. As observed in tendon fibroblasts, PDLFs also secrete autocrine or paracrine growth factors and increase DNA synthesis. Human PDLFs under cyclic equibiaxial mechanical stretching (5%, 0.05Hz, 24 hr) increased DNA synthesis and markedly increased TGF-β1 secretion into the medium (Kimoto et al., 1999). In addition, PDLFs respond to specific magnitudes of tensional loading by differentially altering the ECM protein synthesis. For example, cells exposed to 5% biaxial stretching (0.5 Hz, 24 hr) exhibited significant increases in type I collagen and fibronectin synthesis with a decrease in tropoelastin production compared to unstretched controls (Howard et al., 1998). On the other hand, a 10% stretching exhibited similar responses for fibronectin and tropoelastin, while the amount of type I collagen synthesized by stretched cells did not differ from control levels. Besides, these cells responded differentially to tensile and compressive loads, as manifested by a selective alteration in certain matrix components. A cyclic 10% equibiaxial stretching (0.5Hz, 24hr) exhibited synthetic events with an increase in type I collagen and total protein, Col1A1 mRNA, MMP-2 mRNA, and TIMP-2 mRNA (He et al., 2004). On the contrary, application of 10% cyclic compressional strain accelerated matrix degradation events, decreasing type I collagen, fibronectin protein, and Col1A1 mRNA and increasing total protein, MMP-2 mRNA, and MMP-2 protein, both in latent and active forms. However, TIMP-2 mRNA was unchanged under these conditions, which suggests that matrix equilibrium is tilted towards degradation.

Further experimental evidence is presented to demonstrate how cyclic stretch acts as an inhibitor of the degradation of ECM in PDLFs. For example, cyclic equibiaxial stretching (20%, 0.1Hz, 48hr) of PDLFs increased TIMP-1 and -2 mRNA expressions without affecting MMP-1 and -2 mRNA expressions (Tsuji et al., 2004). Using free-floating and attached 3-D collagen gels, it was shown that mechanical tension is essential for periodontal tissue remodeling. Presence and activity of MMP-2 and MMP-9 were reported in PDLF enriched free-floating collagen gels using gelatin zymography (Von den Hoff, 2003). In addition, histological sections showed matrix degradation around the cells. This reveals that ligament cells in stress-free conditions degrade collagen matrix. The collagen content of attached gels did not change, but after detachment, it rapidly decreased, indicating that mechanical tension can prevent matrix degradation.

Mechanical tension force is known to induce inflammatory mediators, COX-2, PGE_2, and IL-1β in PDLFs (Shimizu et al., 1998; Shimizu et al., 1994; Yamaguchi et al., 1994). In response to mechanical stretching, PDLFs demonstrated variable rates of PGE_2 production at variable stretching magnitudes. For instance, cyclic equibiaxial stretch for 5 days showed 6-fold increase in PGE_2 production when cells were exposed to 9% stretching as opposed to 25-fold production when exposed to 24% stretching (Yamaguchi et al., 1994). In addition, similar stretching conditions at 18% strain induced COX-2 mRNA expression and PGE_2 release in a time-dependent manner without affecting COX-1 expression (Shimizu et al., 1998). Moreover, in presence of selective COX-2 inhibitor, PGE_2 release was completely inhibited, suggesting that tension force-induced COX-2 is responsible for PGE_2 production.

PDL fibroblasts are the particular cells that initiate the bone remodeling process *in vivo* (Middleton et al., 1996). When orthodontic forces are applied to a tooth, the PDL is stretched on one side and compressed on the other. The tension side is characterized by bone synthesis, while the compression side is characterized by bone resorption (Reitan, 1967; Reitan, 1969; Reitan, 1970; Rygh and Reitan, 1972). Therefore, the type of force (i.e., tension vs. compression) applied to tooth support structures plays an important role in bone remodeling of a tooth.

16.4.3 Mechanical Loading Effects in Dermal Fibroblasts

The primary components of skin ECM consist of collagen, elastin, PGs, and fibronectin. The role of mechanical loading in ECM remodeling is underscored in dermal fibroblasts. Both 2-D stretching devices and 3-D collagen gels have been used to study the effect of mechanical loads on ECM synthesis. Mechanical loading promoted procollagen synthesis in presence of TGF-β1 in dermal fibroblasts (Parsons et al., 1999). When human dermal fibroblasts were subjected to cyclic equibiaxial stretching (20%, 1Hz, 48hr) in presence of TGF-β1, procollagen synthesis was increased by 38% over unloaded growth factor controls. Procollagen mRNA levels were also increased by 2-fold in stretched cells. There was also enhanced processing of procollagen to insoluble collagen. Mechanical load acting in synergy with growth factors to enhance procollagen synthesis has direct implications in wound repair and healing.

Furthermore, tenascin-C, an ECM protein that rapidly changes its expression patterns during embryogenesis and the regenerative process, is affected by mechanical loading. Application of cyclic equibiaxial stretches have been shown to induce tenascin-C expression in skin fibroblasts. Cyclic equibiaxial stretching (10%, 0.3 Hz, 6hr) increased tenascin-C mRNA and protein expression 2-fold compared to resting control (Chiquet et al., 2004; Sarasa-Renedo et al., 2006). Cyclic strain caused additional induction in presence of serum, TGF-β, and PDGF (Chiquet et al., 2004). Previously, it was reported that tenascin-C expression was elevated in skin fibroblasts in stressed collagen gels, while expression was decreased in free floating gel (Chiquet-Ehrismann et al., 1994). Similarly, two other ECM proteins, fibronectin and collagen XII, whose expressions are prominently induced in response to wounding (Clark, 1990; Karimbux and Nishimura, 1995), were significantly affected by tensile loading of fibroblasts in collagen gels. Collagen XII and fibronectin productions under tensed or attached conditions were 3- to 5-fold higher than under released or floating conditions (Fluck et al., 2003). Moreover, collagen XII expression could be reversibly regulated in a similar manner as tenascin-C. To demonstrate this finding, collagen gels with attached cells were mounted to movable polyethylene plugs and were relaxed or stretched at 24 and 48 hr intervals (Trachslin et al., 1999).

Dermal fibroblasts in mechanically stressed collagen gels are activated to a "synthetic" phenotype with the induction of several genes encoding matrix proteins such as collagens, fibronectin, and tenascin-C as determined by gene array analysis (Kessler et al., 2001). In addition, tension induced genes coding for growth factors such as connective tissue growth factor (CTGF) and vascular endothelial growth factor (VEGF), protease inhibitors, and proteins involved in cell proliferation. Moreover, Northern blot analysis of cells from stressed gels revealed increased mRNA expression levels of collagens, decorin, TGF-β, and CTGF. Interestingly, collagen expression was further augmented in fibroblasts subjected to equibiaxial cyclic stretching (20% strain, 0.1Hz, 24hr) compared to cells in stressed gels. Therefore, the type of loading and the *in vitro* environment of the cells differentially regulate ECM gene expression.

16.4.4 Mechanical Loading Effects in Tissue Engineered Constructs

Tissue engineering uses 3-D scaffolds such as collagen, cells, and bioactive factors to create viable tissue constructs for damaged tissue repair (Butler, 2003;

Doroski et al., 2006). Besides the common regulatory factors such as growth factors, hormones, and cytokines, mechanical load is an important factor that can not be overlooked while developing successful tissue constructs (Vunjak-Novakovic et al., 2004). Since mechanical loading can regulate matrix production, recent tissue engineering investigations focus on *in vitro* methods of mechanically preconditioning tissue scaffolds to stimulate matrix production and to strengthen tissue engineered constructs. Application of cyclic mechanical strain can enhance the biomechanical properties and stimulate matrix production of bioartificial tissues (BATs) such as tendons. Uniaxial cyclic strain (1%, 1Hz, 1h/day) applied to tendon cell-populated bioreactor system showed that tendon cells could acquire the properties similar to those of native tendons with increased expression of ECM proteins such as collagens, aggrecan, and fibronectin (Garvin et al., 2003). Likewise, mechanically loaded BATs had an ultimate tensile strength that was three times greater than that of non-loaded BATs. Also, cells in loaded BATs assumed morphology of linearly arranged cells aligned with the principal strain direction as in whole tendon fascicles. Strain-induced conditioning of fibroblasts/material constructs has been demonstrated in another study. Fibroblast-seeded polyurethane constructs (FSPC) were subjected to uniaxial cyclic strains (10% strain, 0.25 Hz, 8h/day) with or without ascorbic acid supplementation (Webb et al., 2006). The combination of cyclic strain and ascorbic acid increased elastic modulus ($> 110\%$) of the construct relative to either condition alone. Cyclic strain was sufficient to stimulate significant increases in fibroblast proliferation. In addition, mechanical strengthening was accompanied by increased type I collagen and fibronectin matrix accumulation. Moreover, significant increase in gene expression of type I collagen, TGF-β1, and CTGF was observed.

Mechanical stimulation of cell-collagen sponge constructs improved structural and mechanical properties of healing tendon after surgery (Juncosa-Melvin et al., 2006b). Rabbit mesenchymal stem cells (MSCs) in type I collagen sponges were mechanically stimulated (4% strain, once every 5 min for 8h/day) in flexible silicone dishes for two weeks. *In vitro* testing of the constructs showed that mechanically stimulated constructs had 2.5 times the linear stiffness of non-stimulated constructs. The other stimulated construct and the non-stimulated construct were implanted in the central part of rabbit patellar tendons. Twelve weeks after surgery, structural and mechanical properties significantly differed between stimulated vs. non-stimulated repairs. Mechanical stimulation increased repair maximum force by 15%, linear stiffness by 30%, maximum stress by 20%, and linear modulus by 10%. A particularly encouraging outcome was that the repair tissue exceeded the structural and material properties found in previous studies (Awad et al., 2003; Juncosa-Melvin et al., 2005; Juncosa-Melvin et al., 2006a), and it was able to match normal patellar tendon up to peak *in vivo* forces and displacements. The maximum force, linear stiffness, maximum stress, and linear modulus for the stimulated repairs averaged 70%, 85%, 70%, and 50% of corresponding values for the central part of patellar tendon, respectively.

Mechanical loads play similar roles in the development of tissue-engineered constructs of ligaments. Advanced bioreactor system with medium profusion and mechanical loading was used to engineer ligaments from MSCs (Altman et al.,

2002b). The reactor was designed to apply multiple-dimensional loading: axial stretching, compression, and torsion. The application of axial stretching plus torsion over 21 days of culture induced cell alignment in the direction of force and increased the expression of ligament specific transcripts, type I and III collagen, and tenascin-C (Altman et al., 2002a; Vunjak-Novakovic et al., 2004). Mechanically stimulated gels compared with static controls showed increased amounts of type I collagen fiber bundles, type III collagen, and fibronectin in the direction of load (Altman et al., 2002a).

PCL (poly e-caprolactone), a mechanically stretched synthetic film, has the potential to be used as a matrix for tissue engineering skin. Human dermal fibroblasts seeded on PCL films were biaxially stretched (Ng et al., 2001). Tensile strength of stretched films was increased by more than two times that of unstretched cells. Cell attachment and proliferation were also increased in stretched constructs. In addition, appropriate pre-loading of fibroblasts seeded in collagen constructs has proven to be necessary for the proper development of dermal tissue engineering constructs. Neonatal human dermal fibroblasts seeded in collagen constructs were preloaded to either 2 or 10 mN before subjecting them to cyclic stretching (10%, 1 Hz, 24 hr) (Berry et al., 2003). Collagen synthesis was enhanced by cyclic stretching within constructs preloaded at 2 mN only. Also, the structural stiffness of the construct was enhanced only under this loading protocol. The expression profile of MMPs (MMP-1, -2, -3, and -9) was similar in constructs preloaded at 2 mN with or without the application of cyclic stretching. On the contrary, constructs preloaded at 10 mN and subjected to the same 10% cyclic stretching expressed enhanced levels of both latent and active MMPs compared with other conditioning regimens (Berry et al., 2003). The results indicate that cyclic strain superimposed on a preload can influence both anabolic and catabolic processes in dermal fibroblasts seeded within collagen gel constructs.

16.4.5 Cell-Generated Tensile Forces and Tissue Healing

In addition to the regulation of ECM by external mechanical forces described above, fibroblasts of connective tissues can generate internal tension. This internally-generated tensile force or fibroblast contraction is essential for wound closure during tissue healing. However, when cell contraction is excessive, tissue scarring often results (Desmouliere et al., 2005). During wound healing, fibroblasts acquire myofibroblast phenotype with stress fiber formation and expression of myofibroblast marker, α-SMA. It is demonstrated that mechanical stress is a prerequisite for the expression of α-SMA in stress fibers (Hinz and Gabbiani, 2003; Tomasek et al., 2002).

3-D collagen lattices provide a useful *in vitro* model to study contractile forces generated by fibroblasts (Grinnell, 2003). Endogenous tension can be generated in the collagen matrix by tethering the lattice. Using such a model, the contractility of fetal and adult skin fibroblasts has been investigated. Fetal skin fibroblasts were found to be less contractile with lesser secretion of active TGF-β compared to adult skin fibroblasts (Coleman et al., 1998). The different contractile forces between the fetal and adult skin fibroblasts directly correlate with different healing outcomes: fetal wounds heal without scar formation, whereas adult wounds heal with scar formation (Adzick and Lorenz, 1994; Longaker et al., 1994; Nedelec et al., 2000).

16.5 Cellular Mechanotransduction

The underlying mechanisms of mechanotransduction have been intensively investigated in the past few years because mechanical loading signals critically influence the development, remodeling, and pathogenesis of tissues throughout the body. As a result, various mechanisms have been proposed for mechanotransduction in a variety of cell types such as endothelial cells, fibroblasts, and chondrocytes. Regardless of the cell type, some essential cellular components are involved in mechanotransduction mechanisms, including integrins, cytoskeletal structures, stretch-activated ion channels (SACs), and growth factor receptors, along with numerous other secondary signaling molecules. In general, the earliest responses to mechanical stimuli occur at ECM contacts. These responses include opening of SACs, release of soluble mediators, phosphorylation of focal adhesion kinases (FAKs), and activation of small guanosine triphosphatases (GTPases) (Sadoshima and Izumo, 1997). Subsequent to these responses, numerous intracellular signaling pathways are triggered such as mitogen activated protein kinase (MAPK), PKC, and nuclear factor-κB (NF-κB) (Chiquet, 2001) to regulate ECM gene transcription (MacKenna et al., 2000) **(Fig. 16.3)**. The following section focuses on a brief review of the role of integrins, cytoskeleton, and SACs in mechanotransduction mechanisms. For a more complete and in depth discussion on cellular mechanotransduction mechanisms, interested readers should consult the following excellent reviews (Chiquet et al., 2003; Iqbal and Zaidi, 2005; Lehoux et al., 2006; Lehoux and Tedgui, 2003; MacKenna et al., 2000; Sarasa-Renedo and Chiquet, 2005).

16.5.1 Integrins and Cytoskeleton

Cell-ECM adhesions sense mechanical properties of ECM and alter its environment in connective tissue cells such as fibroblasts (Chiquet, 1999). Fibroblasts anchor to ECM via focal adhesions, which transmit intracellular tension to ECM and also transmit external mechanical stresses from ECM to cells. Both internal and external forces converge on transmembrane integrin receptors that cluster together within focal adhesions. Integrins physically link the ECM to the cytoskeleton (Schwartz et al., 1995). Integrins and cytoskeletal proteins act together as mechanosensors in fibroblasts. These proteins are assumed to be capable of responding to forces through conformational and organizational changes (Nicolas et al., 2004; Nicolas and Safran, 2004). Integrins transmit signals by organizing the cytoskeleton actin filaments through associated molecules including vinculin, paxillin, talin, and

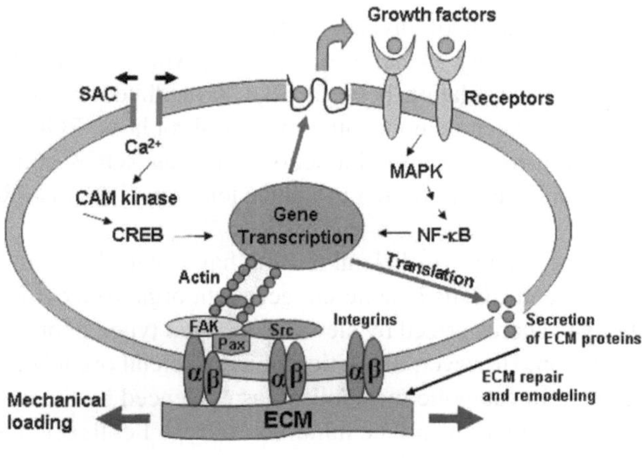

Fig. 16.3 Simplified scheme of how mechanical stress is transduced into changes in gene expression. Tensile stretching of cells causes activation of membrane ion channels elevating cations such as Ca^{2+}, which in turn lead to activation of secondary messengers such as protein calcium/calmodulin-dependent protein kinase (CAM kinase). CAM kinase can phosphorylate and activate transcription factors such as c-AMP response element-binding protein (CREB), whose activation can induce transcription of immediate early genes such as c-fos. Moreover, secondary messenger activation leads to activation of nuclear factors such as NF-κB, leading to gene transcription. Stretching of ECM-integrin contacts leads to cytoskeleton-mediated signals by rearrangement of cytoskeletal components that include actin filaments and recruitment of kinases such as FAK and Src. Activation of FAK and Src further activate secondary signaling pathways to promote selective gene transcription. The altered gene transcription leads to translational and post-translational modification to selectively synthesize and secrete ECM proteins for ECM repair and remodeling

α-actinin (Hynes, 1992; Ingber, 1991). Cyclic equibiaxial strain induced changes in the adhesion sites and the cytoskeleton of skin fibroblasts. In response to stress, cells presented very large cell-matrix adhesions containing vinculin along their margins, and actin stress fibers were distributed parallel to the long axis of cell body (Sarasa-Renedo and Chiquet, 2005). In cells at rest, stress fibers were straight and randomly oriented. This finding suggests that integrins, along with associated proteins and the cytoskeleton, can act as mechanosensors.

Integrins themselves are considered as receptors capable of inducing biochemical signals within the cell that regulate gene expression and cell growth. Upon clustering at focal adhesion sites, integrins are able to recruit kinases such as FAK and Src, and focal adhesion complexes (FACs) (Clark and Brugge, 1995; Juliano and Haskill, 1993). Maturation of FACs requires external mechanical force or internal contractile force (Balaban et al., 2001; Hinz and Gabbiani, 2003). Activation of FAK and Src can further activate various MAPK signaling path ways (Parsons and Parsons, 1997; Plopper et al., 1995).

Cytoskeletal contractility and integrity are also necessary for effective mechanical signal transduction (Bershadsky et al., 2003; Riveline et al., 2001). The actin cytoskeleton and myosin II, the key components of focal adhesion formation, play a major role in force transduction *via* activation of Rho GTPases along with their downstream targets ROCK and mDia1 (Fukata et al., 2001; Nakano et al., 1999; Watanabe et al., 1999). Through direct phosphorylation of myosin light chain or through indirect inactivation of myosin light chain phosphatase, ROCK stimulates actomyosin contractility (Amano et al., 1997; Amano et al., 1996; Kimura et al., 1996). A direct correlation between ROCK-mediated cytoskeletal tension and gene induction has been demonstrated in skin fibroblasts. ROCK inhibitor attenuated stretch-mediated tenascin-C induction, whereas ROCK stimulators acted synergistically with mechanical stretch to induce tenascin-C (Sarasa-Renedo et al., 2006).

In addition, alterations in cytoskeletal tension have been shown to control gene expression in tendon cells. Cells in adherent gels with organized cytoskeleton favor anabolic metabolism, characterized by elevated collagen type I expression and little collagenase expression. However, disruption of cytoskeletal organization using cytochalasin D resulted in catabolic metabolism, as evidenced by the upregulation of collagenase mRNA expression and the inhibition of type I collagen mRNA expression (Lavagnino and Arnoczky, 2005).

F-actin assembly and cellular stiffness are stimulated by Rho via mDia1 (Riveline et al., 2001). The integrity of actin cytoskeleton is also essential for generation of cellular internal tension, as disruption of actin filaments results in the disappearance of such force (Kolodney and Wysolmerski, 1992).

16.5.2 Stretch-Activated Ion Channels (SACs)

Activation of mechanosensitive ion channels has been proposed as the transduction mechanism in various types of cells including fibroblasts (French and Stockbridge, 1988; Stockbridge and French, 1988; Wall and Banes, 2005). Membrane or cytoskeletal stress opens cation channels allowing the passage of ions like Ca^{2+}, Na^{+}, and K^{+}. SACs transfer mechanical signals into elevations in cytosolic Ca^{2+}, thereby activating membrane kinases to specifically phosphorylate other signaling molecules (Ruknudin et al., 1993). Stretch-induced increases in intracellular Ca^{2+} can be blocked by pre-incubation with SACs blockers streptomycin and gadolinium, which inhibit immediate early genes and protein synthesis (Gannier et al., 1994; Lacampagne et al., 1994; Sadoshima and Izumo, 1993; Sadoshima et al., 1992). This finding suggests that SACs are directly involved in mechanotransduction.

It has been reported that ligament and tendon cells responded to stretch and other mechanical loads by increasing their intracellular Ca^{2+} levels (Banes et al., 1994; Banes et al., 2001; Banes et al., 1995b; Hung et al., 1997). Ca^{2+} signaling is proposed as an early response to mechanical load in tendons (Wall and Banes, 2005). Stretching forces applied through flexible substrata have been shown to induce intracellular Ca^{2+} concentration and traction forces of NIH3T3 fibroblasts (Munevar et al., 2004). In addition, MCL fibroblasts that were aligned and subjected to mechanical stretch propagated a calcium wave better than cells that were aligned, but not subjected to stretch (Jones et al., 2005). Furthermore, Ca^{2+} influx through SACs has been reported to be critical for stretch-induced activation and translocation of NF-κB, as well as stretch-induced upregulation of COX-2 expression in human fibroblasts (Inoh et al., 2002; Kato et al., 1998).

The mechanisms of how elevated Ca^{2+} contributes to activation of signaling pathways have been thoroughly reviewed (Iqbal and Zaidi, 2005; Ruwhof and van der Laarse, 2000). Elevated Ca^{2+} may enhance Ca^{2+}-responsive proteins such as protein kinase (PKC), calcium/calmodulin-dependent protein kinase (CAM kinase), and calcineurin. CAM kinase can phosphorylate and activate transcription factors such as c-AMP response element-binding protein (CREB) (Iqbal and Zaidi, 2005; Ruwhof and van der Laarse, 2000). The activated transcription factors can induce transcription of immediate early genes such as c-fos.

SACs are also modulated by other cellular components such as contractile actin cytoskeleton (Kalapesi et al., 2005). SACs can be activated by force application to integrins (Glogauer et al., 1997; Munevar et al., 2004), and the resultant influx of Ca^{2+} into the cytoplasm can modulate contractility influencing both cytoskeletal organization and cell stiffening (Helfman et al., 1999; Sokabe et al., 1997).

It should be noted that signals from mechanical loading initiate a cascade of cellular events through various pathways, resulting in selective gene transcription. This altered gene transcription leads to translational and post-translational modifications such that specific ECM proteins are synthesized and eventually secreted to alter the structural and mechanical properties of ECM. These changes may help to maintain normal tissue functioning or may lead to tissue pathology, depending on the mechanical loading. For instance, tendon ECM turnover is greatly influenced by mechanical loading, with increase in collagen synthesis and activity of MMPs maintaining tissue homeostasis (Kjaer, 2004). High matrix turnover may represent either a tissue repair or maintenance function. However, abnormal loading can alter the net collagen synthesis and deposition, leading to tendon pathological conditions. Similarly, excess deposition of collagen due to mechanical loading in dermal fibroblasts may be responsible for scar formation during wound repair (Parsons et al., 1999).

16.6 Conclusion

Connective tissues in the body perpetually experience mechanical forces. It is known that moderate mechanical forces are essential for connective tissue homeostasis, though abnormal forces induce pathological conditions. Fibroblasts, a major type of cell in the tissues, are mechano-responsive cells. In response to mechanical forces, these cells alter ECM synthesis by favoring anabolic or catabolic metabolism mostly in presence of growth factors and cytokines. Therefore, determining mechanobiological effects on fibroblasts is essential for a better understanding of the physiology and pathology of connective tissues as well as their wound healing mechanisms.

Fundamentally, however, it is necessary to further our understanding of cellular mechanotransduction mechanisms by which cells sense mechanical forces and alter ECM synthesis. Because of intensive research in recent years, the cellular mechanotransduction mechanisms are beginning to be understood. Molecules that mediate mechanotransduction, including ECM, integrins, and cytoskeletal components may represent good targets for the development and implementation of preventive and/or therapeutic strategies. Investigations of mechanotransduction mechanisms at the cellular level are giving precedence to single-molecule studies. These single-molecule investigations are able to reveal significant mechanosensory motifs such as force-regulated recognition, cryptic binding, or phosphorylation sites. However, much future work is needed to elucidate mechanotransduction mechanisms so that we can fully understand how mechanical force regulates ECM composition, structure, and function and thus apply such knowledge to advancing tissue engineering and regenerative medicine.

Acknowledgment We thank Mr. Michael Lin and Ms. Sheena Li for their assistance in preparing this review. We also gratefully acknowledge the funding support of NIH grant AR049921 and the Arthritis Investigator Award (JHW).

References

Adzick, N.S., and H.P. Lorenz. (1994). Cells, matrix, growth factors, and the surgeon. The biology of scarless fetal wound repair. Ann Surg. 220:10–8.

Akeson, W.H., S.L. Woo, D. Amiel, R.D. Coutts, and D. Daniel. (1973). The connective tissue response to immobility: biochemical changes in periarticular connective tissue of the immobilized rabbit knee. Clin Orthop Relat Res:356–62.

Alfredson, H., T. Pietila, P. Jonsson, and R. Lorentzon. (1998). Heavy-load eccentric calf muscle training for the treatment of chronic Achilles tendinosis. Am J Sports Med. 26:360–6.

Altman, G.H., R.L. Horan, I. Martin, J. Farhadi, P.R. Stark, V. Volloch, J.C. Richmond, G. Vunjak-Novakovic, and D.L. Kaplan. (2002a). Cell differentiation by mechanical stress. Faseb J. 16:270–2.

Altman, G.H., H.H. Lu, R.L. Horan, T. Calabro, D. Ryder, D.L. Kaplan, P. Stark, I. Martin, J.C. Richmond, and G. Vunjak-Novakovic. (2002b). Advanced bioreactor with controlled application of multi-dimensional strain for tissue engineering. J Biomech Eng. 124:742–9.

Amano, M., K. Chihara, K. Kimura, Y. Fukata, N. Nakamura, Y. Matsuura, and K. Kaibuchi. (1997). Formation of actin stress fibers and focal adhesions enhanced by Rho-kinase. Science. 275:1308–11.

Amano, M., M. Ito, K. Kimura, Y. Fukata, K. Chihara, T. Nakano, Y. Matsuura, and K. Kaibuchi. (1996). Phosphorylation and activation of myosin by Rho-associated kinase (Rho-kinase). J Biol Chem. 271:20246–9.

Amiel, D., C. Frank, F. Harwood, J. Fronek, and W. Akeson. (1984). Tendons and ligaments: a morphological and biochemical comparison. J Orthop Res. 1:257–65.

Amiel, D., S.L. Woo, F.L. Harwood, and W.H. Akeson. (1982). The effect of immobilization on collagen turnover in connective tissue: a biochemical-biomechanical correlation. Acta Orthop Scand. 53:325–32.

Archambault, J., M. Tsuzaki, W. Herzog, and A.J. Banes. (2002). Stretch and interleukin-1beta induce matrix metalloproteinases in rabbit tendon cells in vitro. J Orthop Res. 20:36–9.

Aumailley, M., and B. Gayraud. (1998). Structure and biological activity of the extracellular matrix. J Mol Med. 76:253–65.

Awad, H.A., G.P. Boivin, M.R. Dressler, F.N. Smith, R.G. Young, and D.L. Butler. (2003). Repair of patellar tendon injuries using a cell-collagen composite. J Orthop Res. 21:420–31.

Balaban, N.Q., U.S. Schwarz, D. Riveline, P. Goichberg, G. Tzur, I. Sabanay, D. Mahalu, S. Safran, A. Bershadsky, L. Addadi, and B. Geiger. (2001). Force and focal adhesion assembly: a close relationship studied using elastic micropatterned substrates. Nat Cell Biol. 3:466–72.

Banes, A., Sanderson, M.,Biotano, S., Hu, P., Brigman, B., Tsuzaki, M., Fischer, T., Lawrence, W. (1994). Cell Mechanics and Celular engineering. 210–232 pp.

Banes, A.J., G. Horesovsky, C. Larson, M. Tsuzaki, S. Judex, J. Archambault, R. Zernicke, W. Herzog, S. Kelley, and L. Miller. (1999). Mechanical load stimulates expression of novel genes in vivo and in vitro in avian flexor tendon cells. Osteoarthritis Cartilage. 7:141–53.

Banes, A.J., Lee, G., Graff, R., Otey, C., Arcahmbault, J., Tsuzaki, M., Elfervig, M., Qi, J. (2001). Mechanical forces and signaling in connective tissue cells: cellular mechanisms of detection, transduction, and response to mechanical deformation. Current opinions in Orthopaedics. 12:389–396.

Banes, A.J., M. Tsuzaki, P. Hu, B. Brigman, T. Brown, L. Almekinders, W.T. Lawrence, and T. Fischer. (1995a). PDGF-BB, IGF-I and mechanical load stimulate DNA synthesis in avian tendon fibroblasts in vitro. J Biomech. 28:1505–13.

Banes, A.J., M. Tsuzaki, J. Yamamoto, T. Fischer, B. Brigman, T. Brown, and L. Miller. (1995b). Mechanoreception at the cellular level: the detection, interpretation, and diversity of responses to mechanical signals. Biochem Cell Biol. 73:349–65.

Barkhausen, T., M. van Griensven, J. Zeichen, and U. Bosch. (2003). Modulation of cell functions of human tendon fibroblasts by different repetitive cyclic mechanical stress patterns. Exp Toxicol Pathol. 55:153–8.

Berry, C.C., J.C. Shelton, D.L. Bader, and D.A. Lee. (2003). Influence of external uniaxial cyclic strain on oriented fibroblast-seeded collagen gels. Tissue Eng. 9:613–24.

Bershadsky, A.D., N.Q. Balaban, and B. Geiger. (2003). Adhesion-dependent cell mechanosensitivity. Annu Rev Cell Dev Biol. 19:677-95.

Bosman, F.T., and I. Stamenkovic. (2003). Functional structure and composition of the extracellular matrix. J Pathol. 200:423–8.

Butler, D.D., M; Awad,H. (2003). Functional tissue engineering:assessment of function in tendon and ligament repair. Springer, New York. 213–226 pp.

Camelliti, P., T.K. Borg, and P. Kohl. (2005). Structural and functional characterisation of cardiac fibroblasts. Cardiovasc Res. 65:40–51.

Canty, E.G., and K.E. Kadler. (2002). Collagen fibril biosynthesis in tendon: a review and recent insights. Comp Biochem Physiol A Mol Integr Physiol. 133:979–85.

Chang, H.Y., J.T. Chi, S. Dudoit, C. Bondre, M. van de Rijn, D. Botstein, and P.O. Brown. (2002). Diversity, topographic differentiation, and positional memory in human fibroblasts. Proc Natl Acad Sci U S A. 99:12877–82.

Chiquet-Ehrismann, R., M. Tannheimer, M. Koch, A. Brunner, J. Spring, D. Martin, S. Baumgartner, and M. Chiquet. (1994). Tenascin-C expression by fibroblasts is elevated in stressed collagen gels. J Cell Biol. 127:2093–101.

Chiquet, M. (1999). Regulation of extracellular matrix gene expression by mechanical stress. Matrix Biol. 18:417–26.

Chiquet, M., Fluck,M. (2001). Early response to mechanical stress: form signals at the surface to altered gene expression. Elsevier, Amsterdam. 97–110 pp.

Chiquet, M., A.S. Renedo, F. Huber, and M. Fluck. (2003). How do fibroblasts translate mechanical signals into changes in extracellular matrix production? Matrix Biol. 22:73–80.

Chiquet, M., A. Sarasa-Renedo, and V. Tunc-Civelek. (2004). Induction of tenascin-C by cyclic tensile strain versus growth factors: distinct contributions by Rho/ROCK and MAPK signaling pathways. Biochim Biophys Acta. 1693:193–204.

Clark, E.A., and J.S. Brugge. (1995). Integrins and signal transduction pathways: the road taken. Science. 268:233–9.

Clark, R.A. (1990). Fibronectin matrix deposition and fibronectin receptor expression in healing and normal skin. J Invest Dermatol. 94:128S–134S.

Coleman, C., T.L. Tuan, S. Buckley, K.D. Anderson, and D. Warburton. (1998). Contractility, transforming growth factor-beta, and plasmin in fetal skin fibroblasts: role in scarless wound healing. Pediatr Res. 43:403–9.

Davidson, C.J., L.R. Ganion, G.M. Gehlsen, B. Verhoestra, J.E. Roepke, and T.L. Sevier. (1997). Rat tendon morphologic and functional changes resulting from soft tissue mobilization. Med Sci Sports Exerc. 29:313–9.

Desmouliere, A., C. Chaponnier, and G. Gabbiani. (2005). Tissue repair, contraction, and the myofibroblast. Wound Repair Regen. 13:7–12.

Doroski, D.M., K.S. Brink, and J.S. Temenoff. (2006). Techniques for biological characterization of tissue-engineered tendon and ligament. Biomaterials.

Fluck, M., M.N. Giraud, V. Tunc, and M. Chiquet. (2003). Tensile stress-dependent collagen XII and fibronectin production by fibroblasts requires separate pathways. Biochim Biophys Acta. 1593:239–48.

French, A.S., and L.L. Stockbridge. (1988). Potassium channels in human and avian fibroblasts. Proc R Soc Lond B Biol Sci. 232:395–412.

Fukata, Y., M. Amano, and K. Kaibuchi. (2001). Rho-Rho-kinase pathway in smooth muscle contraction and cytoskeletal reorganization of non-muscle cells. Trends Pharmacol Sci. 22:32–9.

Gannier, F., E. White, A. Lacampagne, D. Garnier, and J.Y. Le Guennec. (1994). Streptomycin reverses a large stretch induced increases in [Ca2+]i in isolated guinea pig ventricular myocytes. Cardiovasc Res. 28:1193–8.

Garvin, J., J. Qi, M. Maloney, and A.J. Banes. (2003). Novel system for engineering bioartificial tendons and application of mechanical load. Tissue Eng. 9:967–79.

Glogauer, M., P. Arora, G. Yao, I. Sokholov, J. Ferrier, and C.A. McCulloch. (1997). Calcium ions and tyrosine phosphorylation interact coordinately with actin to regulate cytoprotective responses to stretching. J Cell Sci. 110 (Pt 1):11–21.

Grinnell, F. (2000). Fibroblast-collagen-matrix contraction: growth-factor signalling and mechanical loading. Trends Cell Biol. 10:362–5.

Grinnell, F. (2003). Fibroblast biology in three-dimensional collagen matrices. Trends Cell Biol. 13:264–9.

He, Y., E.J. Macarak, J.M. Korostoff, and P.S. Howard. (2004). Compression and tension: differential effects on matrix accumulation by periodontal ligament fibroblasts in vitro. Connect Tissue Res. 45:28–39.

Helfman, D.M., E.T. Levy, C. Berthier, M. Shtutman, D. Riveline, I. Grosheva, A. Lachish-Zalait, M. Elbaum, and A.D. Bershadsky. (1999). Caldesmon inhibits nonmuscle cell contractility and interferes with the formation of focal adhesions. Mol Biol Cell. 10:3097–112.

Hinz, B., and G. Gabbiani. (2003). Mechanisms of force generation and transmission by myofibroblasts. Curr Opin Biotechnol. 14:538–46.

Howard, P.S., U. Kucich, R. Taliwal, and J.M. Korostoff. (1998). Mechanical forces alter extracellular matrix synthesis by human periodontal ligament fibroblasts. J Periodontal Res. 33:500–8.

Hsieh, A.H., C.M. Tsai, Q.J. Ma, T. Lin, A.J. Banes, F.J. Villarreal, W.H. Akeson, and K.L. Sung. (2000). Time-dependent increases in type-III collagen gene expression in medical collateral ligament fibroblasts under cyclic strains. J Orthop Res. 18:220–7.

Hung, C.T., F.D. Allen, S.R. Pollack, E.T. Attia, J.A. Hannafin, and P.A. Torzilli. (1997). Intracellular calcium response of ACL and MCL ligament fibroblasts to fluid-induced shear stress. Cell Signal. 9:587–94.

Hynes, R.O. (1992). Integrins: versatility, modulation, and signaling in cell adhesion. Cell. 69: 11–25.

Ingber, D. (1991). Integrins as mechanochemical transducers. Curr Opin Cell Biol. 3:841–8.

Inoh, H., N. Ishiguro, S. Sawazaki, H. Amma, M. Miyazu, H. Iwata, M. Sokabe, and K. Naruse. (2002). Uni-axial cyclic stretch induces the activation of transcription factor nuclear factor kappaB in human fibroblast cells. Faseb J. 16:405–7.

Iqbal, J., and M. Zaidi. (2005). Molecular regulation of mechanotransduction. Biochem Biophys Res Commun. 328:751–5.

Jones, B.F., M.E. Wall, R.L. Carroll, S. Washburn, and A.J. Banes. (2005). Ligament cells stretch-adapted on a microgrooved substrate increase intercellular communication in response to a mechanical stimulus. J Biomech. 38:1653–64.

Juliano, R.L., and S. Haskill. (1993). Signal transduction from the extracellular matrix. J Cell Biol. 120:577–85.

Juncosa-Melvin, N., G.P. Boivin, M.T. Galloway, C. Gooch, J.R. West, A.M. Sklenka, and D.L. Butler. (2005). Effects of cell-to-collagen ratio in mesenchymal stem cell-seeded implants on tendon repair biomechanics and histology. Tissue Eng. 11:448–57.

Juncosa-Melvin, N., G.P. Boivin, C. Gooch, M.T. Galloway, J.R. West, M.G. Dunn, and D.L. Butler. (2006a). The effect of autologous mesenchymal stem cells on the biomechanics and histology of gel-collagen sponge constructs used for rabbit patellar tendon repair. Tissue Eng. 12:369–79.

Juncosa-Melvin, N., J.T. Shearn, G.P. Boivin, C. Gooch, M.T. Galloway, J.R. West, V.S. Nirmalanandhan, G. Bradica, and D.L. Butler. (2006b). Effects of mechanical stimulation on the biomechanics and histology of stem cell-collagen sponge constructs for rabbit patellar tendon repair. Tissue Eng. 12:2291–300.

Kalapesi, F.B., J.C. Tan, and M.T. Coroneo. (2005). Stretch-activated channels: a mini-review. Are stretch-activated channels an ocular barometer? Clin Experiment Ophthalmol. 33:210–7.

Karimbux, N.Y., and I. Nishimura. (1995). Temporal and spatial expressions of type XII collagen in the remodeling periodontal ligament during experimental tooth movement. J Dent Res. 74:313–8.

Kato, T., N. Ishiguro, H. Iwata, T. Kojima, T. Ito, and K. Naruse. (1998). Up-regulation of COX2 expression by uni-axial cyclic stretch in human lung fibroblast cells. Biochem Biophys Res Commun. 244:615–9.

Kessler, D., S. Dethlefsen, I. Haase, M. Plomann, F. Hirche, T. Krieg, and B. Eckes. (2001). Fibroblasts in mechanically stressed collagen lattices assume a "synthetic" phenotype. J Biol Chem. 276:36575–85.

Kim, S.G., T. Akaike, T. Sasagaw, Y. Atomi, and H. Kurosawa. (2002). Gene expression of type I and type III collagen by mechanical stretch in anterior cruciate ligament cells. Cell Struct Funct. 27:139–44.

Kimoto, S., M. Matsuzawa, S. Matsubara, T. Komatsu, N. Uchimura, T. Kawase, and S. Saito. (1999). Cytokine secretion of periodontal ligament fibroblasts derived from human deciduous teeth: effect of mechanical stress on the secretion of transforming growth factor-beta 1 and macrophage colony stimulating factor. J Periodontal Res. 34:235–43.

Kimura, K., M. Ito, M. Amano, K. Chihara, Y. Fukata, M. Nakafuku, B. Yamamori, J. Feng, T. Nakano, K. Okawa, A. Iwamatsu, and K. Kaibuchi. (1996). Regulation of myosin phosphatase by Rho and Rho-associated kinase (Rho-kinase). Science. 273:245–8.

Kjaer, M. (2004). Role of extracellular matrix in adaptation of tendon and skeletal muscle to mechanical loading. Physiol Rev. 84:649–98.

Kjaer, M., H. Langberg, B.F. Miller, R. Boushel, R. Crameri, S. Koskinen, K. Heinemeier, J.L. Olesen, S. Dossing, M. Hansen, S.G. Pedersen, M.J. Rennie, and P. Magnusson. (2005). Metabolic activity and collagen turnover in human tendon in response to physical activity. J Musculoskelet Neuronal Interact. 5:41–52.

Kolodney, M.S., and R.B. Wysolmerski. (1992). Isometric contraction by fibroblasts and endothelial cells in tissue culture: a quantitative study. J Cell Biol. 117:73–82.

Koob, T.J. (2002). Biomimetic approaches to tendon repair. Comp Biochem Physiol A Mol Integr Physiol. 133:1171–92.

Koob, T.J., P.E. Clark, D.J. Hernandez, F.A. Thurmond, and K.G. Vogel. (1992). Compression loading in vitro regulates proteoglycan synthesis by tendon fibrocartilage. Arch Biochem Biophys. 298:303–12.

Labat-Robert, J., M. Bihari-Varga, and L. Robert. (1990). Extracellular matrix. FEBS Lett. 268:386–93.

Lacampagne, A., F. Gannier, J. Argibay, D. Garnier, and J.Y. Le Guennec. (1994). The stretch-activated ion channel blocker gadolinium also blocks L-type calcium channels in isolated ventricular myocytes of the guinea-pig. Biochim Biophys Acta. 1191:205–8.

Langberg, H., D. Skovgaard, L.J. Petersen, J. Bulow, and M. Kjaer. (1999). Type I collagen synthesis and degradation in peritendinous tissue after exercise determined by microdialysis in humans. J Physiol. 521 Pt 1:299–306.

Lavagnino, M., and S.P. Arnoczky. (2005). In vitro alterations in cytoskeletal tensional homeostasis control gene expression in tendon cells. J Orthop Res. 23:1211–8.

Lee, A.A., T. Delhaas, L.K. Waldman, D.A. MacKenna, F.J. Villarreal, and A.D. McCulloch. (1996). An equibiaxial strain system for cultured cells. Am J Physiol. 271:C1400–8.

Lehoux, S., Y. Castier, and A. Tedgui. (2006). Molecular mechanisms of the vascular responses to haemodynamic forces. J Intern Med. 259:381–92.

Lehoux, S., and A. Tedgui. (2003). Cellular mechanics and gene expression in blood vessels. J Biomech. 36:631–43.

Li, Z., G. Yang, M. Khan, D. Stone, S.L. Woo, and J.H. Wang. (2004). Inflammatory response of human tendon fibroblasts to cyclic mechanical stretching. Am J Sports Med. 32:435–40.

Longaker, M.T., D.J. Whitby, M.W. Ferguson, H.P. Lorenz, M.R. Harrison, and N.S. Adzick. (1994). Adult skin wounds in the fetal environment heal with scar formation. Ann Surg. 219:65–72.

MacKenna, D., S.R. Summerour, and F.J. Villarreal. (2000). Role of mechanical factors in modulating cardiac fibroblast function and extracellular matrix synthesis. Cardiovasc Res. 46:257–63.

Middleton, J., M. Jones, and A. Wilson. (1996). The role of the periodontal ligament in bone modeling: the initial development of a time-dependent finite element model. Am J Orthod Dentofacial Orthop. 109:155–62.

Mostafavi-Pour, Z., J.A. Askari, S.J. Parkinson, P.J. Parker, T.T. Ng, and M.J. Humphries. (2003). Integrin-specific signaling pathways controlling focal adhesion formation and cell migration. J Cell Biol. 161:155–67.

Munevar, S., Y.L. Wang, and M. Dembo. (2004). Regulation of mechanical interactions between fibroblasts and the substratum by stretch-activated Ca2+ entry. J Cell Sci. 117:85–92.

Nakano, K., K. Takaishi, A. Kodama, A. Mammoto, H. Shiozaki, M. Monden, and Y. Takai. (1999). Distinct actions and cooperative roles of ROCK and mDia in Rho small G protein-induced reorganization of the actin cytoskeleton in Madin-Darby canine kidney cells. Mol Biol Cell. 10:2481–91.

Nedelec, B., A. Ghahary, P.G. Scott, and E.E. Tredget. (2000). Control of wound contraction. Basic and clinical features. Hand Clin. 16:289–302.

Ng, K.W., D.W. Hutmacher, J.T. Schantz, C.S. Ng, H.P. Too, T.C. Lim, T.T. Phan, and S.H. Teoh. (2001). Evaluation of ultra-thin poly(epsilon-caprolactone) films for tissue-engineered skin. Tissue Eng. 7:441–55.

Nicolas, A., B. Geiger, and S.A. Safran. (2004). Cell mechanosensitivity controls the anisotropy of focal adhesions. Proc Natl Acad Sci U S A. 101:12520–5.

Nicolas, A., and S.A. Safran. (2004). Elastic deformations of grafted layers with surface stress. Phys Rev E Stat Nonlin Soft Matter Phys. 69:051902.

Olesen, J.L., K.M. Heinemeier, C. Gemmer, M. Kjaer, A. Flyvbjerg, and H. Langberg. (2006). Exercise dependent IGF-I, IGFBPs and type-I collagen changes in human peritendinous connective tissue determined by microdialysis. J Appl Physiol.

Parsons, J.T., and S.J. Parsons. (1997). Src family protein tyrosine kinases: cooperating with growth factor and adhesion signaling pathways. Curr Opin Cell Biol. 9:187–92.

Parsons, M., E. Kessler, G.J. Laurent, R.A. Brown, and J.E. Bishop. (1999). Mechanical load enhances procollagen processing in dermal fibroblasts by regulating levels of procollagen C-proteinase. Exp Cell Res. 252:319–31.

Plopper, G.E., H.P. McNamee, L.E. Dike, K. Bojanowski, and D.E. Ingber. (1995). Convergence of integrin and growth factor receptor signaling pathways within the focal adhesion complex. Mol Biol Cell. 6:1349–65.

Ralphs, J.R., A.D. Waggett, and M. Benjamin. (2002). Actin stress fibres and cell-cell adhesion molecules in tendons: organisation in vivo and response to mechanical loading of tendon cells in vitro. Matrix Biol. 21:67–74.

Reitan, K. (1967). Clinical and histologic observations on tooth movement during and after orthodontic treatment. Am J Orthod. 53:721–45.

Reitan, K. (1969). Principles of retention and avoidance of posttreatment relapse. Am J Orthod. 55:776–90.

Reitan, K. (1970). Evaluation of orthodontic forces as related to histologic and mechanical factors. SSO Schweiz Monatsschr Zahnheilkd. 80:579–96.

Riley, G. (2004). The pathogenesis of tendinopathy. A molecular perspective. Rheumatology (Oxford). 43:131–42.

Riley, G.P. (2005). Gene expression and matrix turnover in overused and damaged tendons. Scand J Med Sci Sports. 15:241–51.

Riveline, D., E. Zamir, N.Q. Balaban, U.S. Schwarz, T. Ishizaki, S. Narumiya, Z. Kam, B. Geiger, and A.D. Bershadsky. (2001). Focal contacts as mechanosensors: externally applied local mechanical force induces growth of focal contacts by an mDia1-dependent and ROCK-independent mechanism. J Cell Biol. 153:1175–86.

Ruknudin, A., F. Sachs, and J.O. Bustamante. (1993). Stretch-activated ion channels in tissue-cultured chick heart. Am J Physiol. 264:H960–72.

Ruoslahti, E. (1988). Structure and biology of proteoglycans. Annu Rev Cell Biol. 4:229–55.

Ruwhof, C., and A. van der Laarse. (2000). Mechanical stress-induced cardiac hypertrophy: mechanisms and signal transduction pathways. Cardiovasc Res. 47:23–37.

Rygh, P., and K. Reitan. (1972). Ultrastructural changes in the periodontal ligament incident to orthodontic tooth movement. Trans Eur Orthod Soc:393–405.

Sadoshima, J., and S. Izumo. (1993). Mechanotransduction in stretch-induced hypertrophy of cardiac myocytes. J Recept Res. 13:777–94.

Sadoshima, J., and S. Izumo. (1997). The cellular and molecular response of cardiac myocytes to mechanical stress. Annu Rev Physiol. 59:551–71.

Sadoshima, J., T. Takahashi, L. Jahn, and S. Izumo. (1992). Roles of mechano-sensitive ion channels, cytoskeleton, and contractile activity in stretch-induced immediate-early gene expression and hypertrophy of cardiac myocytes. Proc Natl Acad Sci U S A. 89:9905–9.

Sarasa-Renedo, A., and M. Chiquet. (2005). Mechanical signals regulating extracellular matrix gene expression in fibroblasts. Scand J Med Sci Sports. 15:223–30.

Sarasa-Renedo, A., V. Tunc-Civelek, and M. Chiquet. (2006). Role of RhoA/ROCK-dependent actin contractility in the induction of tenascin-C by cyclic tensile strain. Exp Cell Res.

Schwartz, M.A., M.D. Schaller, and M.H. Ginsberg. (1995). Integrins: emerging paradigms of signal transduction. Annu Rev Cell Dev Biol. 11:549–99.

Shimizu, N., Y. Ozawa, M. Yamaguchi, T. Goseki, K. Ohzeki, and Y. Abiko. (1998). Induction of COX-2 expression by mechanical tension force in human periodontal ligament cells. J Periodontol. 69:670–7.

Shimizu, N., M. Yamaguchi, T. Goseki, Y. Ozawa, K. Saito, H. Takiguchi, T. Iwasawa, and Y. Abiko. (1994). Cyclic-tension force stimulates interleukin-1 beta production by human periodontal ligament cells. J Periodontal Res. 29:328–33.

Skutek, M., M. van Griensven, J. Zeichen, N. Brauer, and U. Bosch. (2001a). Cyclic mechanical stretching enhances secretion of Interleukin 6 in human tendon fibroblasts. Knee Surg Sports Traumatol Arthrosc. 9:322–6.

Skutek, M., M. van Griensven, J. Zeichen, N. Brauer, and U. Bosch. (2001b). Cyclic mechanical stretching modulates secretion pattern of growth factors in human tendon fibroblasts. Eur J Appl Physiol. 86:48–52.

Sokabe, M., K. Naruse, S. Sai, T. Yamada, K. Kawakami, M. Inoue, K. Murase, and M. Miyazu. (1997). Mechanotransduction and intracellular signaling mechanisms of stretch-induced remodeling in endothelial cells. Heart Vessels. Suppl 12:191–3.

Stockbridge, L.L., and A.S. French. (1988). Stretch-activated cation channels in human fibroblasts. Biophys J. 54:187–90.

Thampatty, B.P., H. Li, H-J. Im, and J. H-C., Wang. (2007). EP4 receptor regulates collagen type-I, MMP-1, and MMP-3 gene expression in human tendon fibroblasts in response to IL-1 beta treatment. Gene. 386: 154–61.

Tomasek, J.J., G. Gabbiani, B. Hinz, C. Chaponnier, and R.A. Brown. (2002). Myofibroblasts and mechano-regulation of connective tissue remodelling. Nat Rev Mol Cell Biol. 3:349–63.

Trachslin, J., M. Koch, and M. Chiquet. (1999). Rapid and reversible regulation of collagen XII expression by changes in tensile stress. Exp Cell Res. 247:320–8.

Tsuji, K., K. Uno, G.X. Zhang, and M. Tamura. (2004). Periodontal ligament cells under intermittent tensile stress regulate mRNA expression of osteoprotegerin and tissue inhibitor of matrix metalloprotease-1 and -2. J Bone Miner Metab. 22:94–103.

Tsuzaki, M., D. Bynum, L. Almekinders, X. Yang, J. Faber, and A.J. Banes. (2003a). ATP modulates load-inducible IL-1beta, COX 2, and MMP-3 gene expression in human tendon cells. J Cell Biochem. 89:556–62.

Tsuzaki, M., G. Guyton, W. Garrett, J.M. Archambault, W. Herzog, L. Almekinders, D. Bynum, X. Yang, and A.J. Banes. (2003b). IL-1 beta induces COX2, MMP-1, -3 and -13, ADAMTS-4, IL-1 beta and IL-6 in human tendon cells. J Orthop Res. 21:256–64.

Visse, R., and H. Nagase. (2003). Matrix metalloproteinases and tissue inhibitors of metalloproteinases: structure, function, and biochemistry. Circ Res. 92:827–39.

Von den Hoff, J.W. (2003). Effects of mechanical tension on matrix degradation by human periodontal ligament cells cultured in collagen gels. J Periodontal Res. 38:449–57.

Vunjak-Novakovic, G., G. Altman, R. Horan, and D.L. Kaplan. (2004). Tissue engineering of ligaments. Annu Rev Biomed Eng. 6:131–56.

Wall, M.E., and A.J. Banes. (2005). Early responses to mechanical load in tendon: role for calcium signaling, gap junctions and intercellular communication. J Musculoskelet Neuronal Interact. 5:70–84.

Wang, J.H., and B.P. Thampatty. (2006). Mechanoregulation of fibroblast function. Encyclopedia of Biomaterials and Biomedical Engineering. pp. 1–11 Taylor and Francis.

Wang, J.H. (2006). Mechanobiology of tendon. J Biomech. 39:1563–82.

Wang, J.H., F. Jia, G. Yang, S. Yang, B.H. Campbell, D. Stone, and S.L. Woo. (2003). Cyclic mechanical stretching of human tendon fibroblasts increases the production of prostaglandin E2 and levels of cyclooxygenase expression: a novel in vitro model study. Connect Tissue Res. 44:128–33.

Wang, J.H., Z. Li, G. Yang, and M. Khan. (2004). Repetitively stretched tendon fibroblasts produce inflammatory mediators. Clin Orthop Relat Res:243–50.

Wang, J.H., G. Yang, and Z. Li. (2005). Controlling cell responses to cyclic mechanical stretching. Ann Biomed Eng. 33:337–42.

Watanabe, N., T. Kato, A. Fujita, T. Ishizaki, and S. Narumiya. (1999). Cooperation between mDia1 and ROCK in Rho-induced actin reorganization. Nat Cell Biol. 1:136–43.

Webb, K., R.W. Hitchcock, R.M. Smeal, W. Li, S.D. Gray, and P.A. Tresco. (2006). Cyclic strain increases fibroblast proliferation, matrix accumulation, and elastic modulus of fibroblast-seeded polyurethane constructs. J Biomech. 39:1136–44.

Woo, S.L., M.A. Gomez, T.J. Sites, P.O. Newton, C.A. Orlando, and W.H. Akeson. (1987). The biomechanical and morphological changes in the medial collateral ligament of the rabbit after immobilization and remobilization. J Bone Joint Surg Am. 69:1200–11.

Woo, S.L., K. Hildebrand, N. Watanabe, J.A. Fenwick, C.D. Papageorgiou, and J.H. Wang. (1999). Tissue engineering of ligament and tendon healing. Clin Orthop Relat Res:S312–23.

Yamaguchi, M., N. Shimizu, T. Goseki, Y. Shibata, H. Takiguchi, T. Iwasawa, and Y. Abiko. (1994). Effect of different magnitudes of tension force on prostaglandin E2 production by human periodontal ligament cells. Arch Oral Biol. 39:877–84.

Yamamoto, E., W. Iwanaga, H. Miyazaki, and K. Hayashi. (2002). Effects of static stress on the mechanical properties of cultured collagen fascicles from the rabbit patellar tendon. J Biomech Eng. 124:85–93.

Yang, G., R.C. Crawford, and J.H. Wang. (2004). Proliferation and collagen production of human patellar tendon fibroblasts in response to cyclic uniaxial stretching in serum-free conditions. J Biomech. 37:1543–50.

Yang, G., H.J. Im, and J.H. Wang. (2005). Repetitive mechanical stretching modulates IL-1beta induced COX-2, MMP-1 expression, and PGE2 production in human patellar tendon fibroblasts. Gene. 363:166–72.

Yasuda, T., M. Kinoshita, M. Abe, and Y. Shibayama. (2000). Unfavorable effect of knee immobilization on Achilles tendon healing in rabbits. Acta Orthop Scand. 71:69–73.

Yost, M.J., D. Simpson, K. Wrona, S. Ridley, H.J. Ploehn, T.K. Borg, and L. Terracio. (2000). Design and construction of a uniaxial cell stretcher. Am J Physiol Heart Circ Physiol. 279:H3124–30.

Zeichen, J., M. van Griensven, and U. Bosch. (2000). The proliferative response of isolated human tendon fibroblasts to cyclic biaxial mechanical strain. Am J Sports Med. 28:888–92.

Zhou, D., H.S. Lee, F. Villarreal, A. Teng, E. Lu, S. Reynolds, C. Qin, J. Smith, and K.L. Sung. (2005). Differential MMP-2 activity of ligament cells under mechanical stretch injury: an in vitro study on human ACL and MCL fibroblasts. J Orthop Res. 23:949–57.

Index

MIX
Papier aus verantwortungsvollen Quellen
Paper from responsible sources
FSC® C105338

If you have any concerns about our products,
you can contact us on
ProductSafety@springernature.com

In case Publisher is established outside the EU,
the EU authorized representative is:
Springer Nature Customer Service Center GmbH
Europaplatz 3, 69115 Heidelberg, Germany

Printed by Libri Plureos GmbH
in Hamburg, Germany